Für Peter

zur Erinnerung an die Heidelberger Physiologie mit allen guten Wünschen.

Jerusalem, im März 1993

Caspar Rüegg

ALLE ZEIT WACH
1842

Johann Caspar Rüegg

Calcium in Muscle Contraction

Cellular and Molecular Physiology

Second Edition

With 110 Figures

Springer-Verlag
Berlin Heidelberg New York
London Paris Tokyo
Hong Kong Barcelona
Budapest

Prof. Dr. JOHANN CASPAR RÜEGG
University of Heidelberg
Department of Physiology II
Im Neuenheimer Feld 326
6900 Heidelberg, FRG

First edition "Calcium in Muscle Activation" published 1986
as Volume 19 of the Series "Zoophysiology". Second printing 1988

ISBN 3-540-55544-7 Springer-Verlag Berlin Heidelberg New York
ISBN 0-387-55544-7 Springer-Verlag New York Berlin Heidelberg

ISBN 3-540-18278-0 1. Auflage Springer-Verlag Berlin Heidelberg New York
ISBN 0-387-18278-0 1st edition Springer-Verlag New York Berlin Heidelberg

Library of Congress Cataloging-in-Publication Data. Rüegg, Johann Caspar, 1930–
Calcium in muscle contraction: cellular and molecular physiology/Johann Caspar Rüegg. – 2nd ed. p. cm. Rev. ed. of: Calcium in muscle activation. c1986. Includes bibliographical references and index. ISBN 3-540-55544-7. – ISBN 0-387-55544-7 1. Muscle contraction – Regulation. 2. Calcium – Physiological effect. 3. Physiology, Comparative. I. Rüegg, Johann Caspar 1930– Calcium in muscle activation. II. Title. [DNLM: 1. Calcium – physiology. 2. Muscle Contraction – drug effects. 3. Physiology, Comparative. WE 500 R919c] QP321.R864 1992 591.1′852 – dc 20

Production Editor: Renate Münzenmayer
Typesetting: Brühlsche Universitätsdruckerei, Giessen
2131/3130-543210

This book is dedicated to
Elvi, Christian and Barbara

Preface to the Second Edition

The years following the appearance of Calcium in Muscle Activation have witnessed an explosion of knowledge on the molecular mechanisms underlying the action of calcium channels, calcium pumps and calcium regulatory proteins. In addition, paradigms concerning calcium activation of contractile proteins have also changed. All of this called for a revision of the text, including a new section on molecular level approaches. This book, then, represents an interdisciplinary approach in describing the role of calcium ions in the regulation of muscle contraction and contractility. The latter topic is new. The reader will appreciate that "contractility" reflects the activity of the molecular force generators, the crossbridges, which is preceded and determined by the processes of activation, but also depends on the properties of the contractile proteins themselves. An understanding of contractility and its regulation by calcium, on the cellular and molecular level, will certainly be of great importance in muscle and heart research. I hope therefore that this book will be a useful reference, not only to the student of musle physiology and biophysics, but also to the clinician in the cardiovascular field. New chapters and sections will highlight the recent progress, leading to a molecular understanding of the structure and function of calcium channels, calcium regulatory proteins and calcium-sensitivity modulation in cardiac, skeletal and smooth muscle.

In revising this monograph I received enormous help from many people, in particular from Werner Melzer who critically read the sections on excitation-contraction coupling and provided ideas, suggestions and figures. Special thanks are also due to Bernhard Brenner, Rainer Fink, and Gabriele Pfitzer for their comments and critical reading of revised sections as well as to C. Ashley, N. M. Green and C. Franzini-Armstrong for providing plates and figures. It is also a pleasure to acknowledge the excellent cooperation of the staff of Springer-Verlag. Last, but not least, I am greatly indebted to Linda Gränz for her skillful assistance in the preparation of the manuscript, to Claudia Zeugner for the artwork as well as to Elke Höflein, Ulrike Weidelich, and John Strauss for their help and assistance in the preparation of the subject index.

Heidelberg, Autumn 1992 JOHANN CASPAR RÜEGG

Preface to the Second Printing

One year after the appearance of the hardcover edition, this text was revised to address graduate students taking advanced level courses in muscle biophysics and physiology. The book may also prove a useful source of reference to scientists in cell biology and pharmacology and other disciplines. Although muscle cell calcium is an area in which rapid progress continues to be made, the main arguments and facts presented here remain valid, and do not require alteration at this time.

In revising the text, I again received the invaluable help of Isolde Berger. The book also benefited from the critical reading by one of my students, Mathias Gautel, and from comments by many readers. I am grateful for this, and I look forward to an ongoing dialogue.

Heidelberg, Autumn 1987 — Johann Caspar Rüegg

Preface to the First Edition

This book offers a comparative and interdisciplinary approach to excitation-contraction-coupling in smooth and striated muscles, including the myocardium. It is an account of the pathways and mechanisms by which cellular calcium is handled and activates the contractile proteins. It also describes how these mechanisms are adapted in various kinds of muscle to meet specific functional requirements, such as speed or economy.

This monography then presents facts, ideas and theories and the evidence on which they are based, and if it stimulates others and furthers research, it will have served its purpose. All of the chapters are self-contained and may be read in any order, but readers unfamiliar with muscle are recommended to start with the introductory chapter on excitation and contraction.

During all the years of writing this book, I received enormous help from Isolde Berger who corrected, edited and transformed my innumerable notes and drafts into a readable manuscript; she also compiled the list of references and the Subject Index. I owe a great debt of gratitude to her and also to Claudia Zeugner, who prepared the figures with expertise and care. Then I would like to thank the Deutsche Forschungsgemeinschaft and the Fritz-Thyssen-Stiftung for supporting the work of my Department which has been reported in this monograph.

A great many people contributed with helpful discussions. Additionally, the following colleagues read parts of the manuscript or provided figures or other materials in advance of publication: Chris Ashley, Giovanni Cecchi, Joe DiSalvo, Clara Franzini-Armstrong, Jean-Marie Gillis, Roger Goody, Peter Griffiths, Wilhelm Hasselbach, Hugh Huxley, Gerrit Isenberg, John Kendrick-Jones, Helmut Langer, Herbert Ludwig, Christoph Lüttgau, Madoka Makinose, Fumi Morita, Richard Murphy, Richard Paul, Gabriele Pfitzer, James Potter, Malcolm Sparrow, John Wray and Mariusz Zydowo. To them and to my colleagues from the Heidelberg II Physiology Institute I am most grateful.

Heidelberg, Autumn 1986 JOHANN CASPAR RÜEGG

Contents

Introduction

When you hold this book in your hands, the activated muscles of your arms and hands receive contraction signals from motoneurons that release acetylcholine, a neurotransmitter, at the nerve-muscle junction. The message conveyed to the muscle is then relayed through a series of ion channels in the cell membrane which ultimately lead to the release of calcium ions from intracellular calcium stores which then activate the contractile proteins. Whereas the events at the nerve-muscle junction are now well understood, much less is known about the mechanism of calcium release and calcium activation of the contractile machinery. Here, we shall consider these problems and how they can be solved by interdisciplinary approaches involving physiology, biophysics and molecular biology. In addition, the strength of the comparative approach will also become apparent, in view of the diversity of the different kinds of striated and smooth muscles. Yet there is unity in this diversity. In all muscle studied so far, contraction appears to be basically an interaction of two proteins, actin and myosin, which is more or less precisely controlled by the intracellular calcium ion concentration. This principle of calcium control is so essential for muscle contraction and cell motility that it has been conserved during evolution for more than a billion years. How diverse, on the other hand, are the mechanisms controlling the intracellular calcium ion concentration, the intracellular calcium target proteins and the way in which calcium ions activate the contractile system! For instance, in fast skeletal muscle calcium is handled entirely within the cell by the sarcoplasmic reticulum, whereas in smooth and cardiac muscle the extracellular calcium supply seems to be quite important. In vertebrate skeletal and heart muscle calcium regulates contraction by operating the protein switch troponin, whereas in the much slower smooth muscles calcium influences the myosin molecule either directly (molluscan smooth muscle) or indirectly through an enzymic cascade involving calmodulin and myosin phosphorylation (vertebrate smooth muscle). Such diversity is a challenge to the comparative physiologist, who attempts to understand its significance. This is no small task considering that the creation of diversity is an essential step in the evolutionary process leading to the adaptation of animals to their environment. The molecular strategies involved have been discussed by Hochachka and Somero (1973). Among these are: diversification of protein structures and variation of enzymic activities within the cell as illustrated by the evolution of the myosin molecule and its diversification into various isoforms exhibiting different ATPase activities. How are such molecular strategies employed in the adaptation of muscles to specific functional requirements such as quick performance?

Of course, one can also look at biological diversity as the outcome of innumerable natural experiments during evolution which scientists might analyze (Schulz and Schirmer 1978). Variations of structure and function, for instance, may help to understand relationships between structure and function as exemplified by A. F. Huxley's analysis of the transversal membrane system (A. F. Huxley 1974). Due to the enormous diversity of biological materials, nature is a "treasure house" that offers a wealth of very different specimens from which a scientist may choose when looking for the most suitable experimental material to solve a specific biological problem. Thus, the Danish physiologist Krogh (1939) advised experimentalists to study the most adapted and specialized animals, tissues or mechanisms since these are most likely to furnish clues. Indeed, much of the progress made in muscle research has been achieved by using highly specialized muscles, such as frog sartorius and insect flight muscle! Rather intuitively, Hoyle predicted already in 1957 that specialized "invertebrate muscles will have much to contribute to the understanding of the coupling processes between the membrane potential changes and the contractile process". Indeed, giant crustacean muscle fibres injected with the calcium probe aequorin proved, in the hands of Hoyle's colleague Ashley, ideal for showing the relation between intracellular calcium ion concentration and contraction many years ahead of similar studies in the more difficult vertebrate muscle preparations. These demonstrated that the intracellular free calcium ion concentration is the link between muscle excitation and contraction (Ashley and Campbell 1979).

The processes involved in the coupling of excitation and contraction of striated muscle consist of many steps, including the spreading of excitation into the interior of the fibre and the intracellular release of calcium ions from the sarcoplasmic reticulum (Chap. 2), the generation of the calcium signal (Chap. 3) and the perception of that signal by the calcium-binding protein troponin (Chap. 4). The subsequent chapters will deal with adaptations of the calcium-activating mechanisms such as those which occur in fast and slow striated muscles, in insect flight muscles and molluscan muscles as well as in vertebrate cardiac and smooth muscles. In the latter as well as in the heart, "contractility" can be increased not only by enhancing myoplasmic free calcium levels, but also by rendering the contractile proteins more responsive to the calcium activator!

On a subcellular and molecular level, the phenomenon of excitation-contraction coupling appears to be basically a question of communication between calcium channel, calcium binding, regulatory, and contractile proteins that function as senders, receivers and transducers of calcium signals, and also as modulators of the calcium responsiveness. However, the cellular calcium signalling system as a whole is complex and has the emergent properties of a communication network subject to feedback and self-regulation. Contractile properties (contractility) and calcium activation, furthermore, are interdependent, as calcium responsiveness affects actin-myosin interaction and vice versa. These cybernetic aspects and molecular level approaches will be discussed in Chapters 9 and 10, respectively. To readers unfamiliar with the physiology of muscle excitation and contraction, some initial facts and basic background information are given in the first chapter.

Chapter 1

Muscle Excitation and Contraction

In the most primitive muscle cells, the myocytes of sponges, contractile phenomena are entirely regulated at a cellular level. In contrast, in all higher phyla, motility is more or less under the control of the nervous system which communicates with muscle cells at the neuromuscular junction. Think, for example, of a frog jumping into a pond. Nerve impulses reaching the motor end-plates of its leg muscles cause the release of acetylcholine from the nerve endings. By opening acetylcholine-gated ion channels, the neurotransmitter released by a nerve impulse then elicits a small potential change in the membrane of the muscle fibres, the end-plate potential, which in turn evokes a propagated action potential exciting the muscle fibres. This event is followed by the processes of excitation-contraction coupling (Sect. 1.2) and contraction (Sect. 1.3) in which muscle cell calcium (Fig. 1.1) plays a pivotal role.

1.1 Muscle Excitation

During an action potential the membrane potential becomes reversed so that for a brief moment the inside of the muscle cell becomes positively charged with respect to the outside; however, within a few milliseconds the initial negativity of the cell interior (about −90 mV) is restored. After a latency of only a few milliseconds, this muscle excitation is followed by a brief contraction of the muscle fibres which reaches a maximum within about 20–100 ms and then gives way to relaxation (Fig. 1.2). Such a phasic contraction response is known as a twitch.

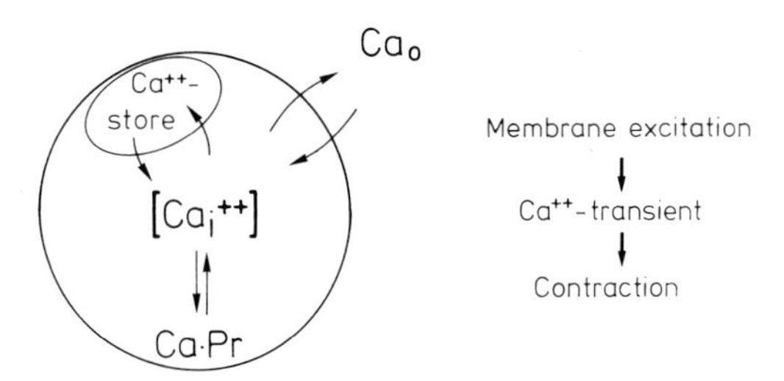

Fig. 1.1. Muscle cell calcium as the link between excitation and contraction. The intracellular ionized calcium (= free calcium) is in dynamic equilibrium with calcium bound to the contractile and regulatory proteins (e.g. myosin, troponin, calmodulin), with extracellular calcium (Ca_0) and with the calcium stored within cell organelles (sarcoplasmic reticulum, mitochondria). In relaxed muscle cells free calcium is approximately 0.1 μM and rises to about 1–10 μM in contraction. This transient increase in $[Ca^{2+}]$ is the link between excitation of the cell membrane and contraction

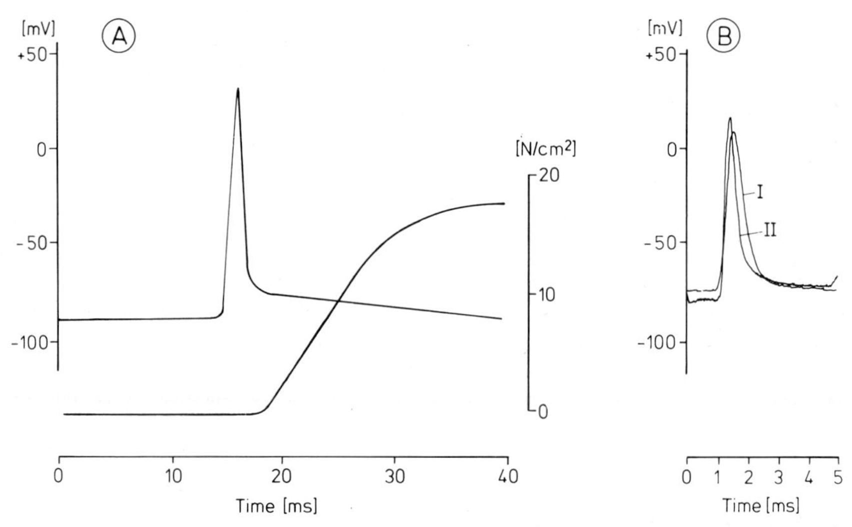

Fig. 1.2. Muscle excitation. *A* Action potential and development of isometric force recorded from an isolated frog skeletal muscle fibre at 18 °C. The upstroke of the action potential is due to a regenerative fast inward Na current, while the "overshoot" of the potential is dependent on the logarithm of the external Na^+ concentration. Repolarization is caused by the inactivation of the Na channels as well as by an outward K current due to a delayed increase in membrane K conductance. The resting potential is −90 mV. Note the brief electromechanical latency between action potential and onset of force development. (From Lüttgau and Stephenson 1986) *B* A comparison between action potentials of (*I*) slow rat soleus fibres (type I fibres) and (*II*) fast rat extensor digitorum longus fibres (type II B). Records were obtained in situ with intracellular electrodes, placed in a fibre of a motor unit stimulated via a nerve axon terminal by means of a wire electrode on the fibre surface. The fibre type was subsequently identified by histochemical techniques. (Wallinga-de Jonge et al. 1985)

The action potential of the excited cell membrane is the "trigger" that elicits a twitch contraction in both invertebrate and vertebrate muscles. In the latter, particularly in mammals, one may distinguish between fast and slow twitch fibres. Both fibre types show a similar cross-striation pattern, but they may be readily distinguished by biochemical and histochemical techniques (Table 1.1). Compared with fast fibres, the resting potential of slow twitch fibres is less negative and the amplitude of the action potential, as well as its rate of depolarization and repolarization, are smaller (Wallinga-de Jonge 1985). In muscles of invertebrates the action potential may be more complex. In scorpion striated muscle, for instance, the initial "overshooting" spike potential may be followed by a burst of smaller spikes which accompany the twitch response (Gilly and Scheuer 1984).

In addition to twitch fibres, many species contain another type of cross-striated muscle fibres which responds to stimulation with a slow "tonic" contraction. The classical representative is the tonic fibre bundle of the frog iliofibularis muscle (Kuffler and Vaughan-Williams 1953a, b) which may serve a postural function. In these tonic fibres, as well as in many slow striated muscle fibre types of insects and crustaceans and in obliquely striated fibres of annelids and nematodes (Hoyle 1983), stimulation by nerves does not cause an action potential followed by a twitch. Instead, repetitive nerve impulses evoke small and slow potential changes

Table 1.1. Diversity of vertebrate striated muscle fibres

Feature	Twitch fibres			Slow tonic fibres
Fibre type	IIB	IIA	I	
Colour	White	Red	Red	Red
Contraction	Fast twitch	Fast twitch	Slow twitch	Slow contracture
Fatigue resistance	Low	Medium	High	High
Metabolism	Glycolytic	Glycolytic and oxidative	Oxidative	Oxidative
Lactic dehydrogenase	High	Medium or high	Low	Low
Succinic dehydrogenase	Low	Medium	High	High
Example	EDL[a] (Rat)	Quadriceps	Soleus (Rat)	Iliofibularis[b] (Frog)

[a] Extensor digitorum longus muscle.
[b] Tonus bundle.

which summate to give a maintained potential change of, for example, 30 mV or more and, accordingly, the membrane becomes less charged: it depolarizes. Increasing depolarization increases, in a graded manner, the extent of tonic contraction. This graded activation contrasts with the "all-or-none" activation in twitch fibres. As soon as the membrane potential (about −90 mV) rises by about 40 mV, a threshold is reached and when it is surpassed, an all-or-none action potential is elicited and a maximal contractile response, the twitch, ensues. Repeated twitches superimpose and, if the stimulation frequency is high enough, fuse into a maintained tetanic contraction. Are such differences between tonic and twitch fibres merely due to a difference in the excitability of the muscle cell membrane or must they be ascribed to a more fundamental difference in the coupling processes between excitation and contraction?

Under certain conditions, the membranes of slow tonic muscle also become excitable (Miledi et al. 1981), so that now the application of an appropriate stimulus causes an action potential followed by a twitch. Conversely, the occurrence of action potentials in twitch fibres can be experimentally prevented. If then the membrane is more and more depolarized, the muscle fibre does not respond in an "all-or-none" manner, but with an increasingly stronger contraction showing that the relation between membrane potential and contraction can be graded. Depolarization can be achieved, for instance, by raising the potassium concentration in the extracellular fluid (Kuffler 1946). From these experiments it became clear that the contractile mechanism of a twitch fibre *is* capable of contracting in a graded manner similar to that of tonic muscle fibres and that the "all-or-none" character of a twitch response is merely a consequence of the "all-or-none" nature of the action potential in a highly excitable membrane.

Membrane excitability depends not only on Na^+ and K^+ channels, but may also be controlled by chloride channels. Thus, hyperexcitability associated with repetitive firing of action potentials and impaired relaxation may occur in myotonic mutants (ADR mice) that lack a functional chloride channel (Steinmeyer et al. 1991).

1.2 Electromechanical Coupling

The intracellular processes that follow the electrical events of the membrane and lead to the activation of the contractile material within the cell are described as excitation-contraction coupling (Sandow 1952) or also as electromechanical coupling. In the last decades, research has gone a long way in determining how the coupling between the membrane potential changes and contraction is brought about and how control mechanisms (Fig. 1.3) evolved to meet the functional requirements in different kinds of muscle. Since the early work of Hodgkin and Huxley, it is well known that action potentials of nerves and muscle involve longitudinal membrane currents as well as currents across the membrane. In vertebrate twitch muscle fibres, as in nerves, these arise mainly from inwardly moving sodium ions and outwardly moving potassium ions (Hodgkin 1964), while "calcium inward currents" are more important in many invertebrate muscles (Fatt and Ginsborg 1958; Gilly and Scheuer 1984). The question arises, therefore: is it the ion current or the change in membrane depolarization per se that elicits contraction?

1.2.1 The Role of Membrane Depolarization

As mentioned already, in tonic frog muscle fibres, graded membrane depolarization causes contraction. In the case of twitch fibres longitudinal currents are only effective in causing contraction to the extent that they cause membrane depolarization (Sten-Knudsen 1960). Furthermore, Orkand (1962) altered the membrane potential in crustacean muscle fibres by injecting current through a microelectrode or by depolarizing the muscle fibre with an increased concentration of potassium chloride. He concluded that the degree of activation was related to the absolute magnitude of the membrane potential rather than to the extent of depolarization or current flow.

What then is the precise relationship between membrane potential and activation of the contractile material in twitch fibres? This problem was addressed by Hodgkin and Horowicz (1960), who attached an isolated, single twitch muscle fibre of the frog semitendinosus at one end to a muscle holder and at the other end to a stiff lever connected to an electromechanical transducer recording the

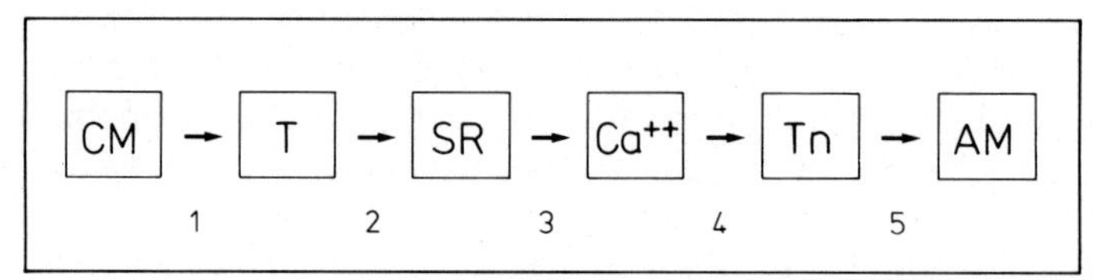

Fig. 1.3. Stages in the excitation-concentration coupling process of striated muscle. *1* Depolarization (excitation) of the cell membrane (*CM*), and inward spread of excitation along T-tubules (*T*); *2* signal transference from the T-system to the sarcoplasmic reticulum (*SR*): T-SR coupling; *3* Ca-release from SR; *4* calcium binding to troponin; *5* activation of actomyosin (*AM*) in the contractile system: relief of inhibition of interaction of actin and myosin leading to the contractile response (activation of actomyosin ATPase, force development, shortening)

force. As long as the fibre was in physiological saline it remained relaxed, but when the potassium ion concentration was increased, to e.g. 20 mM or 30 mM, the fibres developed force while held at constant length and this isometric contraction was soon followed by relaxation as though an "activator" had been removed. A microelectrode inserted into the muscle fibre served to record the membrane potential. It became less and less negative with increasing potassium concentration, the relationship between the membrane potential and the logarithm of the potassium concentration being linear as predicted from the Nernst equation. Force development, on the other hand, was a more complex function of the potential (Fig. 1.4 B). Thus, the muscle fibres remained relaxed until the membrane potential reached a mechanical threshold of about −50 mV at 20 mM K^+. When the potassium concentration was further increased, the force of the fibre rose steeply with the change in potential to reach a maximum at about −20 mV. No additional tension occurred at less negative potentials even when the membrane was completely depolarized at a potassium concentration equal to the intracellular concentration of K^+. In crustacean muscle fibres (Orkand 1962; Zachar and Zacharová 1966) and in fast and slow mammalian twitch muscle fibres (Lorkovíc 1983) force was demonstrated to depend on the potassium concentration and membrane potential in a similar fashion (cf. Fig. 1.4 B). It is noteworthy, however, that in slow muscle the mechanical threshold potential appears to be somewhat more negative than in fast muscle.

In order to investigate the dependence of force over a greater range of membrane potentials, it is convenient to use the voltage clamp technique, whereby two microelectrodes are inserted into a short segment of a muscle fibre. One is used to control the membrane potential, while the other serves to inject or withdraw current until the membrane potential reaches the desired value. This potential is then "held" at a constant value by means of a feedback system which automatically compensates for any flow of depolarizing or repolarizing ionic currents across the membrane. In this way, Léoty and Léauté (1982) compared the force-voltage relationship in fast and slow twitch muscle fibres in the extensor digitorum longus muscle and the soleus of the rat (Fig. 1.4 A). In both fibre types force can be graded over a potential range of −50 mV to −20 mV whereas very little change occurred in the range of −10 mV to +30 mV. This means that a stimulus causing an "all-or-none" type of action potential exceeding +30 mV will always activate the contractile material fully and with a large safety margin and it will, therefore, result in an "all-or-none" twitch response.

It should be noted, however, that the contractile response depends not only on the extent, but also on the duration of membrane depolarization. Thus, as illustrated in Fig. 1.4 C, the relationship between extent and duration of depolarization eliciting a just noticeable "threshold contractile response" may be described by a hyperbolically shaped stimulus-strength-duration curve which is different in invertebrate and vertebrate muscle (Gilly and Scheuer 1984), but remarkably similar in fast and slow tonic vertebrate muscle fibres (Gilly and Hui 1980 a).

Fast and slow twitch fibres and tonic frog muscle fibres, however, differed in the time course of a potassium contracture. When the slow tonic rectus abdominis muscle of a frog or a slow tonic insect leg muscle (Aidley 1965) was immersed in

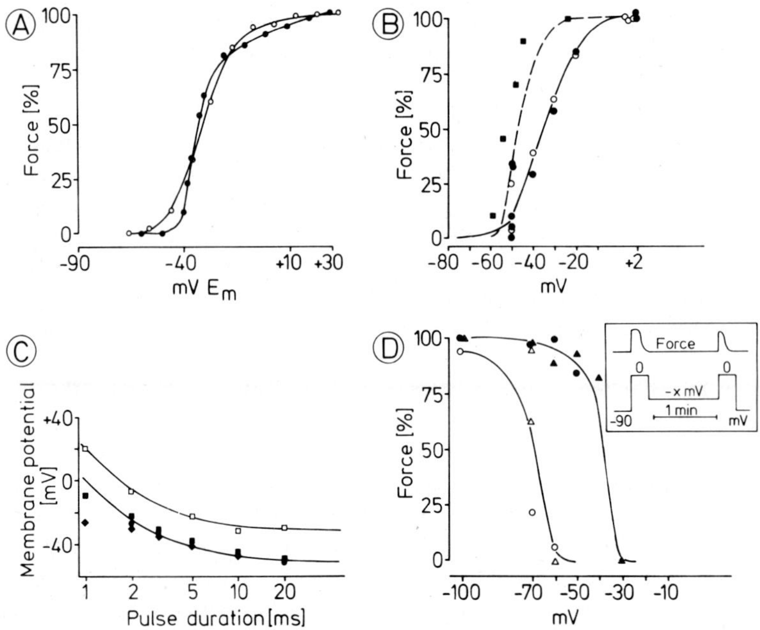

Fig. 1.4. Depolarization-contraction coupling in striated muscles. *A* Relation of peak contracture force and membrane potential in a fast and slow twitch muscle of the rat. ● = EDL (fast); ○ = soleus (slow). The membrane potential was clamped at the values shown on the *abscissa.* (Léoty and Léauté 1982). *B* Potassium contractures. *Ordinate:* Force in % maximum; *abscissa:* membrane potential adjusted by varying the external K^+ concentration. --- Crustacean muscle; — frog twitch muscle at high Ca_0^{2+} (3 mM; ●) and very low (10^{-8} M) external Ca^{2+} concentration (○; based on Lüttgau and Spiecker (1979) and Zachar and Zacharová (1966). For comparison: data from the classical experiments of Hodgkin and Horowicz (1960) with frog muscle ■. *C* The potentials required to elicit a just threshold contraction varies with the duration of a depolarization pulse, as shown in these stimulus-strength-duration curves. □ = scorpion muscle; ●◆ = frog twitch fibres; ■ = frog tonic fibres. (Adapted from Fig. 7 of Gilly and Scheuer 1984) *D* Depolarization-induced inactivation in frog twitch muscle. Loss of contractile force if muscle fibres are partially depolarized prior to maximal K-stimulation. *Ordinate:* Relative force elicited by K-induced depolarization (with 190 mM K) to 0 mV in dependence of the membrane potential (shown on *abscissa*) before contractile stimulation. *Inset:* Experimental procedure showing control contracture followed by test contracture (●▲ = 3 mM Ca_0^{2+}; ○△ = 10^{-8} M Ca_0^{2+}). (Lüttgau and Spiecker 1979)

saline containing high potassium, a contracture developed that was maintained for minutes. When, on the other hand, a fast twitch muscle fibre was depolarized by potassium, the contracture only lasted for a few seconds and somewhat longer in the case of slow twitch muscle. Then the muscle fibre relaxed and the activating system became refractory or inactivated. Unless the muscle fibre had been exposed to a lower potassium concentration for a time sufficient to "reprime" the activating system, a second contracture could not be elicited again by high potassium. The size of the second contracture evoked with high potassium then measured the extent of repriming or restoration of the activating system. It was inversely related to the membrane potential in the "repriming phase". This relationship was described by a sigmoidal curve, the "steady-state inactivation curve". As shown in Fig. 1.4 D, a low external calcium ion concentration shifted this curve to more negative values; it also accelerated potential-dependent inactivation processes, which may be one of the reasons for excitation-contraction uncoupling of twitch muscle fibres at low external calcium (Lüttgau and Spiecker 1979; cf. also Lüttgau et al. 1986).

The coupling between membrane potential and contraction, sometimes called electromechanical coupling or depolarization-contraction coupling, has been found to be remarkably similar in different kinds of muscle, including cardiac and smooth muscle. It can, however, be pharmacologically modified. For instance, potentiators of the twitch contraction, such as perchlorate (Gomolla et al. 1983) and caffeine, shift the curve of the force-potential relationship towards more negative values (Caputo et al. 1981), while local anaesthetics may shift the curve to the right.

The above examples of altered coupling between membrane potential and contraction suggest that membrane depolarization or the electric field cannot, of course, be the ultimate mechanism in the activation of muscular contraction as was once supposed (Szent-Györgyi 1953). This conclusion is supported by a wealth of comparative physiological data. For instance, the myocytes retracting the oscula of sponges are relaxed when depolarized by high potassium, but they can then be stretch-activated without any change in membrane potential (Prosser 1967). In crustacean muscle fibres it is possible to induce abnormally high action potentials that are not followed by contraction (Fatt and Katz 1953), and in many smooth muscles a contracture can be induced by hormonal stimulation without any change in membrane potential: pharmacomechanical coupling (A. P. Somlyo and Somlyo 1968). Many decades ago, Heilbrunn and Wiercinski (1947) showed that injection of calcium salts into muscle fibres would cause them to contract; again, this activation occurs without membrane depolarization (Caldwell and Walster 1963). Moreover, the contraction may even be possible without membrane at all. This is demonstrated in experiments in which the cell membrane is mechanically removed and the myoplasm is replaced by an artificial intracellular medium containing ATP as the only energy source. The demembranated or skinned fibres obtained in this way are relaxed at very low calcium ion concentration, but contract after addition of calcium, suggesting that calcium may be the intracellular messenger responsible for triggering contraction (Bozler 1953).

1.2.2 The Importance of Calcium

The existence of quite different modes of activation that induce contraction led to the question whether there was, in all cases, an ultimate common pathway perhaps involving calcium. Support of this concept came from the calcium injection studies just mentioned and from "calcium depletion experiments". In calcium-free saline tonic frog muscle and insect muscle did not contract after increasing the potassium concentration, but contractility could be restored by readdition of calcium, as shown in Fig. 1.5 (Aidley 1965). The contraction of most invertebrate muscles is absolutely dependent on the extracellular calcium ion concentration (Gilly and Scheuer 1984): calcium ions flowing into the cell during the action potential may trigger the contraction. Cardiac muscle failed to contract when the

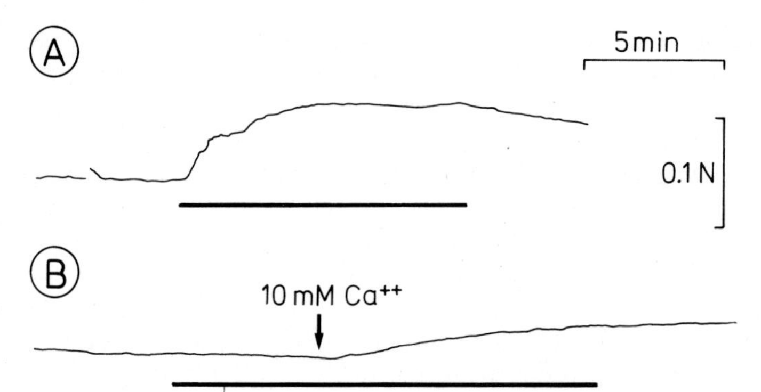

Fig. 1.5. In a slow insect muscle (Extensor tibialis of the mesothoracic leg of the locust *Schistocerca gregaria*) a potassium contracture (164 mM K) is slow and maintained (*A*), but can only be elicited in the presence, and not in the absence, of external calcium (*B*). *Horizontal bar* indicates K-depolarization. (Aidley 1965)

bathing saline was made calcium-free (Ringer 1883), whereas action potentials could still be observed (Mines 1913). Since the removal of calcium uncoupled excitation and contraction, it seemed that these ions must somehow be involved in the excitation-contraction coupling. The question arises, therefore: Is contraction turned on by intracellular release of calcium ions, as suggested by Bailey already in 1942? This idea was, however, met with resistance. How could anything as simple as a calcium ion perform such a specific function, in particular in view of A. V. Hill's (1948) objection that calcium ions cannot diffuse from the excited surface membrane to the contractile substance in the fibre interior within the brief electromechanical latency between stimulating the muscle and the onset of a twitch? This conceptual obstacle was, however, overcome when A. F. Huxley and colleagues showed that excitation of the cell membrane could rapidly spread into the fibre interior along membrane invaginations that form the transversal tubular system (cf. Chap. 2.1 and A. F. Huxley 1974). Depolarization of the tubules then causes the sarcoplasmic reticulum to release calcium ions which activate the contractile machinery, as shown by the above-mentioned skinned-fibre and calcium-injection studies. The ATPase of the contractile proteins is activated and the contractile structures shorten and/or develop force. A detailed discussion of the contractile processes is outside the scope of this book, and the reader is referred to the books of A. F. Huxley (1980), Squire (1981), Woledge et al. (1985), and to Needham (1971) to obtain a historical perspective. Nevertheless, it may be helpful now to look briefly at the contractile structures and mechanisms that are activated by calcium ions in the process of excitation-contraction coupling.

1.3 The Contractile Process

Much of the basic knowledge on muscle contraction (cf. Bagshaw 1982; Woledge et al. 1985) has been obtained from experiments with isolated frog sartorius or even single muscle fibres stimulated with electric current pulses applied with a number of electrodes distributed along the length of the muscle so that all parts of the preparation are activated at the same time. If one end of the muscle is fixed and the other one is attached to a weight, the muscle will shorten in an isotonic contraction, thereby lifting the load and performing mechanical work. If held rigidly fixed at both ends, the stimulated muscle cannot shorten, but contracts isometrically, thus developing force and reaching a peak in about 0.2 s at 2 °C. The muscle is activated!

1.3.1 Characteristics of Activation

The interval between the stimulus and the beginning of isometric tension rise, the latent period, lasts about 10–15 ms in a frog sartorius at 0 °C. But already before the end of latency, the resistance to stretching (stiffness) increases. Soon afterwards, the muscle is able to bear a load comparable to the force developed in tetanus. This rapid mechanical change prompted the suggestion that the transition from resting to active muscle occurs suddenly soon after stimulation, despite the fact that tension rises comparatively slowly (Hill 1950). According to Hill, muscle resembles a system made up of two components, contractile elements in series with elastic elements. Very soon after stimulation the contractile elements are supposedly fully active, but as long as the series elastic elements remain slack, the muscle appears slack as well; force is only developed when the fully activated contractile elements shorten, thereby extending the series elastic components; indeed, a time-consuming effect. Hill's popular concept of the sudden onset of the "active state" can, however, no longer be maintained in its original sense in view of several more recent discoveries:

1. Activation is accompanied by a dramatic increase in "immediate" stiffness which precedes contraction only slightly: it rises with a time course approximately similar to that of isometric tension (Figs. 1.6 and 1.8). This immediate stiffness obeys Hooke's law as shown by A. F. Huxley and Simmons (1973; cf. also Ford et al. 1986; Cecchi et al. 1986). When one end of a single muscle fibre was suddenly released or stretched by various amounts, e.g. 0.2% or 0.4% of the initial muscle length, force (F) changed in synchrony with and in proportion to the length step Δ L. Immediate stiffness (Δ F/Δ L) then is determined by the slope of the nearly linear portion of the force-length diagram, the T_1 curve, as illustrated in Fig. 1.6 (Ambrogi-Lorenzini et al. 1983). Note that force drops to zero, i. e. the curve intercepts the abscissa, when the muscle is released by about 1% of its initial length regardless of whether the release was performed during or at the end of force development. This is not expected on the model proposed by A. V. Hill. Here, the series of elastic components would have been stretched more under high tension than at low tension. According to Hill's model, therefore, the amount of release required to abolish tension should increase during the rise of tetanic tension, in contrast to that which has been found experimentally.

2. It is well known that shortening speed depends on muscle load in a hyperbolic manner (Hill 1938). In contrast to Hill's assumption, however, the establishment of the force-velocity relationship in stimulated muscle is a relatively slow process and may reflect the time course of activation (Cecchi et al. 1978). While

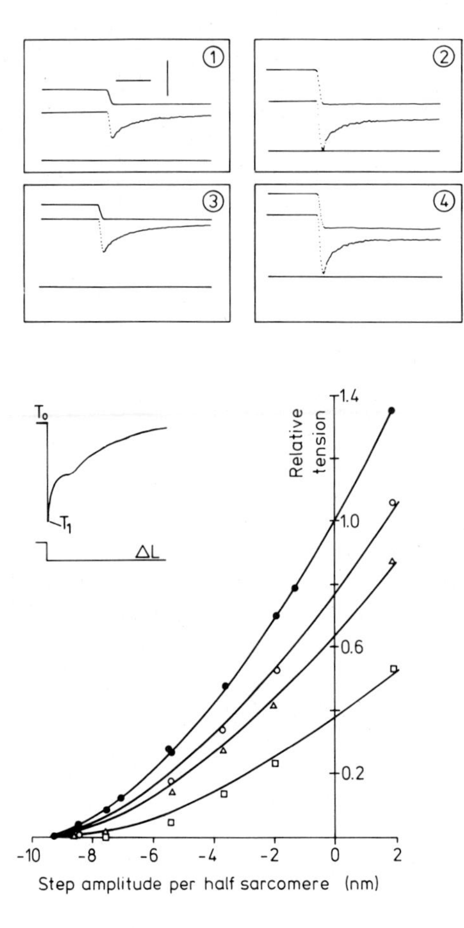

Fig. 1.6. Time course of activation. Force and "immediate stiffness" at different times after onset of tetanus (frog tibialis anterior muscle at 6 °C, stimulated at 15 s^{-1}). *Above:* Method of determining force and stiffness during rise of tetanic tension. Force transients due to quick change in length, applied 80 ms (*upper panels 1,2*) and 200 ms after the onset of tetanic stimulation (*lower panels 3,4*). The length changes (*upper tracings*) were 4 nm/half-sarcomere (*left: 1,3*) or 8 nm/half-sarcomere (*right: 2,4*). *Middle tracings* represent force, *lower tracings* represent resting tension. Vertical force calibration = 15 Ncm^{-2}, horizontal time calibration 1 ms. *Below:* T_1 curves: extreme forces (T_1 values) induced by length change ΔL (cf. *inset*) plotted versus the extent of release or stretch. A step amplitude of 10 nm/half-sarcomere corresponds to about 1% length change. The T_1 curves were obtained at 44 ms (□) 60 ms (△) 80 ms (○) and 240 ms (●) after the onset of stimulation and all intercept the abscissa at about −9 nm/half-sarcomere. Slopes of T_1 curves represent stiffness. The *vertical scale* indicates the relative isometric tetanic force (T_0) as fraction of maximal tetanus force after 200 ms stimulation. (From Ambrogi-Lorenzini et al. 1983 with permission)

unloaded muscle can shorten at its maximal speed very soon after the onset of tetanic stimulation, the capacity to lift a load quickly develops but gradually (Fig. 1.7). Only at a time when force reaches about 50% of its final value is the load-velocity relationship indistinguishable from that of fully tetanized muscle (Fig. 1.8).

3. Activation and contraction are associated with an alteration of the molecular arrangement of the contractile material as determined by X-ray diffraction (H. E. Huxley 1979). These changes (in the intensity ratio of the so-called 1.0 and 1.1 reflections, see Sect. 1.3.3.1) occur gradually and slowly during the rise of force (Fig. 1.12 C).

As the muscle contracts it produces heat which arises from the chemical reactions that take place, in particular the splitting of adenosine triphosphate (ATP) to adenosine diphosphate (ADP) and phosphate (P_i). The hydrolyzed ATP is almost immediately reformed from phosphocreatine which phosphorylates ADP, and breaks down to creatine. Thus, the net concentration of ATP stays practically constant even in fatigued muscle (Dawson et al. 1977). Only if the creatine phosphokinase and other ATP-regenerating reactions are poisoned is it possible to

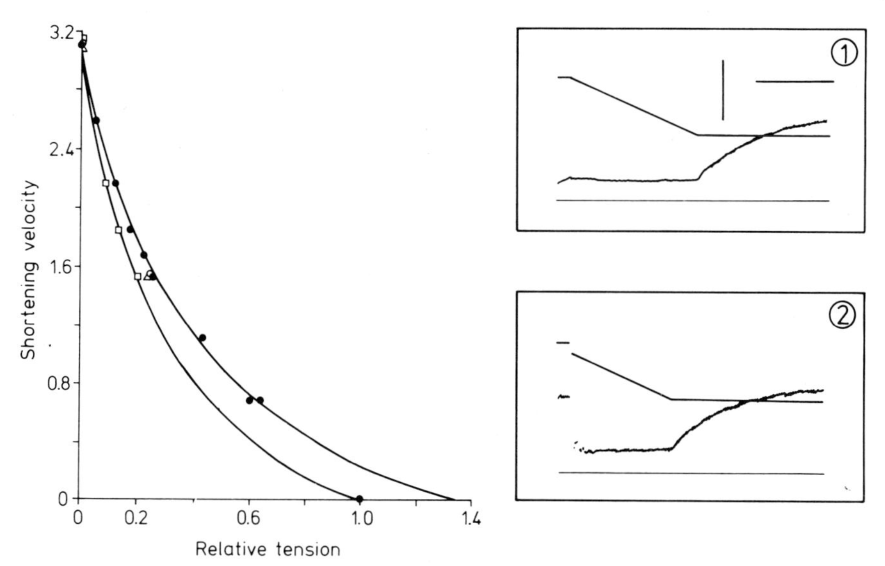

Fig. 1.7. Time course of activation. *Left:* Relationship between force and velocity of shortening (μms^{-1} half-sarcomere) at different times during the development of tetanic force. □ 28 ms, △ 40 ms; and ● 200 ms after beginning of stimulation. Frog tibialis anterior muscle fibre at 6 °C, stimulated at 15 s^{-1}. *Right:* Method of determining force-velocity relation at different times during tetanic rise in force. *1* At 28 ms after the onset of stimulation fibres were released at constant speed (e.g. 1.5 μm s^{-1} per half-sarcomere, as shown in the *upper tracing*) and force exerted during this ramp shortening was recorded (*middle tracing*). *Lower tracing* indicates resting tension. Note that isometric force continues to rise after the cessation of shortening. *2* Ramp shortening 60 ms after onset of tetanus. Horizontal calibration: 20 ms; vertical calibration: 15 N cm^{-2} fibre cross-section. (From Ambrogi-Lorenzini et al. 1983, with permission)

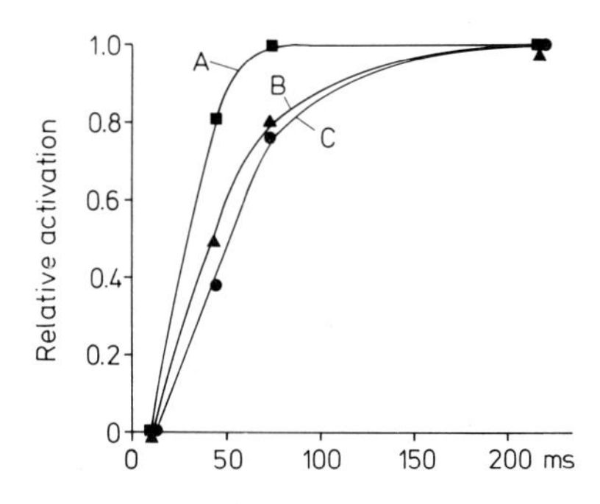

Fig. 1.8. Time course of activation. *A* Establishment of force-velocity relation (activation = force exerted at given speed of ramp shortening). *B* Development of immediate stiffness (force change per change in length). *C* Development of isometric force plotted versus time after onset of tetanic stimulation (on *abscissa*). (Based on Ambrogi-Lorenzini et al. 1983)

demonstrate the usage of ATP during contraction directly (Davies 1964). In a twitch contraction about 0.3 μmol ATP are hydrolized by 1 g muscle in about 200 ms, mainly by the actin-activated, magnesium-dependent ATPase. This means, of course, that during contraction the contractile ATPase activity becomes strongly activated, reaching a level 100–1000 times higher than during rest.

What are the biochemical mechanisms behind this enzymic activation and what are the molecular mechanisms that lead to the gradual increase in stiffness, force and shortening ability during contraction?

1.3.2 The "Contractile" ATPase

The proteins within the contractile organelles, the myofibrils, consist of about 60% myosin and 20% actin. That myosin is involved in contraction seemed apparent when it was shown to hydrolyze ATP (Bailey 1942). Myosin is a long, rod-shaped molecule with a molecular weight of about 500 kDa which bears two globular heads to which two "light" peptide chains are loosely attached (Lowey 1971, Fig. 1.9). One type of light chain is called the essential or alkali light chain since it is easily removed under alkaline conditions. The other one is called the regulatory light chain or P-light chain since it may be phosphorylated. The catalytic site is in the heavy chain of the head portion which can be separated from the remainder of the myosin molecule by papain digestion (Lowey 1971). Many enzymic studies on myosin ATPase have been done using this subfragment S-1 fraction rather than the intact myosin molecules. Under ionic conditions, similar to those of the myoplasm, i.e. at low ionic strength and roughly equimolar concentrations of ATP and Mg ions, the ATP-splitting rate by myosin alone is quite low. However,

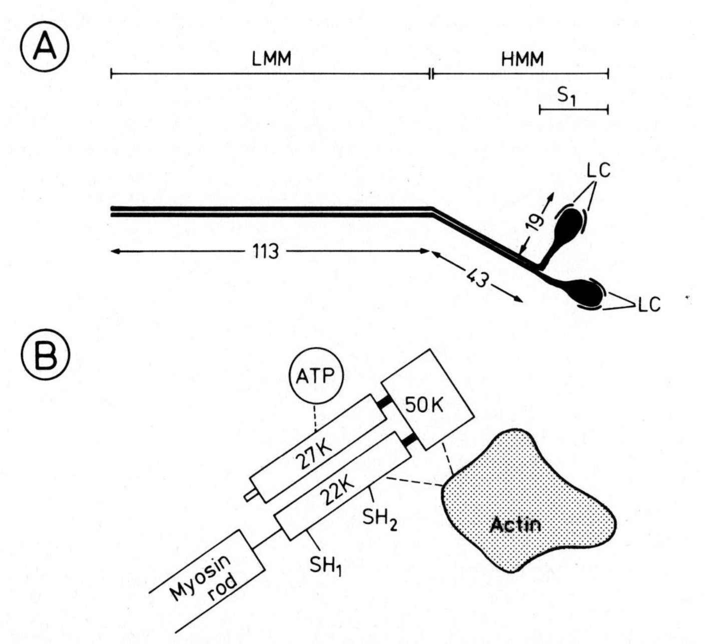

Fig. 1.9. Structure of myosin. *A* The myosin molecule. Composition: two heavy chains consisting of light meromyosin (*LMM*) and heavy meromyosin (*HMM*) bearing two "heads": subfragment-1 (S_1) of heavy meromyosin; four light chains (*LC*), namely two essential light chains (alkali light chains) and two regulatory light chains (synonyms: P-light chains; DTNB-light chains); the latter can be removed with the reagent DTNB and they may be phosphorylated, whereas the so-called essential light chain can be removed under alkaline conditions. Dimensions are given in nm. *B* Domains of myosin subfragment-1, which are involved in the binding of actin, substrate (*ATP*) and alkali light chain. The latter (not shown) extends into the hinge region of the myosin molecule (after Goody and Holmes 1983). The structure of the actin myosin head complex is different in the presence and absence of ATP as shown by electronmicroscopy (cf. Frado and Craig 1992). For the structure of actin and myosin heads, see also Kabsch et al. (1990) and Winkelmann et al. (1985)

it becomes dramatically activated following the addition of the muscle protein actin which has a molecular weight of 42 kDa and exists in muscle in a polymerized form known as F-actin.

The mechanism of ATP hydrolysis has been reviewed by Taylor (1979). Many steps are involved, as shown in Fig. 1.10. It is now well established that the low enzymic activity of purified myosin is not due to the slowness of the hydrolytic step. In fact, when ATP is suddenly mixed with subfragment S-1 in a stopped flow experiment, the enzyme-substrate complex is rapidly formed and the energy-rich bond of ATP is immediately hydrolyzed to form the enzyme-product complex. The latter decomposes, however, very slowly with a halftime of tens of seconds which corresponds to a rate constant of much less than 0.1 s^{-1}. Unless the products are released, a new molecule of ATP cannot be bound and hydrolyzed. Hence, the product release is the slowest reaction during continuous ATP splitting and determines its rate (Bagshaw et al. 1974). By accelerating product release actin activates the Mg-dependent myosin ATPase. Actin combines with myosin to form a ternary actin-myosin product complex which decomposes in a fraction of a second to give an actin-myosin complex (cf. Fig. 1.10 A, Pathway II). After binding a new molecule of ATP, this complex dissociates into actin and myosin ATP, thereby a new cycle of hydrolysis commences (Fig. 1.10 B).

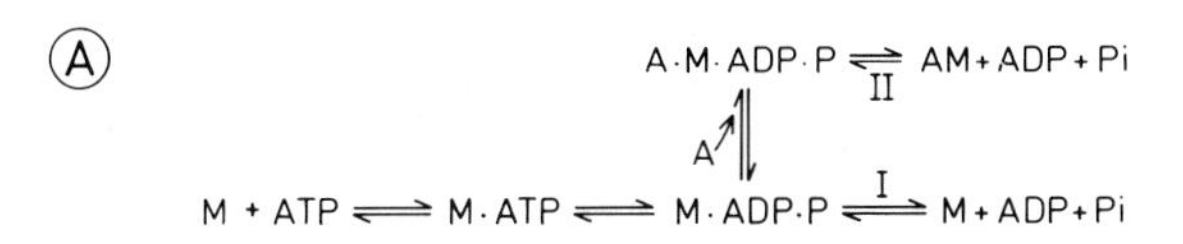

B AM + ATP ⇌ AM·ATP ⇌ AM*·ATP

A

M*·ATP

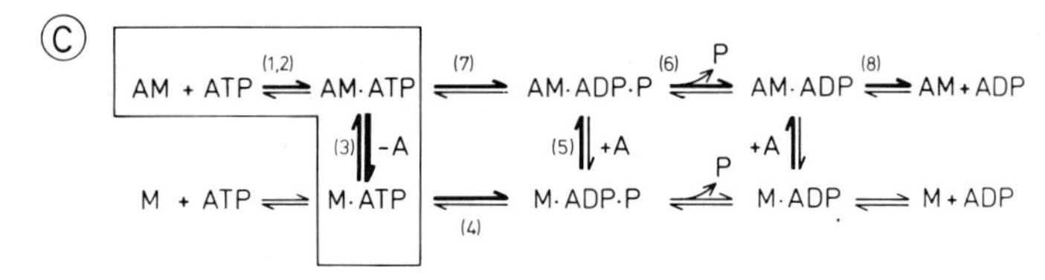

Fig. 1.10. The mechanisms of ATP hydrolysis: *A* ATP hydrolysis by myosin (M) involving the formation and dissociation of the enzyme substrate and enzyme-product complex (*pathway I*). In the presence of actin (A) the reaction proceeds according to *pathway II* since actin accelerates the decomposition of the enzyme-product complex and, hence, increases ATPase activity. *B* Dissociation of actomyosin (AM) into actin and myosin by ATP. *C* A combination of schemes *A* and *B*: actomyosin is dissociated into actin and myosin-ATP which is rapidly hydrolyzed to myosin-ADP · P, which then reacts with actin to form actomyosin ADP · P. This compound may, however, also be formed when the actomyosin-ATP complex is hydrolyzed by the non-dissociating *pathway 7,* see text

Hence, ATP has two actions. Firstly, it serves as a substrate for the actin-activated myosin ATPase to which it supplies energy. Secondly, it breaks the linkages between actin and myosin, whereby dissociating the actomyosin complex. Both actions occur in sequence when the ATP is hydrolyzed by the actin-activated myosin ATPase. The predominant pathway of this reaction is illustrated in Fig. 1.10 C (cf. White and Taylor 1976). Additional steps may be required to describe the release of products, since it would seem that phosphate is released before ADP and presumably it is this step during which most of the free energy, derived from the splitting of the energy-rich phosphate bond of ATP, is set free. This step is relatively easily reversible since labelled inorganic phosphate ($^{32}P_i$) may be reincorporated into ATP, a process that can only occur through the intermediate formation of an energy-rich, actomyosin-ADP complex (Ulbrich and Rüegg 1971; Sleep and Hutton 1980). According to the scheme, the affinity of myosin for actin would depend on the state of the bound nucleotide: it is high for actin and myosin, low for actin and myosin occupied with the products of ATP hydrolysis (ADP and P_i) and also low between actin and myosin occupied with ATP, but perhaps intermediate in the actomyosin-ADP complex detected indirectly by the ^{32}P incorporation experiments.

One gram active frog muscle splits about 1.5 μmol ATP s^{-1} during an isotonic contraction at 0 °C (Davies 1964). Since the content of myosin in the muscle fibre is about 120 nmol g^{-1} muscle, the rate of ATP splitting by subfragment S-1 must be of the order of 6 s^{-1} in situ, assuming that both heads function independently. In close agreement with this prediction, a rate of 4 s^{-1} could be measured in solution when subfragment S-1 was activated with actin, but only about 0.01 s^{-1} when subfragment S-1 was investigated in the absence of actin. Thus, the high ATP-splitting rate in contracting muscle can only be accounted for under the assumption that, under these conditions, myosin interacts with and is activated by actin. Conversely, we must assume that actin-activation is blocked or repressed in resting muscle and the question arises as to how derepression is brought about during activation. As actin-activated ATPase is often regarded as the biochemical correlate of contraction, it may provide an excellent model for studying the molecular processes in muscle activation in vitro. We shall discuss later on in more detail how actin and myosin interaction is controlled by calcium in vertebrate skeletal muscle (Chap. 4), molluscan muscle (Chap. 6) and vertebrate smooth muscle (Chap. 8). The enzymic studies on actomyosin ATPase have propelled the idea that actin-myosin interaction is the basic event in muscle contraction which over the years has been studied on a molecular level by many investigators.

1.3.3 Actin-Myosin Interaction in Muscle Contraction

Ever since the Szent-Györgyi discovery that ATP induces contraction of artificial protein threads made of actin and myosin, these proteins have been referred to as contractile proteins. The way in which the two proteins interact during muscle contraction has been elucidated by a combined approach of structural, biochemical and mechanical studies and is described by the sliding filament theory of contraction.

1.3.3.1 Structural Basis of Contraction

Actin and Myosin Filaments. Within the muscle cell myosin molecules arrange themselves in the form of bipolar filaments approximately 1.6 µm long and 10 nm in diameter (H. E. Huxley 1963). The myosin heads project sideways from the filament shaft, like branch stumps from a tree trunk, and reach across to a neighbouring "thin filament" which consists mainly of polymerized actin (and tropomyosin). In this way the projections form spirally arranged crossbridges. The thick and thin filaments are longitudinally oriented with respect to the fibre axis and the precision of the arrangement resembling a crystal lattice is astonishing (cf. H. E. Huxley 1957). The degree of order and the kind of arrangement, however, vary greatly in different types of muscle (Squire 1981). In vertebrate cross-striated muscle the array of myosin filaments is such that each thick filament is surrounded by six thin filaments, whereas each thin filament is surrounded by three thick filaments in a hexagonal lattice. The ratio of actin filaments to myosin filaments is 2:1, whereas it is 3:1 in insect flight muscle and often much higher in slow invertebrate muscle. In vertebrate striated muscles the myofilaments form bundles, the myofibrils, which are about 1 µm in diameter and constitute the contractile organelles of the fibre. Partitions, the Z-discs, subdivide them into serially arranged compartments of about 2 µm to 3 µm in length, the sarcomeres (Fig. 1.11). Actin filaments, which are fixed to the Z-disc on either side like the bristles of a brush, extend by about 1 µm towards the middle of the sarcomere. Here, they interdigitate with the 1.6-µm-long myosin filaments which appear as a dark anisotropic structure, the A-band, in the light microscope. The bands are flanked by isotropic light zones, the I-bands, which contain only thin filaments. It is this regular arrangement of the filaments that gives the myofibrils their cross-striated appearance (cf. Fig. 2.4).

In many invertebrate muscles the length of filaments and sarcomeres is much longer and the lateral arrangement of myofilaments is different. In all muscles of annelids and nematodes and in many molluscan muscles the sarcomere divisions are not transversely, but diagonally, oriented with respect to the fibre axis and are, therefore, described as obliquely striated. In smooth muscle a regular arrangement of filaments and, hence, cross-striations cannot be recognized.

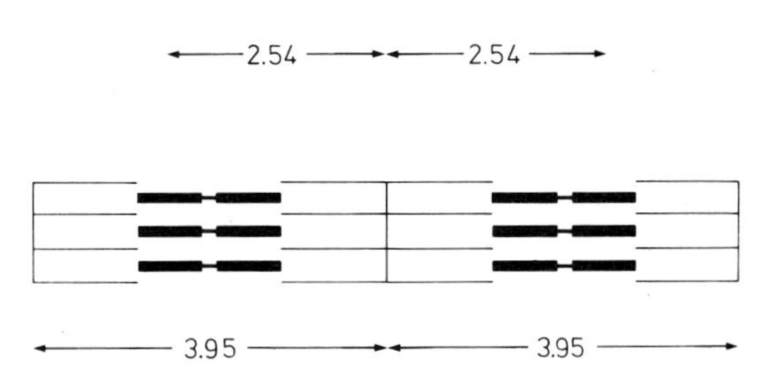

Fig. 1.11. The sliding of thin filaments past thick filaments when the sarcomere length is altered. At the sarcomere length of 3.95 µm no force can be developed in mammalian striated muscle, whereas at 2.54 µm force is maximal (D. G. Stephenson and Williams 1982)

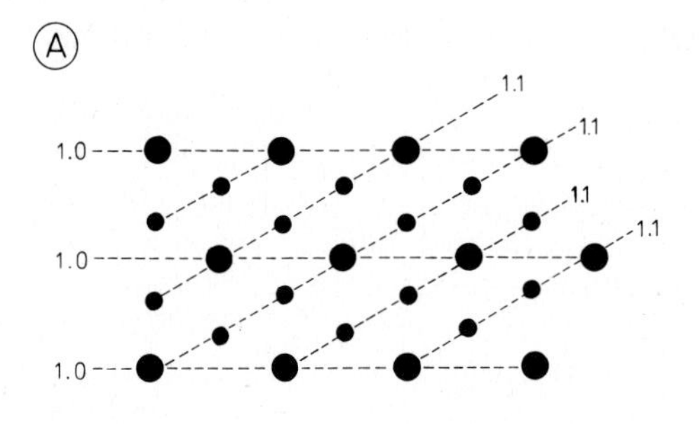

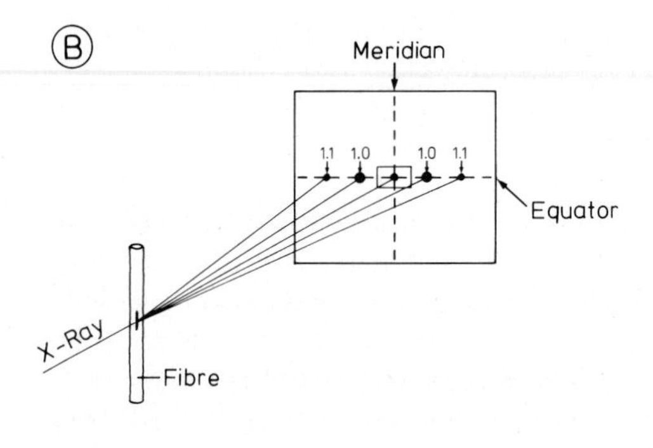

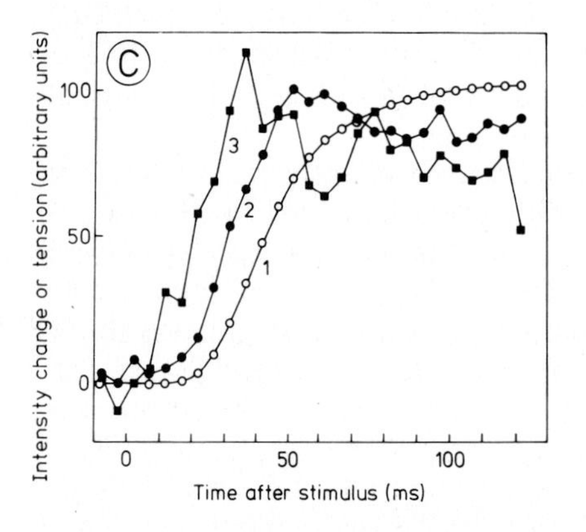

Fig. 1.12 A, B. X-ray diffraction and filament lattice. *A* Schematic diagram of filament lattice that is responsible for the equatorial reflexions. The regular repeating intervals of the 1.0 planes of thick myosin filaments constitute a "diffraction grid". They diffract the X-ray beam by a certain angle, causing 1.0 reflexions when a vertically suspended frog muscle is subjected to a parallel beam of X-rays as shown in *B*. The 1.1 pattern arises from the reflexions of the 1.1 planes containing both thick and thin filaments. *B* Equatorial low angle diffraction by a muscle fibre. The diffracted beams of the X-ray (diffraction angle about 2°) cause a characteristic pattern of spots on a photographic plate. For clarity, only spots corresponding to the 1.1 and 1.0 reflexions are shown. Note that compared with the 1.0 reflexions of the relaxed muscle, the scattering angle of the 1.1 reflexions is larger, while the intensity of the reflexion is weaker, as there is less mass in the 1.1 planes, and the repeating intervals of these planes are smaller than in the case of the 1.0 planes. During isometric contraction, the intensity of the 1.1 reflexions increases due to the azimuthal movement of crossbridges towards the actin filaments, resulting in an apparent increase in the mass of the latter within the 1.1 plane. For time-resolved studies the

Structural Changes During Isotonic Contraction and Stretch. The muscle shortens as the result of the shortening of innumerable sarcomeres connected in series. During shortening, the thin actin filaments slide past the thicker myosin filaments, while the filament length remains constant. Thus, the width of the A-band remains constant, whereas the I-band becomes narrower, as reviewed by Pollack (1983). When sarcomeres lengthen during stretch or relaxation, the actin filaments are pulled out from the array of thick filaments so that the extent of actin-myosin overlap decreases, as shown in Fig. 1.11. At sarcomere lengths where actin and myosin filaments no longer overlap, approximately 3.6 μm in frog muscle and 4.0 μm in mammalian twitch fibres, muscle is unable to develop force. At shorter sarcomere lengths, down to about 2.2 μm, there is a linear relationship between the extent of overlap and isometric force development, suggesting that the interaction of actin and myosin is essential for contraction.

Structural Changes During Isometric Contraction. During an ideal isometric contraction, actin and myosin filaments do not slide past each other, yet the fine structure of the filament's array shows characteristic changes during activation, suggesting an increase in filament interaction.

Figure 1.12 shows that the thick and thin filaments of vertebrate skeletal muscle are hexagonally arranged in two sets of planes. When muscle fibres are exposed to a parallel beam of X-rays, these planes, designated as 1.1 and 1.0, give rise to two diffraction spots on the "equator" of the X-ray diffraction pattern. The 1.0 reflection arises predominantly from the thick myosin filaments, while the intensity of the 1.1 reflections depends on the mass of both thick and thin filaments (Haselgrove and Rodger 1980). Thus, the intensity ratio of both reflections [I (1.1)/I (1.0)] may be taken as a rough measure for the ratio of thin and thick filament mass. During an isometric contraction of muscle fibres this intensity ratio increases, thus indicating that mass has been shifted from thick filaments towards thin filaments (H.E. Huxley 1979; Matsubara and Yagi 1978). Does this mean that crossbridges swing out towards thin filaments to which they become attached in the process of actin-myosin interaction?

Actin and myosin interaction is very strong after complete exhaustion of muscle when ATP stores are depleted and rigor tension develops. Under these conditions, the 1.1/1.0 intensity change is particularly great due both to an increase in the intensity of the 1.1 reflections as well as to a decrease in the 1.0 intensity. Electron micrographs of muscle fibres fixed in rigor show crossbridges attached to actin filaments at an angle of approximately 45°, thus forming an arrowhead structure pointing towards the centre of the sarcomere (Reedy et al.

◄

powerful X-ray beams of synchrotron radiation are used and the intensity of diffracted X-rays is measured by electronic counters. During activation and contraction, the intensity ratio of the 1.0/1.1 reflexions decreases from 2.5 to 0.6. *C* Time course of changes in X-ray diffraction pattern during activation and force development of a frog semitendinosus muscle (6 °C). *1* Increase in isometrically recorded force during tetanic stimulation (20 s^{-1}). In some muscles there is a minute fall in force (latency relaxation) before tension commences to rise (-○-). *2* Increase in the intensity of the equatorial 1.1 reflexions (-●-) indicating an increase in the number of attached crossbridges (cf. Fig. 1.8). *3* Increase in intensity of second layer line (-■-) indicating a change in the structure of the thin filament (tropomyosin movement, cf. Chap. 4). Note that the tropomyosin movement reaches a nearly final value when crossbridge attachment and contraction start. (Adapted from Fig. 2 of Kress et al. 1986)

1965). Similar arrowheads are formed when isolated thin filaments are "decorated" with isolated subfragment S-1 particles in vitro and in the absence of magnesium-ATP. In relaxed muscle, on the other hand, crossbridges exhibit a more random position, indicating that they are probably not attached to actin.

1.3.3.2 The Mechanism of Force Generation

From the structural studies just mentioned, as well as from studying muscle mechanics, a picture of the contractile process emerges (A. F. Huxley 1980; but cf. also Pollack 1983) which in higly simplified form may be summarized as follows (Fig. 1.13). In resting or relaxed muscle actin and myosin are separated. At the onset of contraction the heads of the myosin molecules swing out towards the thin filament where they attach at right angles in the form of crossbridges. Consequently, the contractile structure now forms a highly crosslinked network; it stiffens and hardens. Then the myosin heads bend or rotate through an angle of approximately 45°, presumably as the result of the (electrostatic?) attraction between closely adjacent "charged" sites on the surface of the globular actin molecules and the myosin head. The oblong head, rotating as it were, on the actin filament, stretches the internal elastic structure of the muscle, presumably situated in the myosin necks, like minute molecular levers by about 5 to 10 nm. In this way, rotating crossbridges develop elastic forces and the actin and myosin filaments are, so to speak, braced by forces of traction as long as the ends of the muscle fibre are held so that the sarcomere length remains constant (isometric conditions).

Crossbridge Cycles. Of course, we must not suppose that the individual myosin crossbridge remains in an uninterrupted state of tension when force is produced and maintained, since this would only be true in the case of rigor after depletion of ATP. Rather, a myosin molecule releases the actin filament after only a tenth or a hundredth of a second. After a pause that may be equally short, the crossbridge reattaches again to the thin filament, thus completing one crossbridge cycle associated with the splitting of one molecule of ATP (Fig. 1.13). Despite the rhythmical alteration between attachment and detachment with a frequency, e.g. of 5–50 Hz, force does not oscillate (insect flight muscle may be an exception) be-

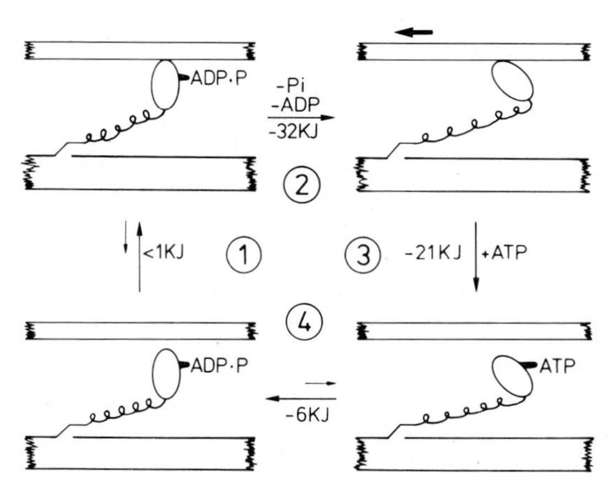

Fig. 1.13. Simplified scheme of a crossbridge cycle. *1* Attachment of crossbridges to actin; *2* power stroke; *3* crossbridge detachment; *4* recovery. The possible chemical states and changes in free energy during the cycle are after Goody and Holmes (1983). Note that ATP binding detaches crossbridges which reattach after ATP hydrolysis to form an "energy-rich" complex of actin, myosin and the products of ATP hydrolysis. The release of products is associated with a large drop of free energy (−32 kJ) which may be converted into mechanical work

cause statistically, at any given moment, there is the same number of bridges attached in the force-generating state.

Mechanical Analysis of Crossbridge Action. If one end of the isometrically contracted muscle fibre is released the attached myosin heads together with the actin filaments snap back because of the spring tension of the myosin necks producing a small movement of some 5–10 nm per half-sarcomere. Whereas a rapid release decreases spring tension in crossbridges, a quick stretch increases it by an amount depending on the stiffness of the crossbridge (Fig. 1.14; cf. A. F. Huxley and Simmons 1973). At least 1000 million of such crossbridges would have to muster their forces in parallel in order to develop the force and the stiffness characteristic of one activated striated muscle fibre. Here, the crossbridges of the myosin filaments pull on the neighbouring actin filaments with (additively) combined forces like a ship's crew pulling a rope.

Since all attached elastic bridges contribute to muscle stiffness, the latter can be used to determine the relative number of crossbridges attached at any one moment and the time course by which the net number of attached crossbridges increases during muscle activation (see Sect. 1.3.1, Fig. 1.8). The results showed that during twitch contraction the stiffness and the number of crossbridges attached at any one moment increase almost with the same time course as force. Evidently, stiffness measurement involving rapid alterations of length and tension recordings must be performed within less than a millisecond, i.e. during the lifetime of an attached state. Otherwise force measurements would be falsified ("truncated")

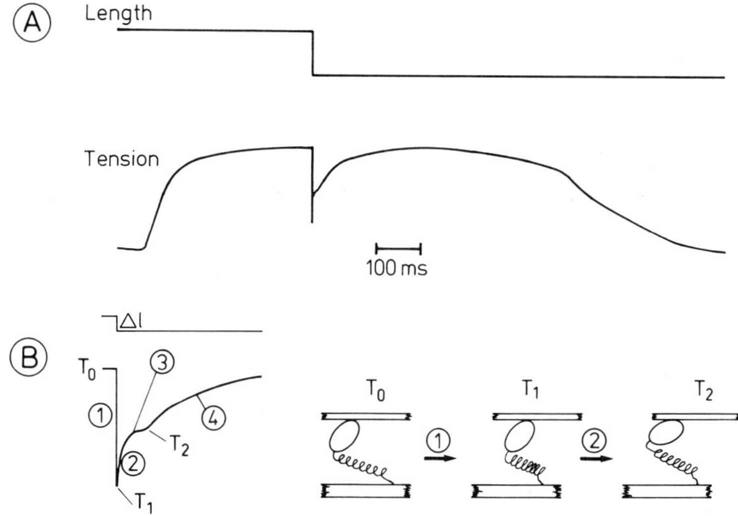

Fig. 1.14. Tension transient observed during and after a sudden change in length (quick release by 1% L_0) applied to a frog semitendinosus muscle during a brief tetanus. (After A. F. Huxley 1980) *A* Time course of tension changes at 2 °C. *B* Tension transient (schematic) showing four phases: *1* the "elastic" tension drop from the initial tetanus tension (T_0) to the extreme tension reached during step (T_1). *2* Quick phase, i.e. the tension adjustment (occurring within 1–2 ms) to the T_2 level corresponding to the plateau phase (*3*) and (*4*) the tension recovery to the initial value. *Right:* Molecular interpretation of the tension transient in terms of the Huxley-Simmons crossbridge theory, see Irving et al. (1992)

by secondary crossbridge detachments or tension adjustments ("give") caused by crossbridge rotation. It goes without saying that these requirements made it necessary to develop a sophisticated technique of precision-controlled, ultrafast changes in sarcomere length with rapid vibrators and time-resolved tension recording using sensitive electromechanical transducers of a natural frequency above 20 kHz (Ford et al. 1977).

A. F. Huxley and colleagues were able to partially synchronize the crossbridge movement for a few milliseconds following a quick release of isometrically contracted muscle fibres by approximately 0.5% resting length, corresponding to a step-shortening of half a sarcomere by about 5 nm (A. F. Huxley 1974), as shown in Fig. 1.14. This length step then results in a complex tension transient consisting of four phases: (phase 1) a rapid fall in force in phase with the length change due to the elastic discharge within the crossbridges. Then (phase 2) the force partially recovers within 1–2 ms because now discharged crossbridges, which are vertically attached, have an increased tendency to rotate into the 45°–angled attitude, thereby developing force. This "quick phase" describing the force-generating process is followed by a tension plateau (phase 3) caused by detachment of strained crossbridges and reattachment at new sites. In phase 4 attachment and detachment continue and the final recovery of tension then reflects the net attachment of crossbridges. All these events take a longer time in slow muscle (Heinl et al. 1974; Steiger 1979; Kawai and Schachat 1984).

The rapid mechanical transients, therefore, make the sequential stages of crossbridge attitudes "visible" so that it is possible to develop a general idea of the kinetics of the involved processes under different conditions of activation. Thus, in fast skeletal muscle of vertebrates crossbridge rotation and movement may occur within less than 1 ms, while a cycle of crossbridge attachment and detachment may take hundred milliseconds or even longer.

1.3.3.3 The Mechanism of Shortening

As a result of activation muscle fibres become capable of shortening. The sliding-filament theory of H. E. Huxley and Hanson (1954; cf. A. F. Huxley 1957; H. E. Huxley 1957, 1969; A. F. Huxley and Niedergerke 1954) explains the way in which the molecular movement of the myosin heads (about 10 nm) is transformed into a visible movement of the muscle. Thus, the crossbridges connected in parallel pull with united forces on the actin filaments. Owing to the bipolarity of the myosin filaments the myosin heads in the two halves of the sarcomere pull in opposite directions. In this way, the actin filaments of the two half-sarcomeres always slide in the direction towards the middle of the sarcomere. Now a single rotational crossbridge movement would make an actin filament of each half-sarcomere slide by only about 10 nm towards the centre of the sarcomere so that individual sarcomeres would shorten only by about 1% of their length. The sarcomeres of a lightly loaded frog sartorius muscle may, however, shorten at 0° C by 0.4 μm in three tenths of a second (cf. Jewell and Wilkie 1960). To achieve this the crossbridges would have to perform the pulling action just described not once, but 20 times, in this length of time like a team of men which can only haul in a long rope "hand-over-hand" by repeated heaves. By means of this rope-pulling principle, the molecular movements of countless crossbridges may cause, for instance, a 3-cm-long frog muscle to shorten by 20%, since the shortening effect of about 15000 sarcomeres connected in series would, of course, be additive.

In any muscle, the rate of shortening (Hill 1938) and, hence, the rate of filament sliding depend on the load in a hyperbolic fashion. It is greatest when the muscle fibres are completely unloaded. Under these conditions, the speed of shortening is actually limited by the rate at which crossbridges can detach after the performance of a power stroke (Ferenczi et al. 1982). While unloaded shortening seems to be governed by the rate of crossbridge detachment, it is obviously not dependent on the number of crossbridges attached at any one moment. This has been proven by Gordon et al. (1966), who stretched fibres by various amounts, thus decreasing the extent of overlap between actin and myosin filaments. In this way the number of crossbridges that could attach to actin at any one moment decreased. While this effect caused a corresponding reduction in immediate stiffness and muscle force (A. F. Huxley and Simmons 1973), the capacity for rapid shortening under no load was not impaired. Besides, unloaded muscle is capable of shortening at maximal speed soon after an activating stimulus even before substantial force is developed (Cecchi et al. 1978). On the other hand, an increase in the concentration of calcium activator in the vicinity of the myofilaments may increase the extent of force development and stiffness and perhaps also the rate of unloaded shortening, thus raising the question whether both the rate and extent of actin and myosin interaction are calcium-dependent (cf. Chap. 4).

1.3.3.4 Chemomechanical Coupling and Crossbridge Cycle

We have seen that activation involves an increase in both the mechanical output as well as in the rate of ATP splitting; this raises the question as to how the two processes are related (Bagshaw 1982; Woledge et al. 1985). During each cyclic action, the crossbridges produce force by stretching an internal elastic structure. By doing so they perform, at the expense of chemical energy, mechanical work that is then stored in the bridges in the form of mechanical potential energy. Bridges that have rotated and are attached at an acute angle to actin filaments are thus in a state of higher mechanical potential energy than crossbridges that are vertically attached. This energy is ultimately derived from the hydrolysis of ATP, so that the question arises as to how the transference of energy is brought about from the ATP molecule to the contractile structure, and how the different chemical states of ATP hydrolysis are related to the different crossbridge states. A plausible mechanism has been proposed by A. F. Huxley (1980).

Relation Between Mechanical and Enzymic Crossbridge Steps. Figures 1.13 and 1.15 illustrate a possible relationship between the mechanical and chemical crossbridge cycle. Probably one molecule of ATP will be bound to a myosin head as soon as its rotational movement, the power stroke, is completed. This lowers the affinity between actin and myosin heads which consequently detach. Immediately, the bound ATP is hydrolyzed into the products ADP and P_i with the intermediate formation of the myosin-product complex. The splitting of the "high-energy phosphate bond" is the prerequisite for the reattachment of the crossbridges

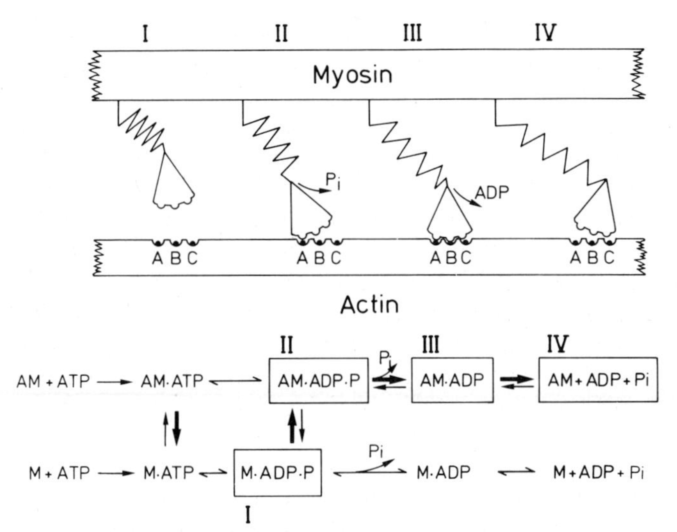

Fig. 1.15. Mechanism of force generation and filament sliding due to the rotation of myosin heads on actin filaments. (Simplified scheme, based on A.F. Huxley 1980) *Above:* Mechanical and structural changes; *below:* corresponding chemical crossbridge states. A = actin; M = myosin; A, B, C = sites on actin with high affinity for either M · P · ADP, M · ADP or M respectively, so that a rotary movement of myosin proceeds from *II* to *IV* when the states of myosin are altered (cf. also E. Eisenberg and Hill 1985). In the relaxed state myosin exists in the M · ADP · P form (*I*). When contraction is initiated myosin heads attach to actin thus forming actomyosin-ADP · P (*II*), which develops force when phosphate and ADP are released (*III, IV*). Crossbridges detach when actomyosin binds ATP and dissociates into actin and myosin. Then ATP is hydrolyzed and a new crossbridge cycle begins. Note that motion may however also be produced by deformation of the myosin head rather than by crossbridge rotation (cf. Huxley and Kress 1985)

to actin with which the myosin-product complex forms a "high-energy" ternary complex. If, on the other hand, the ATPase is inhibited, crossbridges do not reattach and the muscle stays relaxed. The next step following reattachment is the release of bound phosphate which is associated with the liberation of about 50% of the total free energy (60 kJ) available from ATP splitting. The latter may be partly recovered as mechanical energy when the crossbridges rotate into their angled attitude, thereby displacing the actin filament. After this step bound ADP is also released, a new molecule of ATP is bound and a new crossbridge cycle commences.

What is the influence of mechanical and structural constraints on the crossbridge cycle? We have seen how sequential chemical steps in the enzymic hydrolysis of ATP drive the step motions and conformational changes of the cycling crossbridges. As has been pointed out by A.F. Huxley (1980), however, the reverse may also be true. "A particular stage of the mechanical cycle permits the next step of the chemical cycle to occur" so that the rate of the chemical processes may be greatly dependent on the ease with which crossbridges can move. Such a mechanism may explain how it is possible that the total energy liberated during a contraction is much greater when the muscle is allowed to shorten and to do work than under the mechanical constraint of an isometric contraction, even if the muscles were stimulated in an identical manner in both cases. Fenn (1924) summarized these findings by the statement that the amount of work done by a muscle determines the extent of a chemical reaction that takes place.

Obviously, it would be wrong to suppose that the processes of excitation-contraction coupling following an electric stimulus cause the liberation of a definite predetermined amount of energy which would be partly liberated as heat and partly appear as mechanical work. Rather, the rate of ATP hydrolysis and, hence, the rate of crossbridge cycling may be increased when the muscle shortens. As already mentioned, the shortening speed may be rate-limited by crossbridge detachment, a process that has also been shown to be accelerated by shortening (cf. Güth et al. 1981). These few examples may suffice to illustrate the point that the chemomechanical coupling is a two-way affair. Consequently, the ATPase activity of muscle is not solely dependent on the extent of activation, eg. on the free calcium ion concentration in the myoplasm, but may be greatly dependent on the mechanical conditions. It goes without saying, therefore, that the various kinetic schemes for chemical reactions involved in the hydrolysis of ATP by subfragment S-1 and actin in vitro cannot be directly extrapolated to the situation in the intact muscle where the movement of crossbridges is influenced by various structural and mechanical constraints. It would, therefore, be desirable to study the kinetics of actin-myosin interaction in situ in an intact structured system such as skinned fibres.

The Skinned-Fibre Approach. According to A. F. Huxley (1983), a very promising approach to many problems of muscle may be seen in the use of muscle fibre preparations from which the surface membrane had been removed or made highly permeable by membrane skinning, while the contractile system is left intact. Skinned fibres have also been used to analyze the role of ATP in the crossbridge cycle. H. H. Weber and Portzehl (1954) were the first to use this approach in a systematic manner. They removed all the intrinsic ATP from skeletal muscle fibres by extraction with aqueous glycerol solutions which made the cell membrane highly permeable to ATP. After readdition of ATP in the presence of calcium and magnesium ions the fibres contracted, but when ATP was washed out again, they failed to relax. Such ATP-free fibres were in rigor and the crossbridges were firmly attached to actin. After addition of ATP, however, they detached and the fibre became plasticized, provided ATP hydrolysis was inhibited by the addition of poisons or removal of calcium ions. The fibres relaxed from their rigor contraction and the stiffness became very low. The rate of relaxation, however, was very slow since the diffusion of ATP into the fibre required considerable time. This diffusional difficulty could, however, be overcome by letting the ATP-free fibre preequilibrate with a photolabile inert ATP analogue rather than with ATP. This compound, known as "caged ATP", is so photolabile that a brief flash of ultraviolet light transforms it within milliseconds into ATP, which then produces a sudden relaxation associated with a decrease in stiffness. Its time course indicated the rate of crossbridge detachment that depended on the concentration of ATP in a manner similar to that of actomyosin dissociation in vitro (Goldman et al. 1982). With these experiments it could be shown that the mechanical process of actomyosin dissociation in vitro actually corresponds to the mechanical process of relaxation and to crossbridge detachment in situ. Within 20 ms following relaxation one molecule of ATP formed became hydrolyzed, showing that detachment precedes ATP hydrolysis. Addition of the non-hydrolizable ATP analogue AMP-PNP, on the other hand, causes a "pseudorelaxation" which is not due to the detachment of crossbridges, but to their swing back from an acute-angled position into a 90° configuration (Holmes et al. 1976; Beinbrech et al. 1976; Marston et al. 1976). However, crossbridge detachment and reattachment may not necessarily be coupled to ATP hydrolysis (Lombardi et al. 1992).

1.3.3.5 Crossbridge Turnover Kinetics

As we have seen in the preceding section, crossbridge detachment is rather quick and crossbridge attachment may be occurring even faster. Remember that attachment of strongly bound crossbridges to actin and crossbridge detachment are part of the elementary contractile cycle. Thus, during maintained contraction, crossbridges are in a dynamic state, continuously turning over and cycling between detached or weakly attached states and strongly attached conformations. The kinetics of these processes are governed by two apparent rate constants, denoted as f_{app} and g_{app} or simply f and g, that describe the rate at which crossbridges enter and leave the force generating state. The cyclic action of crossbridges hitherto only assumed could recently be demonstrated experimentally by *motility assays* with single myosin molecules as follows (cf. Huxley 1990 for review).

Myosin molecules attached to a solid support, such as a nitrocellulose sheet, were in the presence of ATP interacting with a single actin filament that was attached to a (quasi isometric) extremely sensitive force transducer detecting subpiconewton (pN) force fluctuations related to crossbridge cycling (Ishijima et al. 1991). The size of the crossbridge step motions or of a unitary structural change for that matter could also be deduced (cf. Uyeda et al. 1990, 1991) from observations of the movement of a fluorescently labelled actin filament that was pulled by a single myosin molecule or even a myosin head attached to a nitrocellulose support. With only one myosin molecule pulling with ATP as an energy source, the deduced step motions of the crossbridge were of the order of 10 nm (cf. also Higuchi and Goldman 1991) as predicted from theory (Huxley and Simmons 1973), but this "unitary" shortening velocity was slow. This is because after a powerstroke, the myosin molecule paused before pulling again. However, with two myosin molecules pulling, velocity was doubled as a second molecule was moving, while the first one was pausing, and these repetitive stroke movements occurred with a frequency of about 20/s. More interestingly, even, the sliding speed of the filaments became similar to that in intact muscle fibres during shortening when several myosin molecules were asynchronously pulling on an actin filament (Uyeda et al. 1991).

Surely, this kind of mechanical "single molecule recording" of unitary velocity is quite analogous to the single ion channel recording of unitary conductance by patch clamp methods (cf. Sakmann and Neher 1983). In much the same way as the molecular behaviour of single ion channels is now being discussed in terms of open probability and the probability of being closed, we may want to describe the dynamic properties of the molecular force generators in terms of the rate constants f and g. Thus, the "on" probability or the proportion of time in which the myosin heads are in the on-state will depend on f/f+g according to Ishijima et al. (1991), and it is related of course also to the fraction of crossbridges strongly attached to actin at any one moment (Huxley 1957; cf. also Brenner 1988). This fraction indicates the time during which each myosin molecule actively pulls during the crossbridge cycle or during the hydrolysis of one molecule of ATP and has been called the "duty ratio". It may vary between 0.05 to 0.3, according to conditions (Uyeda et al. 1990; Ishijima et al. 1991), while the unitary force of crossbridges in the "on-state" is constant and of the order of 1 pN. Hence, steady state force is proportional to f/f+g. These rate constants f and g could be deduced from motility assays (Ishijima et al. 1991), but, as we will see below (Sect. 10.2.2), they may be determined by mechanical analysis in skinned fibres as well.

It may well be that during filament sliding crossbridges "beat" more than once during the hydrolysis of one molecule of ATP (Ishijima et al. 1991, cf. also Lombardi et al. 1992) at least under certain conditions. In the (usual) case in which only one molecule of ATP is split in each crossbridge cycle, the crossbridge cycling rate will correspond to the ATP-splitting rate or simply to the molecular turnover number of the actin-activated myosin ATPase, which is proportional to fg/f+g according to A. F. Huxley (1957) and Brenner (1988). The energetic tension cost is then the energy required for maintaining unit force in a given time and is obtained by dividing ATPase activity (proportional to fg/f+g) by the force (which is proportional to f/f+g); thus, it is related to g, the apparent detachment rate constant. Note that in the two-state crossbridge model of A. F. Huxley (1957) the constant g depends on the mechanical strain of the crossbridges. Under isometric conditions, when the strain is high, constant g (then denoted as g_1) is low, e.g. 1/s, while during unloaded shortening velocity, the detachment rate (now denoted as g_2) is high, since crossbridges are not strained, but, on the contrary, may even be compressed, thus producing negative force. According to the modelling of Ishijima et al. (1991), this kind of strain dependence of crossbridge rate constants would best account for the single molecule motility behaviour of myosin. As mentioned already, the strain dependence of crossbridge detachment could also be demonstrated in skinned fibres (Dantzig et al. 1991).

Since both tension cost (Kushmerick and Krasner 1982) and unloaded shortening velocity (Ferenczi et al. 1984) are related to the rate of crossbridge detachment occurring during filament sliding, one might predict that ATPase activity and shortening velocity are related as well in different muscles, developing a similar force. Such a relationship has indeed been found in fast and slow skeletal muscle (Bárány et al. 1967), but also in heart muscle, expressing different isoforms of myosin heavy chain (see Swynghedauw 1986, for review). Because of the different crossbridge detachment rate constants, fast and slowly contracting muscles would of course also be expected to relax under isometric conditions at different rates when contraction ceases. However, it is not known whether the quasi exponential tension decay observed during relaxation of an isometrically contracted muscle is actually rate limited by the detachment rate constant of crossbridges.

1.3.4 The Cessation of Contraction: Relaxation

In activated, force-maintaining muscles, crossbridges are continuously cycling, detaching and reattaching, thereby splitting ATP in each cyclic operation of the bridge. If, however, ATP hydrolysis is inhibited, attached crossbridges still continue to detach, but they are now unable to reattach and perform a "power stroke". Consequently, crossbridge cycling stops and the muscle fibre loses force and stiffness, i.e. it relaxes with a rate corresponding to that of net crossbridge detachment. As we have learned from the skinned-fibre experiments mentioned in the preceding section, this process requires Mg ATP which becomes bound to the myosin heads. In the absence of ATP, bridges remain in a rigidly attached position, typical for muscle in rigor (cf. H. H. Weber and Portzehl 1954).

The physiological inhibitor of the actin-activated ATPase is the regulatory muscle protein troponin in conjunction with tropomyosin if its calcium-specific binding sites are not occupied with calcium (see Chap. 4). This is the case in resting relaxed muscle when the calcium ion concentration in the myoplasm is very low, whereas in an activated muscle troponin-induced inhibition is depressed by the elevated calcium ion concentration in the sarcoplasm (Ebashi and Endo 1968). The free calcium ion concentration in the myoplasm is, therefore, the major determinant of contractility (Chap. 3) and is itself regulated by the sarcoplasmic reticulum (Chap. 2). If, after cessation of the stimulation, the activating calcium ions are returned into the tubules of the sarcoplasmic reticulum by the calcium pump (Hasselbach 1964), the actomyosin ATPase becomes inhibited and the fibre relaxes because the crossbridges detach from actin.

Chapter 2

The Sarcoplasmic Reticulum: Storage and Release of Calcium

It is now well known that the contractile system of muscle fibres becomes activated when the intracellular calcium ion concentration increases to about 1 to 10 μM and that it is turned off when it is lowered to below about 0.1 μM. Here, we shall describe the structures and mechanisms that raise and lower the intracellular free calcium ion concentration quickly during the contraction-relaxation cycle of various types of muscle. In very thin muscle fibres, such as those of the frog heart (Fabiato 1983), vertebrate smooth muscle (Deth and van Breemen 1974), and in the myotome of *Branchiostoma lanceolatum* (Hagiwara et al. 1971; but cf. Melzer 1982a, b), the activating calcium ions seem to originate, at least partly, from the extracellular fluid and enter the cell through calcium channels of the cell membrane during activation. Bianchi and Shanes (1959) detected a small calcium influx even in stimulated skeletal muscle, but Sandow (1965) calculated that the amount of calcium entering the stimulated muscle cell is too small and its diffusion into the interior of the fibre too slow (cf. also Hill 1948) to account for the rapid onset of contraction. In these fast and wide-fibred muscles the ac-

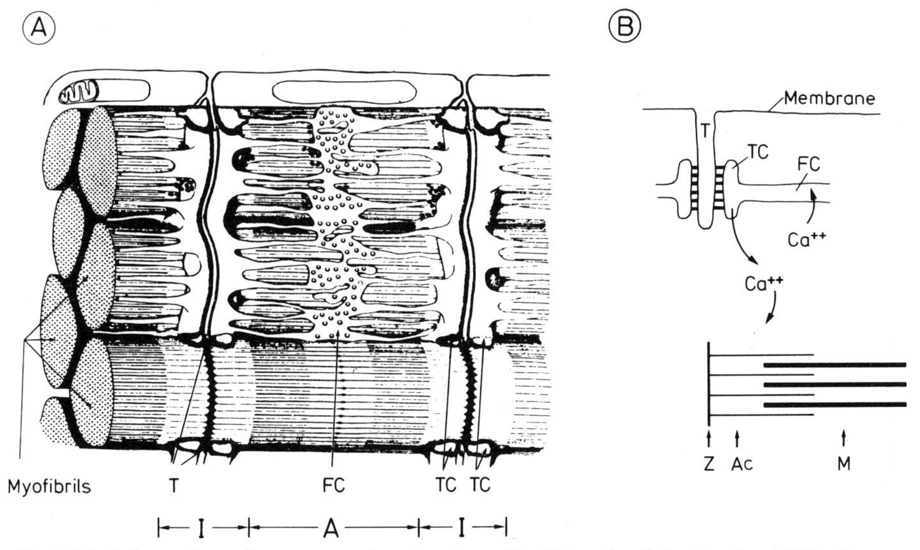

Fig. 2.1 A, B. Internal membrane system in striated muscle. *A* Drawing (after Peachey 1965) of a part of a frog twitch muscle. *T* = transverse tubuli invaginating from the cell membrane; *TC* = terminal cisternae and *FC* = fenestrated collars of the sarcoplasmic reticulum (SR). Myofibrils with I-bands (*I*) and A-bands (*A*). *B* Simplified, schematic representation of a triad (*TC-T-TC*) and of the relationship between cell membrane, T-tubules, sarcoplasmic reticulum and myofilaments. *Ac* = actin; *M* = myosin filaments; *Z* = Z-line

tivator calcium is released from and recuperated by the sarcoplasmic reticulum (Hasselbach 1964) during contraction and relaxation (Sects. 2.2 and 2.3).

As shown in Fig. 2.1 schematically, the sarcoplasmic reticulum is an internal membrane system which forms a network of tubules lying longitudinally, with respect to the fibre axis, between the myofibrils (Peachey 1965). These longitudinal tubules end in blind sacs filled with calcium, the lateral or terminal cisternae. These vesicular structures are closely apposed to the membranes of other tubules that are oriented perpendicularly to the longitudinal system and usually surround the myofibrils at the level of the Z-disc (amphibian muscle) or in the region of the I-band (crustacean muscle, fish muscle, muscles of higher vertebrates). This system of transverse tubules, the T-system, originates from invaginations of the outer cell membrane and, unlike the longitudinal system (L-system), actually communicates with the extracellular space. In a structural and functional sense, the T-tubules bridge the gap between the sarcoplasmic reticulum and the cell membrane (Fig. 2.2 A). The T-system is primarily involved in mediating the excitation signal from the cell membrane to the calcium release sites at the terminal

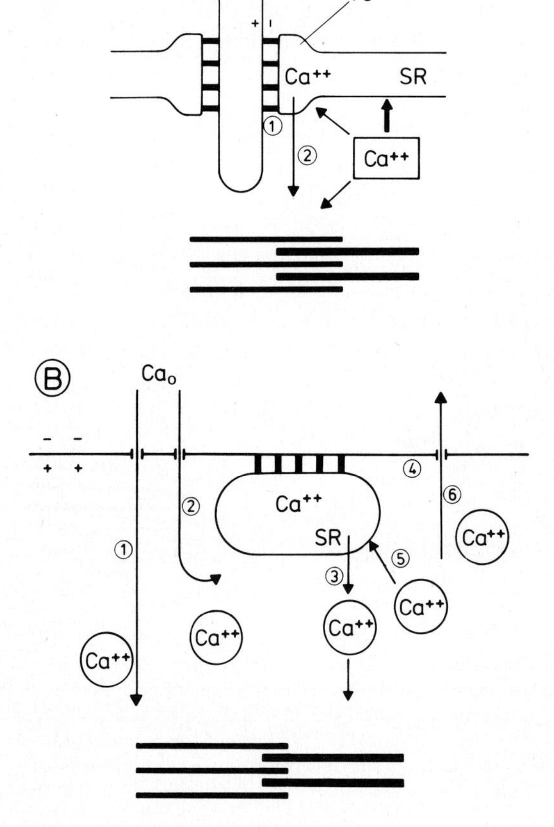

Fig. 2.2 A, B. A Calcium movements in vertebrate striated muscle. Depolarization of T-tubular membrane (*T*) leads via T-SR couplings (*1*) to the release of calcium from the terminal cisternae (*TC*) of the sarcoplasmic reticulum (*SR*). *2* Released myoplasmic Ca^{2+} influences myofilaments and is recuperated by the SR. *B* Calcium movements in muscles without T-tubular system (e.g. striated muscle fibres of *Amphioxus*). Calcium entering the cell via calcium channels may activate myofilaments directly (*1*) or trigger calcium-induced calcium release (*2*) from subsarcolemmal sarcoplasmic reticulum (*SR*) (forming a diad with the cell membrane). *3* Calcium release; *4* depolarized cell membrane; *5* calcium sequestration by sarcoplasmic reticulum; *6* calcium extrusion through the cell membrane

Table 2.1. Internal membrane systems in muscle cells

Animal	Muscle	T-system (%)	SR (%)	Ref.
Frog	Sartorius	0.3	13	Peachey (1965)
Frog	Cardiac (ventricle)	0	0.5	Sommer and Johnson (1979)
Rat	Cardiac (ventricle)	1	3.5	Sommer and Johnson (1979)
Rabbit	Aorta (vascular smooth)	0	6	Devine et al. (1972)
Mnemiopsis[a]	Sagittal muscle (smooth)	0	0.9	Hernandez-Nicaise et al. (1984)

[a] *Mnemiopsis Leydii* (phylum: Coelenterata, order: ctenophores). T-system (transverse tubuli) and SR (sarcoplasmic reticulum) in % muscle cell volume (cf. also Table 5.1).

cisternae (Sect. 2.1). In muscles devoid of the T-system (see Table 2.1) activator calcium may, as already mentioned, enter the cell via membrane calcium channels or it may be released from subsarcolemmal vesicles which form “diadic junctions” with the cell membrane (Fig. 2.2B). This calcium may then be extruded again from the cell by active transport or Na-Ca exchange or it may be recuperated by the vesicles.

2.1 Inward Spread of Excitation in the Transverse System (T-System)

A depolarization of the outer membrane during an action potential may induce contraction of the myofibrils of the fibre interior within a few milliseconds. Think, for example, of a fast muscle such as the cricoarytenoideus of the rat throat which contracts maximally within 3 ms and requires only 5–10 ms for relaxation (Hinrichsen and Dulhunty 1982). How is it possible in this case to transmit a message in such a short time from the excited cell membrane to the innermost myofibrils of a muscle fibre? As mentioned already, a chemical transmitter, such as calcium ions, would not reach the fibre interior in time since, according to Hill (1948), a diffusional process is slow compared with the time required for activation. A. F. Huxley and Taylor (1958) addressed this problem in their famous “local activation experiments” when they discovered a fast electrical inward-conducting mechanism associated with the T-tubular system.

2.1.1 Local Activation Experiments

A. F. Huxley and Taylor (1955, 1958) viewed single muscle fibres of the frog with a polarizing microscope and examined the changes in the striation pattern after application of weak current pulses to the muscle fibre. By using an electrode in

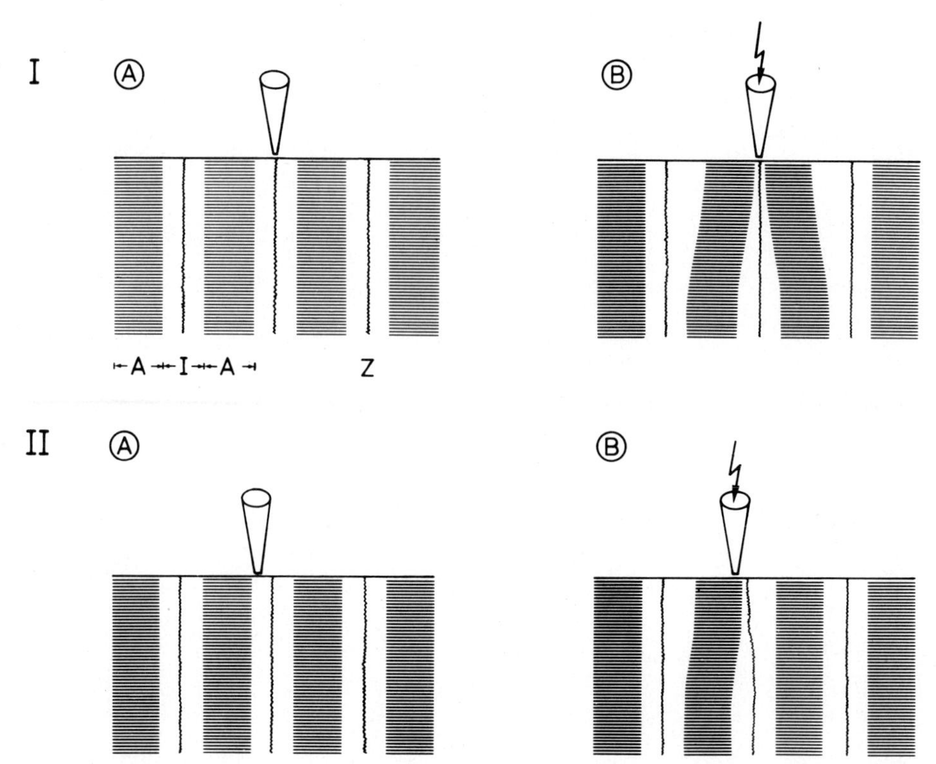

Fig. 2.3. Classical experiment of A. F. Huxley and colleagues to demonstrate local activation and role of T-tubules in excitation-contraction coupling in frog muscle fibres (*I*) and lizard muscle (*II*). In both cases a local contraction is induced only when the stimulating electrode is placed over the orifice of T-tubules (near the Z-line in *I* and near the A-I border line in *II*). Note the change in striation pattern when the resting fibre (*A*) is weakly stimulated by a depolarizing pulse (*B*). (After A. F. Huxley and Taylor 1958; A. F. Huxley and Straub 1958)

form of a micropipette, they were able to depolarize the membrane over an area of the surface so small that it just covered the contact of the Z-line with the sarcolemma. When the depolarization was very weak, no propagated action potential was elicited and correspondingly no generalized contraction occurred. Instead, as shown in Fig. 2.3 I, a local contraction could be observed in the two half-sarcomeres adjoining the Z-line on either side, thus causing the width of the corresponding I-band to become smaller. But when the stimulating electrode was placed elsewhere, for instance opposite the junction of the I-band and A-band, no contraction occurred. At first, A. F. Huxley and Taylor (1955) concluded tentatively from their experiments that the sensitive spot of the membrane was perhaps the Z-line itself and that the latter, also known as Krause's membrane, was a structure capable of conducting membrane excitation into the interior of the fibre. This working hypothesis, however, could not be verified in subsequent comparative experiments on muscle fibres of lizard and crabs.

In lizard muscle (A. F. Huxley and Straub 1958) as well as in crab muscle (Peachey and Huxley 1964) a local contraction did not occur when the stimulating electrode was placed over the Z-line region. But when the pipette was placed over the I-band near the junction of the A-band and I-band, a weak stimulus caused shortening only of the stimulated half-sarcomere. The actin filaments on the other side of the Z-disc in the neighbouring, unstimulated half-sarcomere were pulled over to the stimulating electrode, hence causing a widening of that half-sarcomere (as shown in Fig. 2.3 II). At the time when Huxley and Straub presented their results, Robertson (1956) published an electron microscope investigation of the ultrastructure of muscle fibres from the lizard *Lacerta viridis*. These studies showed a tubular or vesicular structure in the middle of the I-band running from the membrane transversely into the interior of the fibre. This structure, later known as T-tubular system, was, therefore, just at the right place to account for the fast inward-conducting mechanism discovered by Huxley and Straub in lizard muscle. In frog muscle a similar transverse tubular system was found, not at the A-I junction as in lizard, but near the Z-line (Porter and Palade 1957; Peachey 1965), corresponding to the sensitive spot for local activation in these fibres. Thus, from comparative studies in different muscles, A. F. Huxley and colleagues were able to conclude that fibres responded with a local contraction only when stimulated in a membrane area located precisely over a transverse tubule. These investigators also noted that a very weak stimulus caused a contraction limited to the corresponding half-sarcomere of the superficial myofibrils while, with stronger stimulation, the deeper myofibrils within the fibre were also brought into action. Obviously, the T-tubules must be capable of directing the depolarization from the muscle membrane to the innermost myofibrils and they must, therefore, provide an important link in the process of signal transmission between the excitable cell membrane and the calcium-release sites near the myofibrils.

This concept was later verified by the finding that a disruption of the T-tubular system by brief glycerol treatment also impaired excitation-contraction coupling (B. R. Eisenberg and Eisenberg 1968). Could it be possible that an action potential travelling along the muscle membrane would actually penetrate into the fibre along the T-tubules to reach the calcium-release sites in the interior of the fibre? In order to fulfil the proposed function, such a conducting mechanism would require excitability of the T-tubular membrane and its coherence with the surface membrane. Let us see whether this is so by considering the structure and its function in more detail.

2.1.2 Structure of the Transverse System

There is a great variation in the abundance of the T-tubular system and of the longitudinal system (sarcoplasmic reticulum) in different muscles (Table 2.1).

Vertebrate Twitch Muscle. Here, the T-tubules comprise about 0.5% of the fibre volume; they are hardly noticeable in the muscles of neonatal rats, but evolve later from invaginations of the plasma membrane (Schiaffino and Margreth 1969). These eventually form tubules, 30 to 80 nm in diameter, entering the fibre near the border line of the A-I junction or near the Z-disc. They penetrate into the fibre along a line vertical to the fibre axis into the spaces between myofibrils, thereby surrounding or even totally enwrapping each fibril and finally ending in a blind sac. Hundreds of myofibrils lying within a muscle fibre are interconnected in this way at the level of the Z-line or near the A-I border line. In some cases, T-tubules may form a three-dimensional network. Near the cell membrane they often follow a tortuous path and form ramifications and caveolae which open to the extracellular space (Peachey 1981). In this way, the tubules of the T-system communicate with the extracellular medium as demonstrated by showing that large extracellular "marker molecules," such as the protein ferritin, could penetrate into the lumen of the T-tubules where they can be localized by electron microscope techniques (H. E. Huxley 1964). These findings also show that the membranes of the T-tubular system may indeed be regarded as an extension of the cell membrane, a structural feature of considerable functional significance, since it allows the rapid spread of excitation from the outer membrane to the T-tubules; this feature also accounts for the unusually high membrane capacity of the muscle cell membrane (Hodgkin and Nakajima 1972).

According to Peachey (1981), about 30% of the transverse system may consist of small circular elements which do not form junctions with the sarcoplasmic reticulum, while the remaining 60% to 70% of tubules are junctional. This means that their membrane areas are in intimate contact with the membrane of the closely apposed terminal cisternae, the blind sacs at which the longitudinal tubules of the sarcoplasmic reticulum end. A T-tubule and the two terminal cisternae flanking it laterally on either side form three swellings, first described as *"triad"* by Porter and Palade (1957). This is the structure at which membrane excitation influences the calcium-release sites of the sarcoplasmic reticulum. Let us recall in this connection that upon local stimulation of a lizard muscle, only the half-sarcomere penetrated by the stimulated T-tubule shortened. Would this finding not suggest that the T-tubular depolarization affects only the calcium-release sites at the two terminal cisternae, which flank it and lie in the respective half-sarcomere, and not the whole of the longitudinal tubule extending over the entire sarcomere?

Comparison of Fast and Slow Fibres. The structural features just described obviously enable the junctional T-tubules to serve as a rapid communication system between the cell membrane and the calcium-release sites of the sarcoplasmic reticulum. Clearly, such a fast-conducting system would not appear to be necessary in very slow and/or very narrow-fibred muscles where the activator calcium may stem either from the external medium or from sarcoplasmic reticulum vesicles (caveolae) lying underneath and in contact with the surface cell membrane. Well-studied examples are the thin muscle sheets of *Branchiostoma lanceolatum* (Grocki 1982), avian cardiac muscle (E. Page and Surdyk-Droske 1979) and vertebrate smooth muscle (A. P. Somlyo et al. 1982). The other extremes are fast and wide-fibred muscles, e.g. the laryngeal muscles of the rat. One of these muscles, the cricoarytenoideus, contains an extremely abundant and elaborate T-tubular system (Hinrichsen and Dulhunty 1982). As pointed out by Peachey (1981), greater speed of muscle contraction usually appears to be correlated with abundance of the junctional T-tubular system. Thus, chicken fast-twitch muscle contains up to eight triads or diads per sarcomere, while chicken slow-twitch muscle contains only one (S. G. Page 1969) and in slow tonic fibres of frogs only every sixth sarcomere is touched by a T-tubule (S. G. Page 1965). Is it surprising that no special sensitive point could be detected in these slow muscles from which it was possible to induce a highly localized contraction?

A correlation between speed and abundance of T-tubules has also been found in comparative studies on fast and slow crustacean muscle fibres (Eastwood et al. 1982), in which T-tubules usually contact only a terminal cistern on one side, thus forming *diads* rather than triads. For instance, in the fast tail flexor of the crayfish, the transverse system is much more abundant than in the slow leg flexor. The latter, on the other hand, contains in addition to the T-tubular system, which is located near the borders of the A-band, a different type of tubules near the Z-lines, the Z-tubules. These, however, do not seem to be involved in excitation-contraction coupling since, in contrast to stimulation of the A-band borderline region, depolarization of the cell membrane precisely over the Z-disc did not induce a local contraction (Peachey and Huxley 1964).

With the exception of the body wall muscle of the nematode *Ascaris lumbricoides* (Rosenbluth 1965), the obliquely striated muscles of nematodes, annelids and molluscs do not contain T-tubules. Instead, diadic junctions are formed between the cell membrane and subsarcolemmal vesicles communicating with the sarcoplasmic reticulum (SR). These may be analogous to the diadic T-SR junction of crustacean muscle (Rosenbluth 1972). Since the core of many types of obliquely striated muscle fibres is free of contractile material, a fast inward-conducting T-system may not be required for rapid activation.

2.1.3 Conduction of Excitation in T-Tubules

For many years the question had remained open whether the inward spread of excitation along the T-tubules is an active, regenerative process or whether it occurs in a passive, electrotonic fashion. In the experiments of A. F. Huxley and Taylor (1958) the muscle membrane was stimulated with weak current pulses at sensitive spots near the openings of the T-tubules, and the resulting membrane depolarization spread along the T-tubules in an electrotonic and decremental manner into the fibre interior. Since the contraction of fibrils depended on the extent of membrane depolarization (cf. Chap. 1), the innermost myofibrils of the fibre contracted very weakly or not at all, while the fibrils just below the surface membrane contracted strongly. These differential responses of superficial and deep myofibrils were particularly pronounced when the action potential mechanism of the membrane was inactivated by blocking the Na channels with tetrodotoxin (Mathias et al. 1980). However, an increase in stimulation intensity caused the innermost fibrils of the corresponding half-sarcomere to contract strongly.

When strong stimuli were applied to sensitive spots of the fibre membrane in the absence of tetrodotoxin, the elicited action potential travelled along the T-tubule (Costantin 1970). Nakajima and Gilai (1980) visualized the location, size and time course of the potential change in the T-tubule under the microscope after staining the internal membrane system of the muscle fibre with the potential-sensitive dye merocyanine. The rate of inward spread was found to be about

6,5 cm s^{-1} (25 °C) in *Xenopus* muscle fibres and increased by a factor of two when the temperature was increased by 10 °C. Such a high temperature dependence would not be expected if the conduction mechanism were electrotonic. In order to determine the rate of conduction, González-Serratos (1971) used high-speed cinematography and measured the time that elapsed from the onset of shortening in the outermost myofibrils to the onset of shortening in the innermost myofibrils of a muscle fibre:7 cms^{-1} (20 °C).

Thus, in conclusion, it must be assumed that action potentials travelling along the surface membrane of a muscle fibre penetrate into the fibre interior along the T-tubular system, but the safety factor of this action-potential mechanism does not appear to be very high (Adrian and Peachey 1973). Radial propagation into the fibre may even fail when the fibre diameter increases during shortening. Under these conditions, the innermost fibrils often are not activated at all during electrical stimulation, so that force production of the fibre is impaired, especially during muscle fatigue (Garcia et al. 1991; cf. also Westerblad et al. 1990; S. R. Taylor and Rüdel 1970). Whereas a passive electrotonic inward-conducting mechanism, therefore, seems to be of little importance in the physiological activation of twitch muscle, such a mechanism would, of course, be expected to play a major role during graded activation of crustacean muscle and amphibian slow tonic muscle fibres.

2.2 Calcium Release from the Sarcoplasmic Reticulum (SR)

Excitation of the membranes of the transverse system triggers, in a way as yet unexplained, the release of calcium ions from the sarcoplasmic reticulum (SR). Of course, we would expect the coupling process between the two internal membrane systems (T-SR coupling) to occur in the triad at the junction of the membranes of the terminal cisternae of the longitudinal SR and the transverse system, which are separated by a gap of only 15 nm. What is the evidence for the function of the triad in the course of muscle activation?

2.2.1 Calcium Storage and Release Sites

In order to localize calcium intracellularly A. V. Somlyo and colleagues (1977 b) applied electron-probe microanalysis to a preparation with an extremely well-developed sarcoplasmic reticulum: the ultrafast muscle fibres of the toadfish swim bladder containing about 2–3 μmol calcium g^{-1} muscle. Muscle fibres were frozen in the relaxed state and cut into thin cryosections that were irradiated with an electron beam (the probe) in an electron microscope. When the beam was focussed just on a small area corresponding to the terminal cisternae, the energy spectrum of the X-rays emitted from the radiated spot indicated the presence of calcium in a local concentration of about 50 μmol ml^{-1}. Very little calcium could be detected in the myoplasm, in the mitochondria or in the intermediate cisternae of the longitudinal tubules of the SR. Since the volume of the terminal cisternae is only about 4% of the total fibre volume (Mobley and Eisenberg 1975), its measured calcium content fully accounts for the total cell calcium. These results also confirmed Winegrad's pioneering autoradiographic calcium location studies (1965, 1970) that showed qualitatively the loss of stored, radioactively labelled calcium from the terminal cisternae upon stimulation. The extent of calcium release was determined quantitatively by A. V. Somlyo et al. (1981), who showed by electron-probe analysis that the calcium content of the terminal cisternae decreased from 50 μmol in resting muscle to 30 $\mu mol\ ml^{-1}$ in fibres frozen during tetanic stimulation.

Since the calcium released into the myoplasm (20 $\mu mol\ ml^{-1}$ SR) is diluted about 25 times, the total calcium ion concentration of the myoplasm would be expected to increase by about 0.8 $\mu mol\ ml^{-1}$ cell water during a tetanic contraction, while in a twitch contraction, this release was only 0.2 $\mu mol\ ml^{-1}$. As we shall see in Chap. 3, however, the free calcium ion concentration will only increase from 0.1 to 10 μM under these circumstances, since most of the released calcium becomes protein-bound. In the terminal cisternae the free calcium ion concentration is estimated to be about 1 mM. The large concentration gradient of calcium ions between the sarcoplasmic reticulum and myoplasm provides the "driving force" for a purely passive calcium outflux when the calcium permeability of the membranes of the lateral cisternae increases as a result of activation. This passive outflux of calcium could be blocked by dantrolene sodium (Hainaut and Desmedt 1974). This drug also inhibits passive "depolarization-induced" calcium release from the "heavy fraction," but not from the "light fraction" of fragmented sarcoplasmic reticulum isolated by differential centrifugation (Campbell and Shamoo 1980). The heavy fraction containing the bound calcium could be shown to correspond to the isolated terminal cisternae, while the light fraction containing little calcium was identified with the intermediate cisternae of the longitudinal system (Campbell et al. 1980). In conjunction, therefore, these biochemical findings complement the electron-probe studies of A. V. Somlyo et al. (1981), providing further support for their conclusion that upon activation calcium is released passively through the membranes of the terminal cisternae. How then does depolarization of the T-tubules cause an increase in calcium permeability of the cisternal membranes and calcium release? This is a key question, and one way of determining the mechanism is to investigate the structures involved in calcium release in more detail.

2.2.2 Structure of T-SR Junction

The information flow between the T-system and SR occurs within the triad which may be regarded as an "intracellular synapse." The fine structure of these intracellular membrane systems is beautifully shown in electron micrographs (Fig. 2.4) of longitudinal thin sections through a very fast fish muscle (Franzini-Armstrong and Peachey 1981). Note the electron-dense particles in the T-SR gap. In frog skeletal muscle the 15-nm gap between the T-tubular membrane and the junctional membrane of the terminal cisternae is also bridged by such "foot"-like structures which form a regular tetragonal lattice of about 800 "feet" μm^{-2} junctional membrane. The latter comprises roughly 27% of the cisternal membrane

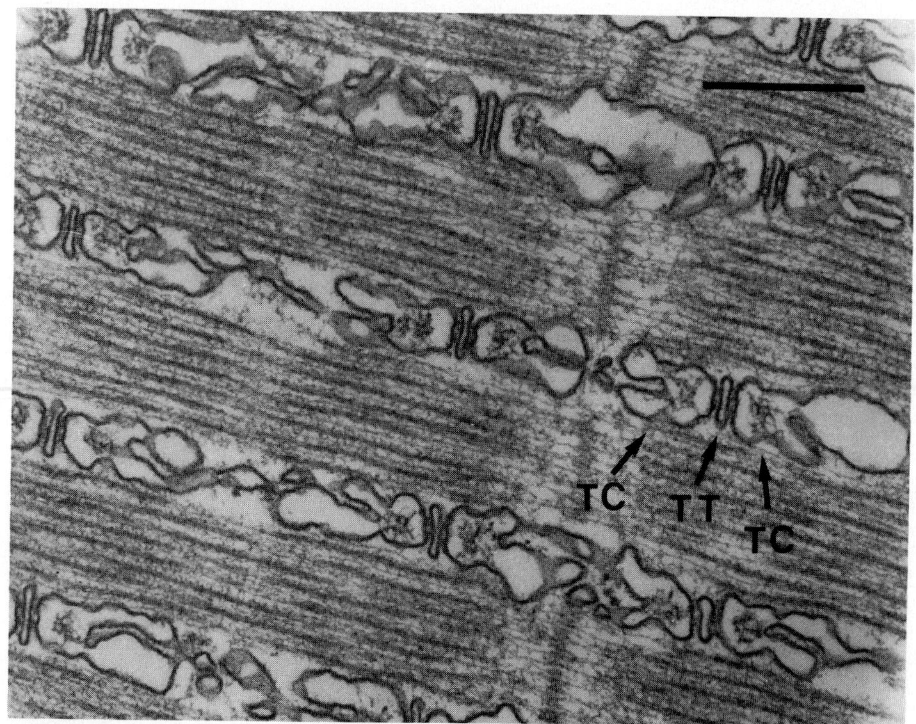

Fig. 2.4. Electron micrograph of longitudinal section from ultrafast fish swim bladder muscle of *Opsanus tau* (Franzini-Armstrong and Peachey 1981). The sarcoplasmic reticulum is very abundant and parallel to the myofibrils in which thick and thin filaments, with crossbridges between them, are clearly recognized. Note the connecting material ("feet") between transverse tubules (TT) and terminal cisternae (TC), within a *triad* (TC-TT-TC), calibration *bar* = 0.4 μm. (Courtesy of Dr. C. Franzini-Armstrong)

and amounts to about 1,500 cm^2 g^{-1} muscle (Franzini-Armstrong 1970, 1975; Franzini-Armstrong and Peachey 1981). These "feet" may be involved in T-SR coupling and may be linked to moving charged particles (Adrian et al. 1976). In fact, the feet structures represent the calcium release channels of the sarcoplasmic reticulum (for review of the evidence, cf. Chap. 10 and Rios and Pizarro 1991). As the rate of calcium release from the sarcoplasmic reticulum is extremely high (Melzer et al. 1984), it is clear that it could never have been accounted for by assuming a carrier-coupled calcium flux as once supposed (cf. Oetliker 1982).

Interestingly, the junctional membrane area and the density of "feet" is much reduced in slow muscles, where the calcium release and activation of contraction are also slow processes. Nonetheless, the fine structure of the T-SR junction, in particular the regular arrangement of "feet," is extremely constant in different muscles throughout the animal kingdom, suggesting that this structural feature has been conserved through evolution. For instance, a regular arrangement of "feet" was not only observed in slow and fast frog muscles, but also in the muscles of insects, spiders, in fish fast muscle and crustacean fast and slow muscle (Franzini-Armstrong and Nunzi 1983).

The detailed structure of the feet is similar in various kinds of striated muscle (see Castellani et al. 1989). On freeze-fracture electron micrographs they may appear on the face of the sarcoplasmic reticulum membrane as square particles consisting of four subunits each according to Block et al. (1988). These units contain pores functioning as calcium channels and extending up to 12 nm from the surface of the SR-membrane into the gap between the T-tubular membrane and the sarcoplasmic reticulum. On the face of the former and opposing the feet, there are groups of four particles called tetrads that are probably identical to the T-tubular calcium channels formerly described by Almers et al. (1981 b) and Sanchez and Stefani (1983). These two kinds of channels are involved in the coupling and signal transmission between the excitable T-tubular membrane and the calcium release site of the sarcoplasmic reticulum during excitation-contraction coupling (Ebashi 1991; see also Sects. 10.11 and 10.12).

2.2.3 Coupling of T-System and Sarcoplasmic Reticulum

In a sense, the junction between the T-tubular membrane and the sarcoplasmic reticulum within a "triad" may be considered as a kind of intracellular synapse. Clearly, the signal transmission across the gap might then in principle be either electrical or chemical including, for instance, calcium ions and inositol-trisphosphate as messengers. Alternatively, there may be a direct "mechanical" linkage mediated by "charge movement" and conformation changes in interacting proteins, e. g. the calcium channels of the T-tubules (dihydro-pyridine receptors) and the calcium release channel (ryanodine receptor). What then is the evidence for and against these proposed coupling mechanisms in different kinds of muscle?

2.2.3.1 Is Calcium Released by SR Depolarization?

Skeletal muscle fibres of frogs were carefully stripped free of their cell membrane and placed in oil; rather surprisingly, stimulation of such demembranated or skinned muscle fibres by small current pulses induced contraction (Natori 1954). Costantin et al. (1967) confirmed Natori's findings and attributed the response to a calcium release from the sarcoplasmic reticulum following the depolarization of internal membrane structures. This interpretation was questioned, however, by Peachey and Eisenberg (1978), cf. also Lamb and Stephenson (1989).

Substituting non-permeable anions, such as propionate, with chloride was said to render the inside of the sarcoplasmic reticulum of skinned fibres more negative and this "SR depolarization" was followed by calcium release (Nakajima and Endo 1973). Replacement of potassium ions by choline also caused a "depolarization-induced" calcium release from skinned fibres, and even from the isolated, heavy fraction of the fragmented, isolated sarcoplasmic reticulum, consisting mainly of terminal cisternae (Kim and Ikemoto 1983). More recent experiments by Ma et al. (1988) and Stein and Palade (1988) suggested that the calcium release channels are voltage-dependent and might therefore be gated by depolarization of the sarcoplasmic reticulum membrane. Yet, it is difficult to conceive how such depolarization may be brought about in vivo. Perhaps, a kind of direct electrostatic coupling exists between the membrane of the T-tubules and the closely apposed feet structures (cf. Rios and Pizarro 1991). As an alternative, transient opening of a gap-channel-like pathway has been proposed between T-tubules and sarcoplasmic reticulum (Mathias et al. 1980; Eisenberg and Eisenberg 1982) which might allow sufficient ionic current to depolarize the SR-membrane and cause depolarization-induced calcium release. This seems now unlikely. Thus, Oetliker et al. (1982) doubted the validity of earlier experiments with potential-sensitive dyes that suggested a change in SR-membrane potential during activation (cf. Oetliker et al. 1975) and A. V. Somlyo et al. (1981) falsified the depolarization theory directly. Using electron-probe microanalysis these investigators determined the elemental composition of the sarcoplasmic reticulum and the myoplasm during rest and activity.

The chloride concentration was similar in the myoplasm and in the sarcoplasmic reticulum both during rest and activation. Since the SR membrane is very permeable to chloride (Kometani and Kasai 1978), these findings argue strongly against a diffusion potential across the sarcoplasmic reticulum membrane of resting muscle and they also exclude the occurrence of long-lasting potential changes during activation.

If the outflow of calcium from the SR were electrogenic, a small potential change across the SR membrane would be expected. This, however, would be transient, as it may cause an influx of Mg^{2+}, K^+ and H^+ (Fink et al. 1988), which would rapidly restore the ion balance within the SR space.

2.2.3.2 Charge Movement in the T-Tubular Membrane

Schneider and Chandler (1973) proposed that mobile, positive charges within the T-tubular membrane may be mechanically connected to rod-like structures, e.g. the "feet", which span the gap between the T-system and the sarcoplasmic reticulum and may be attached to "plugs" in the calcium channels of the cisternal membrane. During activation, the outside of the T-tubular membrane may become negatively charged and would thus attract the positive charges, as illustrated schematically in Fig. 2.5. This charge movement then would unplug the calcium channel, thus causing the release of calcium into the myoplasm. The theory predicts that the charge distribution in the T-tubular membrane would depend on the membrane potential in an S-shaped manner (cf. Lüttgau and Stephenson 1986), and this has been verified experimentally. This S-shaped dependence of charge movement (Q) on membrane potential (V) obeys the equation:

$$Q(V) = \frac{Q_{max}}{1 + \exp[-(V - V')/k]} \quad (2.1)$$

where k is a constant and V′ the potential at which an even distribution of charges within the membrane is reached (cf. Lüttgau and Stephenson 1986).

In the experiments shown in Fig. 2.6, Dulhunty and Gage (1983) compared the slow soleus muscle fibres and the fast extensor digitorum longus fibres of the

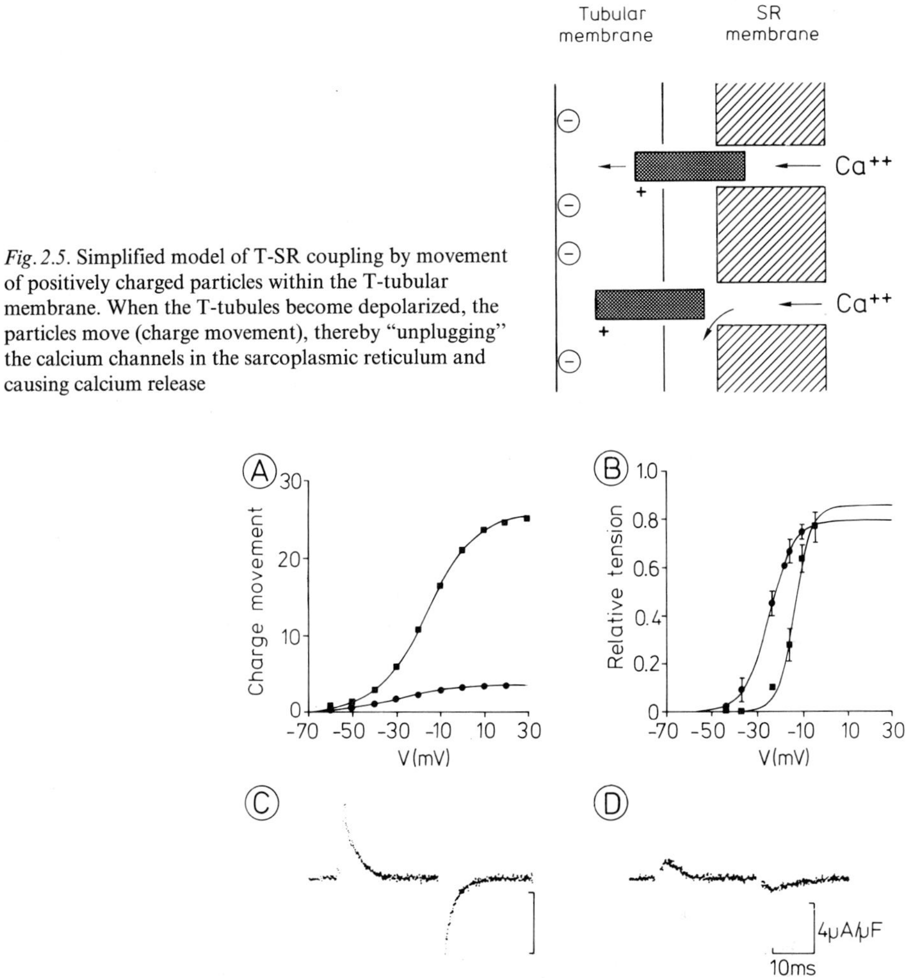

Fig. 2.5. Simplified model of T-SR coupling by movement of positively charged particles within the T-tubular membrane. When the T-tubules become depolarized, the particles move (charge movement), thereby "unplugging" the calcium channels in the sarcoplasmic reticulum and causing calcium release

Fig. 2.6 A–D. Relation between membrane depolarization, charge movement and contraction in skeletal muscle. *A* Charge movements (nC/μF) induced by membrane potential changes to values given on *abscissa*; ■ = fast twitch muscle (rat EDL); ● = slow twitch muscle (rat soleus). *B* Voltage dependency of force development in fast (■) and slow twitch muscle (●). *C, D* Recording of asymmetric charge movement induced by depolarization to +20 mV and subsequent repolarization. *C* = Rat EDL; *D* = soleus. (Dulhunty and Gage 1983)

rat. In both muscles, the extent of charge movement and contractile force depended on the membrane potential reached after depolarization in a similar (S-shaped) manner. In the slow muscle, however, the threshold depolarization was shifted to more negative values, the relationship between charge movement and potential was less steep and the extent of charge movement smaller. Dulhunty and Gage related this to the smaller number of charged particles in the T-C junction rather than to a difference in the tubular junctional membrane area. Incidentally in frog muscle the number of "feet" per μm^2 junctional membrane happens to coincide with the density of charged particles as calculated from electrophysiological studies.

Charge movements may be induced by lowering the membrane potential from a "holding potential" of for instance −70 to −20 mV, using microelectrodes in a voltage clamp setup. Usually, two electrodes are used to control the membrane potential in different places within the fibre and to measure its cable constants, and one is used to inject or withdraw the current for changing or maintaining the membrane potential. The potential change induces a shift of positive charges from the inside of the T-tubular membrane facing the cisternal membrane towards the outside of the T-tubular membrane causing a tiny and transient "outward" capacity current. Its time integral, i.e. the area under the current transient (cf. Fig. 2.6C and D), is related to the number of charges moved. This current, however, is so small (of the order of a few μamp per μF of membrane capacitance) that it can only be recorded if special precautions are taken. For instance, movement artifacts must be avoided and ionic currents should be suppressed by blocking K and Na channels. Even so, currents due to charge movements may be masked by other (capacity) currents. To eliminate these large currents, investigators make use of the observation that charge movement may be elicited, for instance, by a voltage step from −70 to −20 mV, but not if the step occurs between −150 and −100 mV. The current associated with charge movement may, therefore, be simply obtained by recording the difference between the capacity currents obtained under these two conditions. Figure 2.6C and D also show that restoration of the original membrane potential causes a capacity current of a similar time integral, but in the opposite direction, suggesting a reversal of charge movement.

The currents originating from charge movements (asymmetry currents) are no longer obtained after destroying the structure of the T-tubules by brief glycerination (Adrian et al. 1976), showing that charge movement is indeed associated with the T-tubular membrane. Its correlation with activation of contraction under a wide variety of conditions (Adrian et al. 1976) strongly supports its role in the tight coupling of membrane depolarization and contraction (cf. also Sect. 3.2.2).

2.2.3.3 Are Calcium and Inositol-Trisphosphate Messengers Between T-System and Sarcoplasmic Reticulum?

The intramembrane charge movements associated with force activation are probably associated with the gating of voltage-dependent calcium channels (Feldmeyer et al. 1990 b) shown to exist in the T-tubular membrane by Almers and Palade (1981). Thus, calcium ions entering the muscle cell during excitation from the extracellular space (Cota and Stefani 1981) may well be involved in triggering calcium release from the sarcoplasmic reticulum as originally suggested by Frank (1982) and also by Fabiato (1983). However, depolarization-contraction coupling can function normally in the absence of external calcium (Lüttgau and Spieker 1979), and when calcium influx is blocked by impermeable cations like cobalt or cadmium. Thus, calcium influx does not seem to be necessary for signal transmission in activation of skeletal muscle contraction. On the other hand, excitation-contraction coupling may be affected by calcium channel blockers under certain conditions (Eisenberg et al. 1983). The paralyzing effect of these drugs on charge movement (Hui and Milton 1987), calcium release and consequently force generation can be reversed by strong hyperpolarization (Feldmeyer et al. 1990 a). Thus, the receptor protein for calcium antagonists seems to have the pharmacology of a calcium channel that binds dihydropyridines (DHP) and other calcium antagonists and is perhaps even a functional channel (cf. Campbell et al. 1988 for review). However, the DHP-receptor does not send its information on the voltage of the T-tubules to the sarcoplasmic reticulum by opening a pore for the influx of external calcium, but in some other ways. These might involve conformational changes in proteins bridging the gap between the T-system and the sarcoplasmic reticulum or the generation of diffusable messengers like inositol-trisphosphate (cf. Rios and Pizarro 1991). The latter is formed from membrane phospholipids following the activation of a phospholipase C as illustrated in Fig. 2.11, and has already been shown to play a role in excitation-contraction coupling of smooth muscle (cf. Hashimoto et al. 1986 and Chap. 8 for further discussion).

In contrast to vertebrate striated muscle, in many invertebrate muscles (cf. Gilly and Scheuer 1984) and in vertebrate cardiac muscle (Fabiato 1983) the evidence for a trigger function of calcium is much more convincing as shown by skinned fibre experiments.

2.2.4 Calcium Release from "Skinned Fibres"

Muscle fibres from which the external cell membrane has been mechanically removed (skinned fibres) contain, in addition to the contractile mechanism, an intact, internal membrane system (longitudinal system and parts of the transverse system). This experimental model has the advantage that the sarcoplasmic reticulum and the myofibrils can be exposed to a well-defined ionic medium. For instance, the concentration of the energy donator ATP, as well as the free calcium ion concentration, can be accurately controlled as discussed by Fabiato (1983) and Endo (1977). Such fibres accumulate radioactively labelled calcium ions from an appropriate physiological salt solution containing also ATP, magnesium, and potassium propionate and store them in the terminal cisternae (Ford and Podolsky 1972 a). It is possible then to study the amount and the rate of calcium release triggered by the appropriate chemical environment, as well as the optimal conditions for release.

2.2.4.1 "Depolarization" and Calcium-Induced Calcium Release

Figure 2.7 shows an experiment by E. W. Stephenson (1978) in which a calcium-loaded, skinned skeletal muscle fibre is resting in an ATP salt solution containing K-propionate and a low concentration of EGTA to keep the free calcium ion concentration below 10^{-8} M. When the propionate ions were replaced by chloride

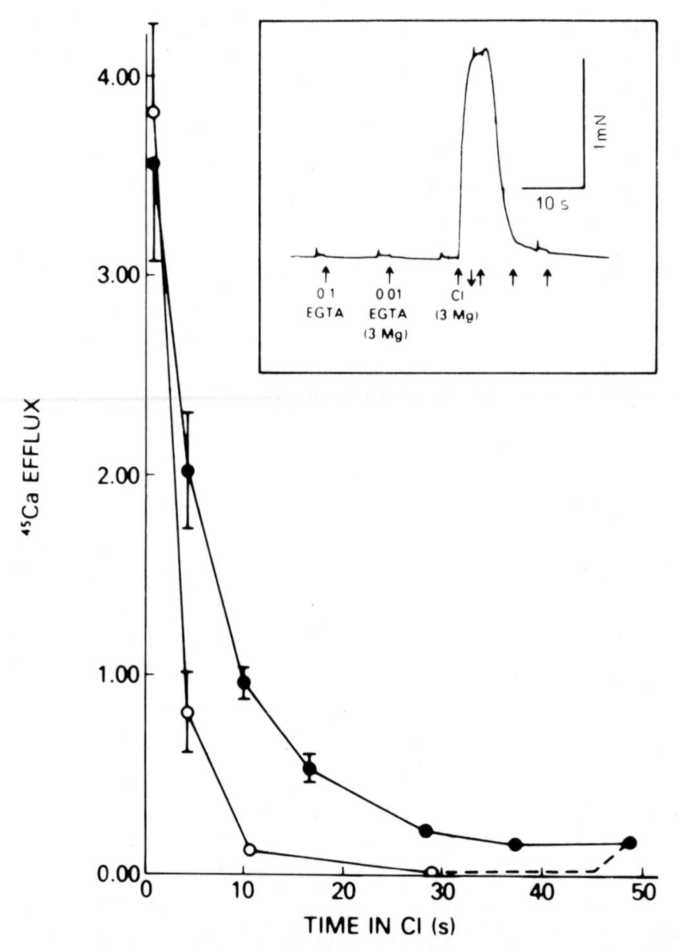

Fig. 2.7. Evidence for calcium release from sarcoplasmic reticulum in skinned fibres. *Ordinate:* Rate of ^{45}Ca efflux (fraction lost per second × 10^2) uncorrected (○) or corrected (●) for calcium reuptake by the sarcoplasmic reticulum. *Inset:* Tracing showing transient contraction of saponin skinned rabbit psoas fibres suspended in ATP salt solution elicited by calcium release from the sarcoplasmic reticulum. Note that calcium release occurs when propionate is replaced by chloride. (From E. W. Stephenson 1978)

ions to "depolarize" the membrane of the sarcoplasmic reticulum, labelled calcium that had been stored in the skinned fibre was rapidly released into the medium and caused a contraction which was soon followed by relaxation as the released calcium was taken up again by the sarcoplasmic reticulum of the fibre. This calcium reuptake, of course, would have led to an underestimate of the released calcium which, however, could be avoided by trapping the calcium (with EGTA in a high concentration). In these experiments the calcium of the terminal cisternae is not totally released, but the remainder could be liberated by caffeine which has been known for a long time to liberate calcium from isolated, fragmented sarcoplasmic reticulum (cf. A. Weber 1968). Without prior depolarization-induced calcium release the amount of caffeine-released calcium, of course, would have been much larger, amounting to maximally 200 nmol g^{-1} muscle (cf. Endo 1977), a value comparable to the calcium release estimated in intact muscle fibres by electron-probe microanalysis (A. V. Somlyo et al. 1981). This is in remarkable agreement, considering that in Endo's (1977) experiments calcium release was measured indirectly from the size of the caffeine-induced force response of the skinned fibre.

By taking the difference between caffeine-induced calcium release with and without prior depolarization, Endo was able to estimate "depolarization-induced" release. It was very small at low calcium ion concentration and became larger as the calcium ion concentration in the medium increased to values of 1 to 10 μM which are similar to those reached in the myoplasm during activation. These experiments indicated that Ca^{2+} could induce the release of more calcium (cf. also E. W. Stephenson 1982, 1985). Is it possible that the release of calcium from the sarcoplasmic reticulum during activation induces a regenerative release of calcium, as originally proposed by Ford and Podolsky (1972b)? Are calcium ions merely modulating T-SR coupling or do they induce per se calcium release, as suggested by Fabiato (1983)?

2.2.4.2 Comparative Aspects

Endo et al. (1983) compared mechanically skinned skeletal muscle fibres from healthy persons and from patients suffering from malignant hyperthermia. This is a rare, but extremely severe condition arising in approximately 1 of 50,000 patients during general anaesthesia. The patients' muscles stiffen spontaneously and contract supermaximally, thus developing so much heat that the body temperature rises to lethal levels. Sometimes the situation may, however, improve after treatment with dantrolene sodium which, according to Hainaut and Desmedt (1974), blocks the release of calcium from the sarcoplasmic reticulum. As shown by Endo and colleagues (1983), the calcium stores of the sarcoplasmic reticulum may be less "tight" in the diseased state since skinned fibres obtained from the patients' muscles were found to be unusually sensitive to calcium ions inducing the release of calcium, due to a mutation of the calcium release channel (Fill et al. 1990).

Calcium-induced calcium release has also been demonstrated in skinned arthropod muscle fibres (D. G. Stephenson and Williams 1980) as well as in mammalian cardiac muscle fibres (Fabiato 1983). Fabiato (1982a) skinned single cells of cardiac ventricle by mechanically removing the cell membrane with microtools; he then attached them to a force transducer and monitored with a differential microspectrometer the amount of calcium bound to the inner face of the sarcoplasmic reticulum membrane as well as the "myoplasmic" calcium resulting from calcium release. When a solution containing 0.1 μM free calcium was very rapidly applied, calcium was released from the sarcoplasmic reticulum in the skinned fibres and caused the myoplasmic free calcium to rise to 10 μM, a concentration sufficient to activate maximally the contractile system. Preceding the calcium release, Fabiato observed in the presence of chlorotetracycline a small fluorescence signal in the membrane of the sarcoplasmic reticulum monitoring perhaps the calcium-induced change in the SR-membrane responsible for the gating of calcium channels. The extent of calcium release increased in a graded manner with increasing concentration of rapidly applied calcium, but at higher calcium ion concentration, the release was inhibited. Interestingly, a slow application of calcium-containing solutions promoted calcium reuptake by the membranes of the sarcoplasmic reticulum and the further filling of the sarcoplasmic reticulum with calcium rather than inducing a calcium release. Calcium-induced calcium release could also be demonstrated in the thin muscle fibres of the cat caudofemoralis muscle, but not in myocytes from frog ventricle. This may not be surprising, since in these small myocytes the sarcoplasmic reticulum is scarce. Since the fibres are only 3–5 μm thick calcium ions entering the cell through the membrane during the action potential will activate the contractile mechanism directly.

Unlike the longitudinal system of the sarcoplasmic reticulum, the transverse system does not appear to be necessary for calcium-induced calcium release, because the release phenomenon could also be observed in skinned fibres from mammalian atrium and avian ventricle (Fabiato 1982b) which are devoid of transverse systems. In these muscles the calcium-containing vesicles of the sarcoplasmic reticulum are to a large extent just beneath the cell membrane with which they form diads. These are structures formed by the vesicles that are attached to the outer membrane by small particles resembling the "feet" previously described in the T-SR junction (Sect. 2.2.2). Calcium ions entering these 20-μm-wide cells during an action potential may rapidly diffuse to the superficial vesicles, thereby causing calcium-induced calcium release. In this case, calcium-induced calcium release may serve to amplify the activating effect of calcium entering the cell during excitation. In skeletal muscle fibres having a diameter of up to 100 μm, calcium entering through the membranes of the T-tubules may serve a similar function, according to Fabiato (1982b). Future work will have to show whether this kind of mechanism or a "charge movement coupled" gating mechanism will be better suited to describe the tight coupling between the membrane potential and the calcium release sites.

2.3 Calcium Reuptake by the Sarcoplasmic Reticulum

We have seen in the preceding sections that the activator calcium required for contraction is released from the terminal cisternae of the sarcoplasmic reticulum; however, in some types of muscle it may also originate partly from the extracellular space. Here, we are concerned with the mechanisms required to remove the calcium activator from the myoplasm during muscle relaxation. The most important calcium sequestration system is the sarcoplasmic reticulum itself (Hasselbach 1964), but supplementary mechanisms will also be briefly considered.

2.3.1 Sarcoplasmic Reticulum in Fast and Slow Muscle

The sarcoplasmic reticulum of intact muscle is particularly well developed in those muscle fibres that relax very rapidly and where, therefore, the activator calcium must be quickly sequestered. Consider the striated muscle fibres of the swim bladder from the toadfish *Opsanus tau* (Fawcett and Revel 1961; Franzini-Armstrong and Peachey 1981) which, by contracting with a frequency of about 100 Hz, produce a gurring noise. Further examples of superfast muscles are the extraocular muscles of fish (Kilarski 1967) or the cricothyreoideus muscle of the bat (Revel 1962). In these muscles the sarcoplasmic reticulum takes up almost one-third of the fibre volume and consists of a network of tubules, arranged longitudinally with respect to the fibre axis and forming curtain-like structures and plexiform sheets surrounding and even totally enwrapping each myofibril (Fig. 2.4). Near the middle of the sarcomere the tubules form swellings, the intermediate cisternae, as well as very elaborate structures, the fenestrated collars. Near the junction of the A-band and I-band these tubules form large swellings, the terminal cisternae, which are in close apposition with the transversal system with which they form the triad structure. In fast vertebrate skeletal muscle, for instance frog sartorius, with a relaxation time of about 40 ms, the sarcoplasmic reticulum is not quite so abundant and elaborate. Here, the volumes of the T-system, the terminal cisternae and the intermediate cisternae, amount to 0.4, 5 and 8% of the fibre volume respectively (Mobley and Eisenberg 1975), whereas in slow twitch muscles of mammals these volumes are about halved (B. R. Eisenberg and Kuda 1976; cf. also Luff and Atwood 1971). In frog slow tonic muscle, in the fibres of some invertebrate smooth muscles, as well as in frog ventricle and in the muscle fibres of *Amphioxus* (Melzer 1982b) the sarcoplasmic reticulum is scarce (cf. Table 2.1).

Evidence for calcium sequestration by the sarcoplasmic reticulum in vivo was first obtained by Portzehl and colleagues (1964), who experimented with giant crustacean muscle fibres from the crab *Maia squinado:* following injection of calcium salts, the fibres first contracted and then, after rabout 30 s, relaxed, suggesting that the injected calcium had been removed again. According to Hasselbach (1964), skinned fibres could take up calcium in the presence of ATP and sodium oxalate and store it in the sarcoplasmic vesicles as calcium oxalate. The uptake rate of labelled calcium into these vesicles may quantitatively account for the rate of relaxation in frog muscle (Ford and Podolsky 1972a). Therefore, the question arises whether, quite generally, differences in relaxation rate of muscle may be paralleled by differences in the rate of calcium accumulation into the sarcoplasmic reticulum. This problem has been addressed in experiments with isolated fragmented sarcoplasmic reticulum.

2.3.2 Fragmented Sarcoplasmic Reticulum

Marsh (1951) and Bendall (1952) discovered a "factor" in muscle juice that was capable of relaxing isolated myofibrils and demembranated muscle fibres. However, the importance of the sarcoplasmic reticulum for this relaxation became apparent only much later when Kumagai et al. (1955) and Portzehl (1957) discovered that the relaxing activity of muscle juice was really due to a "microsomal fraction" that could be separated from inactive sarcoplasm by high-speed differential centrifugation. These vesicles were found to consist mainly of fragmented sarcoplasmic reticulum. The real breakthrough came in 1961 with the discovery of Hasselbach and Makinose (cf. also Ebashi and Lipmann 1962), who showed that preparations of vesicles inhibited myofibrillar contraction and ATP-ase activity by accumulating and storing calcium ions. In the presence of activating Mg ions, ATP as energy source and sodium oxalate, calcium ions were actively pumped into the sarcoplasmic vesicles, where they were deposited as a calcium oxalate precipitate. In this way, the vesicles lowered the free calcium of the medium to below 10^{-8} M, a concentration at which ATP-contracted myofibrils would relax. Two calcium ions could be transported for each molecule of ATP split in the active transport of calcium into the sarcoplasmic vesicles. According to Meissner (1975), ATP splitting and active calcium transport occur predominantly in the tubules from the intermediate parts of the sarcoplasmic reticulum rather than in the terminal cisternae which take up calcium at a somewhat lower rate.

If calcium uptake of the sarcoplasmic reticulum is responsible for muscle relaxation, one would expect this process to be slower in slow muscle fibres than in fast muscle fibres. This question was studied by Heilmann et al. (1977) and more recently by Zubrzycka-Gaarn et al. (1982), who incubated vesicles from the sarcoplasmic reticulum of rabbit fast twitch muscle and slow twitch muscle in solutions containing ATP as energy source, magnesium ions, sodium oxalate and radioactively labelled calcium ions. After incubation for various times, the vesicles were "killed" by addition of the protein-denaturing agent SDS and the reaction mixtures were filtered through Millipore filters. Figure 2.8 shows the amount of ^{45}Ca trapped in the filtered vesicles and plotted as a function of incubation time. Note that with vesicles from fast muscle both the rate and the extent of calcium uptake are many times larger than in the case of slow muscle. During

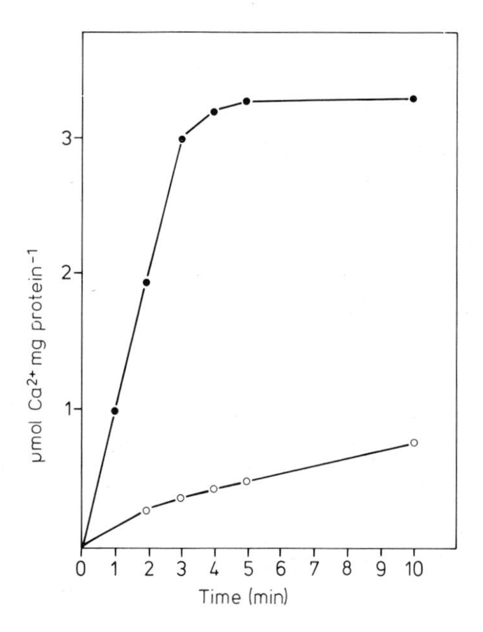

Fig. 2.8. Ca^{2+}-dependent calcium uptake by sarcoplasmic vesicles of rabbit fast twitch muscle (back muscle, *filled symbols*) and slow twitch muscle (soleus, *open symbols*). (From Zubrzycka-Gaarn et al. 1982)

incubation ATP is hydrolyzed progressively by the vesicles. The rate of ATP hydrolysis is much higher in vesicles of fast twitch muscle than in sarcoplasmic reticulum of slow twitch muscle. The higher pumping activity by sarcoplasmic reticulum of fast twitch muscle is, according to Zubrzycka-Gaarn et al. (1982), not due to a higher turnover number of the ATP-splitting, calcium-transport enzyme, but rather to the fact that in sarcoplasmic reticulum of fast muscle the transport ATPase amounted to 80% of the total protein, but only to 20% in the case of slow muscle. Since fast muscle contains twice as much sarcoplasmic reticulum as slow muscle (B. R. Eisenberg and Kuda 1976), the rates of calcium uptake of these fibres would differ by an order of magnitude, more than enough to account for the differences in the relaxation rate (cf. Ferguson and Franzini-Armstrong 1988). When fast muscles are transformed into slow muscles by electrostimulation, its sarcoplasmic reticulum acquires the low ATPase activity and transport capacity typical of slow muscle (Heilmann und Pette 1979).

Rather disappointingly, the absolute rates of calcium uptake, about 5 nmol min^{-1} in the SR from 1 g frog fast twitch muscle at 0 °C, are far too low to account for the relaxation which is completed within about 0.15 s (Ogawa et al. 1981). By that time, the amount of calcium released in a twitch (200 nmol g^{-1}) should be recuperated by the sarcoplasmic reticulum (about 10 mg g^{-1} muscle). Therefore, it seemed questionable whether the rate of calcium uptake during relaxation is reflected by the calcium uptake rate of fragmented sarcoplasmic reticulum under steady-state conditions. However, Ogawa et al. (1981) also discovered that when calcium ions, ATP and sarcoplasmic reticulum are suddenly mixed in a "stopped-flow machine," an initial burst of a very rapid calcium uptake occurs at a rate of 45 nmol mg^{-1} protein in 0.15 s. This amounts to more than 200 nmol g^{-1} muscle and signifies that actually all the calcium released in a twitch could be sequestered by the sarcoplasmic reticulum during the time of mechanical relaxation. In Ogawa's experiments the rapid initial phase was followed by a much slower steady-state rate of calcium uptake, and these two phases of calcium sequestration may be understood in terms of elementary processes occurring during the translocation of calcium through the membranes.

2.3.3 Mechanism of Calcium Transport

Calcium transport is carried out by translocator proteins which are embedded in the lipid phase of the sarcoplasmic reticulum membrane into which they were inserted during embryonic development (Volpe et al. 1982). Because of the fluidity of the membrane, Tada et al. (1978) proposed a rotary carrier model according to which translocator molecules were assumed to float and rotate freely and at random. More recently, however, Martonosi (1984) described a "two-dimensional crystalline" arrangement of translocator proteins within the lipid phase. This highly ordered structure suggests that the calcium movement might be associated with discrete intramolecular configurational changes rather than with translational or rotational movements. Perhaps the calcium translocator protein is forming a channel in which a calcium-binding site is either oriented towards the

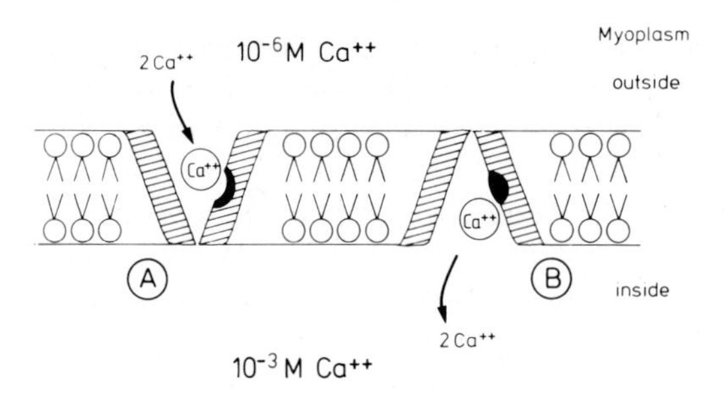

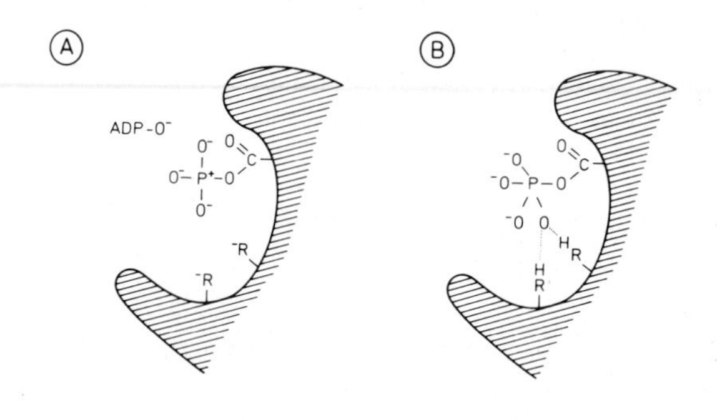

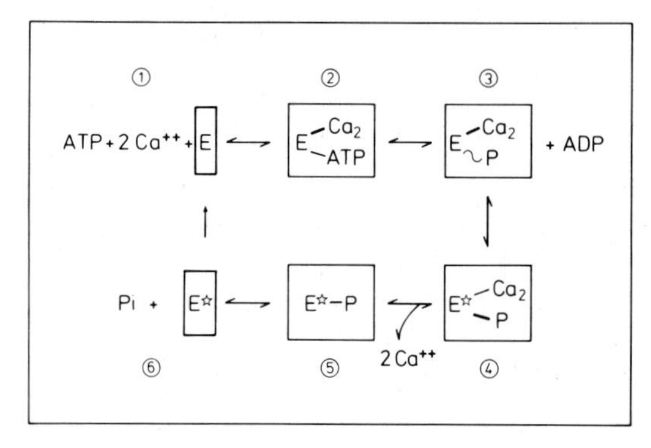

Fig. 2.9. Calcium transport across the sarcoplasmic reticulum membrane. *Upper diagram:* An outwardly oriented high affinity calcium-binding site (*A*) is transformed into an inwardly oriented low affinity site (*B*). *Middle:* Hypothetical structure of phosphorylated intermediates in the calcium transport system and "nucleophilic" attack of H_2O. *Below:* Reaction sequence involved in calcium transport. Formation of the enzyme-substrate complex (*2*) is followed by the phosphorylation of the enzyme (*3*) having a high calcium affinity. As the bonds between the phosphate group and the enzyme become stronger, the affinity for calcium is lowered (*4*) and calcium is released into the interior of the sarcoplasmic reticulum vesicle (*5*), while phosphate is released (*6*). (Adapted from Inesi 1985; E. Eisenberg and Hill 1985; but cf. also Brandl et al. 1986)

myoplasm or towards the interior of the vesicle as depicted in Fig. 2.9. According to Inesi (1985), the first step in calcium transport may then involve rapid calcium binding to the outwardly oriented high-affinity binding sites which is then followed by a ten times slower translocation of calcium into the sarcoplasmic reticulum involving the hydrolysis of ATP.

2.3.3.1 Calcium Binding and Translocation

Calcium Binding. Calcium ions are recognized and bound by the outwardly oriented translocator protein with high affinity. The stability constant of the calcium-protein complex is about $10^6 M^{-1}$ at physiological pH, which is sufficient to allow calcium binding even from solutions containing as little as $10^{-8}M\ Ca^{2+}$.

The higher the Ca^{2+} concentration, the greater the extent of complex formation of calcium and carrier protein and, hence, the rate of transport. The increase in the rate of calcium uptake with increasing Ca^{2+} concentration has been demonstrated by Hasselbach and Makinose (1963) and by A. Weber and colleagues (1966), who found that half-maximum activation occurred at 10^{-6}M Ca^{2+}. Although still controversial, according to Inesi (1985), the high-affinity binding of two calcium ions to the translocator seems to require the (previous) binding of Mg ions and ATP.

Calcium Translocation. In the course of the transport process, the outwardly oriented sites binding two calcium ions become exposed to the inside of the sarcoplasmic reticulum vesicles, where the Ca^{2+} concentration is of the order of 1 mM. Because of this high concentration, calcium ions will only be released into the vesicles when the calcium affinity of the translocator is lowered by several orders of magnitude, a process occurring in several steps (Fig. 2.9): (1) ATP is hydrolyzed, whereby an energy-rich, phosphorylated intermediate of the enzyme is formed, while ADP is released (Makinose 1969). It is this phosphorylation of the translocator protein that enables the subsequent transformation of the outwardly oriented, high-affinity calcium-binding site into an inwardly oriented, low-affinity site. (2) During the transport of Ca^{2+} against a concentration gradient, the energy-rich, protein phosphointermediate of the translocator may be converted into the low-energy form of the phosphoprotein which has a different conformation. Though calcium is now held with low affinity, it may at first be occluded within the translocator protein (Takisawa and Makinose 1981), but then the two bound Ca-ions are released into the vesicles of the sarcoplasmic reticulum and replaced by H^+ and possibly also by K^+. (3) The protein is then dephosphorylated and the original outwardly oriented protein conformation is restored. In this way, H-ions and K-ions bound to the calcium-binding sites are counter-transported across the membrane and released into the myoplasm. It is because of this counter-movement that calcium transport is an electroneutral rather than an electrogenic process (Hasselbach and Oetliker 1983).

In conclusion, the translocator is an ATPase which by hydrolyzing one molecule of ATP moves two molecules of calcium against a concentration gradient of several orders of magnitude.

2.3.3.2 Role of Calsequestrin and Phospholipids

The calcium ions accumulated in the tubules of the sarcoplasmic reticulum diffuse into the terminal cisternae where they are trapped by the membrane protein calsequestrin, discovered by MacLennan and Wong (1971). Since the stability constant of the calsequestrin-calcium complex is rather low, the free Ca^{2+} concentration within the tubules remains comparatively high, in the range of 0.5–1 mM (cf. Hasselbach and Oetliker 1983).

The phospholipids of the sarcoplasmic reticulum, for instance phosphatidylcholine, form a barrier preventing the passive, outward diffusion of calcium. If the lipids are digested by phospholipases or dissolved by treatment with the detergent Triton X-100, the sarcoplasmic reticulum vesicles become so leaky that they are unable to accumulate calcium, although they still hydrolyze ATP. The transport function can be restored by readdition of lipids, and it was even possible to reconstitute artificial, calcium-pumping vesicles from phospholipids and purified transport ATPase (Racker 1972). These studies showed that active calcium transport may in essence be understood in terms of the properties of a single enzyme, the calcium-transport ATPase with a molecular weight of approximately 100 kDa. The molecular structure will be discussed in Section 2.3.6.

2.3.3.3 Calcium-Transport ATPase and Muscle Relaxation

As mentioned already, one molecule of ATP is split to "pay" for the transport of two calcium ions.

Energetics. The energy derived from ATP hydrolysis ($\simeq 46$ kJ mol^{-1}) may be almost entirely converted into osmotic work or concentration work (Hasselbach and Oetliker 1983; Inesi 1985). The maximal work done (ΔG) is then given by the equation:

$$\Delta G = 2\,RT \ln \frac{[Ca^{2+}]_i}{[Ca^{2+}]_o} = 2\,RT \ln \frac{K_o}{K_i} \simeq 46\ \text{kJ} \qquad (2.2)$$

where $[Ca^{2+}]_i$ and $[Ca^{2+}]_o$ are the free Ca^{2+} concentrations inside and outside of the vesicles respectively (10^{-3} M and 10^{-7} M), R and T have their usual meaning, while K_i and K_o are the calcium affinity constants of the inwardly and outwardly oriented binding sites. It is clear then that the free energy available from ATP splitting depends on the ratio of the calcium-affinity constants of the outwardly and inwardly oriented binding sites and may be used entirely to pay for creating a maximal concentration gradient ($1:10^4$) between calcium outside and inside the sarcoplasmic reticulum.

The Mechanism of Calcium Transport ATPase as well as the supporting evidence have been reviewed by Tada et al. (1978), Haynes (1983), and Inesi (1985) (cf. Fig. 2.9). A milestone in the elucidation of the mechanism was the discovery of the acid-stable, phosphorylated intermediate of transport ATPase by Makinose (1969). This had been foreshadowed many years earlier by the finding that an at that time hypothetical protein phosphate intermediate could transfer the terminal phosphate group from ATP to inosine diphosphate by causing the formation of inosine triphosphate or to adenosine diphosphate, thereby forming ATP (Hasselbach and Makinose 1962). This ATP-ADP exchange, as the reaction was called, gave important clues concerning the kinetics of the calcium-transport ATPase.

Kinetics. By studying the ATP-ADP exchange reaction and the overall ATPase reaction, Ogawa and Kurebayashi (1982) compared the transport mechanisms of the sarcoplasmic reticulum of bull frog leg and rabbit back muscle. Interestingly, the steady-state rates of ATP splitting and calcium uptake were quite similar and corresponded to the rate-limiting process, the decomposition of the phosphorylated, high-energy intermediate. In both muscles, the rate of formation of the phosphorylated intermediate was much faster than its decomposition. This finding may account for the mentioned high initial rate of calcium uptake by the sarcoplasmic reticulum, since calcium translocation has been shown to be coupled with the formation of the high-energy intermediate. Recall that one gram muscle contains about 5–10 mg sarcoplasmic reticulum or 50–100 nmol transport ATPase. Simply by binding calcium, the enzyme could therefore recuperate within less than 100 ms the 200 nmol of calcium released into the myoplasm at the onset of the twitch. Accordingly, the rate of initial calcium uptake by fragmented sarcoplasmic reticulum and its temperature dependence were indeed found to be quite similar to that of mechanical muscle relaxation and to the rate of the concomitant reduction in myoplasmic free calcium, i.e. about 10 s^{-1}.

The correlation between relaxation and calcium uptake suggests that during the rapid relaxation of a twitch contraction, the rapid binding of calcium, associated with formation of a high-energy intermediate, may be the decisive process, whereas the steady-state rate of calcium uptake coupled with ATP hydrolysis may be more important during and after a prolonged tetanic contraction. Under these conditions, a large fraction, perhaps a quarter of the total energy expended during a tetanus, would actually be required for the continued calcium pumping into the sarcoplasmic reticulum. This energy, measured as heat of activation, is much smaller in slow muscles as these contain much less calcium transport ATPase (I. C. H. Smith 1972).

2.3.4 Regulation of Calcium Uptake

Calcium. The physiologically most important "activator" of the calcium pump is the internal calcium ion concentration of the myoplasm (Hasselbach and Makinose 1963). Figure 2.10 shows an experiment by A. Weber et al. (1966) on the effect of Ca^{2+} concentration on the calcium-uptake rate. Note that the pump is already active below the Ca^{2+} concentration existing in resting muscle, 0.1 μM, and that it reaches maximal activity at about 1 μM free calcium. However, the pumping activity decreases with time as the sarcoplasmic reticulum vesicles become filled with calcium (A. Weber 1971). High internal calcium presumably inhibits the decomposition of the high-energy phosphate intermediate and the release of inorganic phosphate (Yamada and Tonomura 1972). Such an inhibition would slow down the rate of calcium uptake and ATP splitting by the sarcoplasmic reticulum during a maintained tetanic contraction.

In addition to having a direct effect, calcium ions may also exert a more indirect effect on the sarcoplasmic reticulum. In cardiac and smooth muscle in particular (see Chaps. 7 and 8) calcium interacts with the calcium-binding protein *calmodulin,* which then activates the calcium-transport ATPase (Pifl et al. 1984). Besides calcium, *magnesium ions* are also required for activation of the calcium transport (Hasselbach 1964), since Mg-ATP is the substrate of the calcium-transport ATPase. However, the Mg ion concentration is maintained at such a high level (3 mM) during contraction and relaxation that it cannot be assumed to play a regulatory role.

Cyclic AMP. Hormonal stimulation of muscle with adrenaline and noradrenaline leads to the intracellular formation of cyclic adenosine monophosphate (cAMP) from ATP as precursor (cf. Fig. 2.11). This cyclic nucleotide then acts, via activa-

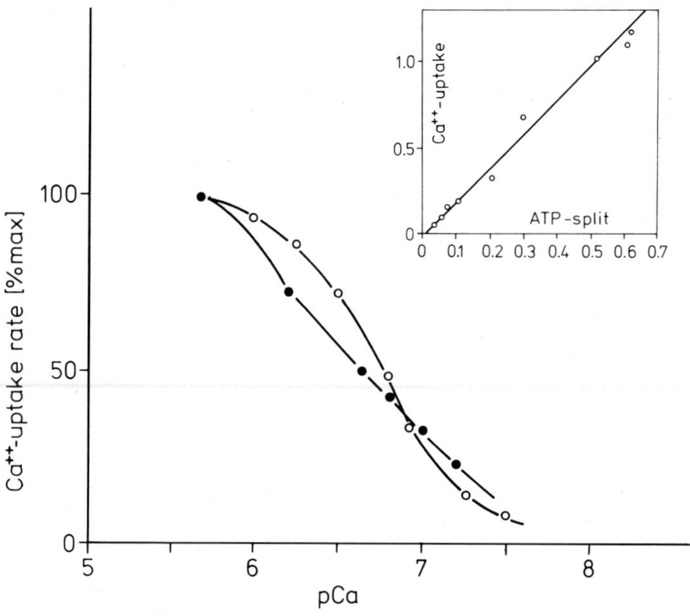

Fig. 2.10. The calcium uptake of fragmented sarcoplasmic reticulum as a function of the calcium ion concentration in the presence (*open circles*) and absence (*closed circles*) of Na oxalate (used to trap calcium inside the vesicles of the sarcoplasmic reticulum). *Inset:* Relationship of calcium uptake and ATP hydrolysis at various Ca concentrations. *Abscissa:* ATP hydrolysis; *ordinate:* Ca uptake in μmol mg^{-1} protein min^{-1}. (From A. Weber et al. 1966)

tion of specific protein kinases, as an intracellular messenger between the agonist-activated, adrenergic β-receptor at the cell membrane and intracellular targets. One of these targets is the sarcoplasmic reticulum membrane. Kirchberger and Tada (1976) found that cAMP and a cAMP-dependent protein kinase increased the rate of calcium uptake of sarcoplasmic reticulum in slow twitch muscle by over 30%, while that of fast muscle was little affected (cf. also Salviati et al. 1982). This activation appeared to be due to a phosphorylation of the 22 kDa protein phospholamban (Katz 1979) in the sarcoplasmic reticulum membrane. Phosphorylated phospholamban activated best at submaximal calcium levels and was, therefore, found to decrease the Ca^{2+} concentration required for half-maximal activation of calcium transport, while maximal activity was little affected (Tada et al. 1974 a, b). An even stronger effect of phospholamban phosphorylation was observed in cardiac muscle (see Chap. 7; Kirchberger et al. 1974; Bárány and Bárány 1981; Mandel et al. 1983). It seems likely that unphosphorylated phospholamban is actually an inhibitor of the calcium transport ATPase (Kim et al. 1990), which is derepressed by phosphorylation.

Katz (1979) argued that cAMP-induced acceleration of calcium uptake by the sarcoplasmic reticulum may well account for an adrenaline-induced increase in the rate of mechanical relaxation of slow twitch fibres as well as that of cardiac muscle, as adrenaline activates adenylate cyclase which, in turn, causes an increase in the intracellular level of cAMP.

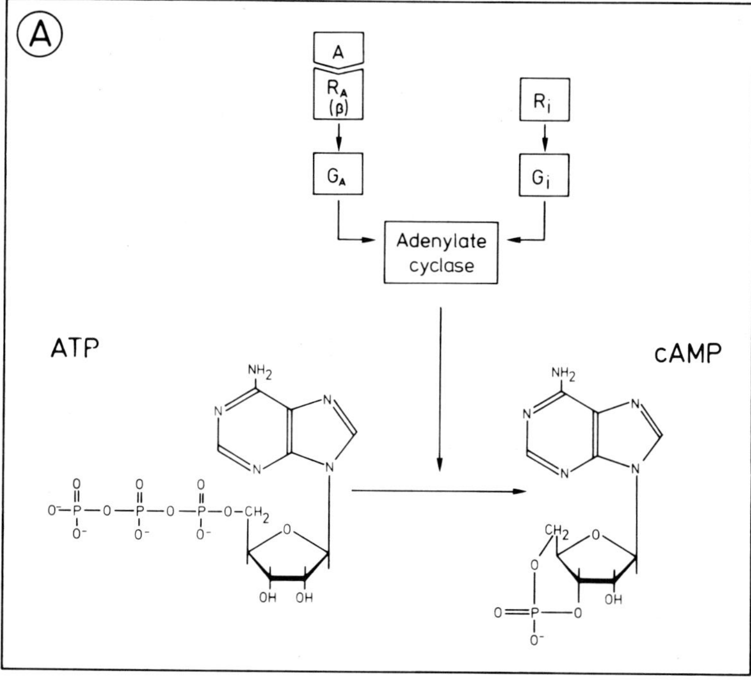

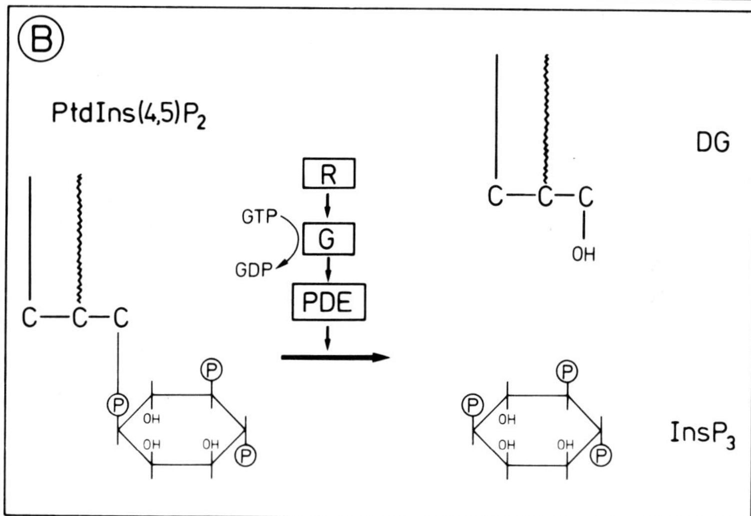

Fig. 2.11 A, B. Formation of "second messengers" influencing the sarcoplasmic reticulum. *A* Cyclic AMP: Agonist-activated *β*-receptor R_A communicates with adenylate cyclase by way of a stimulatory GTP-dependent transducer protein (G_A, G-protein). Activated cyclase catalyzes the transformation of ATP (precursor) into pyrophosphate and cyclic AMP (second messenger), which activates the cAMP-dependent protein kinase phosphorylating phospholamban associated with the calcium pump of the sarcoplasmic reticulum or Ca channels of the cell membrane. *B* The formation of the second messengers diacylglycerol (*DG*) and inositol trisphosphate (*$InsP_3$*) from the membrane-bound precursor phosphatidylinositoldiphosphate (*PtdIns* (4,5) *P_2*). *R* = membrane receptor for agonist, e.g. noradrenaline or histamine; *G* = GTP-binding protein; G_i = inhibitory G-protein; *PDE* = phosphodiesterase (phospholipase C). (After Berridge 1985, cf. also Berridge 1987)

2.3.5 Supplementary Calcium-Sequestering Mechanisms

In non-muscle cells, mitochondria play a major role as controllers of the cytoplasmic-free calcium level (Borle 1981), yet their role in muscle is open to debate.

2.3.5.1 Role of Mitochondria

One milligram respiring mitochondria may accumulate up to 3 μmol calcium ions which are driven across the mitochondrial membrane by an electric gradient generated by a proton ejection mechanism. For each pair of electrons transported along the respiratory chain, two calcium ions are carried, while in the absence of calcium transport, one molecule of ATP may be synthesized by oxidative phosphorylation. When the latter is blocked, however, the transport of two calcium ions requires the hydrolysis of one molecule of ATP. Since mitochondria are able to accumulate and store calcium, many investigators have suggested that they may also play a role in the regulation of the free calcium in muscle, in particular, in cardiac muscle (Carafoli et al. 1972). However, mitochondria cannot really reduce the calcium level of the surrounding medium to below 1 μM and because of their slow pumping activity, it seems unlikely that they play a role in rapid calcium uptake that is responsible for muscle relaxation (Scarpa and Graziotti 1973). According to A. P. Somlyo et al. (1982), even in smooth muscle, mitochondria do not accumulate calcium unless the Ca^{2+} concentration in the myoplasm rises abnormally. This may, however, be the case under anoxic conditions of cardiac cells or when the membranes are damaged so that the cells become flooded with calcium (Nayler et al. 1979). In this situation, mitochondria may effectively buffer a net calcium uptake by the cell (Tsokos et al. 1977).

2.3.5.2 Calcium Efflux: Na-Ca Exchange and Calcium Pump

Ashley and colleagues (1972) injected ^{45}Ca into the giant muscle fibres of the sedentary crustacean *Balanus* and measured the loss of this radiocalcium from the muscle cell with time. They noted that the efflux was depressed when the sodium of the surrounding seawater was substituted by choline. Increasing the Na concentration increased the rate of calcium efflux. Sodium-activated efflux was also found in cardiac muscle by Reuter and Seitz (1968) as well as in frog muscle (Caputo and Bolanos 1978) and in smooth muscle (Reuter et al. 1973; Pritchard and Ashley 1986). Since much of this efflux persisted even after metabolic poisoning of the muscle tissue, the energy for transporting the calcium against a concentration gradient out of the cell cannot be derived from ATP hydrolysis. According to the model proposed by Mullins (1976), it is the Na gradient across the membrane that drives the outward movement of calcium, thus trading, so to speak, Ca for Na. Three sodium ions are exchanged for one calcium ion, thus the exchanger is electrogenic, as it needs for its transport cycle one positive charge that moves across the cell membrane thereby generating a measurable current (cf. Mechmann and Pott 1986). Each time it moves the charge, the exchanger molecule undergoes a conformational change (about 2500 times per second) (Niggli and Lederer 1991). A molecular understanding of these processes would, however, require much more detailed information on the structure of the exchanger protein (cf. Nicoll et al. 1990).

In addition to the calcium exchanger just discussed, ATP-driven sarcolemmal *calcium pumps* are also involved in calcium extrusion from the cell, in particular in cardiac and smooth muscle (cf. Schatzmann 1985; Chaps. 7 and 8).

2.3.6 Molecular Structure and Function of Calcium Pumps

The structures of the calcium pump of the sarcolemma and of the sarcoplasmic reticulum are quite similar and have been reviewed by Carafoli (1991) and Inesi et al. (1990). All Ca^{2+}-transport ATPase proteins belong to a family of P-type ion motive pumps that form a covalent phosphoenzyme in an intermediate step of the elementary process. As the primary structure of these proteins is known, a detailed model of the consensus structure could be proposed (Fig. 2.12; cf. Stokes 1991). The domains involved in ATP splitting seem to be situated in a loop region

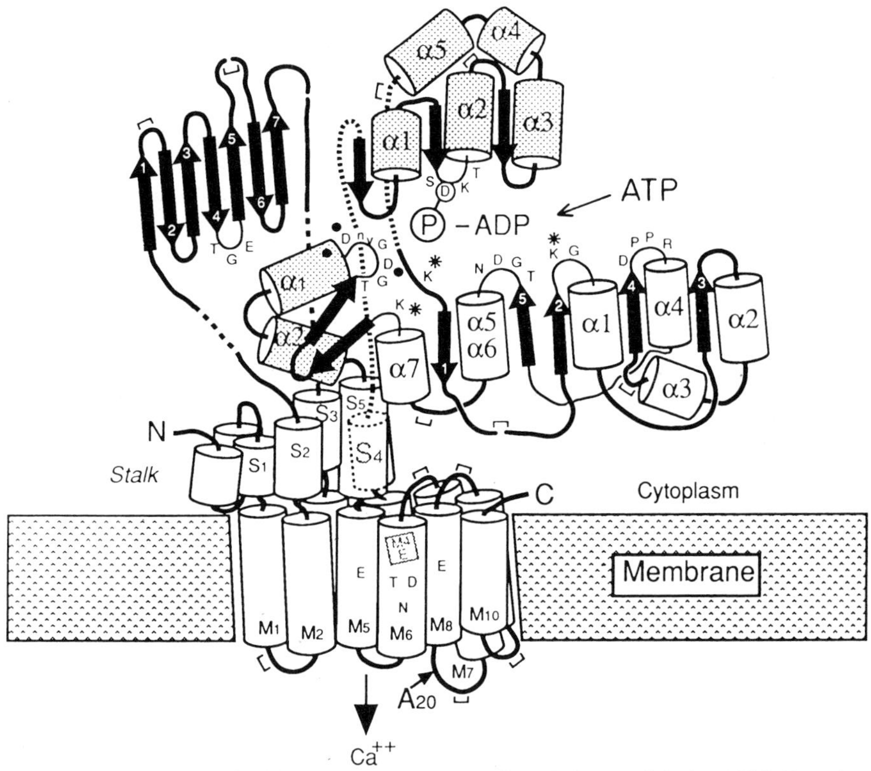

Fig. 2.12. Consensus model of calcium ATPases (courtesy of N. M. Green; cf. Stokes 1991). Note that the ATPase site (phosphorylation domain) is quite apart from the transmembrane-spanning channel regions containing the α-helices M_1 to M_{10} (represented by *cylinders*). *Upper case letters* represent amino acid residues (single letter code, cf. legend to Fig. 4.14) that, when mutated, affect ion pumping. For instance, the substitution of glutamic acid residue (E771) in helix M_5 by glycine will reduce the rate of calcium transport from 6 μmol mg^{-1} protein min^{-1} to less than 0.1 μmol (cf. Clarke et al. 1989)

of the cytoplasmic site of the sarcoplasmic reticulum pump, whereas the calcium binding residues are located in the transmembrane-spanning channel regions. The experiments of Clarke et al. (1989) using oligonucleotide-directed site-specific mutagenesis clearly showed that oxygen-containing residues within the centre of the membrane-spanning channel region provide the ligands of the high affinity calcium binding sites (see Fig. 2.12).

The regulation of the sarcolemmal calcium pumps involves modulator proteins; thus, the calcium pump of the T-tubules and surface membrane contains a domain that binds calmodulin molecules, and which, in the absence of calmodulin, acts as an endogenous inhibitor of the enzyme. In certain pathological states, such as hypertension, the activity of the pump may be decreased (Carafoli 1991). The sarcoplasmic reticulum pump is also affected by calmodulin, though indirectly via its stimulation of phospholamban phosphorylation. When dephosphorylated, phospholamban interacts with the Ca-pump, and inhibits its ATPase activity (Kim et al. 1990). As mentioned already in Section 2.3.4, this pump-associated protein may also be phosphorylated by a cAMP-dependent protein kinase.

Chapter 3

The Dependence of Muscle Contraction and Relaxation on the Intracellular Concentration of Free Calcium Ions

Calcium released from the sarcoplasmic reticulum into the sarcoplasm at the onset of activation triggers contraction in which intracellular free calcium is the major determinant of contractile activity. What is the evidence? In 1966, Jöbsis and O'Connor made an important discovery. Toad muscle fibres loaded with the calcium indicator murexide turned faintly red upon electrical stimulation just before the fibre contracted, indicating the formation of a calcium-murexide complex and a rise in Ca^{2+} concentration. Since then calcium-measuring techniques have been greatly improved and applied to many types of muscle, as reviewed by Ashley and Campbell (1979), Blinks et al. (1982), and Thomas (1982). Now the methodology for the determination of the intracellular Ca^{2+} concentration has become a biological discipline in its own right as new methods are being developed by which the Ca^{2+} concentration can be determined more precisely and conveniently. In the following sections we shall see that the quantitative relationship between intracellular Ca^{2+} concentration and contraction is similar in different kinds of invertebrate muscle (Sect. 3.1) and vertebrate muscle (Sect. 3.2), whereas the temporal relationships may vary widely.

3.1 Crustacean Muscle

Giant crustacean muscle fibres offer a unique opportunity to study problems of excitation-contraction coupling. As they are more than 1 mm wide, these fibres may be easily injected with calcium indicators or calcium salts. In this way, it is possible to determine the intracellular Ca^{2+} concentration during activation or to determine the amount of calcium required for inducing a contraction.

3.1.1 Microinjection Experiments

Before analyzing the role of calcium in "physiological contractions," Caldwell and Walster studied the mechanism of caffeine contracture (1963), for good reasons. Caffeine had long been suspected of acting intracellularly, possibly by releasing calcium from the sarcoplasmic reticulum. At last, there was a possibility to test this hypothesis. Using a fibre cannulation technique, Caldwell and Walster injected caffeine into giant muscle fibres of *Maia squinado*. This caused a contracture followed by relaxation. Obviously, the contracture was associated with and possibly due to a release of calcium from internal calcium stores for several reasons: (1) the intracellular Ca^{2+} concentration rose; (2) the contracture could be prevented by simultaneous injection of the calcium-chelating agent EGTA [ethylene-bis(oxyethylenenitrilo)-N,N'-tetraacetic acid], which would bind any calcium released by caffeine (Ashley 1967); and (3) the effect of caffeine could be mimicked by the microinjection of calcium chloride in concentrations of about 1 mM.

Calcium injection experiments showed that, in principle, the increase in Ca^{2+} concentration inside a muscle fibre could cause contraction. However, the calcium concentrations which had to be used were rather high compared with the micromolar concentrations of Ca^{2+} that are known to activate the contractile proteins in vitro. Yet we should not forget that the injected calcium solutions were diluted by the myoplasm and partly sequestered by the sarcoplasmic reticulum. The free Ca^{2+} concentration inside the cell is only a very small fraction of the total calcium, namely the quantity obtained when the calcium sequestered by mitochondria and by the sarcoplasmic reticulum as well as the protein-bound calcium are subtracted from the total cell calcium. Therefore, to determine the minimum concentration of myoplasmic free calcium needed for evoking a contractile response, special precautions must be taken. Rather than injecting calcium chloride solutions, Portzehl, Caldwell and Rüegg (1964), therefore, injected calcium buffers containing low, stabilized Ca^{2+} concentrations. If a calcium buffer containing calcium and EGTA is injected into a muscle fibre, the small concentration of Ca^{2+} which it contains is maintained, although the buffer will be diluted by the myoplasm, and despite the fact that some of the calcium will be protein-bound or sequestered. The principle of calcium buffering is the same as that of H-ion buffering, as will be discussed in more detail later (Sect. 3.1.6.2). Briefly, the Ca^{2+} concentration is proportional to the ratio of [CaEGTA] to [EGTA], where the proportionality constant is the apparent dissociation constant of the CaEGTA complex, just under 1 μM, under the conditions used. Thus, injection of calcium buffers with equal ratios of [CaEGTA] to [EGTA] would establish an intracellular Ca^{2+} concentration of probably about 1 μM. Injection also caused the fibre to shorten by 20% within 5 s. Half a minute later the fibre relaxed again as the calcium was pumped back into the sarcoplasmic reticulum. In relaxed muscle, the free Ca^{2+} concentration was presumably less than 0.3 μM since injection of calcium EGTA buffers which raised the intracellular Ca^{2+} concentration to that value induced a threshold contraction and fragmented crustacean sarcoplasmic reticulum is known to lower the Ca^{2+} concentration to this level in vitro (Portzehl et al. 1965).

3.1.2 Determination of Ca^{2+} Concentration in Resting and Contracting Muscle with Calcium Electrodes

The injection experiments just described gave only a more or less quantitative idea of the free Ca^{2+} concentration in resting muscle. A much more accurate value could be obtained by using calcium-sensitive electrodes. Ashley and colleagues (1978) found the giant fibres of the retractor muscle of the sedentary crustacean

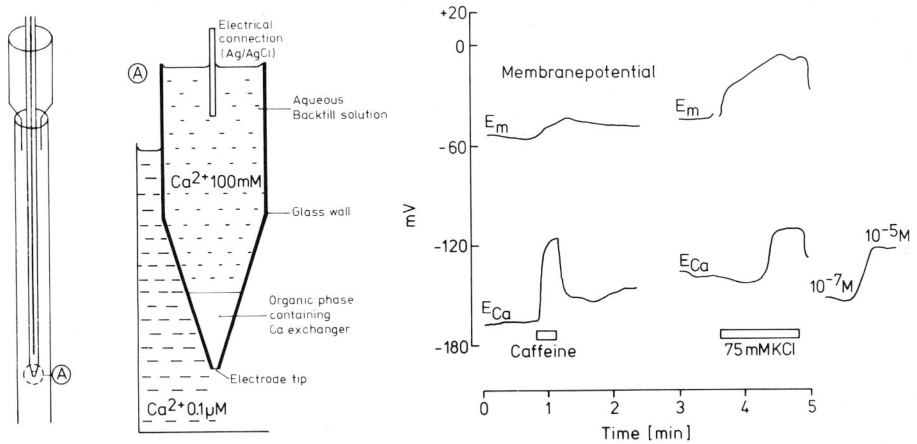

Fig. 3.1. Recording of myoplasmic Ca^{2+} concentration using a calcium selective electrode. *Left:* Electrode inserted onto cannulated giant barnacle fibre (*Balanus nubilus*). Electrode *tip A* enlarged to show basic design of electrode. *Right:* Membrane potential changes and Ca^{2+} concentration of myoplasm during caffeine and KCl contracture. The fibre was stimulated by 10 mM caffeine or 75 mM KCl. E_m = Membrane potential; E_{Ca} = output of calcium electrode (corrected for membrane potential), calibrated by immersion into saline containing 0.1 or 10 μM Ca^{2+} (*right*). (After Ashley et al. 1978)

Balanus nubilus (the giant barnacle) very suitable for such experiments. Figure 3.1 shows the setup whereby electrodes were inserted into the 3-mm-wide muscle fibres. The measured potential depended on both the free Ca^{2+} concentration in the myoplasm and the membrane potential. The latter must, therefore, be recorded simultaneously with a conventional microelectrode and should be subtracted from the total potential measured by the calcium electrode to obtain the potential dependent on the free Ca^{2+} concentration. The principle of this calcium-measuring technique is simple. The tip of the calcium-sensitive electrode has a diameter of about 20 μm and is in direct contact with the test solution. An organic phase which is selectively permeable to Ca^{2+} separates, as a calcium-permeable membrane, the test solution at the tip from the backfill solution in the electrode containing 100 mM calcium chloride. Selective calcium permeability of the "liquid membrane" is achieved by complementing the organic phase with a calcium-specific carrier, such as the "EGTA-type," neutral calcium-binding ligand ETH 1001 (Simon et al. 1977). Because of the concentration gradient of Ca^{2+} between the backfill solution of the electrode and the test solution, a calcium diffusion potential arises across the calcium-permeable electrode membrane which is equal to:

$$E = E_0 + \left(\frac{RT}{zF} \cdot 2.3\right) \cdot \log a, \tag{3.1}$$

where a is the calcium activity in the test solution, E_o is a constant and the other symbols have their usual meaning. The bracketed term of the equation is about 30 mV at room temperature, so that for a tenfold change of the Ca^{2+} concentration, the potential should change by 30 mV. Note that the electrode measures the Ca^{2+} concentration in the myoplasm locally just underneath the tip of the microelectrode in the concentration range of 10^{-8} to 10^{-3}M. Unfortunately, the electrode also responds to other ions to some degree and these effects have to be corrected for by using a selectivity coefficient. Strictly speaking, the electrode measures calcium activity rather than concentration, but concentration values can be obtained if the measured calcium potential is compared with that obtained with the same electrode in calcium buffers of known Ca^{2+} concentration. For calibration it is necessary, of course, that the ionic composition and, in particular, the ionic strength and, hence, the activity coefficient for calcium is similar to that of the myoplasm.

Table 3.1 shows that the resting levels of the free Ca^{2+} concentration of quite different vertebrate and invertebrate muscles all lie in a narrow range between 0.1 and 0.3 μM. The constancy of these values in the animal kingdom is remarkable. Perhaps the low calcium levels may be nature's way of allowing rapid activation: only a small quantity of calcium would have to be released from the sarcoplasmic reticulum in order to cause a hundredfold increase in calcium ion concentration and, thus, a rapid and dramatic activation of calcium-dependent enzymes. As

Table 3.1. Calcium ion concentration in resting muscle cells

Animal	Muscle	Ca^{2+} (μM)	Ref.
Barnacle (*Balanus nubilus*)	Retractor	0,1[a]	Ashley et al. (1978)
Frog (*Rana pipiens*)	Semitendinosus	0.16[a]	Tsien and Rink (1980)
Ferret	Cardiac (ventricular)	0,26[a]	Marban et al. (1980)
Sheep	Cardiac (Purkinje fibre)	0.3[a]	Coray et al. (1980)
Toad (*Bufo marinus*)	Stomach smooth muscle	0.14[b]	Williams et al. (1985)

[a] Ca^{2+} concentration was determined with Ca-sensitive microelectrodes.
[b] Ca^{2+} concentration was determined with Ca-sensitive Fura-2 fluorescence: fluorescence imaging techniques demonstrated compartmentalization of intracellular free calcium.

Table 3.2. Probes for detecting intracellular free calcium in muscle

Probe	MW	app.KD	Stoichiometry	Detection of Ca binding	Ref.[a]
Aequorin	20,000	100 μM	2 Ca[b]	Luminescence (460 nm)	1, 4
Murexide	284	3 mM	Ca L	Absorbance (540–507 nm)	2
Arsenazo III	776	10–50 μM	Ca L_2	Absorbance (660–690 nm)	2, 4
Antipyrylazo III	746	~100 μM	Ca L_2	Absorbance (660–690 nm)	2, 4
Quin-2	541	0.1 μM	Ca L	Fluorescence at 520 nm (excitation 340 nm)	3, 5
Fura-2	758	0.2 μM	Ca L	Fluorescence at 500–530 nm (excitation at 350 and 385 nm)	3

Note that Ca measurements by metallochromic and fluorescence indicators may be disturbed by movement artifacts (e.g. due to changes in fibre geometry) or by the Ca buffering of the ligand (L).
[a] References: (1) Ashley and Campbell (1979); (2) Thomas (1982); (3) Tsien et al. (1985); (4) Blinks et al. (1982); (5) Harvey et al. (1985).
[b] Per molecule of aequorin.

shown in Fig. 3.1, such an increase in Ca^{2+} concentration could be measured in giant muscle fibres of *Balanus nubilus* stimulated by caffeine (Ashley et al. 1978).

During maximum contraction, a calcium concentration of 10 μM was determined by comparing the electrode response in the fibre with the response obtained in test solution containing calcium buffers of known free calcium ion concentration. Using electrodes with a smaller tip diameter of less than 1 μm, similar free Ca^{2+} concentrations could be determined during a contracture of thin muscle fibres of the mammalian heart (Marban et al. 1980) and frog skeletal muscle (Tsien and Rink 1980). Since the response time of calcium-selective electrodes is rather slow, they cannot be used, however, for recording rapid calcium transients and for establishing the exact temporal relationship between membrane changes, free Ca^{2+} concentration and contractile activity. In order to do this, other calcium sensors (listed in Table 3.2), such as the photoprotein aequorin, and calcium-sensitive dyes, such as arsenazo III and antipyrylazo III, should be used, as discussed below.

3.1.3 Calcium Transients Determined by Aequorin

So far we have seen that the intracellular Ca^{2+} concentration rises during a contracture and that an increase in intracellular calcium induced by injection of calcium buffers causes a contracture. The next step then would be to establish that free calcium actually does rise during a normal activation, and that this rise precedes the onset of contraction, while a fall in the intracellular Ca^{2+} concentration would lead to relaxation. In 1962, Shimomura and colleagues reported that a protein could be extracted from the light organs of the jellyfish (*Aequorea forskalea*) which would emit light when it reacted with calcium; the latter catalyzes the conversion of the protein into its oxidized form, a process associated with a bluish luminescence. When they heard about this finding, Ashley and Ridgway (1970a) had the idea of microinjecting *aequorin* into giant barnacle fibres and using it as an intracellular calcium sensor. Thousands of jellyfish had to be sacrificed in order to prepare, according to Shimomura's procedure, sufficient protein for the experiments. By microinjection, Ashley and Ridgway obtained an intracellular aequorin concentration of about 1 μM and then suspended the muscle fibre in a small Perspex bath in front of a photomultiplier tube in a vertical position so that the light from the entire fibre length was received by the photomultiplier during the experiment, as illustrated in Fig. 3.2. In addition to the light signal, force and membrane potential were also recorded simultaneously. Even in the resting state, the fibre emitted some light and this "resting glow" could be partly ascribed to a "dark reaction" and partly to the effect of a very low Ca^{2+} concentration (about 10^{-7}M) that is present in the relaxed muscle. When the muscle fibre was stimulated electrically or placed in a solution containing a high concentration of K^{+}, it contracted and glowed more intensely, suggesting that the internal Ca^{2+} concentration rose. In experiments with giant fibres from the spider crab (Ashley 1969), the attained Ca^{2+} concentration was similar to that obtained in earlier experiments in which contraction was elicited by injecting calcium buffers.

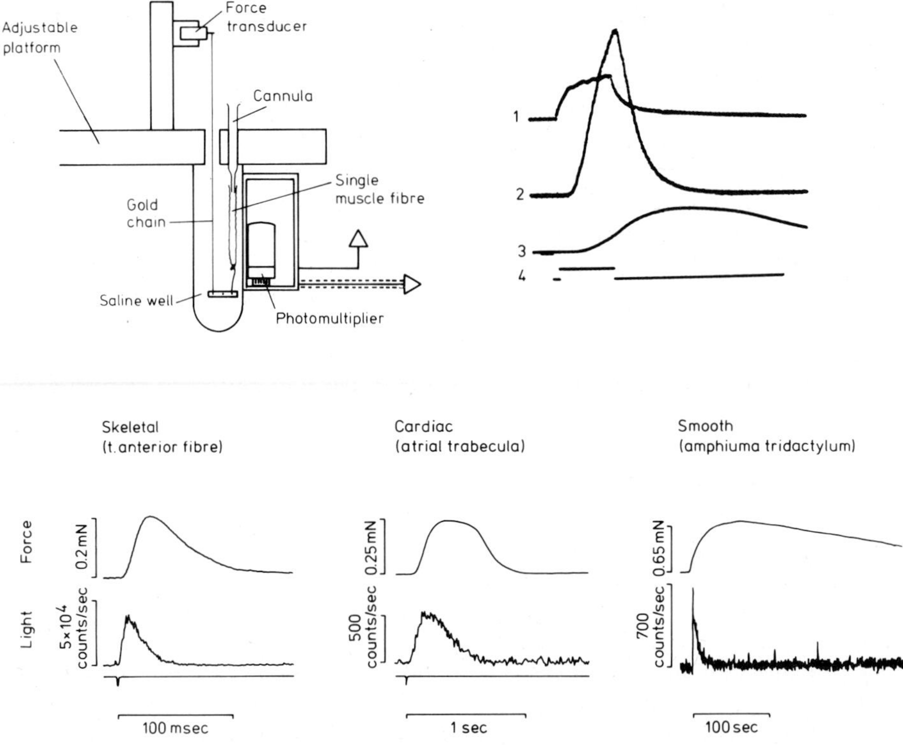

Fig. 3.2. Recording the calcium signal with aequorin. *Above:* Experimental setup (*left*) for recording tension and calcium-dependent light emission from cannulated barnacle fibres which were microinjected with aequorin (*right*). *1* Membrane depolarization (from -56 to -26 mV); *2* calcium signal (light emission); *3* force (0.1 N cm^{-2}); and *4* depolarizing pulse (3.5 V, duration 0.2 s) (From Ashley and Ridgway 1970b). *Below:* Calcium transient as measured by the aequorin method in frog striated muscle (tibialis anterior), atrial frog trabecula and from a large cell from an amphibian stomach (*Amphiuma tridactylum*). Note the difference in time scale. (After Allen and Blinks 1979 and Morgan and Morgan 1982)

To estimate quantitatively the rise in Ca^{2+} concentration during the contraction in a single fibre, Ashley and Ridgway (1970a) injected, in addition to aequorin, a solution containing a calcium buffer of a stabilized Ca^{2+} concentration. In this way, they obtained a linear relationship between the logarithm of the free Ca^{2+} concentration of the buffer within the fibre and the logarithm of the light emission with a slope of just under 2, suggesting a square root relationship between light intensity and Ca^{2+} concentration. Correspondingly, the 100-fold increase in light emission evoked by stimulating the fibre with high K^+ saline was found to correspond to a tenfold increase in the free Ca^{2+} concentration from approximately 0.1 to about 1 μM. Figure 3.2 shows the calcium-dependent light emission obtained by Ashley and Ridgway (1970b) from a giant fibre injected with aequorin and stimulated with a depolarizing electrical pulse. Note that membrane depolarization, e.g. by 30 mV, causes after about 50 ms the intensity of the emitted light to increase, but tension begins to rise much later, after about 100 ms, and much more slowly. Interestingly, the peak of the transient was obtained while tension was still rising and maximum force was achieved at a time when aequorin luminescence was already declining. This decline of the calcium transient started after repolarization of the muscle cell membrane and occurred with a time constant of about 50 ms, while muscle relaxation occurred at a time when the Ca^{2+} concentration had reached almost resting value. Such responses could be repeated many times since very little aequorin was used up when it reacted with Ca^{2+} during the course of the experiment.

3.1.4 Calcium Transients: Comparative Aspects

Due to an impressive methodological development, the aequorin technique could subsequently be applied to very thin muscle fibres with which it was much more difficult to experiment. Figure 3.2 shows that the time course of the rise and fall of the intracellular Ca^{2+} concentration during contraction is extremely different in various types of fast and slow muscle. A comparison of cardiac muscle (Allen and Blinks 1979), skeletal muscle (Blinks et al. 1978), smooth muscle and (contractile cells) of protozoa (Ettienne 1970) suggests that the different contraction speeds may be related in part to the difference in the kinetics of the calcium transients. For instance, in the rat soleus muscle, the calcium transient during a twitch is approximately three times slower than the calcium transient of a fast twitch fibre of the extensor digitorum longus muscle of the same animal (Eusebi et al. 1980). In smooth muscle, the changes in free Ca^{2+} concentration are particularly slow (Morgan and Morgan 1982; Neering and Morgan 1980; Kometani and Sugi 1978; Fay et al. 1979).

All transients, however, have the following in common: the peak is ahead of the force maximum and most of the transient is completed at a time when maximum force is developed. Therefore, we feel justified in generalizing results obtained from barnacle muscle which, because of its large fibre size, was found to be unusually suitable for measuring calcium transients quantitatively by the aequorin technique at a time when experiments with "thin-fibred" muscles were too difficult to perform. This illustrates the Krogh principle (quoted by Krebs 1975) that "for many problems there is an animal on which it can be most conveniently studied." Many comparative physiologists have been guided consciously or unconsciously by this principle.

3.1.5 Graded Activation

Unlike vertebrate twitch muscle, electric stimulation of barnacle fibres does not usually elicit an action potential or an "all-or-none" contractile response; instead, graded responses are elicited. Ashley and Ridgway (1970 b) demonstrated that an increase in stimulus strength caused an increased depolarization, an increase in the rate and the extent of the *aequorin* light response, as well as an increase in force of contraction. A similar increase was produced by increasing the duration of stimulation. In this case, a maximum intensity of the aequorin light response was obtained, and the duration of the calcium transient was prolonged as the stimulus duration was increased. Obviously, the force obtained not only depended on the level of the intracellular Ca^{2+} concentration reached during contraction, but also on the length of time during which the myofilaments were exposed to this increased Ca^{2+} concentration. If the exposure to calcium was too brief, then there would not be enough time to activate all the crossbridges that would have been activated under equilibrium conditions after prolonged exposure.

However, in interpreting aequorin transients some caution may be required, since immediately after stimulation the Ca^{2+} concentration may be overestimated for the following reason. After a short stimulation, Ca^{2+} released from the sarcoplasmic reticulum may not have reached uniform distribution. For instance, the Ca^{2+} concentration could be much higher in the vicinity of the terminal cisternae (Palade and Vergara 1982) and non-uniformity of the Ca^{2+} distribution may also be caused by the negative charges of the myofilaments (D. G. Stephenson et al. 1981). Because of the square root relationship between Ca^{2+} concentration and light intensity, a region with a slightly increased Ca^{2+} concentration would give rise to a more than proportional increase in light emission, whereas a slight decrease in the Ca^{2+} concentration in another region would cause very little reduction in light intensity. Consequently, the average light intensity would be dominated by the regions in the neighbourhood of the terminal cisternae with a high Ca^{2+} concentration, and it would be much higher than in the case where calcium ions were evenly distributed within the fibre.

An overestimation of free calcium due to non-uniform calcium distribution is less likely to occur when the metallochromic dye, *arsenazo III*, is used as a calcium indicator (Blinks et al. 1982), for in this case an exactly linear relationship between the differential absorbence at wave lengths 673 and 713 nm (measured with a multiwavelength spectrophotometer) and the free Ca^{2+} concentration was obtained in vitro under conditions similar to those occurring in the barnacle fibre injected with arsenazo. Using this calibration method, Dubyak and Scarpa (1982) estimated that the free Ca^{2+} concentration rose from below 1 μM to nearly 10 μM during a maximal contraction (cf. Fig. 3.4 A). The transient of absorbence reflected the time course of the calcium transient obtained after stimulation and exhibited a rapid phase completed in about 50 ms followed by a much slower phase. As in the case in which aequorin was used as a calcium reporter, the peak of the calcium transient was reached before maximal force was obtained. Subsequently, the calcium signal declined, but force still continued to increase for a while. Thereafter, the intracellular Ca^{2+} concentration and tension fell more or less in parallel during the relaxation phase in a quasi-exponential fashion. The decay of tension may lag behind the decline of the free Ca^{2+} concentration because the latter is in equilibrium with calcium bound to the myofibrils, and bound calcium may dissociate comparatively slowly from these binding places when the free Ca^{2+} concentration is lowered by the calcium pump of the sarcoplasmic reticulum.

Increased depolarization increased the calcium transient (measured by arsenazo III) and force developed. The relationship between peak force and peak calcium levels (expressed as pCa) was curvi-linear, as shown in Fig. 3.4 A, the mechanical threshold being around 2 μM. At 3 μM Ca^{2+} the force development was nearly 10 N cm^{-2}, while maximum contraction occurred at 10 μM Ca^{2+}. Significantly, these concentrations of Ca^{2+}, occurring in the living muscle during activation, were also similar to the ones found previously to cause a contraction of crab muscle fibres when injected intracellularly (cf. Portzehl et al. 1964) or when applied to demembranated or skinned muscle fibres in vitro.

3.1.6 Ca^{2+} and Contraction of Skinned Fibres

After showing that the Ca^{2+} concentration of the myoplasm increases as a result of activation, the next step is to determine whether the elevated Ca^{2+} concentration is actually capable of inducing a contraction of the contractile machinery. The latter is functionally isolated in skinned fibres after mechanically removing the sarcolemma (Fig. 3.3). The contractile structure of the myofilaments is then accessible to and may be influenced by the free Ca^{2+} concentration of the bathing medium. The latter must, of course, contain ATP as an energy source since skinned fibres are unable to synthesize it in the absence of added substrates. The ATP-containing bathing medium should also resemble the myoplasm as closely as possible since pH, Mg^{2+} concentration, ionic strength and the temperature all affect the responsiveness of the contractile structure to Ca^{2+} (Ashley and Moisescu 1977; Godt and Lindley 1982).

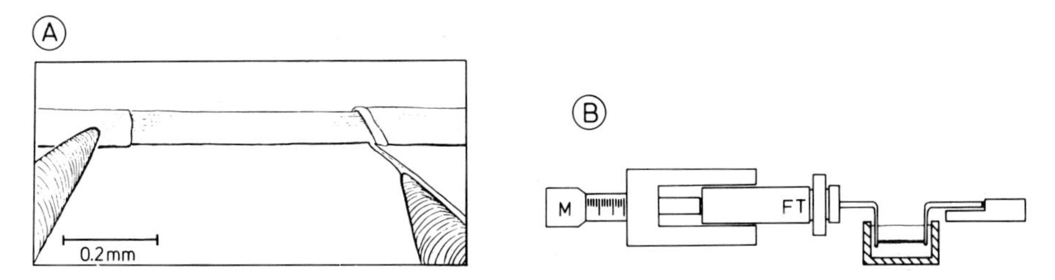

Fig. 3.3. Skinned fibre technique. *A* Mechanical skinning by stripping muscle fibres in relaxing solution or paraffin oil. The muscle cell membrane and a superficial layer of myofibrils are teased off with sharp steel needles (Podolsky 1968). Alternatively, the cell membrane may be removed chemically by means of non-ionic detergents, such as triton X-100 (1%), or it may be rendered hyper-permeable by treatment with saponin (50 μg ml^{-1}), by "freeze drying" or glycerol extraction. *B* Simple apparatus for recording isometric tension from skinned fibres. To the left, the fibre is glued or attached with clips to a hook or rod connected with the arm of the force transducer (*FT*) mounted on a micrometer drive (*M*). Suitable transducers are the semiconductor strain gage 800 series obtained from SensoNor (Horten, Norway) or the Gould Statham type UC-2 transducer. After mounting the fibre is immersed into ATP relaxing solution. For crustacean fibres (cf. D. G. Stephenson and Williams 1980) the composition is as follows (mM): ATP (total) 8, EGTA 50, TES buffer 60, pH 7.1, total Mg concentration 8.5, Mg^{2+} 1, K 117, Na 36, creatine phosphate 10, creatine kinase 150 U/ml. For inducing contraction EGTA is partly replaced by Ca EGTA and the pCa (neg. log of free Ca^{2+} concentration) is determined from the ratio of Ca EGTA and EGTA, see text

3.1.6.1 Calcium-Force Relation

In experiments with skinned barnacle fibres, the Ca^{2+} threshold concentration was approximately 1 µM and maximal isometric force was achieved at 10 µM (D. G. Stephenson and Williams 1980; cf. Fig. 3.4). It is noteworthy that these values are similar to the intracellular Ca^{2+} levels established in vivo during contraction. Note also that the calcium force relationship is less steep than in the case of all-or-none-type fast skeletal muscle fibre, thus facilitating a graded activation by Ca^{2+}. Figure 3.4 B shows the relationship between free Ca^{2+} concentration (expressed as the negative logarithm, pCa) and force development in various types of skinned fibres. Note that in crustacean muscle and also in skinned fibres of vertebrate slow muscle (D. G. Stephenson and Williams 1983), the Ca^{2+} concentration must be increased by a factor of 10 in order to increase the relative

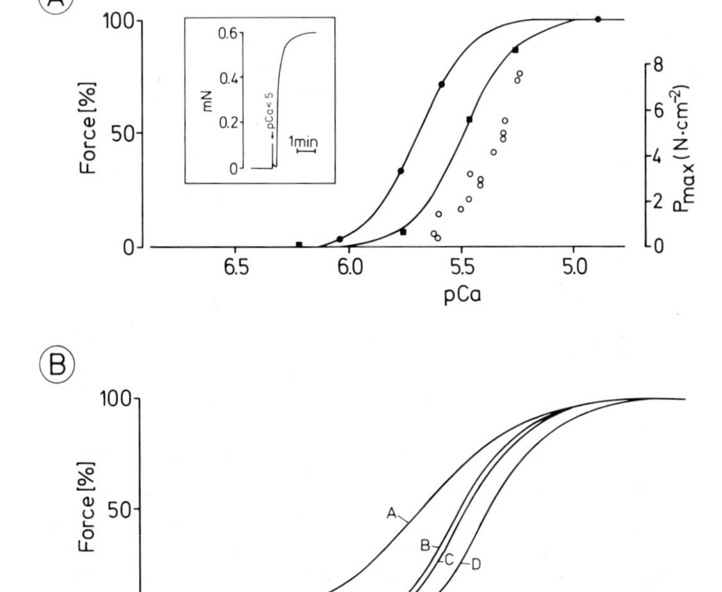

Fig. 3.4. A Relationship between Ca^{2+} concentration (pCa) in the vicinity of the myofilaments and the development of force in crustacean muscle. ● = Skinned barnacle fibre; ■ = skinned crab muscle fibre (force on *left ordinate*) compared to the responses of living barnacle muscle (○, *right ordinate*). In the skinned fibre the Ca^{2+} concentration of the bathing medium is adjusted and "clamped" with CaEGTA buffers. In the living muscle the myoplasmic calcium was varied by varying the membrane depolarization and it was determined by the arsenazo method. *Inset:* Isometric contraction of skinned barnacle fibre after raising the Ca^{2+} concentration from 10^{-8}M to 10^{-5}M with CaEGTA. The bathing medium is given in Fig. 3.3. (Based on data of Dubyak and Scarpa 1982; D. G. Stephenson and Williams 1980.) *B* Relationship between Ca^{2+} concentration (neg. log of free calcium) and force development in skinned fibres from different vertebrate muscles suspended in ATP-containing bathing solution. *A* Rat slow twitch muscle (soleus); *B* frog slow (tonic muscle); *C* rat fast twitch muscle (EDL); *D* frog fast twitch muscle. (Based on D. G. Stephenson and Wendt 1984)

force from 10% to 90%. In fast twitch muscles of amphibians (Hellam and Podolsky 1969) and mammals (D. G. Stephenson and Williams 1981), the calcium-force relationship is much steeper.

The skinned fibres of barnacles respond to a stepwise increase in free Ca^{2+} concentration as strongly and as fast as living fibres when the intracellular Ca^{2+} concentration rises after a depolarization. In experiments with skinned fibres in vitro, however, it is quite difficult to establish a high Ca^{2+} concentration quickly in a step-like fashion within the fibre, since the calcium diffusion is comparatively slow. To overcome these experimental difficulties, Ashley, Moisescu and Rose (1974 b) introduced the Ca jump technique, with which it is possible to increase the free Ca^{2+} concentration within 0.2 s (Moisescu 1976). This is almost as fast as during the Ca transient of living muscle fibres. Skinned fibres were first kept in ATP solution containing not more than 0.2 mM EGTA in order to keep the free Ca^{2+} concentration below 10^{-8}M and the fibre relaxed. This bathing solution was then suddenly exchanged with a solution containing a high concentration (e.g. 50 mM) of Ca buffer stabilizing the Ca^{2+} concentration at 3 μM. Since the concentration gradient of the Ca buffer between the inside and the outside of the fibre is so large, the buffer penetrates rapidly into the interior of the fibre, thus quickly establishing the desired free Ca^{2+} concentration.

3.1.6.2 Use of Calcium Buffers to "Clamp" Ionized Ca^{2+}

As already briefly mentioned, the buffered free Ca^{2+} concentration is given by the expression $[Ca^{2+}] = K \cdot \frac{[CaEGTA]}{[EGTA]}$, where K is the apparent dissociation constant (i.e. the reciprocal value of the stability constant) at a given pH, CaEGTA is the calcium complex of EGTA, and EGTA is the free ligand. At pH 6.8 the logarithm of the apparent stability constants equals 6 (cf. Thomas 1982), which is much lower than the log value of the true affinity constant of 10.4 determined by Schwarzenbach et al. (1975). The difference arises because at neutral pH only a small fraction of the uncomplexed EGTA is in the $EGTA^{4-}$ form which has the mentioned high affinity for Ca^{2+}, the majority being in the less ionized form H_2EGTA^{2-} with which calcium binds with a much lower affinity. The ratio of H_2EGTA to EGTA is then virtually equal to the ratio of the true and apparent stability constant, and may be used for calculating the former. If the H^+ concentration is increased by 0.1 log units, the fraction of $EGTA^{4-}$ is decreased, thus leading to a decrease in the apparent stability constant by 0.2 log units. It must be realized, however, that the effective H^+ concentration at physiological ionic strength (the H^+ activity corresponding to the pH) and the nominal H^+ concentration differ by about 0.1 log units. The H^+ concentration is higher and thus the solution is more acid than suggested by the pH (cf. Tsien and Rink 1980, Blinks et al. 1982).

As Mg^{2+} competes with Ca^{2+} for the metal-binding sites in EGTA, its concentration has to be taken into account as well. Clearly, a neglect of such considerations and the use of different stability constants of the buffer systems in the past may have been partly responsible for apparent diversities between calcium-force relationships in skinned fibres of various types of muscle that have been reported from different laboratories. Fortunately, the situation changed with the acceptance of true stability constants of the EGTA-calcium buffer system (cf. Blinks et al. 1982) and with the advent of a useful calculator programme for the determination of the free Ca^{2+} concentration at different pH, temperature, Mg^{2+} concentration and ionic strength (Fabiato and Fabiato 1979).

Since then it has become increasingly clear that the force-calcium relationship of the contractile system is not unique, but may vary somewhat among different muscles and even in the same muscle under different conditions, suggesting altered calcium sensitivity. In skinned barnacle fibres, for instance, the calcium-force relationship may be obtained by either cumulatively raising the free Ca^{2+} concentration or, alternatively, by lowering it again, but in the latter case the pCa required for half-maximal force production, i.e. the calcium sensitivity is said to be considerably higher (Ridgway et al. 1983). On the other hand, the calcium sensitivity is diminished at increased concentrations of Mg^{2+} or phosphate or at lowered pH. The latter effects may at least partly be responsible for the loss of contractility in fatigued muscle. When measuring calcium responsiveness of skinned muscle fibres, it is important to control, however, not only the ionic environment of the myofilaments, but also the mechanical conditions. Of particular importance is sarcomere length, which may have a profound effect on calcium sensitivity, as reviewed by D. G. Stephenson and Wendt (1984).

3.1.7 Excitation-Contraction Coupling in Crustacean Muscle

The experiments described in the preceding section have shown that in crustacean muscle fibres the intracellular concentration of activator calcium increases from 0.1 μM up to 10 μM during graded activation, depending on the extent of membrane depolarization. Experiments with skinned fibres demonstrated that this increase in Ca^{2+} concentration does, indeed, activate the contractile structures in a graded manner. But how is the Ca^{2+} concentration itself controlled in crustacean muscle fibres? Each of these fibres is multiply innervated by an inhibitory neuron and by both fast and slow motor neurons, each forming many synaptic contacts along the muscle fibres: multineuronal and multiterminal innervation (Grundfest 1966). Muscle contraction is initiated by the repetitive impulses of a motor nerve which, by releasing glutamate as a transmitter, gives rise to a series of excitatory junction potentials which summate and lead to a graded membrane depolarization and calcium release (Ashley and Campbell 1978). The change in membrane potential is dependent on the frequency of the junction potentials and it may induce calcium gating currents (Ashley et al. 1986). Suarez-Kurtz (1982) pointed out that a transmembranal influx of calcium, occurring during depolarization, may cause a subsequent calcium-induced calcium release from the sarcoplasmic reticulum. Indeed, when the calcium inward current is blocked by calcium entry blockers such as verapamil, depolarization does not induce a contraction because calcium-induced calcium release cannot occur then; contraction is also inhibited by procaine, which blocks the calcium release from the sarcoplasmic reticulum rather than the calcium current through the membrane. This shows that the calcium entering the cell through the cell membrane is not sufficient to activate the contractile machinery, but it could, as demonstrated in the case of skinned crustacean fibres by D. G. Stephenson and Williams (1980), induce a release of calcium from the sarcoplasmic reticulum. Contraction is terminated when the calcium ions released are taken up again by the sarcoplasmic reticulum or extruded through the cell membrane via the Na-Ca exchange mechanism (Ashley et al. 1972, 1974a).

3.2 Vertebrate Skeletal Muscle

In contrast to the crustacean fibres described in the preceding section, most vertebrate muscles twitch in response to a single depolarizing stimulus. Twitch muscle fibres of birds (Ginsborg 1960) and mammals (Close 1972) can be divided into a fast and a slow type according to the speed of contraction and relaxation. As already briefly mentioned (Sect. 3.1.4), stimulation of fast twitch fibres causes a fast and transient increase in the intracellular Ca^{2+} concentration, whereas in slow twitch fibres, these transients are comparatively slow (Eusebi et al. 1980). The skeletal muscles of amphibians (Kuffler and Vaughan Williams 1953a, b), fish (Bone 1964) and reptiles (A. Hess 1970) also contain two clearly distinct types of muscle fibres: fast twitch and slow or tonic fibres. The latter usually serve a postural function and are normally unable to twitch. Incidentally, such tonic fibres have also been found within the extraocular muscle of the mammalian eye (A. Hess and Pilar 1963). The mode of excitation of these two types of fibres is very different: twitch fibres are innervated by one motor neuron only; they contain only one motor endplate and the endplate potential, following a nervous impulse by the motor nerve, is usually sufficiently large to elicit an action potential, a large transient increase in the intracellular Ca^{2+} concentration (Blinks et al. 1978) and a twitch. Tonic muscle fibres are also innervated by one motor nerve which, however, branches and innervates the muscle fibres at many points: mononeural, multiterminal innervation (A. Hess 1970). Like the muscle cell membranes of slow crustacean muscle, the membranes of these tonic fibres are usually not excitable and respond to a nervous stimulus by graded depolarization followed by contraction rather than by an action potential.

3.2.1 Amphibian Tonic Fibres

In tonic fibres, a single stimulus usually results in a small localized endplate potential which is followed by a small contracture. But when a fibre is stimulated repeatedly, the endplate potentials summate so that, depending on the frequency of stimulation, depolarization increases. This causes a graded increase in contraction due to a graded increase of the intracellular Ca^{2+} concentration as a result of graded depolarization (Miledi et al. 1981). A small depolarization of the membrane induces, within 100 ms, a small and slow increase in the free Ca^{2+} concentration which, after the end of stimulation, decays very slowly with a half-time

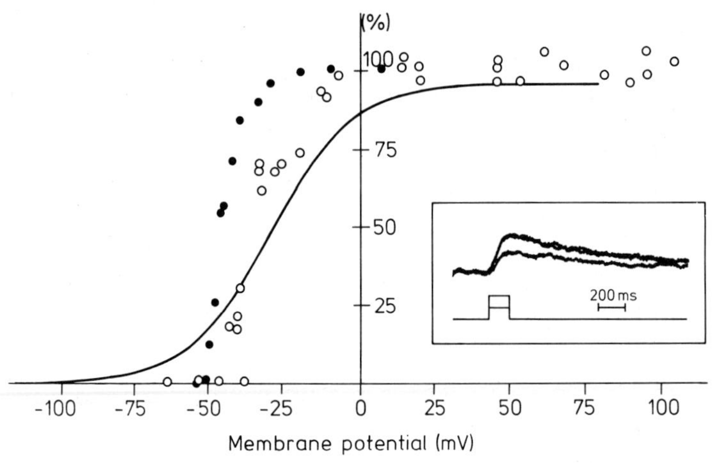

Fig. 3.5. The dependence of relative intracellular Ca^{2+} concentration, charge movement and force on the membrane potential in frog slow muscle fibres. *Abscissa:* Membrane potential; *ordinate:* force (o—o), charge movement (———) or peak size of arsenazo signal (●—●) (Gilly and Hui 1980a, b; Miledi et al. 1981). *Inset:* Time course of calcium signal (measured with arsenazo at two different depolarization pulses)

of about 200 ms. Figure 3.5 shows that the free Ca^{2+} concentration measured with arsenazo III increases only if the potential is less negative than approximately −50 mV. As the membrane becomes more depolarized, there is a gradual increase in the free Ca^{2+} concentration which maximizes when the membrane is depolarized to 0 mV. By prolonged stimulation and prolonged depolarization of the membrane, the intracellular calcium level is usually maintained at a high level as long as the external Ca^{2+} concentration is also sufficiently high (about 10 mM).

At a lower external Ca^{2+} concentration, on the other hand, excitation-contraction coupling is altered: immediately after stimulation, the intracellular Ca^{2+} concentration and force increase as much as with high external calcium, but only transiently; then tension and free calcium decline again, implying that a high external Ca^{2+} concentration is required for tension maintenance. This behaviour is even more pronounced when muscle fibres are denervated and subsequently reinnervated by the nerve fibres normally supplying a twitch muscle. Under these conditions, the muscle acquires many characteristics of twitch muscle, including the capacity to generate propagated action potentials and the loss of ability to maintain a contracture for a prolonged period.

3.2.2 Relation of Membrane Potential and Calcium Release in Twitch Fibres

As in slow muscle, depolarization of the cell membrane causes calcium release and contraction. In twitch fibres, depolarization of the membrane could be induced by increasing the external K^+ concentration or by depolarizing currents, a procedure which also causes a graded contracture (Sect. 1.2.1). This technique, known as the voltage clamp method, allows one to alter the membrane potential suddenly to a preset value and to study the force development (cf. Léoty and Léauté 1982) or the calcium transients as the function of the membrane potential. For such experiments, Kovács et al. (1979) used fibres that were cut at both ends, so as to allow the inward diffusion of a calcium-sensitive dye, antipyrylazo-III, into the muscle fibre. In these experiments, the free calcium ion concentration within the fibre increased with increasing membrane depolarization in much the same way as in tonic skeletal muscle fibres and in crustacean fibres. The slope of the calcium transient was also enhanced with increasing membrane potential suggesting that the latter affects the rate of calcium release (Rakowski et al. 1985). It was of interest then to compare the depolarization-induced calcium release quantitatively with the voltage dependence of the movement of charges within the T-tubular membrane (cf. Melzer et al. 1986). Figure 3.6 shows that both charge movement and the rate of calcium release change in a similar sigmoidal manner when the membrane is depolarized. This suggests that the movement of charges may be coupled to the gating of the calcium release channel as originally suggested by Schneider and Chandler (1973). Note that some charge is already displaced by a very small depolarization, even below the voltage threshold for calcium release.

The calcium release during a prolonged depolarization lasting 200 ms showed a phasic time course. Thus, the rate of release exhibited a peak only a few milliseconds after the onset of the (sudden) potential change and subsequently declined to a much smaller value, as also shown in Fig. 3.6. This phenomenon is

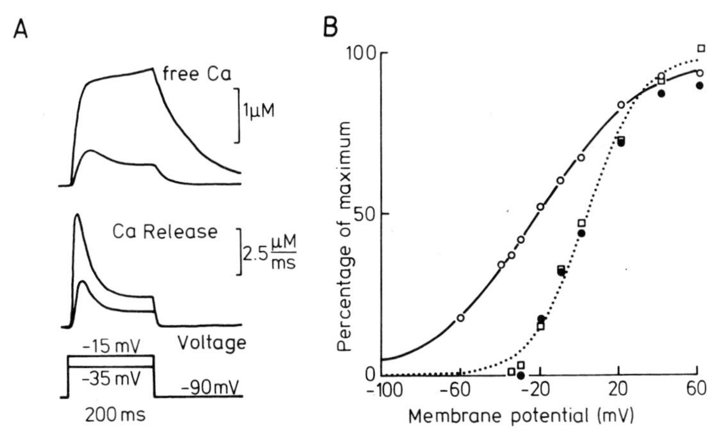

Fig. 3.6. A, B. Calcium transients, calcium release and intramembrane charge movement as affected by the membrane potential in frog muscle. *A* Membrane depolarizations (of 200 ms duration to −35 and −15 mV, respectively). The transient increase in free calcium was determined photometrically using the metallochromic indicator antipyrylazo-III (*top panel*). The calculated rate of calcium release expressed as a change in Ca^{2+} concentration per unit time in the myoplasmic volume (*bottom panel*) exhibits an early peak followed by a decline to a steady value (according to Melzer et al. 1984). *B* Voltage dependence of intramembrane charge movements (*open circles*) and peak calcium release (*open squares*) determined in a cut muscle fibre. The *closed circles* represent the extent of charge movement above threshold (Melzer et al. 1986; courtesy of the authors)

obviously part of a negative feedback mechanism, designed to generate large, but short-lasting calcium transients in fast skeletal twitch muscle fibres. In addition, the twitch muscle fibres also exhibit a slower spontaneous relaxation of force during a prolonged depolarization. This is due to a complete inactivation of the calcium release mechanism. When the muscle fibre is repolarized, the charge moves in the reverse direction and the capacity for calcium release and force production upon depolarization recovers.

In summary then, all these experiments on depolarization-contraction coupling clearly demonstrated that the intracellular or myoplasmic calcium level and, hence, contraction can be graded and controlled by the membrane potential in the range of −50 to −10 mV. Calcium release also depends on the duration of depolarization, but at values more positive than +20 mV even brief depolarizations will cause maximal calcium release (Miledi et al. 1983 b). It will be shown below that, following an action potential, the rise in myoplasmic free calcium is probably supramaximal causing an all-or-none twitch response.

3.2.3 Twitch Contraction

In isolated frog twitch muscle fibres held isometrically, a single stimulus elicits an action potential and after a short time (electromechanical latency) it causes a rapid rise in force followed by relaxation. This mechanical response, called a twitch, is preceded by a minute fall in tension, the latency relaxation (Mulieri and Alpert 1982), a change in birefringence (Kovács et al. 1983) and above all by a transient rise in free calcium. The latter could be measured by using the photoprotein aequorin (Blinks et al. 1978; Eusebi et al. 1983) or by employing the metallochromic dye arsenazo III (Miledi et al. 1982, 1983 a). The peak of luminescence coincides with the steepest part of tension rise, while the maximum force is developed during the declining phase of the light signal presumably caused by reaccumulation of calcium into the sarcoplasmic reticulum (Fig. 3.2).

If a second stimulus is applied after completion of the first twitch, e.g. after 200 ms, a second twitch of equal size is initiated, although the calcium transient associated with it may be smaller than the first transient. Despite this “negative staircase effect”, force does usually not decline during a sequence of twitches unless the muscle fatigues. This suggests that calcium release is supramaximal. Furthermore, the declining phase of the transient, reflecting the decrease in the intracellular Ca^{2+} concentration during relaxation, becomes considerably slower during repeated twitches, a finding that may partly account for the slowing of relaxation under these conditions.

3.2.4 Twitch Superposition and Tetanus

A second stimulus, applied before the first twitch is completed, e.g. after 100 ms at 2 °C, elicits a second twitch which is mechanically superimposed on the first, thus giving a larger maximum tension. The increase in total force does not seem to be due to an increase in the intracellular free Ca^{2+} concentration, since the cal-

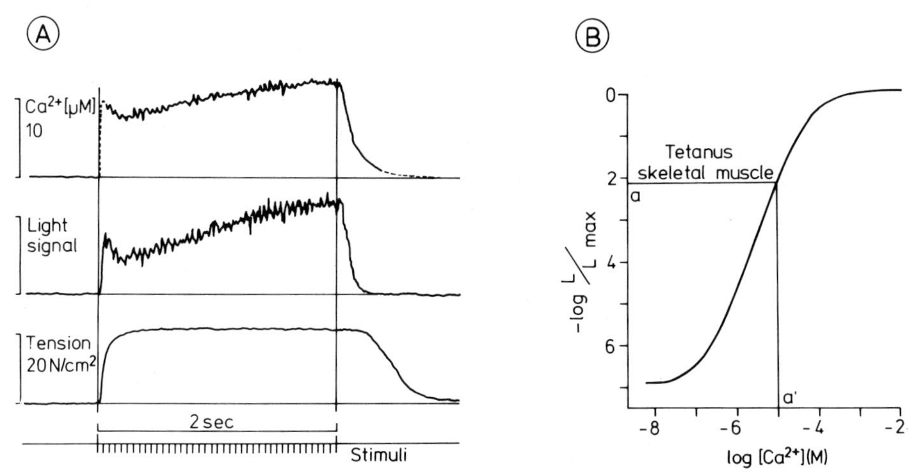

Fig. 3.7 A, B. Intracellular free calcium in frog striated muscle (From Allen and Blinks 1979.) *A* Aequorin light signal and estimated Ca^{2+} concentration during isometric tetanus of frog tibialis anterior. (From Allen and Blinks 1979.) *B* In vitro calibration curve: neg. log of L/L_{max} (*ordinate*) in dependence of log $[Ca^{2+}]$. Estimation of fibre Ca^{2+}: L_{max} is the aequorin luminescence after permeabilizing the fibre with triton X-100 (pCa 4). From the relative luminescence (*a*) the actual pCa (*a'*) can be obtained with the help of the in vitro calibration curve (since it is also valid in situ, see text)

cium transient preceding the second twitch is of similar or even smaller amplitude than that preceding the first twitch. During repetitive stimulation at intervals of 100 ms, twitches fuse, though not completely, so that tension fluctuates with the frequency of stimulation in this incomplete tetanus. The light emission signal of aequorin, on the other hand, fluctuates between high and resting values, indicating a corresponding fluctuation of free calcium (Blinks et al. 1978). When the frequency of stimulation is increased still further, the intracellular Ca^{2+} concentration is maintained at a high level since the calcium released from the sarcoplasmic reticulum cannot entirely be pumped back into the longitudinal system in the short intervals between stimuli. As each stimulus released more calcium, the intracellular Ca^{2+} concentration even rises and may eventually reach a much higher value than during a twitch, as shown in Fig. 3.7. The twitches have now fused to a "smooth" tetanus which may be defined as a state of prolonged contraction maintained by intermittent stimulation and membrane excitation (action potentials) at a frequency exceeding the fusion frequency.

Twitches fuse to a tetanus if the interval between stimuli is shorter than about one-third of the twitch duration; therefore, fusion frequency would be expected to depend on temperature and on the intrinsic speed of contraction. For instance, a fast muscle with a highly developed sarcoplasmic reticulum would require a much higher tetanus fusion frequency than a slow muscle in which the rate of calcium release and reuptake is also slow. In smooth muscle, where the sarcoplasmic reticulum is very scarce and where the calcium transients are exceedingly slow (Morgan and Morgan 1982; Neering and Morgan 1980), the fusion frequency may be below $0.1\ s^{-1}$.

After cessation of tetanic stimulation, the size of the aequorin signal decreases, mainly because of the reaccumulation of calcium by the sarcoplasmic reticulum.

Interestingly, the decrease of the aequorin light signal occurs much more rapidly than tension in contrast to the calcium signal measured by arsenazo III, which is better correlated with relaxation (Miledi et al. 1982). The discrepancy of these methods, as pointed out before (Sect. 3.1.5), is partly due to the fact that the arsenazo signal, but not the aequorin signal, is linearly related to the free Ca^{2+} concentration. It should also be remembered, however, that the change in absorbency of arsenazo caused by a change in the Ca^{2+} concentration is critically affected by the pH and by the Mg^{2+} concentration within the myoplasm. Therefore it was important that Baylor et al. (1982) and P. Hess et al. (1982) were able to show that the pH of the myoplasm (7.0) and the Mg^{2+} concentration (about 3 mM) remained constant during contraction and relaxation.

3.2.5 Quantitative Estimation of Intracellular Calcium Ion Concentration During Twitch and Tetanus

The maximum calcium levels achieved in a twitch or tetanus may be as high as 8 μM when determined by the arsenazo method (Miledi et al. 1982) or 10 μM when determined by the aequorin method (Blinks et al. 1978). Thus, the two- or fourfold difference in tetanic and twitch force of frog fibres cannot be simply ascribed to a difference in intracellular free calcium.

Methodology. For calibration of the aequorin method in the case of thin skeletal muscle fibres and cardiac muscle fibres, Allen and Blinks (1979) adopted an ingenious and simple procedure. First, they dissolved aequorin in vitro in a solution of similar ionic composition as the myoplasm and added calcium buffers of various stabilized Ca^{2+} concentrations. In this way, a straight line was obtained when the logarithm of the luminescence intensity was plotted as a function of the negative logarithm of the Ca^{2+} concentration (pCa). For convenience, the negative logarithm of relative luminescence L/L_{max} was plotted on the ordinate in order to normalize the light response with respect to the maximal emission (L_{max}) obtained at a Ca^{2+} concentration of 0.1 mM or higher (Fig. 3.7). Since the relative light emission should be independent of the concentration of aequorin used and of the solution volume or the length of the light path through the solution, one should, in principle, obtain a very similar response curve in aequorin-injected, single muscle fibres at various myoplasmic free Ca^{2+} concentrations. Conversely, if the fractional luminescence in a fibre is known, the free Ca^{2+} concentration can be derived from the calibration curve just described. For instance, a ratio (fractional luminescence) of 1 : 100 was found between the luminescence during tetanic stimulation and that obtained following treatment of fibres with the detergent triton X-100. The latter caused a maximal calcium release into the myoplasm so that the free Ca^{2+} concentration exceeded 0.1 mM. As seen from Fig. 3.7, this ratio ($-\log=2$) corresponds to a free Ca^{2+} concentration of 10^{-5}M in skeletal muscle. In cardiac muscle, in comparison, Allen and Blinks (1979) found a free Ca^{2+} concentration of only 1 μM and the rate of calcium release was also much slower. This difference is presumably due to the fact that the atrium of the frog heart contains very little, if any, sarcoplasmic reticulum and that the rise in intracellular Ca^{2+} concentration is, therefore, due to Ca^{2+} supplied by the slow transmembranal inward diffusion of calcium during the action potential (Fabiato 1983).

Time Course of the Calcium Ion Concentration Change in a Twitch. The $d[Ca^{2+}]/dt$ function determined by calcium indicators may be used to establish the quantitative and temporal force-calcium relationship as well as to estimate the rate of calcium release from the sarcoplasmic reticulum. This is because the rate of calcium increase in the myoplasm may be predicted from a model according to which the temporal changes in the intracellular Ca^{2+} concentration reflect the balance between calcium release from and calcium uptake by the sarcoplasmic reticulum and calcium-binding proteins (Gillis 1985). According to the model, the rate of Ca release would have to reach its maximum value (30 mM s^{-1} l^{-1} cell water) within 2–3 ms after stimulation to account for the observation that the transient calcium peaks at 10 ms with a concentration of 9 μM, as shown in Fig. 4.12. Therefore, the time course of calcium release is by no means identical with the "upstroke" of the calcium transient, but precedes it (cf. Melzer et al. 1984). The rate of decline of the intracellular Ca^{2+} concentration during the descending phase of the calcium transient, on the other hand, may be largely accounted for by the rate of calcium uptake by the sarcoplasmic reticulum which is a function of myoplasmic Ca^{2+}:

$$\frac{dCa^{2+}}{dt} = \frac{V_{max} \cdot [Ca^{2+}]^n}{[Ca^{2+}]^n + K_m^n}, \tag{3.2}$$

where V_{max} is the maximal calcium pumping rate (1.5 mmol s^{-1} per l fibre volume, n = 2 (since two calcium ions are transported per ATP split, cf. Fig. 2.9), and K_m is the Ca^{2+} concentration giving half-maximal activity (approximately 1 μM, according to Ogawa 1981).

As it takes time to return the released calcium to the terminal cisternae, their calcium content declines during repeated twitches. Consequently, the rate and extent of calcium release and the size of the calcium transients also decrease, particularly in shortened fibres. These are, therefore, not completely activated during a twitch, but activation may be enhanced by nitrate and zinc (Taylor et al. 1982). During twitches at rest length of the muscle, however, twitch potentiators do not enhance but prolong the calcium transient (Matsumura and Ochi 1983), suggesting that then the intracellular Ca^{2+} concentration may be already maximal.

Is Calcium Release Supramaximal in Twitch and Tetanus? The free Ca^{2+} concentration established during a twitch (10 μM) is so high that it activates the actomyosin system fully, as suggested by skinned-fibre experiments. Figure 3.4 illustrates that skinned skeletal muscle fibres exhibit an extremely steep dependency of force on the free Ca^{2+} concentration in the range between 1 and 3 μM Ca^{2+}. Above these values, force remains maximal and constant up to Ca^{2+} concentrations of 20 μM. One might conclude, therefore, that following an action potential the actomyosin system ought to be nearly maximally activated, thus giving rise to an all-or-none twitch contraction. It seems clear then why twitch tension does not decrease during repeated stimulation at a low frequency despite the fact that the size of the calcium transient may decrease considerably.

As already mentioned, superposition of frequent all-or-none twitches causes a maintained tetanic contraction of much greater strength. Clearly, the gradual rise of force during the "build-up" of tetanic force cannot be due to an increase in the myoplasmic free calcium level, since the latter is already maximal in the twitch. Rather it may be ascribed to the longer time during which the myofilaments are exposed to a high Ca^{2+} concentration in a tetanus. While the activation of crossbridges by calcium may be over 80–90% complete in a twitch contraction (Gillis 1985; cf. Matsubara and Yagi 1978; Kress et al. 1986), the crossbridge attachment and force generation seems to take considerable time after a rise in free calcium (Moisescu 1976) and may indeed take much longer than the time available in a twitch. This striking conclusion seems to be unavoidable when considering the experiments of A. F. Huxley and Simmons (1973), Ford et al. (1986), and Cecchi and colleagues (1986), who showed that immediate stiffness (a measure for the number of crossbridges attached at any one moment) increases gradually during the "build-up" of tetanic force. Incidentally, these results are also at variance with Hill's concept (1950) of a sudden onset of active state. Hill supposed that the contractile material was fully activated very early in the twitch and ascribed the subsequent force development in twitch or tetanus to the shortening of fully activated contractile elements stretching the series elastic elements.

3.2.6 Control of Force in Intact Twitch Muscle

Since the concentration of calcium activator is always supramaximal in an all-or-none twitch response, force cannot normally be graded in these muscles by a cellular control mechanism. Instead, a different strategy of a more "centralized" control has been realized. In the intact muscle within the body, force can be gradually increased by increasing the number of "recruited" active motor units, each unit comprising one motor neuron and the group of muscle fibres that it innervates. However, at a low frequency of stimulation by motor neurons, only a fraction of the asynchronously twitching muscle fibres contribute to force at any one time. If the frequency of impulses of motor neurons increases, twitches fuse into tetanic contraction and the muscle tension produced is much larger than in a twitch for two reasons: (1) *all* fibres of a motor unit are in a force-producing state at any one moment. (2) Due to mechanical summation, the tetanic force of each fibre is greater than its twitch force. Hence, force depends on the number of motor neurons "recruited" and on the frequency with which they fire (cf. Carlson and Wilkie 1974 for a brief review). This regulatory mechanism contrasts with that in many invertebrate muscles, vertebrate tonic muscles and cardiac and smooth muscles, where force is usually graded by varying the intracellular Ca^{2+} concentration.

3.2.7 Intracellular Free Calcium and Heat of Activation

As shown in Fig. 3.8, a transient of intracellular free calcium induced by the application of a single shock to the muscle fibre is followed by the production of heat of activation. Unlike the initial heat associated with actin-myosin interaction, activation heat can be observed even in fibres which have been stretched to such an extent that actin and myosin filaments no longer overlap (Fig. 3.9). The initial phase of this heat closely follows the calcium transient and may indeed be attributable to the reaction of the released calcium with the calcium-binding protein troponin on the thin filaments, since this reaction is exothermic. During the decline of the calcium transient, a second, much larger quantity of heat is produced, about 4 mJ g^{-1} muscle, which, according to Mulieri and Alpert (1982), may be attributable to the action of the calcium pump. Assuming that the pumping of 2 mol calcium requires the splitting of 1 mol ATP with an enthalpy of 46 kJ mol^{-1}, the observed activation heat of 4 mJ g^{-1} could be accounted for by the release and reuptake of 200 nmol calcium g^{-1} muscle; this is actually the amount released in a twitch (Lüttgau and Stephenson 1986). In cardiac muscle the free

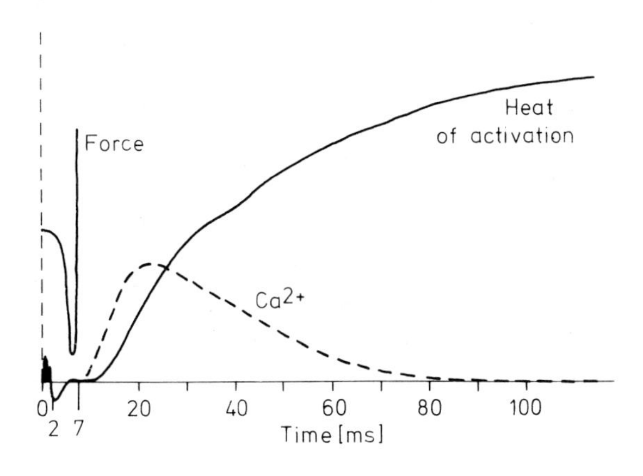

Fig. 3.8. Schematic diagram of time course of heat production, latency relaxation (force record interrupted) and calcium transient in the twitch of a frog muscle at 15 °C and 3.6 μm sarcomere length. (After Mulieri and Alpert 1982)

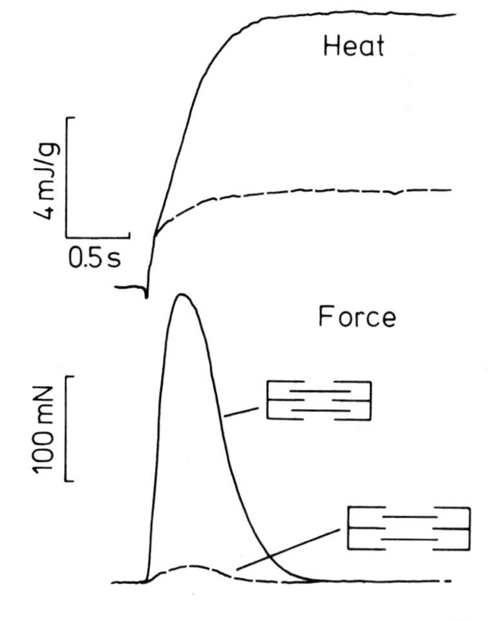

Fig. 3.9. Evolution of heat in twitch of frog sartorius (at 4 °C). *Continuous line* Response at optimal filament overlap (2.2 μm sarcomere length). *Broken line* Twitch force and heat at 3.6 μm sarcomere length where actin and myosin no longer overlap. Note that this heat of activation is about one-third of the total initial heat produced in a twitch and is associated mainly with intracellular "calcium cycling." The difference in heat production at the 2 sarcomere lengths corresponds to the energy required for the contractile process (actin-myosin cycling). (Rall 1982)

Ca^{2+} concentration following an action potential is much smaller, so that the calcium-binding sites on the filaments are only partly saturated. This means, of course, that the quantity of calcium released and taken up by the sarcoplasmic reticulum must also be much smaller, with the consequence that the amount of activation heat is smaller, too.

3.2.8 Coordination of Metabolism and Contraction by the Intracellular Free Calcium Ion Concentration

The ATP split during contraction is rapidly resynthesized in the course of glycogen breakdown. In twitch fibres of the fast, glycolytic type (Close 1972) the mobilization of the energy store glycogen, which occurs within seconds at the onset of muscle activity, is itself started by an increase in the concentration of myoplasmic free calcium (Fisher et al. 1976). Calcium-activated (and cAMP-dependent) phosphorylase kinase causes the conversion of the inactive *b*-form of phosphorylase into a phosphorylated active *a*-form which catalyzes the breakdown of glycogen to glucose-1 phosphate units. Interestingly, the calcium dependency of the phosphorylase kinase and, hence, of the phosphorylase *b*-*a*-conversion is rather similar to that of contractile activation despite the fact that the calcium-binding proteins are different in the two cases (Fig. 3.10). This suggests that the intracellular free calcium plays a major role in integrating cellular functions. It coordinates in a concerted action such different activities as glycogen metabolism and actin-myosin interaction, thus providing a link between metabolism and contractility. As the concentration of Ca^{2+} increases in the sarcoplasm as a result of ac-

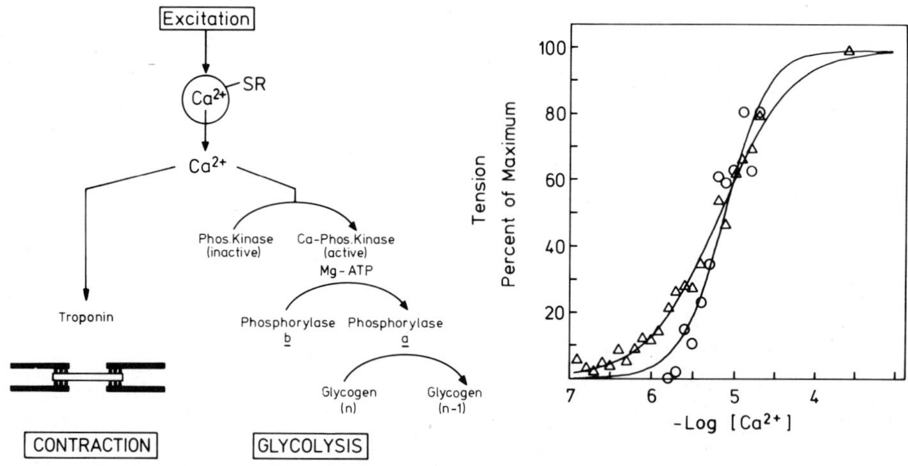

Fig. 3.10. Concerted calcium action in muscle. *Left* Ca^{2+} released from the sarcoplasmic reticulum (*SR*) activates muscle contraction and energy provision by glycolysis. Cyclic AMP-dependent protein kinase and calcium activate phosphorylase kinase to convert phosphorylase-b into phosphorylase-a, which is phosphorylated and which degrades glycogen. *Right* Calcium dependency of phosphorylase kinase activation (△) and force generation (○) of skinned rabbit psoas fibres suspended in ATP salt solution. (Fisher et al. 1976). In addition, some mitochondrial enzymes are also dependent on cytoplasmic free calcium (cf. Hansford 1987).

tivation, the metal ions will bind to troponin and to the calmodulin subunit in phosphorylase kinase and will trigger contraction and glycogen breakdown simultaneously. The interaction of calcium with these various calcium-binding proteins will be discussed more fully in the following Chapter 4.

3.2.9 Calcium Homeostasis: Physiology and Pathophysiology

Calcium imaging techniques with excellent space and time resolution of the imaging system and the use of a new generation of calcium-sensitive fluorescent dyes (such as fluo-3, see Vergara et al. 1991; or fura-2, see Westerblad et al. 1990) are now being used to determine the level and the spatial distribution of intracellular free calcium. In healthy resting muscle fibres free calcium levels are kept low, probably around 100 nM, as the tonic slow influx of calcium through leakage channels is completely compensated by the activity of the membrane calcium pump. This is not so in muscle dystrophy, where calcium influx is enhanced (Turner et al. 1991) and intracellular calcium is greatly elevated (Duncan 1978; Dunn and Radda 1991). Myoplasmic free calcium is also elevated in muscle fatigue (Lee et al. 1991) and in patients suffering from *malignant hyperthermia* (Lopez et al. 1985). In the latter the calcium release channel is apparently mutated (Sects. 2.2.4.2 and 10.1.1.4).

Duchenne muscle dystrophy is a very severe inherited disease that is linked to a defect in the muscle dystrophy gene located at the chromosome XP 21 locus (Hoffman and Kunkel 1989). The gene product is a very large protein (MW 500 kDa), called *dystrophin* that appears to be associated with other cytoskeletal proteins and glycoproteins of the cell membrane (Ibraghimov-Bekrovnaya et al. 1992); its function is not known. Thus, it has been suggested to confer mechanical strength to the cell membrane (Menke and Jockusch 1991) or it may be implicated in the regulation of a stretch-dependent calcium channel (Lansmann and Franco 1991). When dystrophin is lacking, these channels would not readily close in the absence of mechanical tension, but would remain open, thus allowing a continuous tonic influx of calcium leading to an increase in cytosolic calcium. Indeed, in an animal model of muscle dystrophy, the mdx-mouse, it has been found that these mechanosensitive channels have a high resting activity (Franco and Lansmann 1990), and may well account for the rise in intracellular free calcium and the muscle degeneration observed (cf. Stedman et al. 1991). In fact, the calcium imaging studies of Williams et al. (1990) showed that free calcium is uniformly distributed within a quiescent muscle cell, but may reach comparatively high values (up to 300 nM) in dystrophic fibres.

The causal linkage between disease and the high intracellular free calcium levels is still not understood. It is clear, however, that the calcium levels observed are below the mechanical threshold known to increase force in muscle fibres or skinned fibres from diseased human tissue (Fink et al. 1990). On the other hand, the increased free calcium ion concentration of the myoplasm may well be causally involved in an enhanced proteolysis and local contractures (cf. Duncan 1978).

As the dystrophin gene is lacking in dystrophic muscle fibre, its substitution by a cloned complementary DNA of the dystrophin gene is a challenge to molecular biologists. These experiments are very promising as they led to the expression of dystrophin protein in these muscle fibres (Ascadi et al. 1991) and this raises several questions: Does the reexpression of dystrophin cure the dystrophic fibre? In particular, does it normalize calcium channel activity and intracellular free calcium levels, and if so, what is the functional involvement of other cytoskeletal proteins, such as α-actinin, that interact with dystrophin (cf. Perry et al. 1991)?

In summary then, the development of calcium imaging techniques and molecular genetic approaches has led to an understanding of calcium homeostasis in protecting the muscle cell from an invasion of high toxic amounts of calcium which might otherwise cause muscle damage and even dystrophy.

Chapter 4

Calcium Binding and Regulatory Proteins

While it is now well known that a rise in the intracellular free Ca^{2+} concentration precedes muscle contraction, it is still puzzling how exactly the contractile machinery is switched on by calcium. Much progress was achieved, however, by studying the calcium-binding protein troponin, which proved to be the intracellular calcium switch regulating the activity of the contractile machinery in vertebrate striated muscle and many types of invertebrate muscle. Thus, troponin senses the free Ca^{2+} concentration in the myoplasm and transmits the calcium signal to the actin filaments (Sect. 4.1). By influencing the filament structure troponin regulates potential crossbridge-binding sites on a segment of actin monomers and, hence, the activity of the contractile ATPase (Sect. 4.2; cf. also Sect. 10.2). This calcium activation may be modulated indirectly by ancillary calcium-binding proteins, such as calmodulin and parvalbumin and by the regulatory myosin light chains (Sect. 4.3).

4.1 Structure and Function of Troponin

The first clue as to the involvement of calcium and troponin in the regulation of muscle contraction came from studies on the activation of the actomyosin ATPase. A. Weber and Winicur (1961) showed that in physiological concentrations of Mg^{2+} the rate of ATP splitting by crude actomyosin from skeletal muscle was considerable even if no other divalent cation was added. After addition of 1 mM of the metal chelator EGTA, however, the ATPase activity was drastically reduced, though the Mg^{2+} concentration was barely affected. Activity could be restored by addition of calcium salts, thereby increasing the free Ca^{2+} concentration to 1 or 10 μM. These calcium-activating effects were, however, not observed with all actomyosin preparations. For instance, a synthetic actomyosin formed from highly purified skeletal muscle actin and myosin exhibited, in the presence of trace concentrations of Ca^{2+}, a high ATPase activity that could not be inhibited when the calcium was bound by EGTA. The calcium sensitivity of crude actomyosin was also lost when it was repeatedly washed with solutions of low ionic strength, raising the question whether the washing procedure removed a factor conferring calcium sensitivity to actomyosin. The breakthrough came with the discovery of Ebashi in 1963 that a protein factor isolated either from the washing solution or from the whole muscle could restore calcium sensitivity. This factor inhibited "supercontraction" of actomyosin gels as well as the ATP-splitting rate in the absence of Ca^{2+}, but not at micromolar concentrations of free calcium. Thus, in the presence of the factor, ATPase activity could again be controlled by the free Ca^{2+} concentration. As the calcium-sensitizing factor contained mostly tropomyosin, Ebashi and Kodama (1966) initially called it "native tropomyosin," in contrast to pure tropomyosin which per se had no calcium-regulating activity. Unlike pure tropomyosin, native tropomyosin contained an additional protein which could be separated from tropomyosin by isoelectric precipitation. This protein, called troponin (Ebashi et al. 1968), alone was capable of conferring calcium sensitivity to purified actomyosin, but its regulatory effect could be considerably improved by the addition of tropomyosin.

4.1.1 The Subunits of Troponin

Obviously, troponin must have at least three functions. First, it would have to sense the free Ca^{2+} concentration. Secondly, it should inhibit actomyosin ATPase activity and contraction depending on the Ca^{2+} concentration. Thirdly, it would be expected to collaborate and interact with tropomyosin. These functions are located in different subunit parts of the molecule designated as troponin-C, troponin-I and troponin-T (Greaser and Gergely 1971, 1973), or simply Tn C, Tn I, and Tn T.

The Ca^{2+} concentration is sensed by troponin-C, a peptide chain of 159 amino acid residues and having a molecular weight of 19 kDa which binds four calcium ions per molecule (Fig. 4.1). Troponin-I is a globular protein with a molecular weight of 23 kDa. It binds to actin and inhibits actomyosin-ATPase activity, but is "told" by troponin-C when to do so (Hartshorne et al. 1969). Thus, the inhibitory action of troponin-I is relieved by troponin-C when the latter is occupied by calcium, whereas by itself troponin-I is always inhibitory. Only together, therefore, do the two subunits of troponin form a calcium switch that regulates actomyosin ATPase in a calcium-dependent manner. The third component, troponin-T, has a molecular weight of 30 kDa when estimated from the amino acid sequence of 259 residues, but apparently 39 kDa when determined by gel electrophoresis. Three isoforms have been described, restricted to fast skeletal, slow skeletal and cardiac muscle cells (Dhoot et al. 1979; Dhoot and Perry 1980). The elongated molecule consists of a single peptide chain of 259 residues shaped in the form of a hairpin, and it connects the other two troponin subunits to tropomyosin, with which it interacts strongly.

Tropomyosin is a double-stranded, 40-nm-long protein with a molecular weight of 70 kDa; each α-helical peptide chain is wrapped around its partner to form a linearly coiled structure with the dimensions of 2×40 nm. In a solution of low ionic strength tropomyosin molecules aggregate to form two linear polymers; this self-association is promoted by troponin-T, which binds to tropomyosin and extends over 10 nm at the C-terminal end of tropomyosin. Discovered already in 1948 by Bailey, tropomyosin remained for nearly 2 decades without any known function until Ebashi et al. (1969) discovered that it amplified the action of troponin. The amount of troponin required to regulate skeletal muscle actomyosin may be considerably reduced after addition of tropomyosin (cf. Perry et al. 1973).

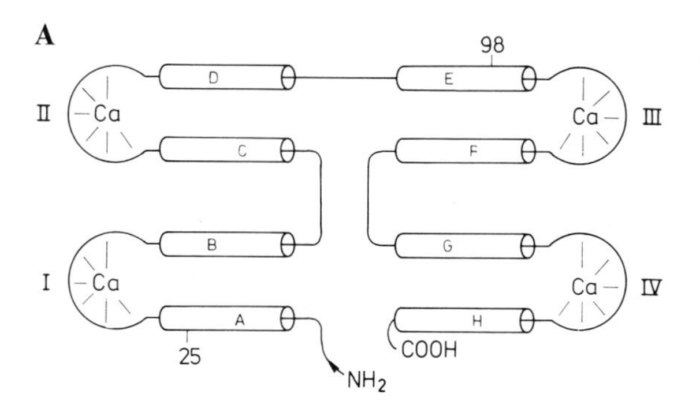

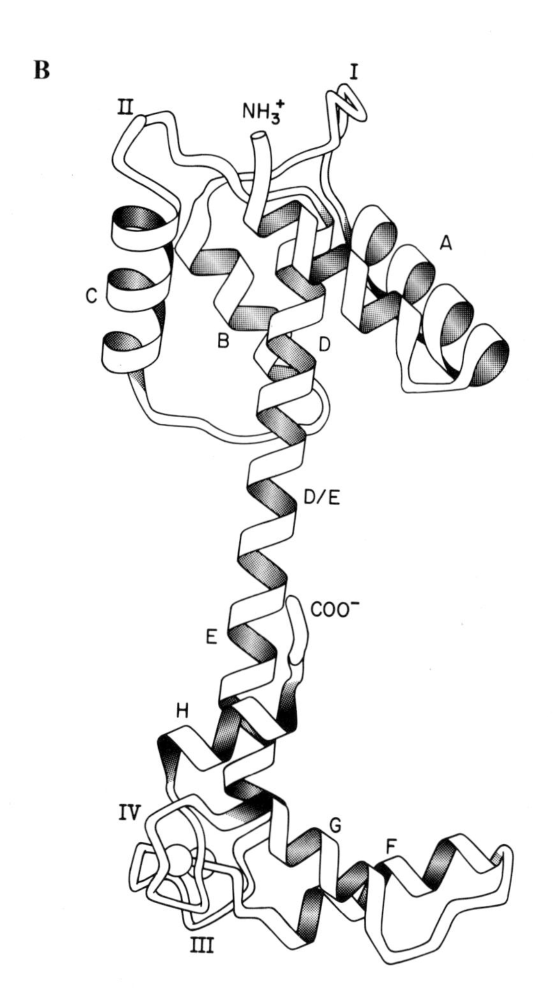

Fig. 4.1. A: Schematic diagram of predicted troponin-C structure with two low-affinity specific calcium-binding domains (*I* and *II*) and two high-affinity Ca- and Mg-binding sites (*III* and *IV*). *A–H* are helical regions of the peptide chain (after Potter and Johnson 1982). *B:* Structure of troponin-C (ribbon representation) as derived from X-ray crystallographic studies at 2.8 Å resolution. The upper helical domain is the calcium regulatory part containing the low-affinity binding sites. The lower domain contains the unspecific Ca/Mg sites. (Herzberg and James 1985; courtesy of the authors)

To summarize, the three subunit designations C, I, and T stand for calcium-binding, inhibiting action and tropomyosin binding. The troponin molecule (Mw 80 kDa) reconstituted from these three subunits has all the properties of the native protein: in conjunction with tropomyosin it switches the actomyosin-ATPase on and off, depending on whether the troponin-C subunit is or is not occupied by calcium.

4.1.2 Troponin-C is the Intracellular Calcium Receptor

The rise in the intracellular Ca^{2+} concentration to 1 or 10 μM during muscle activation is sensed by troponin-C as it binds Ca^{2+} in this range of concentration. Calcium binding to troponin-C, therefore, is the first step in the sequence of events associated with calcium activation of the contractile proteins.

4.1.2.1 Calcium-Binding Sites

Troponin-C of vertebrate skeletal muscle contains four calcium-binding domains, two on each side of the molecule (Fig. 4.1), and, by convention, numbered I to IV from the N-terminus of the peptide chain. Each domain contains a loop of 12 amino acids flanked on either side by α-helical peptide chains lettered sequentially

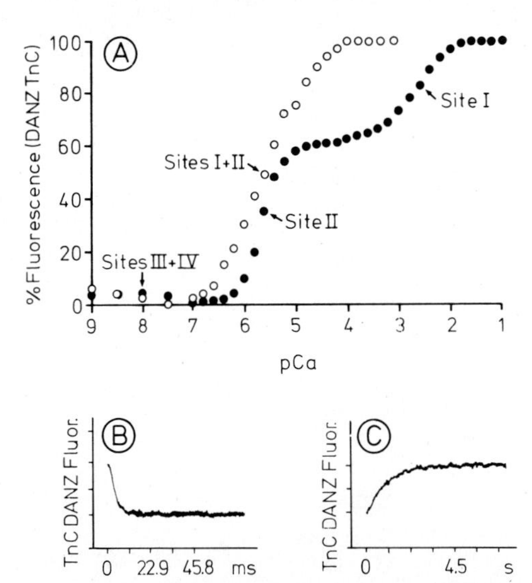

Fig. 4.2 A–C. Ca binding to troponin-C determined by measuring the fluorescence change of the dansylaziridine derivative of troponin (TnC-DANZ). *A* Ca binding to sites *II* and *I* of both skeletal muscle (○) and cardiac troponin-C (●) causes an approximately twofold fluorescence increase, but the calcium affinity of site *I* of cardiac and slow twitch muscle troponin-C is greatly reduced. Ca binding to sites *III* and *IV* (the unspecific high-affinity divalent metal-binding sites) causes a slight decrease in fluorescence (Potter and Johnson 1982). *B* Fast time course of the decomposition of the calcium complex of the Ca-specific binding site. *C* Slow time course of the dissociation of the calcium complex with the high-affinity binding sites. The tracings represent the fluorescence changes obtained after rapidly mixing Ca^{2+} TnC-DANZ with EGTA in a "stopped flow" machine (Johnson et al. 1981)

A to H. The two flanking α-helices are set at right angles and form, together with the connecting loop, a distinct structure known as EF-hand (Kretsinger 1980, Fig. 4.14). The amino acids of the loop region containing carboxylic or hydroxylic groups grip, as it were, a calcium ion. Since the oxygen atoms of these groups form a tetrahedral coordinating loop, they provide the dendates for chelating the calcium ion (Zot et al. 1983). In general, the structure of troponin-C has been conserved ever since it first appeared in evolution, probably some 500 million years ago. There is, however, some variation in structure between fast-twitch muscle troponin and the troponin of slow skeletal muscle and cardiac muscle (Dhoot et al. 1979). Differences also exist with regard to the number of calcium-binding sites of different troponin-C molecules (Goodman 1980). In troponin of vertebrate fast-twitch muscle the sites I and II on the N-terminus of the troponin molecule bind calcium with low affinity and as they do not bind Mg^{2+}, these sites are called calcium-specific sites. In contrast, sites III and IV on the other side of the troponin molecule bind calcium with much higher affinity, but not specifically as they also bind Mg^{2+}; these sites are, therefore, called Ca-Mg sites (Potter and Gergely 1975). Unlike troponin of fast-twitch muscle, cardiac troponin has only one functional, specific calcium site in addition to its two Ca-Mg sites, whereas the other one is rendered non-functional by gene mutation (Holroyde et al. 1980). Crustacean troponin has only one calcium-binding site (Wnuk et al. 1984).

Our knowledge of the stoichiometry and affinity of calcium-binding sites has been obtained largely by determining the calcium occupancy of sites after equilibrium dialysis against varying Ca^{2+} concentrations. As shown on Fig. 4.2. A, calcium binding to a binding site (Y) occurs over a wide range of calcium concentrations which may be described in terms of the simple equilibrium:

$$Y + Ca^{2+} \underset{k_{-1}}{\overset{k_{+1}}{\rightleftarrows}} Y\text{–}Ca\,, \tag{4.1}$$

where the equilibrium constant for the dissociation of the calcium-bound state (Y–Ca) is equal to:

$$K = \frac{[Y]\,[Ca^{2+}]}{[Y\text{–}Ca]} = \frac{k_{-1}}{k_{+1}}, \tag{4.2}$$

where k_{+1} and k_{-1} are the rate constants of complex formation and dissociation. The occupancy or fractional saturation equals

$$Z = \frac{[Y\text{–}Ca]}{[Y\text{–}Ca] + [Y]}. \tag{4.3}$$

By substitution one obtains the calcium occupancy at a given calcium ion concentration

$$Z = \frac{[Ca^{2+}]}{[Ca^{2+}] + K}. \tag{4.4}$$

The dissociation constant K, therefore, is the Ca^{2+} concentration allowing for 50% occupancy and is obtained by determining the midpoint of calcium titration. It is interesting to note that the dissociation constant of the calcium-specific sites of troponin-C is approximately 10^{-6}M, i.e. it is well within the physiological range of Ca^{2+} concentrations occurring in the living muscle cell after calcium release from the sarcoplasmic reticulum. Note that within myofibrils the calcium dissociation constant of the specific sites is very similar, whereas it is decreased by a factor of 10 in isolated troponin (Zot et al. 1983). The reciprocal value of the dissociation constants is a measure of calcium affinity. Incidentally, these values could also be independently obtained by kinetic calcium-binding measurements since the calcium affinity is simply the ratio of the rate constants of complex formation ("on-rate"; k_{+1}) and the rate of dissociation ("off-rate"; k_{-1}).

4.1.2.2 Kinetics of Calcium Binding and Conformational Changes

A knowledge of the temporal relationship between calcium binding to troponin and contraction is essential for our understanding of muscle activation. Clearly, the calcium occupancy of troponin-C should precede the mechanical event of contraction which it initiates. Conversely, relaxation of contracted muscle would have to be preceded by the dissociation of the calcium-troponin complex. The rates of calcium binding and dissociation, therefore, must be higher than the rates of contraction and relaxation. Is this so? Determination of calcium occupancy by the slow process of equilibrium dialysis is unsuitable for kinetic studies. It is possible, however, to measure the calcium occupancy of troponin-C from moment to moment using the fluorescent probe dansylaziridine (DANZ). DANZ-labelled troponin exhibits a strong fluorescence in UV light when calcium occupies the calcium-specific sites (Johnson et al. 1978), whereas the calcium occupancy of the high-affinity sites may cause a slight decrease in fluorescence. Thus, by following the fluorescence intensity as a function of free calcium these sites can be titrated (Fig. 4.2 A).

To measure the time course and rate constants of calcium binding to specific sites, Johnson et al. (1979) mixed troponin-C or troponin and calcium solutions very rapidly, within 2 to 3 ms, in a specially designed "stopped-flow" machine and measured the change in fluorescence, as illustrated in Fig. 4.2 B, C. Significantly, troponin became occupied with calcium almost immediately, i.e. within the mixing time of the instrument. When, on the other hand, troponin occupied with calcium was suddenly mixed with a calcium solution containing the calcium-chelating substance EGTA, fluorescence intensity decreased in a quasi-exponential manner within about 40 ms to 50% of the initial value and even faster in the case of TnC. From the halftime of fluorescence decay it can be calculated that the calcium-troponin complex decomposes with a rate constant (the calcium off-rate) of approximately 23 s^{-1}, while the calcium off-rate from troponin-C is 300 s^{-1}. Now, a moderately fast twitching muscle, the fast extensor digitorum longus muscle (EDL) of the rat contracts within 15 ms and relaxes in about 50 ms (Close 1965). This means that the rate of calcium binding and the rate of calcium dissociation from the calcium-specific sites would be fast enough to account for the speed of contraction and relaxation in a fast twitch muscle. The measured rates of calcium dissociation from the calcium high-affinity sites (the Ca-Mg sites) are, however, by orders of magnitude slower, about 3 s^{-1}, which is too slow to account for relaxation. Though calcium binding to the high affinity site would also be a fast process, we should recall, however, that in the presence of a physiological Mg^{2+} concentration (about 3 mM) these sites are occupied by Mg^{2+}, at least in resting muscle. When the Ca^{2+} concentration rises at the onset of a twitch, Ca^{2+} would have to displace Mg^{2+} first before it can be bound to the high-affinity sites. It is because of the slow dissociation of Mg^{2+}, therefore, that calcium binding to high-affinity sites is slow in the living muscle cell.

In conclusion, it is clear that only the low-affinity calcium-specific sites can bind and release calcium faster than the process of muscle contraction and relaxation. The high-affinity Ca-Mg binding sites, on the other hand, are too slow to play any role in switching the contractile machinery on and off. Their function is still unknown. Although these considerations were originally based on measurements of calcium binding to isolated troponin-C, they are probably also valid for troponin in situ within the intact, structured system (Zot et al. 1983).

In living fast skeletal muscle the Ca^{2+} concentration increases after stimulation almost, but not quite, as rapidly as in the case of the stopped-flow experiments just described. When the time course of the calcium transient in living muscle and the rate constant of calcium binding are known, it is then possible to calculate the calcium occupancy of troponin at any one time during a twitch. Such calculations show that the binding of calcium to and the calcium dissociation from troponin-C precede the rising and falling phase of the twitch both in skeletal and cardiac muscle (Fig. 4.12). These calculations have recently been confirmed by measuring fluorescence changes of DANZ-labelled troponin injected into muscle fibres (cf. Ashley et al. 1985).

When the muscle is activated, the binding of calcium to the calcium-specific sites of troponin-C induces a small conformational change in the troponin-C subunit. This is the message that is then passed on to the neighbouring troponin-I subunit within the troponin molecule. To measure the speed of this signal propagation, Johnson et al. (1981) attached a fluorescent label (IAANS) to troponin-I. When the labelled troponin was suddenly exposed to calcium in the stopped-flow machine, the fluorescence intensity of the probe decreased with a halftime of about 6 ms, indicating that the signal transmission from troponin-C to troponin-I must have been very rapid indeed. Next we shall discuss in more detail the calcium-induced structural changes in troponin and their role in the transmission of the regulatory signal to the actin filament.

4.1.3 The Calcium Signal Alters Thin-Filament Protein Interactions

How does the calcium signal received by troponin-C propagate along the thin filaments, and how is this message translated into an activation of the contractile proteins? As mentioned already (Sect. 4.1.1), troponin-I inhibits, in conjunction with tropomyosin, the actomyosin-ATPase when the Ca^{2+} concentration is low. This inhibition is relieved when calcium ions bind to troponin-C. Calcium activation, therefore, seems to be essentially a disinhibition of the contractile process. It is obviously due to some structural changes in the thin filaments enabling actin to activate the myosin ATPase. For this calcium-activated state, or "on-state," of the thin filaments may be preserved by structural fixation with glutaraldehyde (Mikawa 1979) and such permanently "turned on" filaments are then able to activate the myosin ATPase in vitro even in the absence of Ca^{2+}. In the following section we shall describe the structure and the calcium-induced structural changes of the thin filaments which may be responsible for the initiation of contraction.

4.1.3.1 Location of Troponin and Tropomyosin in Thin Filaments

The backbone of thin filaments consists of two intertwining helical chains of polymerized 5.5-nm-thick, boot-shaped actin monomers (Fig. 4.3). On electron micrographs crossover points of the chains may be seen at regular intervals of approximately 38 nm. The actin helix is not quite integral since there are 13 to 14 actin monomers per half-turn (38 nm). Two linear chains of tropomyosin molecules lie along the groove between the actin monomers on both sides of the helix. The chains arise from head-to-tail aggregation of slightly overlapping rod-shaped tropomyosin molecules, each having a length of just over 40 nm. Antibodies directed against troponin label actin filaments in muscle sections at a repeating distance of just under 40 nm (Ohtsuki 1975), and this axial periodicity is defined by the interaction of the troponin subunit-T with the head-to-tail binding sites of tropomyosin. Since there is about one troponin molecule for one molecule of tropomyosin and seven actin subunits, the thin filament may be visualized as chains of polymerized actin and tropomyosin bearing two troponin molecules at regular intervals of approximately 40 nm (Squire 1981), as illustrated in Fig. 4.3. Accepting this structural model, the question naturally arises as to how it is possible for troponin to regulate a segment of seven actin monomers so that they either can or cannot interact with myosin crossbridges. Briefly, the calcium signal alters the interaction between troponin subunits which, in turn, causes a movement of tropomyosin from a peripheral inhibitory position into a position that permits the attachment of crossbridges and their subsequent movement (Fig. 4.4 A). Calcium binding, however, may also affect the structure of actin monomers, which conversely affects myosin binding (Zot et al. 1983). What is the evidence for the proposed function of tropomyosin?

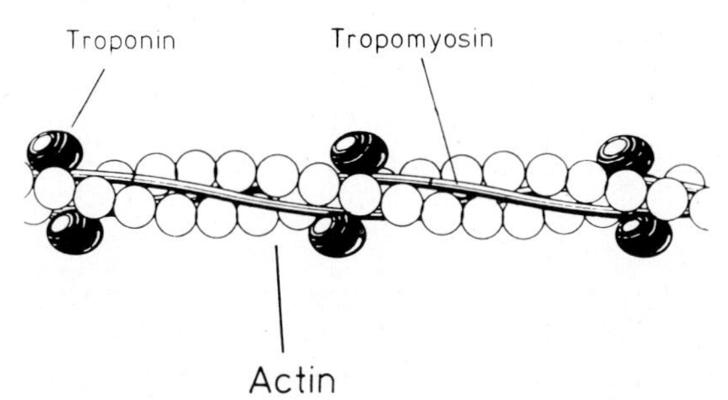

Fig. 4.3. Structure of thin filaments based on Ohtsuki's model showing extended tropomyosin molecules along the grooves of the actin structure and helical arrangement of troponin, generating seven actin functional units in the thin filament. The actin double helix contains 28 actin monomers in every repeating unit and is labelled by troponin molecules at repeating intervals of about 40 nm. Note that the paired troponin molecules are not exactly opposite (cf. Ebashi et al. 1969), but axially displaced by 2.7 nm (Maéda et al. 1979). The figure was kindly provided by Dr. Maéda (cf. Maéda 1979)

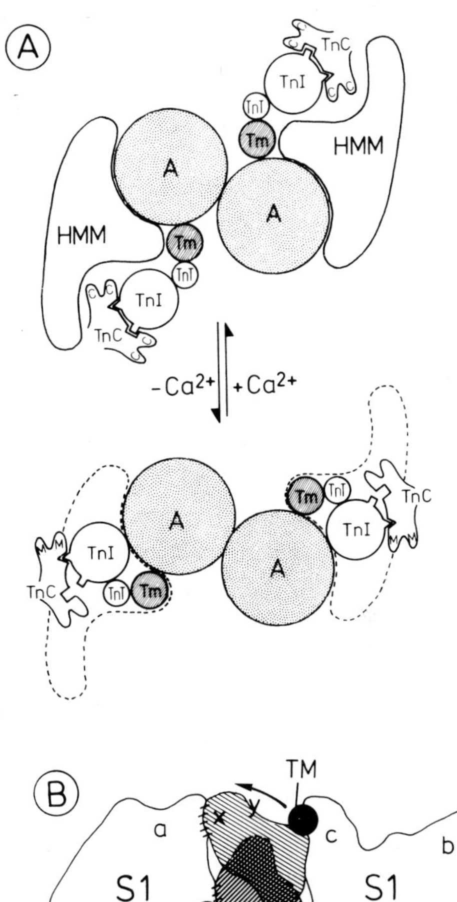

Fig. 4.4 A Model of regulation of skeletal muscle contraction showing the different interactions of troponin subunits, tropomyosin, HMM and actin at high (10^{-5}M) and low (10^{-7}M) concentration of Ca^{2+}. Upon Ca^{2+} activation troponin-C interacts more strongly with troponin-I, while bonds between troponin-I and actin are loosened. At the same time, tropomyosin moves towards the groove between the two actin strands, and crossbridges (HMM) attach to actin. Note that troponin-T interacts with both tropomyosin and troponin-I and possibly also with troponin-C (not shown). (After El-Saleh et al. 1986). *B* A more realistic diagram of the "cross-section" through a thin filament showing the interaction of subfragment-1 (S_1), tropomyosin (*TM*) and actin (*A*) in active muscle. *Arrows* indicate the direction of tropomyosin movement into the "off-position" (*x* or *y*) when the muscle relaxes. The subfragment-1 on the *left* lies 2.7 nm above S1 on the *right*. (After Egelman 1985)

4.1.3.2 Tropomyosin Movement

Tropomyosin has been assumed to have a regulatory function ever since it was found to improve the stoichiometry of the inhibitory effect of troponin-I on actin. As mentioned already, actomyosin ATPase may be inhibited by troponin-I when the troponin-actin ratio is 1:1, whereas in the presence of tropomyosin the required ratio is 1:4 (Perry et al. 1973). In which way, however, tropomyosin plays this regulatory role is still uncertain, despite a wealth of structural information on conformational changes of tropomyosin in situ.

Originally, tropomyosin function was explained in terms of a merely steric blocking model (H.E. Huxley 1973; Haselgrove 1973). In resting or relaxed muscle tropomyosin would lie at the periphery of the actin filament in a position where it would block the attachment of myosin crossbridges to actin, as shown

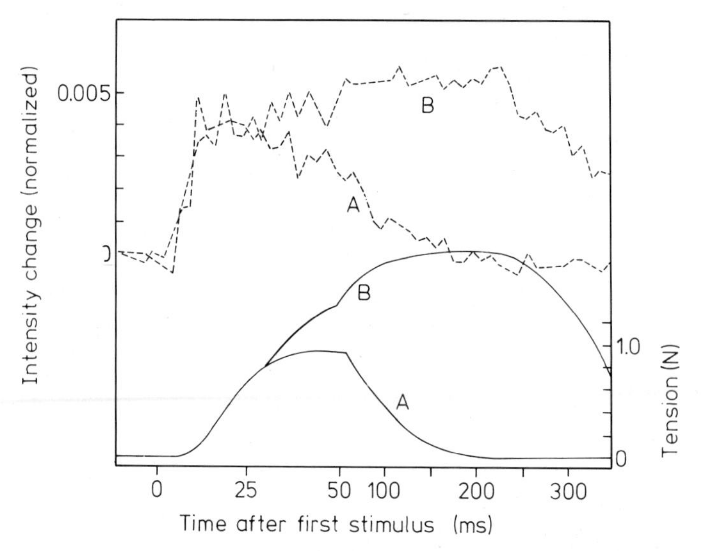

Fig. 4.5 A, B. Time course of tropomyosin movement and force development in a muscle twitch (*A*) and tetanus (*B*) of frog semitendinosus, at 14 °C. *Broken lines:* Intensity change in "second layer line" of X-ray diffraction pattern. *Continuous lines:* Force. Note that tropomyosin movement precedes crossbridge attachment and force generation, but may not be as complete in a twitch as in a tetanic contraction, thus suggesting that in the twitch a striated muscle may not be completely activated at 14 °C (whereas it is completely activated at 6 °C; cf. Fig. 1.12 C). (After Kress et al. 1986)

in Fig. 4 A. When activator calcium combines with troponin, tropomyosin would slip into the grooves between the actin strands, thereby relieving the steric blockage of crossbridge attachment. The theory requires, of course, that tropomyosin strands are indeed located on the same side of the actin filaments, and Taylor and Amos (1981) finally proved that this is so (cf. also Egelman 1985, Fig. 4.4 B). The slippage of tropomyosin into the actin grooves could be seen by X-ray diffraction investigations of stimulated striated muscle. Thus, the studies of H. E. Huxley (1973) showed a change in the relative intensities of the actin layer lines. When resting muscle fibres of the frog are hit by a parallel monochromatic beam of X-rays, the periodicity on the actin filaments (38 nm) gives rise to a clear first-order and third-order reflection (actin layer lines) of 1/380 $Å^{-1}$ (first layer line) and 1/127 $Å^{-1}$ (third layer line), while the second layer line (1/190 $Å^{-1}$) is hardly visible. Upon activation, however, the X-ray diffraction pattern changes dramatically. Now the intensity of the second layer line increases at the expense of the third layer line due to the movement of tropomyosin. Interestingly, these structural changes precede the changes in the X-ray diffraction pattern associated with the attachment of crossbridges to actin (Kress et al. 1986; cf. Fig. 4.5). This finding is clear evidence that the movement of tropomyosin on the actin filaments may be causally involved in the process of muscle activation rather than being its consequence. The latter possibility had been raised as tropomyosin movement has also been observed during activation of smooth muscle lacking troponin (Vibert et al. 1972; see also Squire 1981 for a more detailed discussion of these structural changes).

4.1.3.3 Interaction of Troponin Subunits

After showing that tropomyosin moves, the next step was to determine how this movement is brought about by troponin after receiving a calcium signal. According to the model proposed by Potter and Johnson (1982), a sequence of alterations in the subunit interaction of troponin may play a critical role (Fig. 4.4). Troponin is anchored to tropomyosin by its subunit-T, a highly elongated molecule extending over 10 nm (90 amino acid residues) over the C-terminal end of tropomyosin. Troponin-T also binds troponin-I in the middle region of the molecule. In this way, tropomyosin is held in its peripheral position when the troponin-I subunit binds to actin in the absence of calcium. Binding of calcium to troponin-C then has the following consequences: its bonds with troponin-I are strengthened with the result that the ties of troponin-I to actin become loosened. Troponin-I then no longer grips the actin filament so that the whole complex of tropomyosin and troponin drops, figuratively speaking, into the groove between the actin strands. This is the activating position of tropomyosin in which it would no longer block the attachment of crossbridges. Now the actin filament is "turned on" and it is possible that the actin structure itself is also affected. According to this scheme, it is, therefore, the regulation of the structure of the actin filament that determines the extent of actin-myosin interaction and the state of activation. In contrast to this actin-linked or thin filament-linked regulation of vertebrate skeletal muscle, a "myosin-linked regulation" functions in many types of invertebrate muscle and vertebrate smooth muscle where calcium ions influence the myosin molecule and its interaction with actin directly or indirectly (see Chaps. 6 and 8).

The model of actin-linked regulation just described was developed on the basis of studying the interaction of troponin subunits with each other as well as with tropomyosin and actin, both in the presence and absence of Ca^{2+} (cf. Hitchcock 1975). Gel filtration provides a powerful tool to determine whether two different kinds of small protein molecules, such as the troponin subunits, interact with each other to form larger molecules (Hincke et al. 1979). The results of such studies are summarized in Table 4.1. Note that troponin-I can interact with troponin-T as well as with troponin-C. Since it is not clear whether the latter interacts with troponin-T, the primary troponin interactions probably involve the sequence TnC – TnI – TnT rather than TnI – TnC – TnT, as previously assumed.

Of great functional importance, of course, are the interactions of troponin-I with troponin-C and actin, since these are altered when troponin-C is occupied by Ca^{2+}. It is not astonishing, therefore, that the part of the troponin-I molecule

Table 4.1. Alteration of thin filament protein interaction by calcium

Ca^{2+}	Thin filament proteins[a]
10^{-8} M	A ——— TnI ----- TnC ······· TnT ——— TM
10^{-5} M	A ----- TnI ——— TnC ······· TnT ——— TM

[a] A, actin; TnC, troponin-C; TnI, troponin-I; TnT, troponin-T; TM, tropomyosin.
Strong interaction ———, weak binding -----, interaction doubtful ······. (Based on Potter et al. 1982, see also Zot and Potter 1987).

involved in the interaction (in the middle region of the peptide chain containing the residues 96 to 117) is the structurally most conserved one, while the remainder of the molecule may vary greatly in troponin-I derived from different muscles, such as vertebrate fast twitch, slow twitch and cardiac muscle (Wilkinson and Grand 1978). Troponin-I from insect flight muscle is particularly large (Bullard 1984). As pointed out by Perry et al. (1982), "the differences in troponin-I are no doubt responsible, in part at least, for the characteristic features of the contraction relaxation cycles in different kinds of striated muscle." For instance, cardiac troponin-I contains an extra serine residue in position 20 of the amino acid sequence which can be readily phosphorylated by cAMP-dependent protein kinase, and if this site is phosphorylated, the calcium affinity of troponin-C is reduced (Solaro et al. 1981). Obviously, the interaction of troponin-I and troponin-C is a two-way affair. On the one hand, calcium interaction with troponin-C causes a reduction in the affinity of troponin-I with actin while, on the other hand, phosphorylation of troponin-I influences the calcium-binding site of troponin-C. The peptide chain of troponin-I from fast-twitch and slow-twitch muscle is much shorter than cardiac troponin and lacks, therefore, the phosphorylatable serine residue. Here, additional phosphorylation sites exist which, however, do not seem to have a regulatory function since they are not phosphorylated in the intact troponin complex.

In summary, calcium binding to troponin initiates a change in the interaction of thin-filament proteins and in the position of tropomyosin. We shall now consider how these changes initiate contraction.

4.2 Alterations of Thin Filaments Trigger Contraction

As mentioned in the preceding section, signal transmission in the thin filaments was found to involve the following stages: (1) calcium binds to troponin-C after which (2) the inhibitory troponin-I-actin interaction is loosened and (3) tropomyosin shifts from a peripheral inhibitory position on the actin filaments into an activating position near the central groove of the thin filament thereby (4) "switching the thin-filament structure on." The thin filament is now capable of interacting with myosin and activating the contractile process in a manner that is still not thoroughly understood.

Basically, muscle contraction, as we have seen (Chap. 1), may be discussed in terms of a cyclic interaction of actin and myosin crossbridges driven by the continued hydrolysis of ATP. Activation of contraction, therefore, requires first of all a stimulation of the actomyosin ATPase.

4.2.1 Calcium-Dependent Regulation of Actomyosin-ATPase

The myosin subfragment S-1 moiety of the crossbridges contains both the ATP-splitting site and the actin-binding site (Lowey 1971), but not in the same domain (Fig. 1.9). The catalytic ATP-binding site is located in the 27-kDa fragment of myosin near the N-terminal region of the myosin heavy chain covering S-1 (Yamamoto and Sekine 1979). The part of subfragment S-1 interacting with two globular subunits of actin is located on the 50 kDa fragment further away from the N-terminus (Mornet et al. 1981). Despite being quite apart, however, the nucleotide-binding site and the actin-binding site are influencing each other. For instance, as soon as myosin binds ATP the actin-binding site dissociates from actin, and ATP is hydrolyzed. Conversely, interaction with actin accelerates the rate-limiting step of the ATP-splitting mechanism, namely the release of ADP+P_i from the nucleotide-binding site (step 4 in the reaction scheme):

$$\underset{(1)}{AM+ATP \rightarrow A+M \cdot ATP} \underset{(2)}{\rightarrow} A+M \cdot ADP \cdot P \underset{(3)}{\leftrightarrows} AM \cdot ADP \cdot P \underset{(4)}{\rightarrow} AM+ADP+P_i . \qquad (4.5)$$

The decomposition of the actomyosin-ADP-P complex is much faster than that of the myosin-ADP-P complex.
Three mechanisms of ATPase regulation have been proposed:

1. Steric Hindrance Theory. According to H. E. Huxley (1973) and Haselgrove (1973), in the absence of Ca^{2+} tropomyosin is located at the periphery of the thin filament and prevents the attachment of myosin crossbridges to actin which, in turn, becomes possible when tropomyosin moves into the centre of the groove between two actin strands after adding Ca^{2+}. A steric blockage of the actin-myosin interaction would block step 3 in the reaction scheme [Eq. (4.5)] and would, therefore, also prevent step 4, the rapid dissociation of the enzyme-product complex thereby inhibiting the ATPase.

2. Blockage of Product Release. According to the hypothesis of Chalovich et al. (1981), the regulatory proteins might, in the absence of Ca^{2+}, inhibit the ATPase by directly blocking step 4 of the reaction scheme [Eq. (4.5)], the dissociation of the enzyme-product complex. In order to determine whether attachment of crossbridges (isolated myosin subfragment S-1) to actin was inhibited by the regulatory protein system, Chalovich et al. (1981) performed the following experiment. First, they mixed subfragment S-1 with actin filaments containing the regulatory proteins troponin and tropomyosin in the presence of ATP and magnesium ions. At high Ca^{2+} concentration, the myosin subfragment interacted as expected with the regulated actin filaments forming a complex that could be spun down in the ultracentrifuge. Very little subfragment S-1 remained unbound in the supernatant after spinning. Next, the experiment was repeated in the absence of calcium. Surprisingly, however, the same quantity of subfragment could be spun down together with actin, leaving again very little unbound subfragment S-1 in the supernatant. These experiments were carried out at quite different concentrations of myosin subfragment in order to determine the concentration required for 50% binding corresponding to the dissociation constant of the actomyosin S-1 complex. Through such experiments it became clear that the affinity of actin for myosin was the same in the presence and absence of activator calcium, though in the latter case the ATPase was inhibited (cf. Fig. 4.7 B). If the regulatory proteins inhibited the ATPase at low Ca^{2+} concentration, this inhibition could not be ascribed to an inhibition of actin and myosin interaction. Possibly, therefore, the regulatory proteins interfered in the absence of Ca^{2+} with one of the later steps in the reaction scheme, for instance, with step 4, the dissociation of phosphate from the ternary actin-myosin complex. Such an inhibitory effect could, according to the hypothesis of Trueblood et al. (1982), however, also be brought about by a steric effect of tropomyosin.

3. Modified Steric Hindrance Hypothesis. Consider the relation of the various enzymic steps of the ATP-splitting mechanism to the structural states of the crossbridge cycle already discussed in Chap. 1. Here, it may suffice to recall a few key points: (1) crossbridges attach to actin filaments in a vertical position while they are in the M-ADP-P state. (2) Following attachment, crossbridges rotate into an angled position when phosphate is released; but as pointed out by A. F. Huxley (1980), the reverse is also true. Phosphate would not be released unless the crossbridges are allowed to move into the angled position. Now suppose that the regulatory proteins, although not blocking the attachment of the myosin ADP-P to actin, are nonetheless preventing the interaction of the myosin-ADP crossbridge with the corresponding recipient site on the actin filament (site b in Fig. 4.6). In this way, crossbridges would attach weakly to actin, but crossbridge rotation as well as the concomitant product release [reaction 4 in Eq. (4.5)] would be prevented. Consequently, continued ATP splitting would be inhibited as well since myosin would, for a comparatively long time, be unable to dispose of its ADP and bind and hydrolyze a new molecule of ATP. At elevated Ca^{2+} concentrations, however, reaction 4, representing a dissociation of the enzyme-product complex, would be accelerated, thus enhancing the rate of ATP splitting.

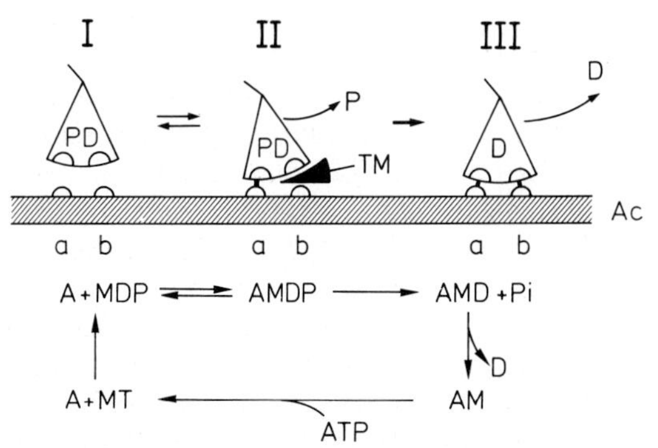

Fig. 4.6. Regulation of force and ATPase by Ca^{2+}. *Above:* Possible steric blockage of crossbridge rotation by tropomyosin (*TM*) in the absence of Ca^{2+}. *I* = Detached state; *II* = weakly attached state (AM · ADP · P state) [myosin binds only to one site of actin (*a*)]; *III* = strongly attached state (AM · ADP state): actin binds at two sites (*a* and *b*). In the absence of Ca^{2+} tropomyosin may block the interaction site *b*, thus preventing the myosin heads to move from state *II* to the force generating state *III*. The release of the ATPase product P_i is inhibited. Calcium ions bound to troponin release this inhibition. *D* = ADP; *P* = phosphate; *Ac* = actin filament. *Below:* Catalytic cycle involved in ATP hydrolysis. ATP dissociates actomyosin (*AM*). Myosin-ATP (*MT*) is hydrolyzed to form the myosin-product complex (*MDP*) which reacts with actin (*A*) to the complex *AMDP*. Its decomposition is blocked by tropomyosin (*TM*) in the absence of Ca^{2+}, so that a repetition of the hydrolytic cycle and, hence, continued ATP hydrolysis is inhibited.

4.2.2 Regulation of Muscle Force, Stiffness and Shortening Velocity

The dissociation of the enzyme-ADP-P complex (cf. Fig. 4.6) is probably directly involved in the generation of force, since the myosin-ADP-P state is weakly bound to actin, whereas myosin-ADP is strongly bound and force-generating. Suppose now that in the absence of calcium the regulatory proteins inhibit the formation of strong bonds between myosin-ADP and actin, but not the formation of weak bonds between myosin-ADP-P and actin. Under these circumstances, crossbridge rotation and force generation will then be sterically hindered, as illustrated in Fig. 4.6. Blockage of site b by tropomyosin would prevent crossbridges attached to site a (low-affinity site) from rotating and attaching to site b, but crossbridge rotation and force generation would be possible when tropomyosin moves out of the way in the presence of calcium (Trueblood et al. 1982). In the following section, we discuss the evidence for and against this proposal.

4.2.2.1 Skinned Fibre Experiments at Low Ionic Strength

According to Trueblood's hypothesis, it is not a steric hindrance of crossbridge attachment per se, but steric impediment of crossbridge rotation that may keep muscle in the relaxed state. In support of this idea, X-ray diffraction experiments (Matsuda and Podolsky 1984) showed that, at least at low ionic strength, crossbridges may also be attached to actin in the relaxed state, although in a different manner than crossbridge attachment in the contracted state and in rigor. Evidence for loose attachment of crossbridges in the relaxed state was also obtained from the determination of muscle fibre stiffness (Brenner et al. 1982). In these experiments Brenner and colleagues used skinned (membrane-free) fibres so that the contractile structures could be equilibrated with salt solutions of low ionic strength and with the desired concentration of calcium and ATP. One end of the fibre was attached to a very sensitive force transducer and the other to a rod connected to the membrane of a loudspeaker with which very rapid, small amplitude changes in length could be imposed on the muscle fibre, thereby stretching it within less than 100 μs by only a fraction of a percent. These stretches produced dramatic increases in tension even when the fibre was held under relaxing conditions in the absence of calcium, suggesting that stiffness was high (Fig. 4.7 A). The tension increase was, however, extremely short-lived; it decayed within a fraction of a millisecond, and it is clear that it would not have been detected had the stretch been performed more slowly, as was the case in previous experiments. From the increment in tension per unit change in length, a value of immediate stiffness was obtained which was almost identical in relaxed fibres at low Ca^{2+} concentration and in activated fibres in the presence of high Ca^{2+} concentrations. Moreover, the stiffness decreased when the sarcomere length was increased to values above 2.5 μm suggesting that it was due to attached elastic crossbridges. At low Ca^{2+} concentration these crossbridges must have been non-cycling since ATPase activity was low, and their conformation differed from that in contraction (Yu and Brenner 1989; cf. Xu et al. 1987).

From these observations Brenner and colleagues (1982) concluded, therefore, that while tropomyosin and troponin did not block crossbridge attachment, they

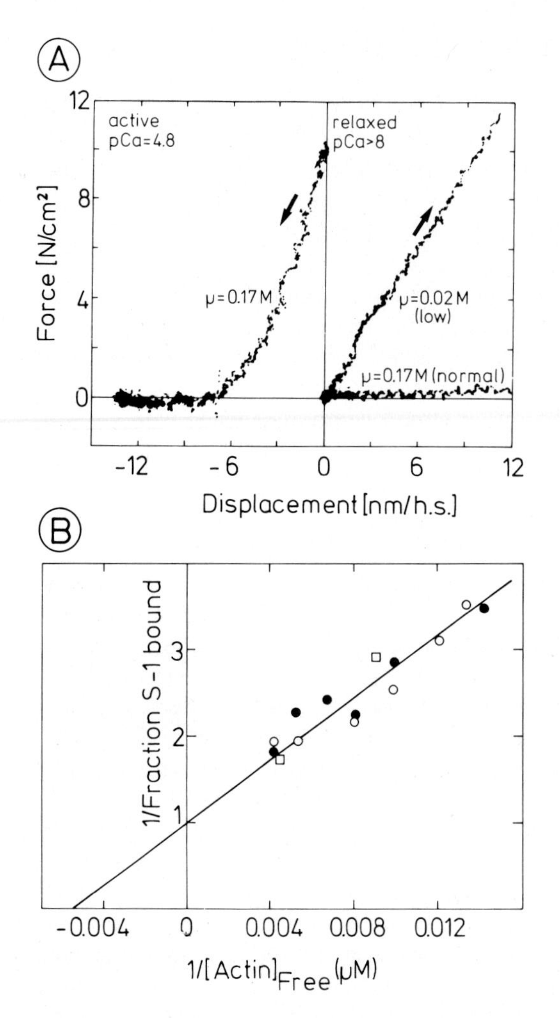

Fig. 4.7 A, B. Evidence for actin-myosin interaction at low ionic strength in the absence of calcium. *A* Immediate stiffness of skinned rabbit psoas fibres (suspended in ATP salt solution) determined by plotting the change in force (*ordinate*) as a function of the length change (release, *left*; or stretch, *right*) performed within 0.3 ms. The stiffness, i.e. the slope of the force extension curve, is a measure of the relative number of myosin crossbridges attached to actin at any one time. Note that it is high in calcium-activated contracted fibres (*left*) and in relaxed fibres at low ionic strength, but negligibly small in relaxed fibres at physiological ionic strength (0.17). *Arrows* indicate direction of length change per half-sarcomere (h.s.) (stretch or release) (Brenner et al. 1982). *B* Double reciprocal plot of actin binding to subfragment-1 of myosin as a function of actin concentration in the presence (●) and in the absence of 0.1 mM Ca^{2+} (○, □). Note that the dissociation constant of the actin-S-1 complex (the reciprocal value of the abscissa intercept) is identical in the presence and absence of Ca^{2+}, even though the rate of ATP splitting is greatly calcium-dependent. (Brenner et al. 1982)

did prevent the crossbridges from proceeding to the next step leading to the generation of force (cf. Fig. 4.6).

The lifetime of attached, non-force generating crossbridges seems, however, to be very short, as judged from the mentioned extreme shortness of the stretch-induced tension transient. This shortness is in accordance with the concept of Stein et al. (1979), that attached crossbridges are in rapid equilibrium with detached ones. The reaction of myosin-ADP·P and actin to AM·ADP·P is an equilibrium in which the forward and back reaction rate constants are extremely high.

In conclusion, all experiments reported in this section seem to indicate that it is possible, at least in principle, to regulate contraction by controlling the product-release step [4 in Eq. (4.5)]. But whether this mechanism, investigated at extremely low ionic strength, is of biological importance requires further investigations under more physiological conditions.

4.2.2.2 Mechanical Experiments at Physiological Ionic Strength

Membrane-Skinned Fibres. These are also an ideal preparation for investigating the graded effect of the free Ca^{2+} concentration on both the immediate stiffness and force at physiological ionic strength. Figure 4.8 shows the increase in immediate stiffness and force of the skinned semitendinosus muscle fibre from frog in ATP salt solution as the free Ca^{2+} concentration surrounding the contractile structures is increased (Herzig et al. 1981 c, cf. also Fig. 4.7). Since immediate stiffness measures the number of crossbridges attached at any one moment, these results are relevant for our understanding of the activation mechanism. Increasing calcium activation obviously causes the attachment of more and more force-generating crossbridges rather than increasing the force development by a constant number of attached crossbridges. Indeed, further experiments (Brenner 1988) proved that an increase in free calcium levels enhances the probability of crossbridge attachment in a strongly attached force-generating mode.

Living Muscle. Experiments carried out in skinned fibres at very low ionic strength are at variance with the experiments of A. F. Huxley and Simmons (1973) and others showing that under physiological conditions the immediate stiffness of living resting muscle is extremely small. At physiological ionic strength obviously crossbridges are predominantly detached in the relaxed state, as also shown by X-ray diffraction experiments (Kress et al. 1986, cf. Fig. 1.12C). Do these results suggest that the regulatory mechanisms for contraction are different at low and high ionic strengths? It seems that under physiological conditions the regulatory proteins may block the reaction step 3 of Eq. (4.5) (the attachment of myosin to actin; cf. Wagner and Giniger 1981), and in particular step 4, i.e. the

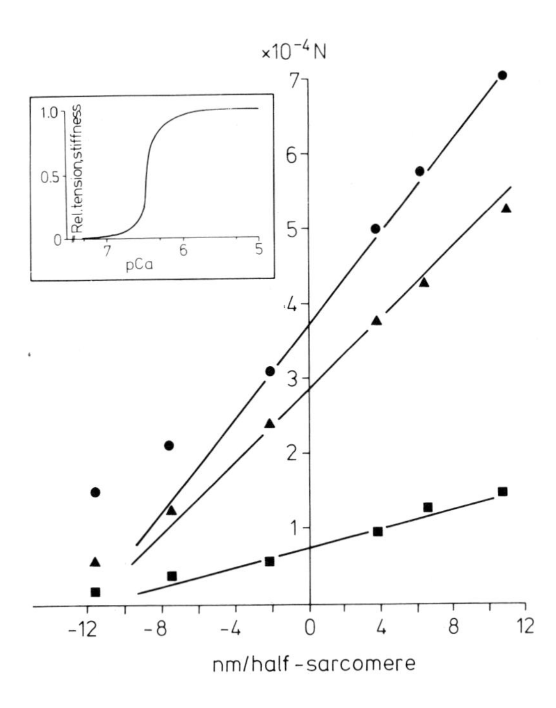

Fig. 4.8. Ca^{2+} dependence of force extension diagrams (T_1-curves) of frog semitendinosus skinned fibre. Immediate stiffness indicated by the slope of the T_1-curves obtained at three different Ca^{2+} concentrations (■, pCa 6.7; ▲, pCa 6; ●, pCa 5). *Symbols* in T-curves represent extreme tensions reached during quick stretch or release (within 1 ms) by an amount given on the *abscissa* (in nm per half-sarcomere). Ordinate intercepts of curves are isometric tensions at different Ca^{2+} concentrations (see Fig. 1.6 for comparison with living fibres). Solutions contained (in mM): ATP (total) 10, Mg (total) 10, CaEGTA buffer 4, pH 6.7. *Inset:* Force or stiffness as function of pCa. (Herzig et al. 1981 c)

dissociation of phosphate from the ternary enzyme-product complex. The immediate consequence of such an interruption of a metabolic pathway is an accumulation of the products of reactions 2 and 3 in Eq. (4.5) which are "upstream" to the blockage. These would include the attached AM · ADP · P state as well as the detached M · ADP · P state. The latter state accumulates since the attachment of crossbridges is reversible. Any accumulation of the AM · ADP · P state is bound to drive the reactions backwards, leading to the dissociation of crossbridges. Now suppose that calcium activation causes a shift in tropomyosin and relieves the blockage of reaction 4 [Eq. (4.5)]. Then attached crossbridges would proceed into the angled, force-generating state, and would thus complete the cycle. Consequently, the myosin-ADP · P state and the AM · ADP · P state would no longer accumulate as the reactions proceed in the foreward direction: the muscle develops force.

The time course of changes in crossbridge-attachment can be detected by the change in the intensity ratio of the 1.0/1.1 reflections in the X-ray diffraction diagram (Kress et al. 1986, cf. Fig. 1.12C) and by a more or less parallel increase in force and stiffness. Following activation, structural changes (Kress et al. 1986) and stiffness (Cecchi et al. 1986; Ford et al. 1986, cf. Sect. 1.3.1) actually slightly precede the increase in force and reach a maximum before tension reaches maximum values in a tetanic contraction.

At physiological ionic strength, therefore, calcium seems to activate muscle by promoting the formation of crossbridges in a strongly bound state, which gives rise to increased stiffness. Still, these force-generating bridges might be generated from weakly bound ones, rather than by crossbridge attachment (cf. Brenner et al. 1990; and Sect. 10.2 for further discussion).

4.2.2.3 Regulation of Crossbridge Turnover and Shortening Velocity

According to the contraction theory proposed by A. F. Huxley (1980), the shortening speed of unloaded muscle fibres should be independent of the proportion of crossbridges that are turned on and which can interact with actin. Indeed, the maximal unloaded shortening velocity is almost identical in the early rise of a tetanic contraction and at the height of the tetanus (Lombardi and Menchetti 1984). Thames et al. (1974) determined the shortening velocity in unloaded skinned fibres at high and intermediate Ca^{2+} concentration. Although the maximum force that could be developed or the number of attached crossbridges was quite different under the two conditions, the maximal unloaded shortening velocity was identical. These results, however, are at variance with experiments of Julian and Moss (1981) and Moss et al. (1986), who found that unloaded shortening velocity depended on the Ca^{2+} concentration and, hence, on the degree of activation. In these experiments velocity was found to be time-dependent. Thus, at intermediate levels of calcium activation, two distinct phases of shortening velocity were observed, namely an initial calcium-independent phase of high velocity, followed by a low velocity phase that was calcium-dependent. The lowering of the velocity at late phases was probably due to the presence of an internal load in the fibre possibly associated with attached crossbridges that hinder sliding. Maximal shortening velocity might then depend on the rate at which these crossbridges detach. Under isometric conditions, crossbridge detachment – unlike crossbridge attachment – does not seem to depend on the free calcium ion concentration (cf. Brenner 1988; cf. also Kawai et al. 1981; Kerrick et al. 1990).

4.2.3 Relation of Muscle Force and Calcium Occupancy of Troponin

How is the number of strongly bound crossbridges related to the calcium occupancy of troponin? In order to answer this question it is important to measure force development and calcium binding in the same preparation since force and calcium binding may be interdependent (cf. Ridgway and Gordon 1984). Skinned fibres are once again the preparation of choice for such studies. Using a double isotope technique, Fuchs and Fox (1982) and Fuchs (1985) determined in these isolated contractile systems the ^{45}Ca binding to troponin as well as force development as a function of the free Ca^{2+} concentration to which the regulatory proteins of the thin filaments were exposed. Over the free Ca^{2+} concentration from 6×10^{-8}M to 1.2×10^{-5}M, calcium binding increased from 0.25 to 1.65 µmol g^{-1} protein, while force increased over a narrow range of Ca^{2+} concentration, being half-maximal just below 1 µM Ca^{2+} (Fig. 4.9). An analogous result had been previously obtained with isolated myofibrils. Here, calcium binding occurred over a wide range of Ca^{2+} concentrations, whereas the ATPase activity exhibited a steep calcium dependence (Murray et al. 1975; Wnuk et al. 1984; cf. Fig. 4.11). These results are, of course, partly expected since only the low-affinity sites are involved in the calcium regulation of force and ATPase. Therefore, it would be important to measure the calcium binding to these sites concomitantly with force production. Recently, Zot et al. (1985, 1986) were able to replace the endogenous tropo-

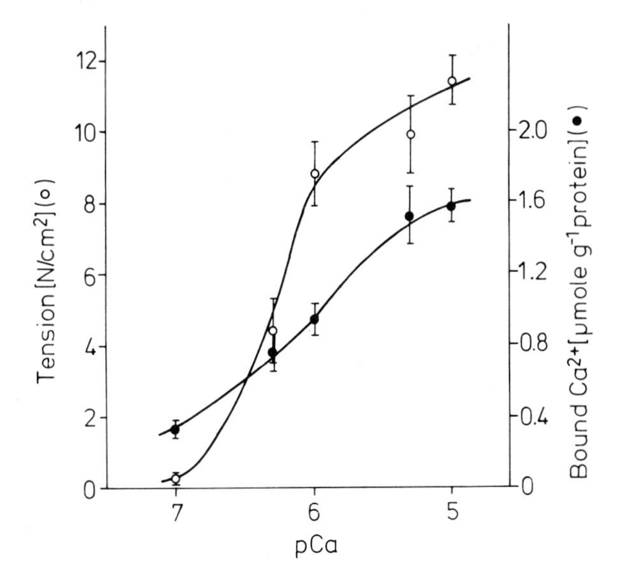

Fig. 4.9. ^{45}Ca binding to thin filaments and force development in skinned rabbit psoas fibres at various Ca^{2+} concentrations (pCa). The skinned fibres were suspended in saline containing 5 mM Mg-ATP (ph 7). (Fuchs 1985)

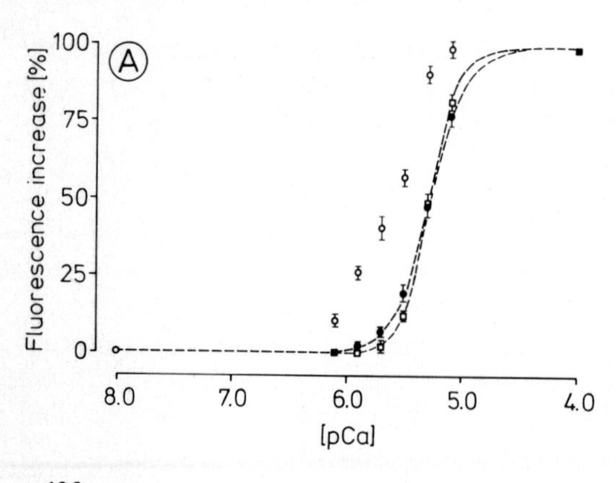

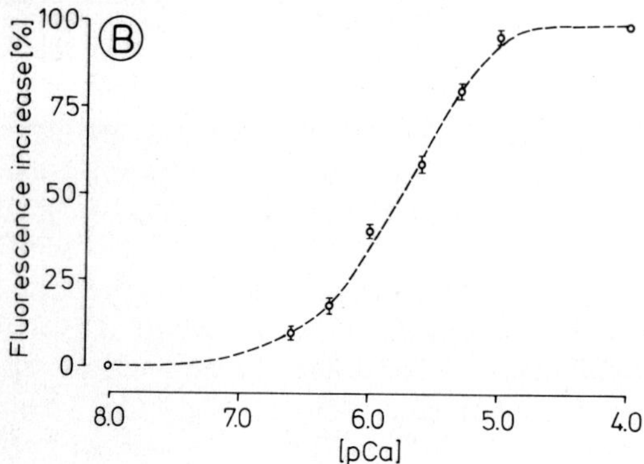

Fig. 4.10 A, B. Force development and calcium binding to troponin in skinned rabbit psoas muscle, determined according to Zot et al. (1985, 1986). *A* Fluorescence of DANZ-labelled troponin-C incorporated into skinned fibres (○) or force (□, ●) in Mg-ATP salt solution (4 mM Mg^{2+}, pCa as indicated). *B* Calcium binding to DANZ-labelled TnC of skinned psoas in ATP-free "rigor solution". (Courtesy of Dr. K. Güth)

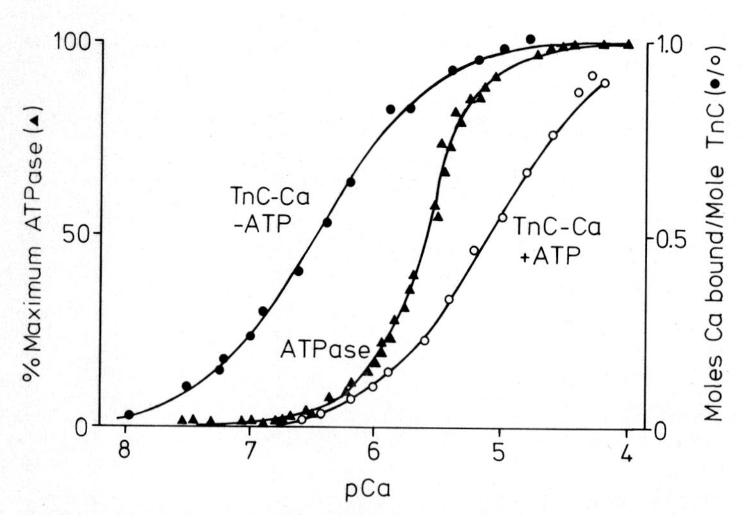

Fig. 4.11. Ca dependency of Ca binding to troponin-C of crayfish tail muscle (●, ○) and myofibrillar ATPase activity or actomyosin-ATPase activity (▲). *Ordinate:* Ca binding to troponin-C containing "regulated" actomyosin in the absence of ATP (*closed circles*) or in the presence of ATP (*open circles*); *abscissa:* neg. log of Ca^{2+}. Note that maximally one molecule of Ca^{2+} is bound per mole troponin-C and that the calcium affinity is different in the presence and absence of ATP ($K = 3.1 \times 10^6 M^{-1}$ and $1.3 \times 10^5 M^{-1}$). (Wnuk et al. 1984)

nin of skinned fibres by troponin which had been labelled by the fluorescent probe DANZ. In this way, they were able to measure, at the same time, calcium binding to the specific sites as well as force generation. Interestingly, both parameters were linearly related, and their relationship to free calcium was extremely steep, indicating cooperativity (Güth and Potter 1987, cf. Fig. 4.10).

4.2.3.1 Cooperativity

The steepness of the calcium force relation may be described by the equation:

$$\frac{T_x}{T_0} = \frac{[Ca_x]^n}{EC_{50}^n + [Ca_x]^n}, \tag{4.6}$$

where T_0 is force or ATPase activity at saturating concentration of Ca^{2+}; T_x is force or activity value at the actual Ca^{2+} concentration $[Ca_x]$ and EC_{50} is the Ca^{2+} concentration giving 50% activation of force or ATPase; n is the Hill coefficient which is a measure of cooperativity. To ease the graphic determination of the Hill coefficient, let us introduce the term U for the fractional force T_x/T_0, then:

$$\log \frac{U}{1-U} = n(\log [Ca_x] - \log EC_{50}), \tag{4.7}$$

where n is therefore simply the slope of the straight line obtained when the expression $\log(U/1-U)$ is plotted versus the logarithm of the Ca^{2+} concentration. In this way, a Hill coefficient of at least 2 or even much higher values was obtained, suggesting that the response of tension or ATPase to calcium must be highly cooperative (see also Sect. 10.2.2.3).

The mechanism of cooperativity is not yet clearly understood. Remember that the calcium binding to isolated troponin may be described in terms of Michaelis-Menten kinetics (Sect. 4.1.2), i.e. calcium binding to specific sites depends on the Ca^{2+} concentration in a hyperbolic manner over a concentration range of two orders of magnitude, suggesting that there is no cooperativity between neighbouring calcium-binding sites of isolated troponin molecules. Therefore, the steepness of calcium binding in skinned fibres must be attributed to mechanisms lying outside the individual troponin molecules, for instance, to interactions of neighbouring troponin molecules which may be mediated between actin monomers bound to neighbouring tropomyosin-troponin complexes (Trueblood et al. 1982). Alternatively, the attachment of crossbridges during force generation may increase the apparent affinity of subfragment S-1 of myosin crossbridges to actin and may, therefore, promote the attachment of neighbouring crossbridges. This effect might, in turn, facilitate calcium binding. An increase of the Ca^{2+} concentration would, therefore, also increase the apparent calcium affinity of troponin with the result that the relationship between the Ca^{2+} concentration and calcium occupancy of troponin or force development would become extremely steep (cf. also Wnuk et al. 1984). In accordance with this proposal, cooperative calcium binding could be observed if skinned fibres were immersed in activating solution but not in ATP-free "rigor solution," when all crossbridges are permanently attached to actin (Fig. 4.10). These findings of Güth and Potter (1987) suggest that attached crossbridges can influence the calcium-binding properties of troponin-C.

Indeed, there is now growing evidence that the attached crossbridges communicate with calcium-binding sites on troponin-C via actin-tropomyosin and troponin-I, thereby changing the calcium affinity of the binding sites. First, it could be shown that addition of troponin-I increases the calcium affinity of troponin-C, but when actin is added, the calcium affinity decreases again (Zot et al. 1983). In the absence of ATP, in the so-called rigor state, nucleotide-free crossbridges attach to actin firmly even at an extremely low Ca^{2+} concentration (less than 10^{-8}M); the affinity of binding sites for calcium is then greatly increased (Bremel and Weber 1972; cf. also Wnuk et al. 1984). Thus, in skinned fibres, more calcium is bound in the absence than in the presence of ATP, but the amount of bound calcium can be decreased again by stretching the fibres to the extent that actin and myosin no longer overlap (Fuchs 1977). This suggests that it is the attaching rigor crossbridges rather than the lack of ATP per se that is the cause of the increased calcium affinity of troponin-C within the myofibrils.

4.2.3.2 Potentiation

Nucleotide-free crossbridges (so-called rigor bridges) may not only exist in the absence, but also at very low concentrations of Mg-ATP, below about 0.1 mM (Bremel and Weber 1972). These bridges would supposedly attach firmly to the actin filaments and once attached they might push the tropomyosin strands into the non-inhibitory position on the periphery of the thin filament, as discussed by Taylor (1979). In this way, rigor bridges would cause a firm, force-generating attachment of nucleotide-loaded "non-rigor bridges" as well, thus causing an increase in force generation and ATPase activity. Under these conditions, ATPase activity and force are both, in the presence and absence of calcium, potentiated well above the activity values observed at physiological Mg-ATP concentration. At the same time, the calcium affinity of troponin-C increases as well (A. Weber and Murray 1973).

Potentiation depends on the presence of tropomyosin which acts as a potentiator at low ATP concentration and also in the presence of a large excess of added subfragment S-1 of myosin (cf. Greene and Eisenberg 1980). At high physiological concentration of Mg-ATP and low myosin concentration tropomyosin does not potentiate, but inhibits actomyosin ATPase of vertebrate skeletal muscle (Bremel et al. 1972), while the actomyosin ATPase of *Limulus* muscle, on the other hand, is activated by tropomyosin even under these conditions (Lehman and Szent-Györgyi 1972). Collectively then, all of these experiments suggest that tropomyosin may not only play a role in the signal transmission from troponin to the actomyosin system, but also in the reverse information flow from crossbridges to troponin-C. It is possible, but not proven, that nearest-neighbour interactions between tropomyosin-troponin-actin complexes may be involved in these cooperative effects (Trueblood et al. 1982).

Feedback Between Mechanical Events and Activation. The idea that crossbridge action influences calcium occupancy of troponin-C emerged from the potentiation experiments just described and is also plausible on theoretical grounds (Shiner and Solaro 1982). Recently, however, it received further support from skinned-fibre studies on barnacle muscle. Ridgway et al. (1983) determined the force of these preparations when the free Ca^{2+} concentration was either increased stepwise or decreased, starting with maximally contracting fibres. In the latter case, the apparent calcium sensitivity (−log of Ca^{2+} concentration required for 50% force generation) was much higher than in the "ascending limb" of the calcium-force response curve (hysteresis). When surviving barnacle muscles were released immediately after stimulation, the calcium occupancy was smaller than under the constraints of an isometric contraction. But when the muscle was stretched, calcium occupancy was larger, suggesting that it may depend on cross-

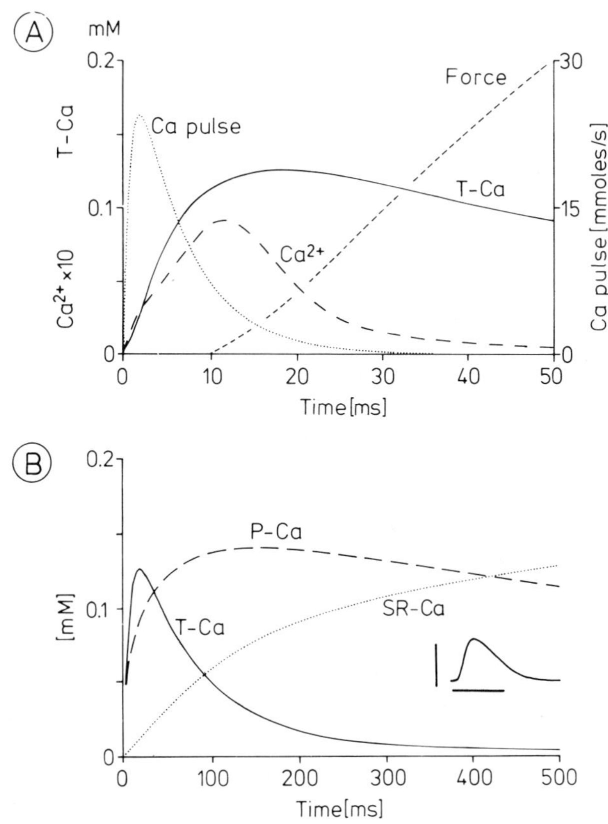

Fig. 4.12 A, B. Time course of calcium release (*Ca pulse*), free Ca^{2+} concentration, calcium occupancy of troponin (*T-Ca*) and parvalbumin (*P-Ca*) as well as calcium uptake into the sarcoplasmic reticulum (*SR-Ca*) during and after a single twitch of frog muscle (computer simulation) assuming a calcium release of 200 μmol l^{-1} myoplasm following a single shock at 4 °C. *A* Note that 10 ms after the stimulus calcium occupancy of troponin is nearly complete, whereas force is just beginning to rise. The rate of calcium release peaks before the calcium transient. *B* Computer simulation showing that calcium binding to parvalbumin precedes calcium uptake into the sarcoplasmic reticulum, which is too slow to account for the rapid decomposition of the calcium-troponin complex. (After Gillis 1985). *Inset:* Actual time course of twitch in frog tibialis anterior. Time calibration 300 ms, force calibration 15 N cm^{-2}. (cf. Cecchi et al. 1984)

bridge strain (Ridgway and Gordon 1984). Clearly, it would be desirable to measure calcium occupancy at a given free Ca^{2+} concentration in the isolated contractile structure of skinned fibres under different mechanical conditions. As it is possible to label the troponin of skinned fibres with calcium-sensitive fluorescent probes (Güth and Potter 1987) such information is now forthcoming (cf. also Brozkovich et al. 1988; Sect. 10.2.2 for further discussion).

4.2.3.3 The Temporal Relationship of Calcium Occupancy of Troponin and Force Development in Living Muscle Fibres

After injecting DANZ-labelled troponin into living muscle cells it is possible to study the temporal relationship of calcium-troponin occupancy and force (Ashley et al. 1985). These measurements may be compared with the time course predicted by various models of excitation-contraction coupling. For instance, as pointed out by Gillis (1985), the regulatory sites of troponin in frog muscle (0 °C) must be at least 90% saturated with calcium within 20 ms after a stimulus to cope with the speed and completeness of activation. Indeed, time-resolved X-ray diffraction measurements (Kress et al. 1986) indicate that the calcium-induced movement of tropomyosin and possibly, therefore, the activation of thin filaments are complete soon after the onset of tetanic stimulation at a time when tension reaches about 20% of its final tetanic value or even before (Fig. 1.12 C).

The possible relationships between activation and calcium occupancy of troponin, myoplasmic-free calcium and muscle force during the rising phase of tetanic tension are schematically shown in Fig. 4.12. It can be seen that activation is complete long before generation of force. Activated crossbridges obviously require considerable time for attachment and subsequent force generation. It is unlikely, therefore, that the difference between twitch tension and tetanic force of frog muscle at 0 °C is due to a difference in the degree of activation. In these experiments, activation (measured by the degree of occupation of troponin by calcium and the extent of tropomyosin movement in the thin filaments) was obviously already maximal in a single twitch. At higher temperatures, however, activation may be less complete in a twitch than in a tetanus (Fig. 4.5) and "twitch potentiators", such as nitrate, accelerate the rate of activation (Cecchi et al. 1978).

4.3 Ancillary Calcium-Binding Proteins: Calmodulin, Parvalbumin and Myosin Light Chains

Besides troponin, other calcium-binding proteins may be involved in the regulation of skeletal muscle contraction.

Calmodulin for instance, the ubiquitous calcium-binding protein of all eukaryotic cells, plays a major role in the calcium regulation of the contractile machinery of vertebrate smooth muscle, whereas in vertebrate skeletal muscle regulation it may have an ancillary function.

Calcium binding to the regulatory light chain of myosin has also been implicated in skeletal muscle regulation since it is homologous to the regulatory light chain of molluscan muscle in which its role as a "calcium switch" is well established.

Parvalbumin, a calcium-binding protein which is soluble in the myoplasm, presumably acts as a soluble relaxing factor that traps the calcium released from troponin. Soluble calcium-binding proteins which are possibly homologous to parvalbumins have also been found in the myoplasm of crayfish muscle (Cox et al. 1977) and in the sandworm *Nereis virens* (Gerday et al. 1981). All these proteins as well as troponin-C belong to a family of calcium-binding proteins which may have evolved from a common ancestor protein, but have retained common amino acid sequences in their peptides. Indeed, the extent of sequence homology is so striking that the design of the calcium-binding structures is probably identical in all cases (Kretsinger 1980).

4.3.1 Is Parvalbumin a Soluble Relaxing Factor?

This protein is a single peptide chain with a molecular weight of 12 kDa that contains two sites binding Ca^{2+} as well as Mg^{2+} with even higher affinity than the Ca-Mg sites of troponin-C (Potter et al. 1978). Its calcium-binding affinity is, however, lower than that of the sarcoplasmic reticulum calcium pump. Thus, as suggested by Gerday and Gillis (1976), parvalbumin might function as a "calcium shuttle" transporting calcium from troponin to the calcium pump during muscle relaxation. Even when the Ca^{2+} concentration of the myoplasm is lowered by the sarcoplasmic reticulum during relaxation, its local concentration would still be high in the neighbourhood of troponin, as it would depend on the ratio of occupied calcium-binding sites to non-occupied sites in the troponin molecule. Near troponin, therefore, parvalbumin would be loaded by calcium, which would then diffuse to the calcium-binding sites of the sarcoplasmic reticulum by a process of carrier-supported or -facilitated diffusion (cf. Demaille et al. 1974). Near the calcium-binding sites of the pump the Ca^{2+} concentration is naturally much reduced so that the parvalbumin-calcium complex decomposes, and calcium will be bound to the translocator protein of the sarcoplasmic reticulum. The result of these various reactions is, of course, a unidirectional flow of calcium from troponin sites to calcium-binding sites of the sarcoplasmic reticulum, as elucidated in Fig. 4.13.

Intuitively, it is easy to visualize in a qualitative manner that parvalbumin would ease calcium sequestration by the sarcoplasmic reticulum and, hence, muscle relaxation. These considerations led to the proposal of a quantitative model of parvalbumin action by Gillis et al. (1982).

When calcium is released from the sarcoplasmic reticulum, parvalbumin does not at once compete with calcium-binding sites of troponin for activator calcium. At first glance, this seems to be a paradox, as the calcium-binding affinity of parvalbumin is much higher than that of troponin calcium-specific sites. Consider, however, that the Mg^{2+} concentration of resting striated muscle is fairly high, of the order of 1–3 mM, so that at low levels of ionized calcium in resting muscle the parvalbumin sites may be almost entirely (up to 92%) occupied by Mg^{2+}. Un-

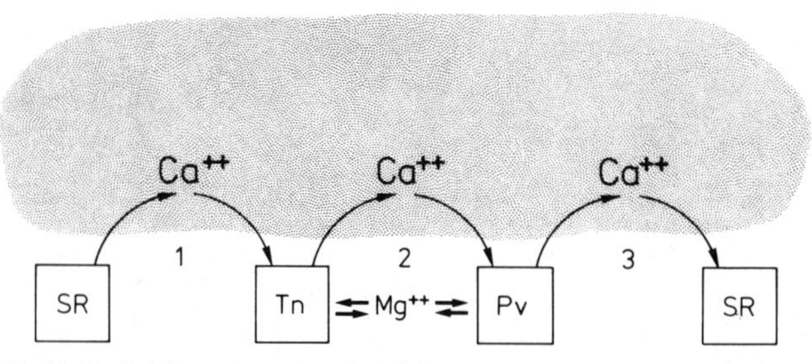

Fig. 4.13. Calcium movement from the Ca-release sites of the terminal cisternae in the sarcoplasmic reticulum (*SR*) to the calcium reuptake sites (calcium pump) of the SR. *1* Ca release from SR and Ca binding to troponin-C (*Tn*); *2* Ca release from troponin-C and binding to parvalbumin (*Pv*); *3* Ca dissociation from parvalbumin and reuptake by the *SR*. Note that the free Ca^{2+} pool is in equilibrium with Ca bound to troponin and parvalbumin

less Mg^{2+} dissociates from its binding sites, parvalbumin cannot bind calcium. The latter is, therefore, bound with a considerable delay following the calcium release from the sarcoplasmic reticulum (Fig. 4.12), in particular since the dissociation of the parvalbumin-magnesium complex is a relatively slow process according to Robertson et al. (1981). These kinetic considerations also predict a biphasic time course of calcium occupancy of the specific binding sites in troponin-C following a calcium release from the sarcoplasmic reticulum during repetitive stimulation in tetanus. At first, the occupancy would increase rapidly, but then it would decline again as some of the troponin-bound calcium is taken over by the high-affinity Ca-Mg binding site of both troponin-C and parvalbumin. In a sense, therefore, parvalbumin may behave like an EGTA-type soluble relaxing factor which, however, exhibits relaxing activity only after dissociation of the Mg-parvalbumin complex. Although still controversial, it seems to follow from these considerations that the rate of this dissociation and the amount of parvalbumin present in the myoplasm may therefore limit the rate of calcium binding to parvalbumin and, hence, the decomposition of the troponin-C-calcium complex and the rate of relaxation. As argued by Gillis et al. (1982), calcium removal by parvalbumin may indeed be faster than the calcium uptake by the sarcoplasmic reticulum, at least in some types of muscle and under special conditions. Thus, calcium binding to parvalbumin rather than to the sarcoplasmic reticulum may ultimately determine the relaxation speed of muscles in cold-blooded animals. This may be particularly true at low ambient temperatures, where the ATPase and pumping activity of the sarcoplasmic reticulum is considerably reduced.

Think, for instance, of trouts or pikes in cold water, which during a predatory prey activity must swim rapidly for very brief periods with their fast muscles twitching only a few times. These fish have obviously adapted by evolving muscles with a particularly high parvalbumin content (up to 0.6 mM in pike, according to Baron et al. 1975) rather than by increasing the maximal rate of calcium pumping in the sarcoplasmic reticulum. The highest known cellular parvalbumin levels, approximately 1.5 mM, have been found in the ultrafast swim bladder muscle of the toadfish *Opsanus tau* (Hamoir et al. 1980). Following a twitch this muscle is able to relax in a few milliseconds, much faster than could be accounted for by the calcium re-uptake by the sarcoplasmic reticulum (Gillis 1980). In muscles of warm-blooded animals the activity of calcium-transport ATPase is greater and the parvalbumin content usually smaller; but here, too, there is a correlation between parvalbumin content and speed of relaxation (Berchtold et al. 1982; Heizmann 1984), suggesting that parvalbumin is implicated in relaxation and may act as a temporary calcium trap.

Following a single twitch, the calcium content of parvalbumin will decrease again with time as calcium is slowly taken up by the sarcoplasmic reticulum. If, however, the muscle is stimulated and twitches repeatedly at fairly high frequency, the interval between calcium-releasing stimuli may be too short to clear parvalbumin from bound calcium. With each stimulus, therefore, the calcium occupancy of parvalbumin would increase until after a few twitches the fully calcium-loaded parvalbumin may be no longer able to act as a calcium-trapping, relaxing factor. For this very reason, burst swimming of fish is, by necessity, limited to a short time. After a few stimuli there is an increase in the relaxation time of twitches which then fuse into maintained tetanic contraction, as parvalbumin becomes saturated with calcium. Attractive as it is this parvalbumin hypothesis of muscle relaxation is, however, still quite controversial (cf. Robertson et al. 1981; Ashley and Griffiths 1983, Stuhlfauth et al. 1984).

4.3.2 Calcium-Calmodulin-Dependent Activation of Myosin Light Chain Kinase

Many of the physiological calcium effects within the cell are exerted by calmodulin (Fig. 4.14) and calmodulin-regulated enzymes (Cheung 1980). One of these enzymes, myosin light-chain kinase, may be important for regulation of contraction. When activated by the calcium-calmodulin complex (Nairn and Perry 1979), it phosphorylates the regulatory light chain of myosin. This may affect the ATPase activity of actomyosin (Pemrick 1980), but does not alter calcium-binding affinity of myosin (Holroyde et al. 1979).

Myosin Light-Chain Phosphorylation. This also occurs in vivo at the onset of a tetanic contraction and eventually leads to the incorporation of up to 0.6 mol phosphate mol^{-1} light chain (Bárány et al. 1979). The rate of phosphorylation may or may not parallel force development and with regard to the extent and rate of phosphorylation, there are large variations among different muscles and species (Butler et al. 1983). There is certainly no quantitative or temporal correlation with force production, in particular, since the phosphorylated state may outlive the end of a tetanic contraction by many seconds: in contrast to relaxation, dephosphorylation by the myosin light-chain phosphatase is a slow process. What then is the role of this covalent modification of the light chains? It has been known for many years that following a tetanic contraction a vertebrate striated muscle may respond to a supramaximal stimulus with a larger twitch than before the tetanus. This phenomenon, called *post-tetanic* potentiation, cannot be ascribed to a greater release of intracellular calcium, but it may well be due to increased phosphorylation of the myosin light chain, as proposed by Stull et al. (1980, cf. Fig. 4.15).

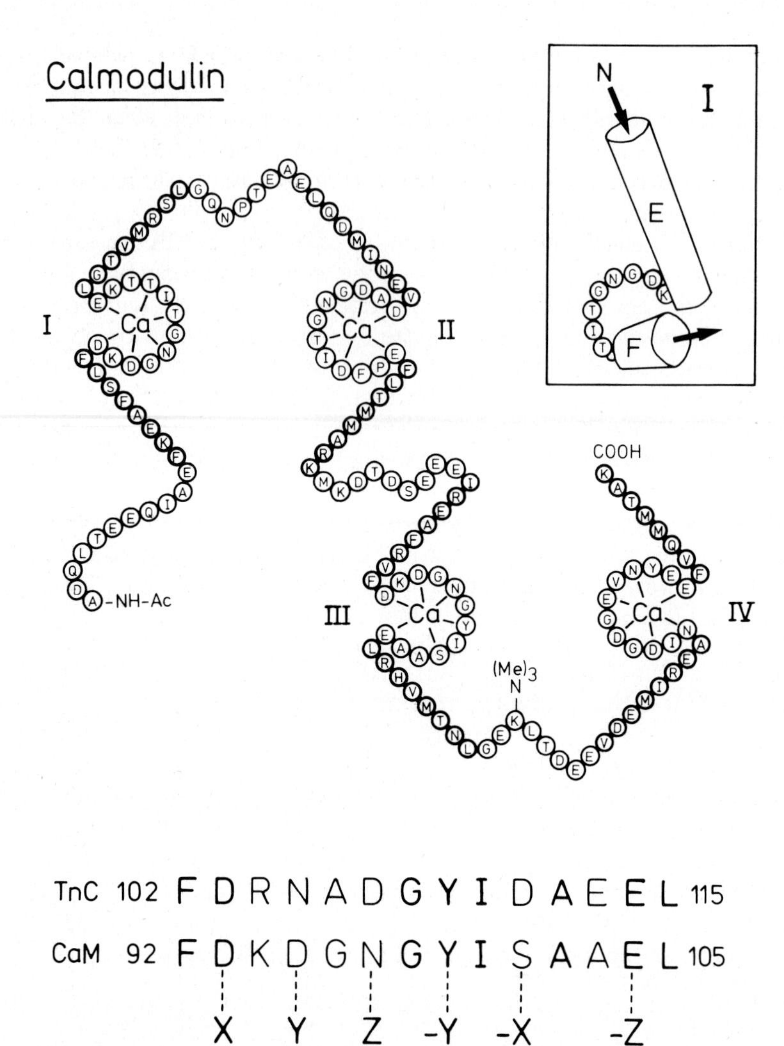

Fig. 4.14. Diagrammatic representation of calmodulin structure showing the amino acid sequence and the calcium-binding domains (EF-hands, cf. Inset). (Partly adapted from Klee et al. 1980, for threedimensional structure see Babu at al. 1985). *Below:* Comparison of the calcium-binding domain III of calmodulin (*CaM*) with that of troponin-C (*TnC*). Ligands involved in the coordinated binding of calcium denoted as *X, Y, Z*. One letter amino acid code: *A* ala; *B* Asx; *C* Cys; *D* asp; *E* glu; *F* phe; *G* gly; *H* his; *I* ile; *K* lys; *L* leu; *M* met; *N* asn; *P* pro; *Q* gln; *R* arg; *S* ser; *T* thr; *V* val; *W* trp; *Y* tyr; and *Z* glx

Alternatively, myosin phosphorylation has been proposed to participate in the regulation of the cycling speed of crossbridges without affecting the number attached at any one moment (Crow and Kushmerick 1982a, but cf. Butler et al. 1983). Suppose that force depends only on the number of crossbridges attached at any one moment and not on the frequency with which they cycle, thereby splitting one molecule of ATP in each cyclic interaction of the crossbridge. Thus, when the cycling rate is reduced, the muscle becomes intrinsically slower, but also more

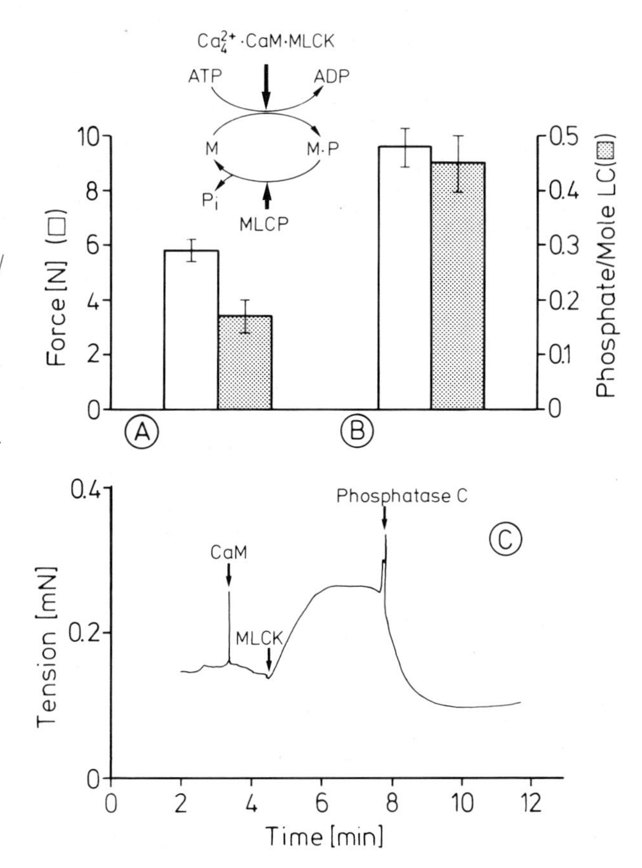

Fig. 4.15 A–C. Role of myosin phosphorylation. *A, B* Calcium/calmodulin-dependent phosphorylation of the myosin P-light chain. *MLCK* Myosin light chain kinase; *CaM* calmodulin; *M* myosin. Correlation of twitch-force and myosin phosphorylation in vivo. *A* Single twitch of rabbit M. plantaris after resting period. *B* Potentiation of the last twitch at the end of 20 s, 5 Hz stimulus train. (Data from Moore et al. 1985). *C* Effect of myosin light chain kinase and phosphatase-C on force development of skinned rabbit psoas fibres in Mg-ATP salt solution at submaximal calcium activation (pCa 6). (After Persechini et al. 1985)

economical, since bridges will cycle less often and, hence, split less ATP in a given time. Cooke et al. (1982) and Rüegg et al. (1984c) attempted to phosphorylate the light chains of myosin in demembranated, skinned fibres of skeletal or cardiac muscle by equilibrating the preparation with a phosphate donor (the ATP analogue ATP-γ-S) with myosin light-chain kinase as a catalyst and calmodulin as a calcium-dependent activator. Interestingly, this treatment did not affect the capacity of skinned fibres to develop force in ATP salt solution, but it did reduce the ATP-splitting rate drastically. It should be noted, however, that light chain phosphorylation by myosin light chain kinase (by using ATP as substrate) and thiophosphorylation (using ATP-γ-S as substrate) seem to have different effects on ATPase activity and mechanical fibre properties (cf. Morano et al. 1990; Persechini and Stull 1984).

The mechanism of post-tetanic potentiation has also been discussed in terms of light chain phosphorylation as mentioned already. Indeed, in skinned fibres, phosphorylation will increase contractile force by enhancing the rate constant of crossbridge attachment (Sweeney and Stull 1990; cf. Sect. 10.2.2). Recall that the fraction of crossbridges existing in a strongly attached force-generating state is related to $f/(f+g)$, where f and g denote the apparent rate constants of crossbridge attachment and detachment respectively. As a consequence, the number of crossbridges in a state of force generation at any one moment would increase and as a result, of course, more force would be developed. In addition, attachment of more crossbridges might also increase the affinity of troponin-C for calcium, as discussed in Sect. 4.2.3, so that calcium sensitivity increases. Indeed, an increased force development has also been observed in partially activated skinned fibres after phosphorylation of the light chain (Fig. 4.15). Increase of calcium sensitivity may well occur in the case of post-tetanic potentiation. In a sense, therefore, post-tetanic potentiation may merely reflect an alteration of crossbridge kinetics induced by a calcium- and calmodulin-dependent phosphorylation of the myosin light chain. Incidentally, at high Ca^{2+} concentration the calcium-binding sites of the regulatory myosin light chain may also become partially occupied with calcium, but it is not known as yet whether this has functional consequences (cf. Moss 1982).

Metabolic and Calcium Homeostatic Effects of Calmodulin. Besides activating myosin light-chain kinase, calmodulin has an enormous repertoire of skills and functions (see Cheung 1980 for a brief review). Here, it may suffice to recall that the calcium-calmodulin complex activates, in conjunction with cAMP-dependent protein kinase, the enzyme phosphorylase kinase that is converting the inactive b-form of phosphorylase into its active a-form. By activating these glycolytic enzymes, calcium stimulates glycogen breakdown, thus coordinating metabolism and calcium-activated contraction. This activation is, however, automatically damped since calcium and calmodulin also stimulate the enzyme cAMP phosphodiesterase which degrades cAMP, one of the activators of the phosphorylase kinase. Furthermore, calcium-calmodulin tends to reduce the internal concentration of the calcium activator since it stimulates the sequestration of Ca^{2+} by calcium pumps of the cell membrane as well as of the sarcoplasmic reticulum and diminishes Ca^{2+} release from the lateral cisternae (Smith et al. 1989). In this way, calmodulin plays an essential part in the intracellular control system involved in the regulation of contraction metabolism and calcium homeostasis. By reducing intracellular Ca^{2+} levels and the energy requirements of the crossbridges and increasing energy provision through glycolysis, it may exert a beneficial "trophotropic" effect on muscle cells.

4.3.3 Role of Myosin Light Chains in Skeletal Muscle

In the preceding section we mentioned already that calmodulin and myosin light chain kinase influence contractility by enhancing the phosphorylation state of the myosin light chains. Phosphorylatable light chains (or P-light chains), essential light chains and myosin heavy chains form together the hexameric myosin molecule (cf. Fig. 1.9). Both types of light chains belong to the superfamily of calcium binding proteins that also includes parvalbumin, calmodulin and troponin-C, and which evolved from a small ancestor protein probably containing four similar calcium binding regions. Each of the regions consisted probably of a pair of helices flanking a central, 12-residue, calcium binding site. Presumably, this primordial calcium binding protein originated from gene duplication and reduplication of a single gene. The evolutionary tree of calcium binding proteins obviously branched early; already in the most primitive eukaryotes, such as slime molds, calmodulin and essential light chains already emerged as distinct proteins though still exhibiting about 40% sequence homology (see Collins 1991 for review). In the further course of evolution then, essential light chains may have completely lost their calcium binding capacity, while the P-light chains (also known as light chain-2, LC-2 or regulatory light chain) still retained one calcium binding site.

In skeletal muscle, unlike vertebrate smooth or molluscan muscle, the *regulatory light chains* seem to play a modulatory rather than a primary regulatory role. For instance, the regulatory light chain of skeletal muscle has a comparatively low calcium affinity. Thus, under conditions of maximal calcium activation and physiological Mg^{2+} concentration, less than 40% of the calcium binding sites is calcium-saturated (Holroyde et al. 1979). However, the binding and unbinding kinetics are probably too slow to play a role in the regulation of twitch contraction in vertebrate muscle (for a different role of these light chains in molluscan muscle and in vertebrate smooth muscle, cf. Chaps. 6 and 8). Yet, it would seem that the light chain is involved in calcium binding to myosin (Wagner 1984) as well as in the calcium sensitivity of myosin (Pulliam et al. 1983). As noted by Hofmann et al. (1990), the regulatory light chains may modulate contraction kinetics in a calcium- and perhaps also phosphorylation-dependent manner. Indeed, if the light chains are partially extracted, the shortening velocity of unloaded skinned fibres of skeletal muscle is reduced by 30% and can be normalized again by readdition of the light chain. Perhaps this light chain interacts with the neck region of the myosin heavy chain to modulate the affinity of myosin for actin and hence the crossbridge detachment rate. In this way, it might cause a modulation of calcium responsiveness of skinned fibres (Hofmann et al. 1990) that is known to be dependent on the crossbridge turnover kinetics and in particular on the detachment rate constant (cf. Brenner 1988; Sect. 10.2.2). By phosphorylation of the serine residue 15 of the light chain, the calcium responsiveness is then further increased as mentioned already.

The other type of light chain is called *alkali light chain*, as it could be removed under alkaline conditions, a procedure which also destroyed the enzymic activity of myosin. For this reason, light chain-1 has been thought to be essential for myosin ATPase activity and has hence also been called "*essential light chain*". Later on, however, this light chain could be removed from myosin even under non-denaturing conditions and without loss of ATPase activity (Weeds 1980). While slow skeletal muscle contains only one type of essential light chain (Mr 27 kDa), fast twitch muscle has two isoforms, LC-1 (Mr 21 kDa) and LC-3 (Mr 17 kDa). Both these essential light chains are products of the same gene, but arise from alternative RNA splicing as tissue-dependent, developmentally regulated gene products (cf. Collins 1991 for review). As a single myosin head contains one regulatory light chain and one alkali light chain (either LC-1 or LC-3), the double-headed myosin molecule may exist in three isoforms, namely as a heterodimer, containing LC-1 and LC-3, and as two homodimers, containing either LC-1 or LC-3. According to Greaser et al. (1988), the isoforms of essential light chains and in particular the ratio of LC-3 and LC-1 may vary between different fast skeletal muscle fibre types, and maximal shortening velocity may actually be correlated with the relative LC-3 content (Greaser et al. 1988). In heart muscle, too, there are light chain isoforms (e.g. Barton et al. 1988) and cardiac hypertrophy may also be associated with different crossbridge kinetics and expression of light chain isoforms (Schaub 1990; Sütsch et al. 1992). Alkali light chains are bound to the neck region of the myosin heavy chain, but they interact also with the C-terminus of actin. Therefore they may well be one of the factors controlling the rate and magnitude of transition from weakly attached to strongly attached crossbridge states (Trayer et al. 1987).

Chapter 5

Diversity of Fast and Slow Striated Muscle

As pointed out by Hoyle (1983), the 10000-fold range of shortening speeds among different kinds of muscle (cf. Prosser 1960) may be possible because of differences in the mode and rate of activation; however, adaptations in the contractile machinery itself could be equally important. In the following we shall be concerned with structures and mechanisms responsible for slowness or fastness.

As described in the preceding chapters, activation involves excitation of the cell membrane in the T-tubular system, calcium release from the sarcoplasmic reticulum, the establishment of an elevated intracellular Ca^{2+} concentration in the myoplasm and eventually the binding of calcium to troponin causing the cycling attachment of crossbridges. All of these processes in the excitation-contraction cascade determine the rate at which calcium is delivered to the contractile system, but as we shall see below, a high rate of calcium delivery would be of little use if it were not matched by an equally fast response of the contractile machinery. An intrinsically slow contractile machinery, on the other hand, would not require a fast activation. It is, therefore, not surprising that slow and fast striated muscles differ not only in the velocity of the contractile mechanism, but also in the rate at which calcium is supplied to and removed from the contractile structures. The one exception to this principle is insect fibrillar muscle which is activated by stretch during a high-frequency oscillatory movement, while the myoplasmic-free calcium level may stay constant.

5.1 Vertebrate Tonic Muscle Fibres

As already mentioned in Chap. 3, amphibian muscles contain two main populations of fibres, the “fast-twitch” type, and another tonic type which does not respond to a single nerve stimulus, but may contract slowly when stimulated repetitively. As reviewed by Morgan and Proske (1984), such tonic fibres responding to prolonged depolarization with a sustained contracture have also been found in fish, reptiles and birds, as well as in the slow extraocular muscles of mammals. In all of these muscles the sarcomere length and the length of the myosin and actin filaments are similar to those in fast-twitch muscles, but the sarcoplasmic reticulum is generally much less extensive in slow than in fast muscles. An extreme example of this principle is the very slow tonic skin muscle of the garter snake, in which the T-tubular system is absent, while the scarce sarcoplasmic reticulum is restricted to a few subsarcolemmal vesicles (Hoyle et al. 1966; cf. also A. Hess 1965).

5.1.1 Amphibian Slow Muscles

In our comparison of fast and slow muscles, frog twitch muscle will serve as a point of reference since its excitation-contraction coupling mechanisms have been extensively studied and described in the preceding chapters. Compared with these twitch fibres, the tonic fibres in the tonus bundle of frog iliofibularis muscle shorten about five times more slowly, but they also maintain tension with approximately four to five times less energy expenditure and apparently without fatigue (Floyd and Smith 1971). Hence, these muscles are ideally suited for a postural function. The activating processes are also slow. Nerve stimuli elicit local depolarizations of the cell membrane which cause a slow contraction only when they summate, but even when the membrane is suddenly depolarized by using the voltage clamp technique, the delivery of activator calcium is still delayed in these fibres for structural reasons. The sarcoplasmic reticulum and the T-tubular systems are much less developed than in twitch fibres (S. G. Page 1965); the latter represents only 0.2% of the fibre volume, about half of which is formed by longitudinal T-tubular elements not engaging in contacts with the terminal cisternae of the sarcoplasmic reticulum. Flitney (1971) calculated that the contact area with the terminal cisternae (T-SR junction) is about five to ten times smaller than in the case of twitch muscle (cf. Fig. 5.1 for a similar difference between fast and slow fish muscle). While the T-tubules in twitch fibres reach the fibre surface at the level of every Z-line, tonic fibres contain T-tubules only at every fifth or sixth Z-line. In tonic fibres the scarce longitudinal tubules of the sarcoplasmic reticulum are oriented in parallel to the fibre axis as they run continuously over a stretch of several sarcomeres, eventually forming a diadic or triadic junction with T-tubules (cf. Sect. 2.1.2). Therefore, the diffusion pathway of activator calcium from its release sites at the triad to the myofilaments may be rather long, perhaps of the order of 6–7 μm, and the calcium released from the cisternae will be considerably diluted by the myoplasm before it reaches the myofilaments. Correspondingly, diffusional delays for calcium activation may be approximately 50 times larger than in twitch fibres where the diffusion path from the triad (at the level of every Z-line) to the region of filament overlap may be less than 1 μm. Consequently, the time course of events in excitation-contraction coupling is quite different in fast and slow muscle.

In fast twitch fibres, depolarization of the T-tubules causes almost immediately the "gating" of the calcium channels in the terminal cisternae, as indicated by a rapid charge movement within these membranes. Calcium permeability may peak after 2 ms, and after this time the gradient of Ca^{2+} concentration is large, ranging from about 50 μM in the vicinity of the triad to nearly zero in the middle of the sarcomere. After 5 ms, however, the average Ca^{2+} concentration of the sarcomere may already have risen to about 5–10 μM. This concentration then is nearly sufficient to saturate the specific calcium-binding sites of troponin evenly. Although the gradient of free Ca^{2+} may still be large at this time, the gradient of troponin occupancy by calcium may nevertheless be small, so that the twitch muscle is maximally and uniformly activated (Cannell and Allen 1984). In tonic fibres, on the other hand, depolarization of the membrane causes a much smaller charge movement (Gilly and Huy 1980b) as there is much less junctional sarcoplasmic reticulum involved in the T-SR coupling between the T-tubules and the terminal cisternae of the sarcoplasmic reticulum. Consequently, the rate of calcium release from the cisternae will be much smaller than in twitch muscle. For this reason, and also because of the long diffusional pathways for Ca^{2+}, the peak of a calcium transient will be reached as late as 100 ms after the onset of stimulation (Miledi et al. 1977, 1981, cf. Sect. 3.2.1). It may even take longer to occupy uniformly the troponin-binding sites with calcium, since the calcium supply is rate-limiting. Therefore, the sarcomeres lying nearer to the tubules may be more highly activated and contracting with greater force than the remaining sarcomeres, which, therefore, become stretched.

The slow rate of activation in frog tonic muscle fibres just described is matched by a slow contractile machinery. Even if all the specific binding sites of troponin-C were to be occupied by calcium instantaneously, the rate of rise of an isometric contraction would still be slow compared to that in twitch muscles, because, as proposed by A. F. Huxley (1957), the rate of rise in tension of isometric contraction depends critically on the rate constants of crossbridge attachment and detachment. These constants may differ by a factor of 10 or 15 between slow and fast muscle, according to Julian and Morgan (1979), who analyzed the tension transients observed after quick changes in length or release in frog muscle. As already mentioned in Chap. 1, these transients may, according to A. F. Huxley and Simmons (1973), reflect the kinetics of crossbridge attachment, movement and detachment, and they may differ widely in fast and slow muscle. Such differences may, of course, reflect differences in crossbridge properties and, in particular, alterations in the structure of myosin light chains and myosin heavy chains forming the head portion of the myosin molecule (Pliszka and Strzelecka-Golaszewska 1981; Rowlerson et al. 1985).

Differences in the properties of the contractile machinery may also account for differences in the unloaded shortening velocity, which is 5.5 lengths s^{-1} in twitch fibres and 1.1 lengths s^{-1} in slow fibres (cf. Lännergren 1978, 1979). Indeed, a similar difference in shortening velocity was also found by Costantin et al. (1967) in their classical skinned-fibre studies. These investigators isolated the contractile mechanism by removing the outer cell membrane and observed by high-speed cinematography the shortening of unloaded sarcomeres after the application of a droplet of calcium solution to the contractile filaments. In this way, they circumvented delays in the excitation-contraction coupling and showed that the contractile machinery of tonic fibres contracted slowly in ATP salt solution even if it is suddenly and maximally activated by an excess of free calcium.

5.1.2 Avian Tonic Fibres

The chicken anterior latissimus dorsi (ALD) is a postural muscle which serves to hold the wings close to the body. Its fibres resemble amphibian tonic fibres in that depolarization induces a prolonged contracture, but unlike frog muscle, twitch responses associated with action potentials may occasionally also be observed (Ginsborg 1960). When stimulated, these muscles take approximately five times longer to develop maximum force than the fast twitch muscle of the posterior latissimus dorsi (PLD) and relax as much as eight times more slowly (Canfield 1971). The two muscles also differ in their unloaded shortening velocities, in the rate of energy expenditure during tension maintenance (Rall and Schottelius 1973) and in the ATPase activity of actomyosin (Bárány 1967) by a factor five or six. Interestingly, the myosin isoforms (Hoh et al. 1976) and, hence, crossbridge properties are also different in avian fast and slow muscle. While force is dependent on the number of crossbridges attaching to actin in a force-generating state, speed may be determined by crossbridge cycling frequency or by the rate at which crossbridges attach, "pull" and detach. Since presumably one molecule of ATP is hydrolyzed in each cyclic operation of a crossbridge, the rate of ATP splitting or ATPase activity and the rate of shortening should also be related. Such a relationship has indeed been established in the now classical investigations of Bárány (1967), who compared the speed and myosin-ATPase activity of a great variety of muscles. Recently, it has even become possible to directly observe the rate of myosin movement along actin filaments. Sheetz and Spudich (1983) obtained long strands of actin filaments from the alga *Nitella*, on which they placed small microspheres coated with fluorescently marked myosin. When they added ATP in amounts sufficient to provide the energy, the myosin beads started to slide along the actin filaments, a movement that could be directly observed under the microscope. Perhaps not suprisingly, myosin of high ATPase activity isolated from fast muscle moved at a higher rate than myosin isolated from slow muscle, thus demonstrating directly that speed of sliding and ATPase activity were causally related (cf. Sellers et al. 1985).

Since during tension maintenance of fast muscle, ATP-splitting crossbridges cycle more often per unit time than in the case of slow muscle, tension maintenance is more costly or, in other words, less economical. The holding economy is equal to the quotient of tension × time/energy used, i.e. the time during which unit force can be maintained by unit energy in a muscle of 1 cm length. Consequently, the holding economy of an ATP-hydrolyzing and force-generating population of cycling crossbridges will depend on how long these bridges remain in the force-generating attached state during each crossbridge cycle. Since the duration of the attached state will increase with increasing cycle time of the crossbridge, muscles with slowly cycling bridges are both slow and economical. As the rate at which crossbridges detach is presumably slow as well, slowly contracting muscles may also require a longer time for relaxation than fast muscles.

The prerequisite for crossbridge detachment in relaxation is, however, the dissociation of the calcium-troponin complex when the Ca^{2+} concentration is lowered by the calcium pump of the sarcoplasmic reticulum. The rate of both of these processes may, therefore, limit the rate of relaxation. There is now indirect evidence suggesting that the rate of dissociation of the calcium-troponin complex (the calcium off-rate) may differ in fast and slow muscle. In skinned fibres from ALD, for instance, the Ca^{2+} concentration required for 50% activation of force is about three to five times lower than the corresponding concentration in skinned fibres of fast muscles (G. Moore et al. 1983). In interpreting these findings, we should remember that the calcium level required for half-maximal activation may correspond to the dissociation constant of the complex formed between Ca^{2+} and the specific, calcium-binding sites on troponin-C and, therefore, to the ratio of calcium off-rate and calcium on-rate. Since the on-rate is high and only diffusion-limited, differences in calcium sensitivity may, therefore, reflect differences in the calcium off-rate. If this is so, calcium will be expected to dissociate more rapidly from the regulatory protein of fast PLD muscle than slow ALD muscle. Furthermore, these calcium ions may be taken up more rapidly by the sarcoplasmic reticulum from the fast muscle, which is more highly developed than in the anterior latissimus dorsi (S. G. Page 1969; Ovalle 1982). In these fast muscles, the sarcoplasmic reticulum does not need to lower the Ca^{2+} concentration as much as in slow muscle, since the calcium threshold concentration eliciting a contraction in isolated contractile structures is comparatively high. In summary, fast and slow muscle differ in many structural and biochemical aspects; all these differences in conjunction will contribute to the more rapid relaxation in avian fast than slow muscle.

5.1.3 Fish Muscle

The discovery of white and red muscle in the fish *Torpedo* by Lorenzini (1678) was the earliest recognition of different muscle fibres in a single species. Fish use their red "tonic" muscles for slow cruising rather than for a postural function, and their muscle activity is graded by the frequency of nervous stimulation. In isolated fast fibres, but not in slow fibres, a single stimulus elicits a twitch, while repetitive stimulation of both fibre types causes a maintained graded tetanus, the extent of contraction depending on the frequency of stimulation (Flitney and Johnston 1979). Red tonic and white muscles of fish differ in the rate of activation and force development and in their speed of contraction and relaxation by a factor of 2 to 3; these differences result, at least partly, from differences in the contractile machinery as revealed by skinned-fibre studies (Altringham and Johnston 1982; Johnston and Brill 1984). The higher rate of contraction and relaxation in white tonic fibres may, however, at least partly be due to differences in the properties of the sarcoplasmic reticulum and in the transverse system, as suggested by the experiments illustrated in Fig. 5.1. In the ultrafast muscles of the swim bladder of the tropical fish *Opsanus tau*, both the T-tubular system and the sarcoplasmic reticulum are extremely abundant, amounting to over 30% to 40% of the fibre volume (Fawcett and Revel 1961). The red body muscles of fish contain as extensive a T-system and sarcoplasmic reticulum as white muscles, but the latter is less effective as a calcium pump. Thus, the calcium-uptake rates measured in tissue slices of fast and slow muscles differ by a factor of nearly 3, and the activity ratios of calcium-activated transport ATPase and basic ATPase may be even more important (McArdle and Johnston 1981). Additionally, the two types of muscle differ also in their parvalbumin content. White fast muscles have particularly large amounts of this calcium-binding protein, which has an important role as a temporary calcium sink at the onset of relaxation and as a shuttle protein between troponin and the sarcoplasmic reticulum as relaxation continues (Heizmann 1984). As mentioned already (Chap. 4), fish fast muscle may contain parvalbumin in a concentration of up to 3 mM, enough to bind the calcium liberated into the myoplasm during five to ten twitches. Differences in the contraction speed of fish

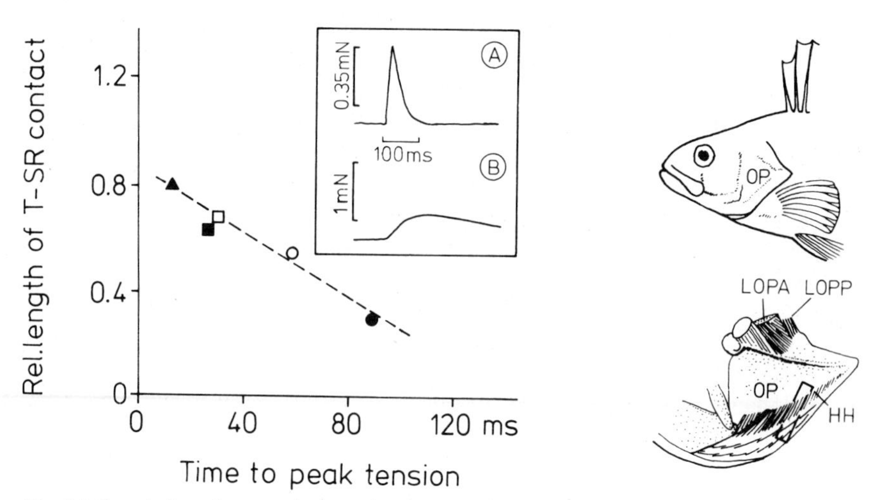

Fig. 5.1. Correlation of contraction speed and extent of T-SR junction in fish muscle. *Ordinate:* Relative extent of contact length between T-tubules and the cisternae of the sarcoplasmic reticulum. *Abscissa:* Time to peak in isometric twitches (at 20 °C) of perch white muscle (▲), intermediate (pink) muscle (■) and slow (red) fibres [m. levator operculi anterior (●)]. Carp intermediate fibre (□); carp slow fibre (○). *Right:* Anatomy of muscles used. Key: *LOPA* = Levator operculi anterior, fast white (▲) or red slow (●); *LOPP* = Levator operculi posterior; *HH* = m. Hyohyoideus, intermediate (■); *OP* = opercular bones. *Inset:* Time course of isometric twitches in perch white fibre (*A*) or slow fibre (*B*) of *LOPA*. (After Akster et al. 1985)

fast and slow muscle may be an expression of altered myosin isozyme composition. Indeed, the myosin heavy chains as well as the light chains of the two kinds of fish muscle can be distinguished by immunological techniques and by gel electrophoresis (Rowlerson et al. 1985).

In fish muscles, adaptation to a warm or cold environment may be of great importance. It is well known that, in general, the speed of muscle contraction and the power output are highly temperature-dependent, the Q_{10} being between 3 and 5. However, individual fish species are adapted to habitats that differ widely in temperature and salinity. *Tulapaia mossambica,* for instance, lives in African lakes at an ambient temperature of over 30 °C, whereas *Notothenia rossi* swims in antarctic waters at temperatures below 0 °C. Even then the power output of its muscles is surprisingly high, amounting to over 60% of that in tropical fish at their ambient temperature. Such adaptation of ectothermic animals to a freezing environment, which has been a challenge to comparative biochemists, would seem to be at least partly accounted for by a biochemical adaptation of the proteins in the contractile machinery and in the sarcoplasmic reticulum. According to McArdle and Johnston (1982), for instance, there is not much difference in the calcium-uptake rate of the sarcoplasmic reticulum from muscles of antarctic and tropical fish at an ambient temperature of 20 °C. In the warm-adapted species, the calcium-transport rate is, however, dramatically reduced when the temperature is lowered to 0 °C, whereas the calcium-uptake rate of a cold-adapted fish is comparatively little affected. In the antarctic fish the rate of relaxation may, therefore, be comparatively high even in a cold environment, in particular since the sarcoplasmic reticulum and the calcium-binding protein parvalbumin are more abundant than in tropical fish.

The speed of fast muscle fibres of tropical and antarctic fish and their myofibrillar ATPase activity are quite similar when measured at 20 °C. At 0 °C, however, the myofibrillar ATPase activity is greatly reduced in the case of tropical fish, but not so much in antarctic fish. Such a difference in the temperature dependence of the contractile ATPase is also seen when warm-adapted and cold-adapted carp are compared. A goldfish, for instance, can acclimatize within a few weeks to an ambient temperature near freezing because it can change the molecular properties of its actomyosin. In the process of such adaptation it synthesizes a myosin ATPase that is comparatively active even at low temperature, but this change in the Q_{10} is paid for by a greater susceptibility to heat denaturation (Johnston 1982). On the other hand, warm-acclimatized goldfish and tropical fish normally living at high environmental temperatures have a relatively heat-stable ATPase which, however, is only active at higher temperature. Force production is also more highly temperature-dependent in the warm-adapted fish than in cold-adapted specimens (Johnston and Brill 1984).

5.2 Comparison of Mammalian Fast- and Slow-Twitch Fibres

A mammalian twitch muscle may respond with a short and powerful effort or with sustained activity, but it does so by using different fibre populations (already listed in Table 1.1). For the prolonged maintenance of tension, such as is required for postural functions, slow twitch fibres (type-I fibres) are recruited. These are non-fatiguing and red since they are rich in mitochondria and myoglobin. The fast-contractile responses, on the other hand, are produced by type-IIB (fast glycolytic) or type-IIA (fast oxidative) fibres. The former are white, rich in glycolytic enzymes and fatigue rapidly, while the latter are equally fast, but pink or red, and do not fatigue easily since they contain an abundance of mitochondria and enzymes involved in oxidative metabolism. The rat extensor digitorum longus (EDL) is a well-studied muscle containing predominantly fast-twitch fibres, whereas the slow soleus muscle is mainly composed of type-I fibres. These two fibres not only differ functionally, but also have different structures and biochemical properties.

5.2.1 Diversity and Plasticity of Fibre Types

At birth, there is little difference in the contraction speed of muscles that will become fast or slow later in life (Close 1964). All fibres are uniformly and multiply innervated, and the muscle protein myosin is of the neonatal type. It is only after muscles have received their postnatal innervation that fast- and slow-fibre types emerge within the first weeks of life. The importance of innervation in differentiation of fibres is clearly illustrated by "cross-innervation experiments" in which

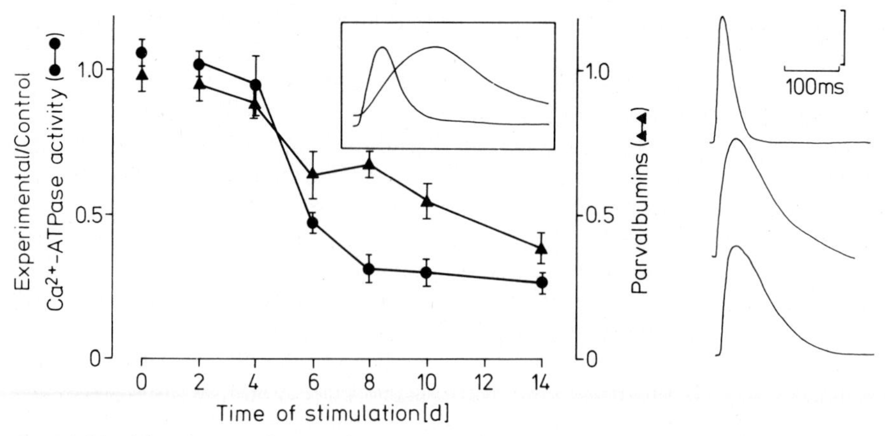

Fig. 5.2. Plasticity of mammalian twitch muscle. Decrease of sarcoplasmic reticulum content, Ca^{2+}-dependent ATPase (●) and parvalbumin content (▲) in rabbit twitch muscle as a consequence of chronic stimulation with 10 Hz impulses. *Inset:* Time course of twitch before stimulation (half-relaxation time 22 ms) and after chronic stimulation (half-relaxation time 130 ms). (After Pette 1984). *Right:* Comparison of time course of isometric twitch in rat EDL muscle before denervation (*top*, calibration bar 2.4 mN) and after 48 days of chronic denervation (*bottom*). The *middle tracing* shows the slow twitch response of innervated rat soleus muscle. (Dulhunty 1985)

fast muscles can be transformed into slow ones and vice versa. Buller et al. (1960) carried out the first experiments of this type when they cut the motor nerve fibres of a slow- and a fast-twitch muscle and exchanged the two nerve endings when reimplanting them into the muscles. After a few weeks, when cross-innervation had become established, the originally slow muscles became fast, whereas the fast muscles contracted more slowly. For a long time researchers speculated that a mysterious trophic factor released from the nerves might be responsible for this transformation, but this idea was eventually abandoned after the importance of the impulse pattern during stimulation had been discovered (Lomo et al. 1974). As shown in Fig. 5.2, denervation or stimulation at a frequency of 10 Hz transformed a fast muscle into a slow one, whereas slow fibres could be made fast when stimulated by a pattern mimicking the impulses signalled by fast motor neurons, for instance 100 Hz impulses applied for brief periods every few hours (Pette 1984). Transformation of fibre type was associated with structural and biochemical changes (cf. Fig. 5.2) as well as with the expression of proteins controlling muscle differentiation (Eftimie et al. 1991).

5.2.2 Differences in Excitation-Contraction Coupling

The rate of rise and fall of twitch tension differs by a factor of 2 to 4 between fast and slow fibres, reflecting similar differences in the rate of muscle activation. The possible relationship between these differences in twitch characteristics and alterations in excitation-contraction coupling have received much attention. Following a single stimulation, both the action potential (Fig. 1.2 B, cf. Wallinga-de Jonge et al. 1985) and the rise in the intracellular free Ca^{2+} concentration (Eusebi et al. 1980) are about three times faster in type-II than in type-I fibres, suggesting that in the former, calcium may be delivered more rapidly to the contractile proteins. These functional differences are matched by an adaptation of the structure of the sarcoplasmic reticulum and of the T-system. Fast-twitch muscles contain two T-tubular networks per sarcomere, namely one at each A-I border zone, forming contacts with the terminal cisternae of the longitudinal tubules which release calcium into the myoplasm. Although the arrangement and architecture of the T-system is qualitatively similar in slow-twitch fibres, important quantitative differences exist (Table 5.1). Thus, the volume of the T-tubules and of the terminal cisternae and the surface area of these membrane systems all differ roughly by a factor of 2 between fast and slow fibres (B. R. Eisenberg et al. 1974; B. R. Eisenberg and Kuda 1975). The importance of such differences is apparent when one considers that the release of Ca^{2+} is probably caused by a movement of charged particles within the membrane of the T-tubules, as discussed in Sect. 2.2.3.2. Significantly, the extent of charge movement following a depolarization and, therefore, presumably the rate of calcium release from the sarcoplasmic reticulum appear to be much less in slow-than in fast-twitch fibres (Dulhunty and Gage 1983, Fig. 2.6). Incidentally, these characteristics of excitation-contraction coupling change concomitantly with the mechanical twitch characteristics within 3 weeks from fast to slow type when a fast-twitch muscle is denervated (Dulhunty and Gage 1985). At the same time, the composition and kind of contractile proteins remain unaltered (Finol et al. 1981), whereas several weeks are required for new myosin isozymes to emerge. This suggests that it is the characteristics of the excitation-contraction coupling and rate of activation rather than the properties of the contractile machinery that principally determine the time course of the isometric twitch.

After fibre transformation, following denervation or chronic stimulation with 10 Hz, an originally fast-twitch muscle relaxes more slowly, a change that is generally attributed to diminution of the sarcoplasmic reticulum (B. R. Eisenberg and Salmons 1981) and its calcium-pumping activity (Heilmann and Pette 1979). As seen in Table 5.1, a fast-twitch muscle contains more sarcoplasmic reticulum (B. R. Eisenberg et al. 1974, B. R. Eisenberg and Kuda 1975) and a more active

Table 5.1. Comparison of mammalian fast and slow twitch muscle[a]

Guinea pig	Slow red (soleus)	Fast white (vastus)
T-system (% fibre volume)	0.14 (1)	0.27 (2)
J-SR (% volume)	0.96 (1)	1.62 (2)
SR total (% volume)	3.15 (1)	4.59 (2)
Maximal rate of calcium uptake (nmol Ca mg^{-1} SR s^{-1})	5.8 (3)	22.5 (3)
Maximal calcium uptake (μmol Ca mg^{-1} SR protein)	2.8 (3)	5.5 (3)
Halftime of relaxation t/2 (ms)	113 (4)	21 (4)
Time to peak contraction (ms)	82 (4)	22 (4)
Actomyosin ATPase (μmol P_i min^{-1} mg^{-1} protein)	0.05 (4)	0.13 (4)

[a] Key: J-SR: junctional sarcoplasmic reticulum; (1) B. R. Eisenberg et al. (1974); (2) B. R. Eisenberg and Kuda (1975); (3) Fiehn and Peter (1971); (4) Barnard et al. (1971).

calcium pump than a slow muscle (Fiehn and Peter 1971; Salviati et al. 1984); it may, therefore, relax much faster. Because of the high calcium-pump ATPase activity the energy expended for calcium cycling (i.e. the energy of activation) is higher than in slow muscles (cf. Sect. 2.3.2).

Note, however, that calcium pump activity of the slow-twitch muscle may be underestimated, since it is only slightly stimulated by calcium, but more extensively by cAMP, which causes phosphorylation of phospholamban, a regulatory protein component of slow-muscle sarcoplasmic reticulum (Salviati et al. 1982; Kirchberger and Tada 1976, cf. Sect. 2.3.4). Gillis (1985), therefore, warns against a too simplistic correlation of relaxation rate and calcium-pumping activity. In his detailed comparison among various kinds of muscle with respect to relaxation, development of the sarcoplasmic reticulum and activity of the calcium pump, he found very little correspondence between the functional and biochemical properties of muscle. In his view, factors other than the rate of calcium sequestration may be more important in determining the speed of relaxation. One such factor may be the rate at which calcium is removed from troponin, a process that is very much dependent on the properties of troponin itself. It is noteworthy, therefore, that the troponin of fast- and slow-twitch muscles is quite distinct (Dhoot et al. 1979), and that there is also evidence for differences in their calcium-binding properties. According to D. G. Stephenson and Williams (1982), for instance, the Ca^{2+} concentration required for 50% activation of skinned fibres of slow-twitch muscle is much lower than in fast-twitch muscle, suggesting that in the latter the calcium-troponin complex may dissociate more rapidly after lowering the Ca^{2+} concentration. In fast-twitch muscle this lowering of the Ca^{2+} concentration may be brought about by calcium-binding to parvalbumin or to SR-bound calcium translocators rather than by the activity of the calcium pump in the sarcoplasmic reticulum (cf. Sects. 2.3.3.3 and 4.3.1):

Parvalbumin may play a role as a soluble-relaxing factor, as discussed already in Chap. 4. It is particularly abundant in fast-twitch muscle, but virtually absent in slow muscle (Heizmann 1984). In certain mutant mice, on the other hand, the parvalbumin content in fast muscle is drastically reduced, and yet, in a twitch contraction, the rate of relaxation is not affected (Stuhlfauth et al. 1984). According to Peckham and Woledge (1986), calcium binding to parvalbumin also causes the "labile heat" observed in single twitches of fast muscle. Thus, the amount of calcium that is handled in twitch contraction is increased in the presence of parvalbumin.

5.2.3 Shortening Velocity and Myosin Isozymes

Whereas the time course of an isometric contraction appears to be largely determined by the rate at which calcium is delivered to and removed from the contractile system, unloaded shortening speed may be related to crossbridge kinetics. When a muscle is fully activated in an isometric tetanus, its shortening speed can be measured by performing an isotonic release experiment. Remember that in an isometric contraction maximal force is equivalent to the load that can be supported without causing a change in muscle length. When this load is suddenly removed, the tetanically stimulated and fully activated muscle will shorten at its maximal speed, the unloaded shortening speed.

Relation Between Shortening Speed and ATPase Activity. As shown in Table 5.2, the unloaded shortening velocity of fast- and slow-twitch skeletal muscle fibres may differ by a factor of 2 to 3 (Close 1964). Similar differences have also been found in actomyosin-ATPase activity (Bárány 1967; Barnard et al. 1971), as well as in the unloaded shortening velocity and crossbridge kinetics of skinned fibres from different types of twitch muscle that have been maximally activated by Ca^{2+} in an ATP salt solution (cf. Kawai and Schachat 1984).

A greater speed in shortening requires a greater rate of filament sliding, of attachment and detachment of crossbridges and of ATP splitting. Since ATP is split at a lower rate in slow than in fast muscle, the latter maintains tension less economically. The heat of maintenance, and the rate of energy expended for maintaining unit tension in a 1-cm length of muscle is roughly three times greater in extensor digitorum longus muscle than in soleus muscle (cf. Crow and Kushmerick 1982c). Energy expenditure and speed of shortening are obviously correlated, since speed is dependent on actomyosin-ATPase activity. Maximal isometric force in tetanus is, however, identical in fast and slow fibres (cf. Table 5.2).

Myosin Polymorphism. The differences in ATPase activity of the contractile proteins in fast and slow muscles may be due to the different structure of the ATP-

Table 5.2. Relationship between shortening speed and "tension cost" in fast- and slow-twitch muscle[a]

Muscle (mouse)	Fibre type	ATP consumption $\mu mol\ g^{-1}$ muscle s^{-1}		Speed of unloaded shortening $L\ s^{-1}$	Rate of relaxation $t_{1/2}$ (ms)	Force ($N\ cm^{-2}$)
		Ca handling[b]	Total[c]			
EDL	IIB	1.2	4	6	7	19
Soleus	I	0.4	1.3	2	22	21

[a] (From Luff 1981; Crow and Kushmerick 1982c, 1983; Kushmerick 1983).
[b] Energy expenditure in muscle stretched to the extent that actin and myosin no longer overlap.
[c] Total energy expenditure or ATP consumption at 2.2 μm sarcomere length (optimal overlap of actin and myosin filaments). The unloaded shortening velocity is expressed in muscle lengths (L) per second, and the rate of relaxation after cessation of isometric tetanus is given as the halftime of the exponential tension decay. Cyclic crossbridge attachment and detachment is associated with ATP-hydrolysis:

$$\text{contractile ATPase activity } (k_{cat}) = k\frac{fg}{f+g}$$

$$\text{and force } = k'\frac{f}{f+g}$$

f and g are rate constants for crossbridge attachment and detachment in the two-state crossbridge model of A. F. Huxley (1957).
By dividing ATPase activity by force, one obtains the "tension cost" which is obviously dependent on g. Since g also determines the rate of shortening, it follows that holding economy (the reciprocal value of tension cost) and speed of contraction will be inversely related. Some muscles may be able to "switch gear" by altering the rate constant g which presumably corresponds to the rate of decomposition of the actomyosin-ADP complex.

splitting enzyme. For instance, a superfast myosin with extremely high ATPase activity is abundant in the powerful jaw muscles of carnivores and in certain fast extraocular muscles. Ordinary type-II fibres contain fast myosin, which may exist in three isoforms differing in the composition of the light chains, whereas type-I fibres contain a slow myosin of low ATPase activity. Structural differences in these myosins have recently been studied in more detail. All myosin molecules are hexamers consisting of two heavy chains of MW 200 kDa and four light chains, namely one "regulatory light chain" (LC-2) and one "essential light chain" (either LC-1 or LC-3) attached to the head of each heavy chain. As shown by "native gel electrophoresis" of the non-denatured proteins in pyrophosphate gels, the heavy chains of fast and slow myosins are quite distinct (Hoh and Yeoh 1979) and the light chains differ as well (Sarkar et al. 1971; Lowey 1980; Weeds 1980). In skeletal muscle containing different fibre types, shortening velocity has been found to be correlated with both the type of myosin heavy chain expression (Reiser et al. 1985) and with the expression of different alkali light chain isoforms (Greaser et al. 1988). In addition, the regulatory or P-light chain may, however, also be important in determining contraction speed. This has been demonstrated by experiments with skinned fibres from which the regulatory light chain had been extracted and subsequently reconstituted (Hofmann et al. 1990; cf. also Moss 1982). Further work is needed, however, to establish more precisley, how the different contractile properties of fast and slow twitch muscles are determined by the type of myosin heavy or light chain.

5.2.4 Fast and Slow Muscle: Calcium Cycling and Energetics

In all types of vertebrate striated muscle fibres, greater contraction speed appears to be generally correlated with a greater extent of the junctional area between the T-system and the sarcoplasmic reticulum as well as with other structural and biochemical characteristics listed in Table 5.3. Notably the rate of calcium release from and the active calcium reuptake by the sarcoplasmic reticulum are both higher in fast than slow muscle. This is also the case in a maintained contraction when the processes of calcium release and reuptake are in a steady state. Compared with slow muscles, fast muscles, therefore, require a higher "energy cost of activation" for calcium cycling during tetanus. But, in addition, their higher crossbridge cycling rate (cf. Table 5.2) also contributes to their lower holding economy. In conclusion, contraction speed and "holding economy" are inversely related. Thus, terrestrial animals will use slow muscles for postural "antigravity" functions in maintaining tension with little fatigue and with as little energy as possible. As explained in Table 5.2 (cf. also Sect. 1.3.3.5), the slow detachment rates of the crossbridges probably account for the low unloaded shortening velocity as well as the lower tension cost. To power maximal movement, both terrestrial and adequatic animals have evolved fast white muscle fibres. However, to also use these muscles for slow movements would not be efficient. This is why fish, for instance, make use of red muscles for slow cruising at slow shortening velocity that is

Table 5.3. Factors determining speed of contraction and relaxation in a twitch

Contraction[a]	Relaxation
1. T-SR junction and -coupling	1. Ca binding to SR-membrane
2. Ca release from SR	2. Ca reuptake into SR
3. Ca diffusion to myofilaments	3. Ca binding to Pv[b]
4. Ca binding to TnC	4. Dissociation of TnC-Ca
5. Crossbridge cycling (ATPase activity of actomyosin)[c]	5. Crossbridge detachment[c]

[a] Numbers indicate the sequence of events in a twitch during activation and subsequent relaxation. Key: T-SR junction: contact between transverse tubuli and terminal cisternae of the sarcoplasmic reticulum (SR); Pv, parvalbumin; TnC, troponin-C.
[b] Optional; may precede step 1, but significance uncertain (cf. Gillis 1985 and Ashley and Griffiths, 1983).
[c] Rate dependent on isoform of myosin.

actually optimal for mechanical power production and efficiency (about one-third of the maximal unloaded shortening velocity; cf. Rome et al. 1988).

5.3 Diversity of Crustacean Muscles

As many crustacean muscle fibres may be more than 1 mm wide, they have been uniquely suitable for investigating problems of excitation-contraction coupling by microinjection (see Sect. 3.1.1). As in other arthropod muscles, e.g. scorpion muscle (cf. Gilly and Scheuer 1984), contraction is initiated by membrane depolarization causing an inward calcium current (Hagiwara et al. 1969), which may trigger the release of calcium from the sarcoplasmic reticulum (D. G. Stephenson and Williams 1980; cf. Sect. 3.1.7). Unlike vertebrate muscle, the calcium-binding protein troponin contains, however, only one rather than four calcium-binding sites (Regenstein and Szent-Györgyi 1975; Wnuk et al. 1984, cf. Sect. 4.1.2, Fig. 4.11).

5.3.1 The Cell Membrane

Crustacean muscles contain a wide variety of fibre types differing greatly in structure and function. For instance, the dactylopodite adductor (closer) of the walking leg in the swimming crab *Portunus sanguinolentus* has been found to contain cell types ranging from fast excitable fibres, exhibiting all-or-none action potentials, to fibres that are slow and unexcitable. These fibres may contract tonically when the membrane is depolarized, the extent and rate of contraction depending on the extent and rate of membrane depolarization (Atwood et al. 1965). Similarly, the dactylopodite flexor of *Cancer magister* may contain at least three different fibres, the fastest ones being activated by an all-or-none action potential inducing twitches of 300 ms duration. On the other hand, the slower fibres, which require about 20 times longer for contraction and relaxation, are not excitable, but respond to graded depolarization with a maintained contraction. When these fibres are stimulated at increasing frequency, they develop a stronger contraction because superimposing excitatory post-synaptic potentials summate under these conditions, resulting in a larger membrane depolarization causing a greater increase in intracellular free calcium. In slow crustacean muscles the resting membrane potential is usually only slightly less negative than the mechanical threshold potential, leading to a just noticeable contractile response so that a small depolarization induced by nerve impulses is sufficient to elicit a slow fibre contraction. Among the most tonic crustacean muscles is the levator of the eyestalk of the crab *Podophthalmus*. This muscle is, most of the time, in a state of continuous contraction since it is depolarized even if it is not stimulated by nerves. The same muscle, however, relaxes when stimulated by inhibitory nerves, since these induce a hyperpolarization of the cell membrane to potentials more negative than −50 mV, the mechanical threshold potential (Hoyle 1968).

In fast fibres, on the other hand, the resting potential of the membrane is much more negative than the mechanical threshold, so that contractions can only be elicited by large potential changes evoked by all-or-none calcium-dependent action potentials. The functional diversity of different crustacean fibre types may, however, only partly be explained by differences in the properties of the cell membrane. Differences in the contractile machinery as well as in the internal membrane system and in the coupling mechanisms linking excitation and contraction may be equally important.

5.3.2 Diversity of Sarcomere Structure and ATPase Activity

Fast-white and slow-red fibre types have recently been described in the closer muscles of the claw of the crab *Eriphia spinifrons* (Rathmayer and Erxleben 1983). Here, the slow fibres appear to contain a myosin of low ATPase activity, whereas the fast types contain a myosin with high ATPase activity (Maier et al. 1984) which, however, is less resistant to alkaline or acid treatment (pH 10.4 or 4.6) than slow myosin. As in vertebrate muscle the slowness of contraction and energy release may, therefore, result at least partly from the slow cycling of ATP-splitting actin-myosin crossbridges. In addition, however, fast and slow crustacean muscles may also differ in their sarcomere length. For example, the slow closer muscle of the crusher claw of a lobster (*Homarus americanus*) contains sarcomeres of approximately 8 μm length, whereas the sarcomeres of the faster cutter claw adductor are only 3–4 μm long (Jahromi and Atwood 1971). The relationship between sarcomere length and function is illustrated schematically in Fig. 5.3.

Let us assume that the rate of filament sliding depends on the rate of crossbridge cycling and, hence, on the ATPase activity, while force development may be a function of the number of crossbridges acting in parallel. With a given rate of cycling, therefore, the schematized contractile unit consisting of two short sarcomeres in series (Fig. 5.3) will, therefore, shorten twice as fast as an equally long contractile unit containing only one long sarcomere. In the long sarcomere, however, the number of crossbridges pulling on one actin filament, and acting therefore in parallel, would be twice as large as in the unit with two short sarcomeres

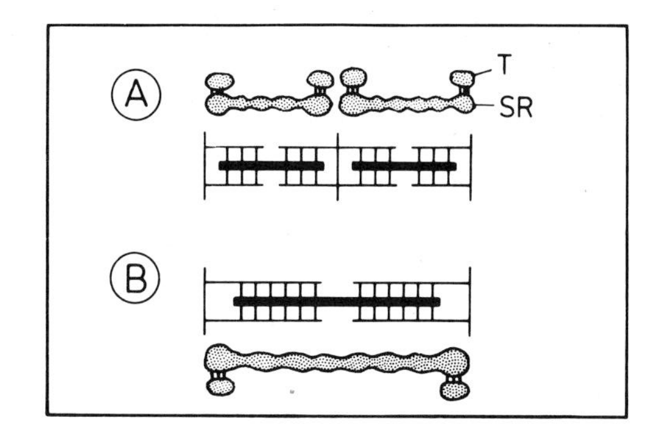

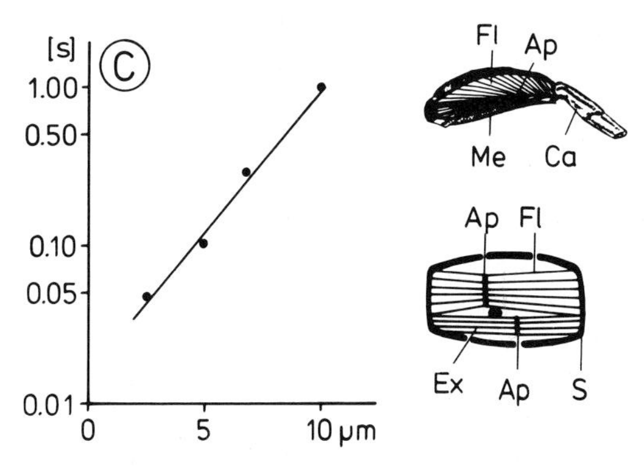

Fig. 5.3 A–C. Relation between sarcomere structure and contraction speed in crustacean muscle fibres. *A, B* Relationship between sarcomere length, diad spacing and relative number of crossbridges acting in series or in parallel. Transverse tubuli (*T*) and cisternae of sarcoplasmic reticulum (*SR*) form diadic junctions (= Ca^{2+} release sites). *C* Twitch contraction time in dependence of sarcomere length in slow and fast fibre types of the carpopodite extensor of the ghost crab *Ocypode*. (After Hoyle 1983). *Right:* Location of the carpopodite extensor (*Ex*) within the meropodite of walking leg; *S* = shell of meropodite (Me), *Ap* = apodome; *Fl* = flexor of carpopodite (*Ca*). Note that the slow fibres with long sarcomeres (*B*) contain less diads or T-SR couplings per unit length than fast fibres with short sarcomeres (*A*). In the latter the rate of Ca^{2+} release and delivery to the contractile system is, therefore, higher than in slow muscles. With a given rate of filament sliding the segment (*A*), containing two short sarcomeres, will shorten twice as fast as an equally long segment (*B*) containing one sarcomere

in series. Force development would, therefore, also be twice as large; it depends on the product of sarcomere length and number of crossbridges per unit fibre length. Conversely, the maintenance of the same force would require only half as many ATP-splitting crossbridges in the one long sarcomere as in the case of two short sarcomeres linked in series, and tension maintenance would, therefore, be less costly. Other factors being equal, the rate of energy expenditure for the maintenance of a given force (tension cost) will, therefore, be related to the speed of shortening as both are dependent on the sarcomere length. When the specific myosin-ATPase activity due to crossbridge cycling is different as well, the following relationship holds (cf. Rüegg 1971):

$$v = k \cdot e \cdot s^{-1}, \tag{5.1}$$

where e is the molecular turnover number or specific ATPase (equivalent to cross-bridge cycling frequency), s is the sarcomere length and v is the unloaded shortening velocity. Variations of sarcomere length and myosin-ATPase activity are, therefore, possible mechanisms of adaptation to the various demands of muscle speed and economy in invertebrate striated muscle. In invertebrates the sarcomere length may range from 0.8 μm (Platyhelminthes) to 30 μm in the striated muscle of the worm *Syllis* (Anderson 1982), and the filament structure also shows an amazing variability (Wray 1979a).

5.3.3 Internal Membrane Systems

In crustacean muscle fibres the cell membrane invaginates into the fibres, thus forming numerous, branching clefts extending far into the fibre interior. Transverse tubules may originate from these clefts as well as from the outer surface membrane (cf. Sect. 2.1.2). These T-tubules are located near the A-I band border zone where they form diadic junctions with the longitudinal tubules of the sarcoplasmic reticulum (Fig. 5.3). Short sarcomeres will, therefore, contain two to four times as many diads per unit length as fibres with long sarcomeres. For instance, lobster abdominal muscles contain 2000 diads in a 1 mm segment of fast fibre, but only 600 diads mm^{-1} in slow tonic fibres (Jahromi and Atwood 1969). Clearly, the diffusional pathways from the calcium-release sites at the diads to the middle of the sarcomeres must, therefore, differ by a factor of 2 to 4 between fast and slow fibres. Since, however, the diffusion time of calcium depends on the square of the diffusion distance, there cannot be a linear relationship between sarcomere length and time required for activation. Instead, Hoyle (1983) showed a relationship between the logarithm of the time required to reach maximum isometric force after a stimulus and the length of the sarcomeres in various crustacean muscles of the ghost crab *Ocypode ceratophtalma* (Fig. 5.3). Similarly, twitch duration has been found to decrease exponentially with increasing content of sarcoplasmic reticulum in different crustacean fibres (Josephson 1975). Among the fastest crustacean muscles are the muscle fibres of lobster antennae which, by contracting and relaxing at a frequency as high as 300 Hz, can produce a buzzing sound. These fibres contain an extremely well-developed sarcoplasmic reticulum occupying up to 70% of the fibre volume (Mendelson 1969, cf. Rosenbluth 1969). Moreover, the transverse tubular system forms longitudinal extensions along the myofilaments, thus making many diadic contacts with calcium-storing cisternal elements of the reticulum which intimately surround the flattened myofibrils along the entire sarcomere. Because of these structural features the time required for calcium diffusion from the calcium-release sites to the myofibrils and from there to the calcium pump will be rather brief.

5.3.4 Comparison with Other Arthropod Muscles

A division of muscle fibres into fast and slow types with attendant physiological, biochemical and ultrastructural differentiation has been recognized not only in all crustacean muscles, but also in those of arachnids and insects. Even muscles of the horseshoe crab *Limulus polyphemus,* a "living fossil" unchanged over 300 million years, apparently contain two fibre types with different thick filament and sarcomere length. Presently, these findings are, however, still controversial, since there also appears to be a variation of sarcomere length within the same fibres and since the thick filaments are not well aligned and apparently themselves able to shorten considerably even after isolation (Dewey et al. 1982). Thick filament "shortening" in *Limulus* muscle is calcium-dependent, but its mechanism is not yet understood. Physiologically, shortening of muscle fibres appears to be accomplished by filament sliding, a process triggered by calcium ions. These activate the actomyosin ATPase via phosphorylation of myosin light chains by means of calcium- and calmodulin-dependent myosin light chain kinase (Sellers 1981), as in the case of vertebrate smooth muscle (Chap. 8). Additionally, however, tropomyosin/troponin-mediated regulation may play a role as well (Lehman 1982).

In insect muscle, calcium regulation of contraction and actomyosin ATPase may also depend on troponin and tropomyosin (Bullard et al. 1973), as well as on the phosphorylation of the regulatory light chain (Winkelman and Bullard 1980). Contractions may be fast or slow depending on the fibre type. The leg muscles of the locust *Schistocerca gregaria,* for instance, contain fast fibres that are rich in sarcoplasmic reticulum and slow fibres in which the sarcoplasmic reticulum is scarce (Hoyle 1978). In these slow muscles, as in most crustacean muscles, about 8 to 12 thin filaments surround each thick filament, the ratio of actin to myosin filaments being about 7:1. Interestingly, this filament ratio is only 3:1 in the structurally most regular muscle of the entire animal kingdom, insect flight muscle.

5.4 Insect Flight Muscle

During flapping flight of many small insects the indirect flight muscles may oscillate several hundred times per second, in certain midges even up to 500–1000 times per second to produce the wing beat by compressing and decompressing the elastic insect thorax. Since the latter acts as a lever system for the wings, oscillatory contractions need not exceed a few percent of the muscle length, yet their power output may be extremely high. The sarcomeres, which are about 2–3 μm in length, exhibit a most regular array of thick and thin filaments which overlap nearly completely: large I-bands are not required since these muscles shorten only by a very small amount during the oscillatory movement.

Flight muscles are of two kinds. One is called non-fibrillar because myofibrils, though present, are thin, and therefore not readily recognized in the light microscope. In these muscles, the sarcoplasmic reticulum is very elaborate, thus allowing a rapid release and reuptake of activator calcium. In fibrillar flight muscles, on the other hand, the rather large myofibrils are separated by rows of huge mitochondria, while the sarcoplasmic reticulum is scarce, since the oscillatory movements of these muscles do not appear to be dependent on the rapid release and reuptake of calcium.

5.4.1 Non-Fibrillar Muscle

The muscle movements causing the wing beat of many lower insects, such as locusts, butterflies and moths, exhibit the characteristics of twitches, each contraction being initiated by a nerve impulse followed by an action potential (Pringle 1957, 1981). Because of the one-to-one relationship of electrical and mechanical activity (cf. Fig. 5.5), non-fibrillar flight muscles are also called "synchronous." In terms of the calcium theory of muscle activation, neurogenic oscillations of synchronous flight muscle require an exactly timed calcium release and its reuptake by the sarcoplasmic reticulum within a few milliseconds. It is, therefore, not surprising that the sarcoplasmic reticulum is unusually extensive, surrounding the narrow myofibrils in tight sheaths (Smith 1966). Neurogenic contractions of this type are, however, only possible as long as the interval between nervous stimuli or action potentials producing a calcium release is larger than that required to pump the released calcium back again into the sarcoplasmic reticulum. As the frequency of stimulation increases to a critical value, the sarcoplasmic Ca^{2+} concentration as well as the calcium occupancy of troponin, will remain high even between stimulations so that twitches fuse to a maintained tetanic contraction. For this reason, synchronous flight muscles of most insects are, as a rule, unable to move the wings at frequencies surpassing approximately 100 Hz, the tetanus fusion frequency of these muscles (Wootton and Newman 1979). The exception to the rule is the synchronous tymbal muscle of the cicada *Okanagana vanduzeei*, which oscillates at a frequency of more than 500 Hz (Josephson and Young 1985), while in other cicada a single twitch contraction may elicit many sound pulses (Huber et al. 1990).

5.4.2 Fibrillar Muscle

Features completely different from those of non-fibrillar muscles are found in the so-called fibrillar flight muscles of small insects with fast wing beats, e.g. bees and flies, and even of some larger insects with slow wing beats, such as some species of bugs. Characteristically, the sarcoplasmic reticulum, which is poorly developed, does not surround the thick myofibrils, thus leaving ample room for the accommodation of large mitochondria (Ashhurst 1967; Smith 1966; cf. Fig. 5.4). Accordingly, the tetanus fusion frequency is quite low, approximately 10 s^{-1}, in isometrically contracting fly muscles. Surprisingly, however, the same muscles are able to contract and relax in situ by about 1–3% of their muscle length at a frequency that is by one order of magnitude higher than the tetanus fusion frequency: i.e. several hundred times per second (cf. Boettiger 1957). These frequent oscillatory contractions are, however, not elicited by equally frequent nerve impulses so that the muscle action potentials are, therefore, no longer in synchrony with the wing-beat movement (Fig. 5.5 A), but occur at random with respect to the oscillations (Nachtigall and Wilson 1967). During the song of the cicada *Platypleura*, about four to six sound pulses are evoked by the fibrillar tymbal muscle by a single motor-nerve impulse, and stimulation at 50 cycles s^{-1} would even induce a myogenic rhythm of about 320 oscillations s^{-1}. Under strictly

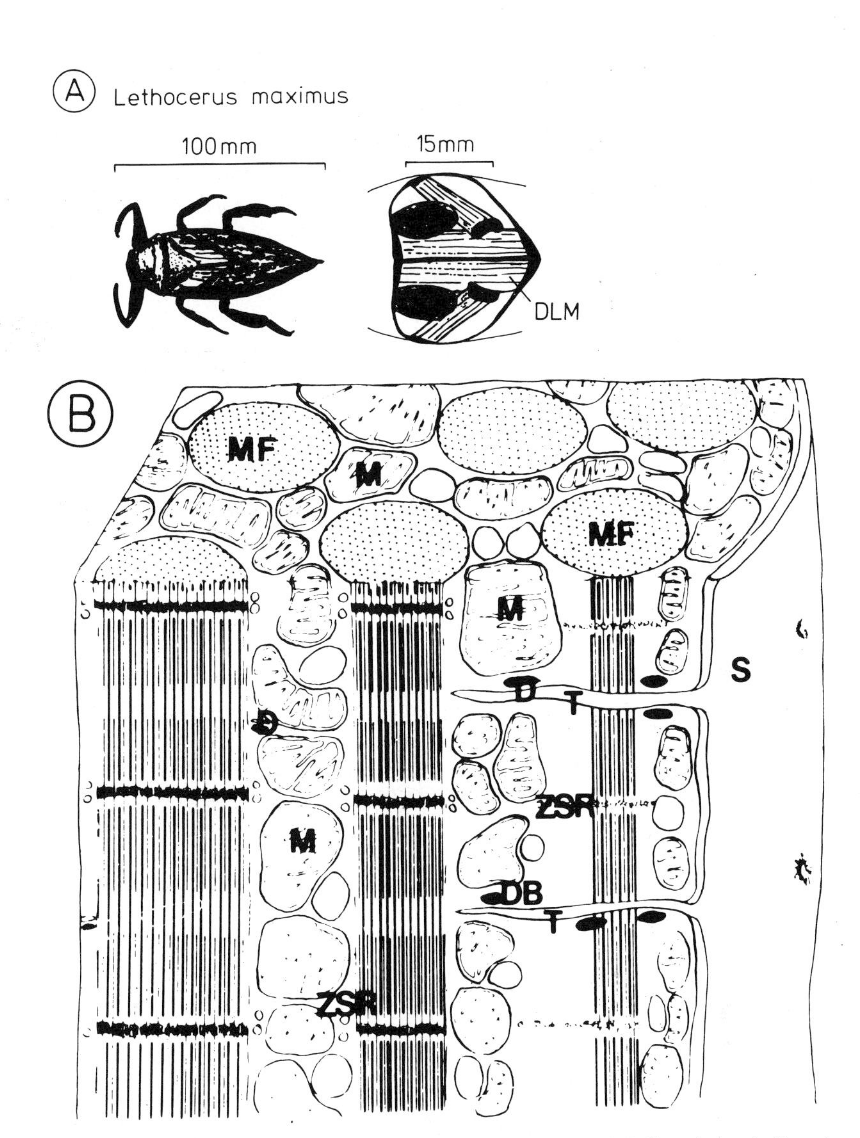

Fig. 5.4 A, B. Lethocerus flight muscle. *A Lethocerus maximus* from Trinidad (*left*) and sketch (dorsal view) showing location of dorsal longitudinal muscle (*DLM*) within the thorax. *B* Schematic drawing of *Lethocerus* DLM muscle fibre. *MF*=myofibrils; *M*=mitochondria; *D*=diad; *DB*=dense body of diad; *S*=sarcolemma; *ZSR*=sarcoplasmatic reticulum; *T*=transverse tubuli, T-system. Note that the sarcoplasmic reticulum is very rudimentary. (Ashhurst 1967)

isometric conditions, on the other hand, the tetanus fusion frequency is only 40 to 60 s^{-1}, and a single twitch was found to last as long as 0.1 s (Pringle 1954). These observations indicate, of course, that oscillations of fibrillar or asynchronous insect muscle cannot be directly controlled by nerve impulses. Oscillation, in other words, is myogenic and not neurogenic (Pringle 1957).

In the following we shall see that the oscillatory movement is not only independent of the functional cell membrane, but also of the "give" and "take" of

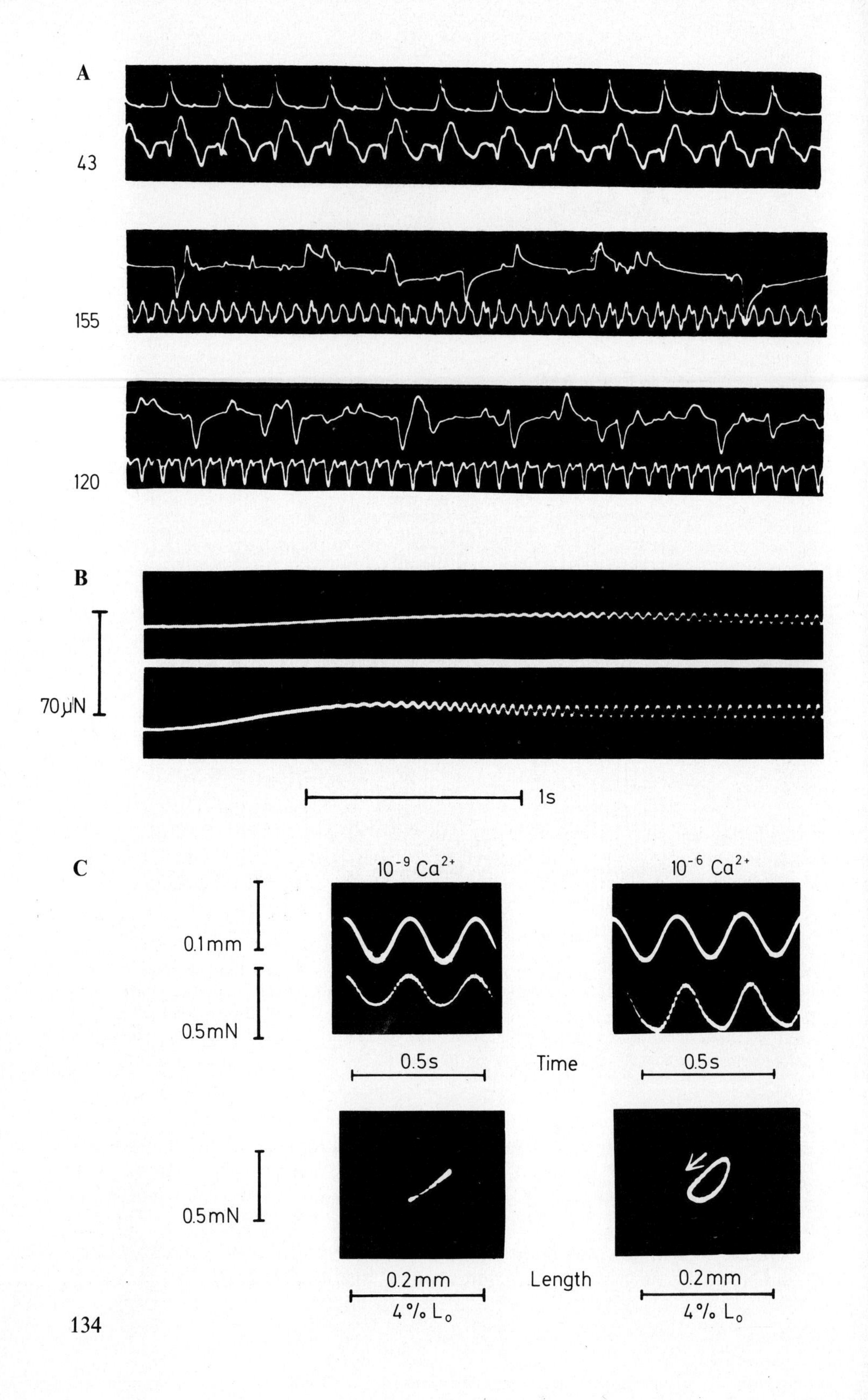
A
43
155
120
B
70 µN
1s
C
10^{-9} Ca^{2+}
10^{-6} Ca^{2+}
0.1mm
0.5mN
0.5s
Time
0.5s
0.5mN
0.2mm
$4\% \ L_o$
Length
0.2mm
$4\% \ L_o$

Ca^{2+} by the sarcoplasmic reticulum. Obviously, oscillation is generated by a myofibrillar "automatism" since oscillatory movement was still possible after removing the cell membrane and the sarcoplasmic reticulum by a chemical skinning procedure.

5.4.3 The Myofibrillar Origin of Myogenic Oscillation: Skinned-Fibre Studies

In order to prove that the oscillatory mechanism resides within the contractile machinery, the latter must be functionally isolated. For this purpose, Jewell and Rüegg (1966) freed the contractile structures of the giant water bug *Lethocerus maximus* of the cell membrane and sarcoplasmic reticulum by extraction with a mixture of glycerol and phosphate buffer. They then attached a single "skinned muscle fibre" or a small fibre bundle to a compliant auxotonic lever system with a natural frequency close to the resonance frequency of the thorax wing system of the intact bug, i.e. 20 cycles s^{-1}. After immersing the preparation in a bathing solution containing Mg-ATP and EGTA to lower the free Ca^{2+} concentration to below 0.01 μM, the skinned fibres were fully relaxed. When, however, the Ca^{2+} concentration was increased to approximately 0.1 μM, the skinned fibre started to oscillate with a frequency of approximately 20 Hz which corresponds to the natural frequency of the lever system. As shown in Fig. 5.5 B, such oscillations could be maintained for many hours with ATP as the only energy source, but stopped as soon as the ATPase activity was inhibited or when ATP was removed. Characteristically, sinusoidal force changes lag behind the sinusoidal length changes in an oscillatory contraction (Fig. 5.5 C).

After raising the free Ca^{2+} concentration stepwise the amplitude of oscillations increased, reaching a maximum of about 1% to 2% L_0, whereas the frequency of oscillation remained unaltered. It could be varied, however, by altering the resonance frequency of the lever system within a certain range. When, however, the lever was stiffened to increase the resonance frequency above 50 Hz, the fibre ceased to oscillate, but contracted isometrically instead. The oscillatory contractions observed in skinned fibres, therefore, appeared to be essentially identical with the oscillations in living beetle flight muscle of *Oryctes* in which the frequency depends on the inertial load of the thorax wing system (cf. Machin and Pringle 1959) and may be increased, for instance, by clipping the wings.

The dependence of frequency of oscillatory contractions on the resonance frequency of the recording lever system suggests that the switching on and off of the contractile machinery may be controlled by the small changes in length that occur

◄

Fig. 5.5 A–C. Oscillation of insect flight muscle. *A* Simultaneous recording of electrical activity (*upper tracings*) and mechanical activity (*lower tracings*) from the thorax of three insects with three different wing-beat frequencies: moth 43 Hz, fly 155 Hz, wasp 120 Hz (From Pringle 1978). *B* Free oscillation of "skinned fibres" from an insect flight muscle (DLM, *Lethocerus maximus*), attached to an auxotonic lever system and suspended in ATP salt solution. The oscillatory contraction starts when the free Ca^{2+} concentration is increased to approximately 1 μM (From Jewell and Rüegg 1966). *C* Driven oscillation of skinned fibres from *Lethocerus* suspended in ATP salt solution of pCa 9 (*left*) or pCa 6 (*right*). In the latter case sinusoidal length changes imposed by a vibrator cause delayed tension changes. The tension length diagram (Lissajou figure) is represented by a loop described in a counterclockwise fashion. (Rüegg 1968 b)

during these oscillations rather than by the changes in the Ca^{2+} concentration which is heavily buffered by CaEGTA. Let us suppose that these changes in length are sensed by the myofibrils and that lengthening switches the contractile machinery on, whereas shortening of the muscle fibre switches the machinery off again with a delay. If this is so, during each oscillatory cycle, the movement of the lever system would stretch the fibre and activate the contractile system; a counterforce would, therefore, be evoked with a delay so that the fibre will shorten again, thereby becoming deactivated and thus relaxing and lengthening under the pull of the restoring force of the lever system. Once more, the resulting elongation would stretch-activate the fibre, commencing a new cycle of oscillation which would be exactly timed by the (restoring) movement of the resonant lever system.

5.4.4 Stretch Activation

Direct evidence for stretch activation of skinned fibres in an ATP salt solution at constant Ca^{2+} concentration was obtained by Jewell and Rüegg (1966). When

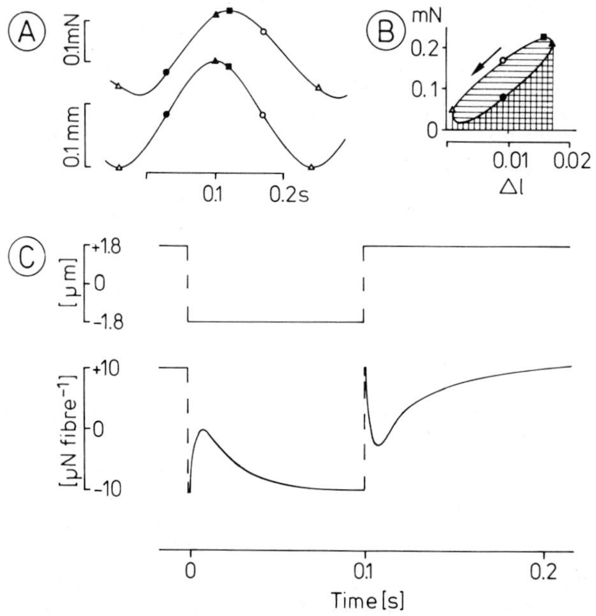

Fig. 5.6 A–C. The oscillatory work cycle of skinned fibres from DLM of *Lethocerus maximus* suspended in ATP salt solution at intermediate level of Ca^{2+}. *A* Sinusoidal length changes cause delayed tension changes. *B* Force changes plotted versus length changes (Lissajou figure). *Vertically hatched area* represents work done on the fibre during stretch, *horizontally hatched area* represents work done by the fibre during release. The difference is the net oscillatory work in a stretch-release cycle. The quadrature tension is the force between *open* and *closed circles* (width of loop) and the ratio of quadrature tension to length change is the viscous modulus. The elastic modulus is the quotient of the tension difference at the largest and smallest length divided by the length change. *C* Tension changes (*lower tracings*) following a quick release and a quick stretch of partly activated skinned fibres (length 0.5 cm) from the water bug *Hydrocyreus columbiae*. Activating solutions contained (in mM) K 65, Na 10, Cl 51, histidine 10, Mg-ATP 5, CaEGTA buffer 2, Ca^{2+} 0.12 μM, pH 7.1. (Jewell and Rüegg 1966)

they suddenly stretched a single skinned fibre by 1% or 2% of its length, there was a sudden and transient increment of passive tension which was followed by a much larger delayed rise of force associated with an activation of the actomyosin ATPase, whereas a quick release to the initial length caused a delayed decrease in tension (Fig. 5.6C). The myofibrillar ATPase activity could also be activated by stretch, as shown in Fig. 5.7 (Rüegg and Tregear 1966).

During flight of many kinds of insects, indirect flight muscles are also suddenly stretched and released in a repetitive manner by a special "click mechanism," while in other insects, such as the tropical water bugs, the muscle length is altered in a more sinusoidal fashion. In this case, sinusoidal tension changes will be delayed with respect to the sinusoidal length changes. If a skinned insect flight-muscle fibre suspended in an ATP salt solution is stretched and released in a sinusoidal fashion, it responds with sinusoidal force changes which lag behind the length changes, indicating that the muscle performs work (Fig. 5.6A). The extent of work done in an oscillatory cycle can be most readily recognized when tension is plotted against length at any one moment of an oscillatory cycle, thereby generating a counter-clockwise Lissajou figure, indicating that the muscle fibre develops more force during shortening than is required to restretch the fibre; thus, the area enclosed by the loop indicates the work done by the fibre in each cycle of this "driven oscillation" (cf. Figs. 5.6B and 5.5C).

The elastic and viscous moduli (Fig. 5.8) and hence the work produced in a cycle of driven oscillation and the power output in the fibre (work times frequency), as well as the rate of ATP splitting depend on the frequency at which the fibre is stretched and released. This relationship is bell-shaped (Fig. 5.9). Interestingly,

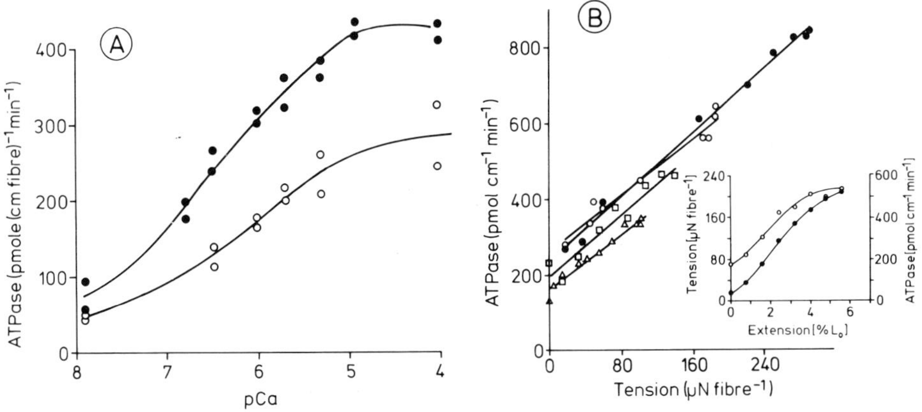

Fig. 5.7 A, B. Stretch activation of ATPase activity in an insect flight muscle. *A* Calcium dependence of ATPase in stretched and unstretched skinned fibres of the dorsal longitudinal muscle of *Lethocerus maximus*. *Open circles:* ATPase activity at resting length (sarcomere length 2.2 μm) and after 4% stretching (*closed circles*). Note that the ATPase activity is increased by stretch, but [Ca^{2+}] for 50% activation does not change. All solutions contained 5 mM Mg-ATP, 20 mM Tris buffer, pH 7.1, 10 mM NaN_3 and 4 mM CaEGTA buffer, to "clamp" the Ca^{2+} concentration, pH 7.1. (Rüegg and Tregear 1966). *B* Linear relationship between stretch-induced active tension and ATPase activity in skinned *Lethocerus* fibre bundles. *Inset:* Increase in ATPase activity (open circles) and force (closed circles) by elongation of fibres at a "clamped" Ca^{2+} concentration. (Loxdale and Tregear 1985)

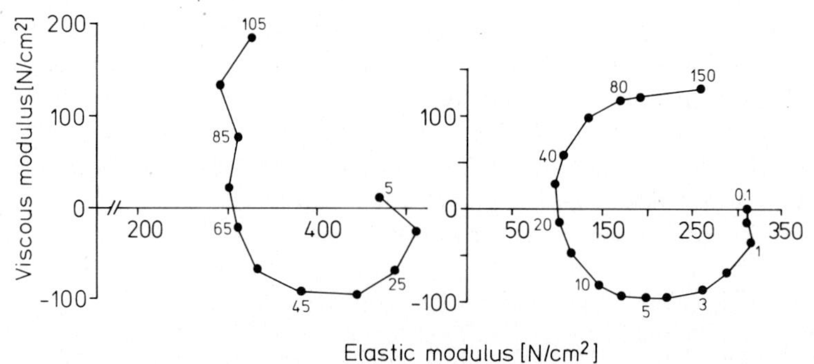

Fig. 5.8. Dynamic modulus plot of viscous versus elastic modulus defined in Fig. 5.6 B in a maximally stimulated living flight muscle of the beetle *Oryctes rhinoceros* (*left*) and in calcium-activated skinned fibres (DLM of *Lethocerus*) suspended in ATP salt solution (*right*). *Figures on curves* give the frequency of the small amplitude oscillation. (Machin and Pringle 1960; Jewell and Rüegg 1966)

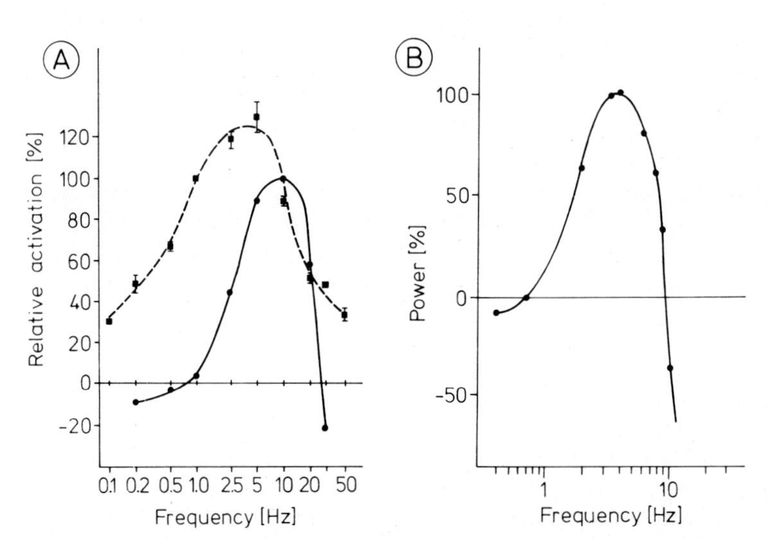

Fig. 5.9 A, B. Comparison of experiment and model of myogenic oscillation. *A* Oscillatory power output of skinned fibre bundles from *Lethocerus* (*circles*) during driven oscillation at various frequencies of sinusoidal stretch and release cycles (length change 3% L_o). Also shown is the relative rate of ^{32}P-phosphate incorporation (*squares*) into ATP in % of ATP-phosphate exchange rate at oscillation frequency of 1 Hz. Fibres in Mg-ATP salt solution (cf. Fig. 5.7) at "clamped" free Ca^{2+} concentration (1 μM). (Ulbrich and Rüegg 1976). *B* Prediction of oscillatory power output of *Lethocerus* fibre bundles by the three-state crossbridge model of Murase et al. (1986)

an optimal rate of ATP splitting and power output may be obtained when the frequency of oscillation is close to the natural wing-beat frequency or the resonance frequency of the thorax wing-system (Steiger and Rüegg 1969). This optimal frequency is about 20 Hz in the giant water bug *Lethocerus maximus* at the physiological temperature of the thorax (40 °C) and about 40–50 Hz in the case of the smaller, but faster, species *Lethocerus annulipes* and *Lethocerus cordofanus*. It is even much higher in skinned fibres of bumble bees and the fruitfly *Drosophila melanogaster*. Thus, it would seem that the frequency characteristics of the oscillatory mechanism whithin the contractile machinery are adapted to the requirements of insect flight in different species. The smaller the delay between the stretch and stretch-activated development of force, the higher the frequency which can be obtained during oscillation! It is, therefore, important from a comparative point of view to know the factors which cause the delay in stretch activation (cf. Molloy et al. 1987 and Drummond et al. 1990 for genetic approaches).

The time course of the delayed increase in tension or decrease therein following a stretch or release will be determined by the rate of attachment and detachment of crossbridges within the contractile system (Thorson and White 1969). Obviously, stretch activation involves the attachment of additional crossbridges, as shown by experiments demonstrating an increase in immediate fibre stiffness following a quick stretch, whereas a quick release would decrease the number of crossbridges attached at any one moment in contracting insect flight muscle (Güth et al. 1981). The additional attachment of crossbridges due to stretching is indicated by a greatly increased fibre stiffness, as well as by an increase in the rate of ATP hydrolysis. In addition, stretch also enhances the exchange reaction leading to incorporation of ^{32}P-labelled phosphate into ATP (Ulbrich and Rüegg 1971; Fig. 5.9). This reaction, as we have seen (Chap. 1), is evidence for the formation of an energy-rich, actin-myosin nucleotide complex which is one of the biochemical intermediates in the crossbridge cycle; it may be intimately linked with the process of force generation. As shown by Beinbrech et al. (1976): crossbridges may attach at right angles to actin in a non-force-generating state when they are occupied by the ATP analogue AMP-PNP. When the nucleotide is removed, the ternary nucleotide-actomyosin complex decomposes, whereby leading to crossbridge rotation and force generation.

The mechanism by which crossbridges are "recruited" during stretch is still open to speculation. As skinned fibres of vertebrate skeletal muscle become more calcium sensitive during stretching, a similar mechanism might be proposed for stretch activation of insect flight muscle. This hypothesis, however, has been disproved by showing that the amount of calcium bound to troponin did not increase during stretch (Marston and Tregear 1974); nor was the Ca^{2+} concentration required for 50% activation of skinned fibres grossly altered by changes in length (Schädler 1967). Moreover, stretch activation was still possible at very high "saturating" Ca^{2+} concentrations (Rüegg and Tregear 1966; Fig. 5.7). For this reason, it seems to be more likely that stretching the contractile structures strains the myosin filaments which are known to be connected to the Z-line by giant proteins (Lakey et al. 1990). Deformation or displacement of filaments might improve the match between myosin heads and actin monomers interacting at the crossbridge level (Wray 1979 b). Perhaps, stretch may cause a displacement of a tropomyosin- or troponin-I-like inhibitory protein, such as troponin-H (Bullard et al. 1988) which might interfere with actin-myosin interaction in the unstretched state only. Troponin-H may be a likely candidate for such a steric blocking role, as it is particularly bulky (MW 80 kDa) and spaced at regular intervals of 39 nm along the thin filament (Bullard et al. 1988). Indeed, Abbot and Cage (1984) were able to demonstrate that the mechanical activity of insect flight muscle increased and decreased over a period of approximately 3% (corresponding to a 38-nm-filament displacement) as the muscle was extended. On the other hand, stretch-activation phenomena may also be found, although to a lesser degree, in non-fibrillar insect flight muscle (Aidley and White 1969), and even in skinned fibres of vertebrate striated and cardiac muscle (Rüegg et al. 1970; Steiger 1977). These findings imply that stretch activation may be a more general property of the contractile mechanism of all muscles. Recently, Murase et al. (1986) proposed a three-state crossbridge model which could account for oscillatory phenomena of muscle in general and, in particular, for the frequency-dependence of oscillatory-power output (cf. Fig. 5.9). However, the physiological significance of stretch activation and oscillatory properties in non-fibrillar muscle remains to be elucidated.

5.5 Obliquely Striated Muscle of Annelids and Nematodes

In the body wall muscles of nematodes and annelids the thick myosin-containing filaments are not in register within the sarcomeres, but arranged in a staggered fashion, each filament being slightly displaced with respect to its neighbour (Rosenbluth 1965; Mill and Knapp 1970). Thus, the border line between A-band and I-band no longer forms a right angle with the fibre axis, but gives rise to an oblique or diagonal striation pattern. Interestingly, however, the degree of obliquity changes as the muscle lengthens or shortens. During extension, the thick filaments slide not only relative to the thin filaments, but also in relation to their thick-filament neighbours. Conversely, shortening may be associated not only with the sliding of thin filaments past thick filaments, but also with shearing of thick filaments. Shearing phenomena have even been observed without sliding of actin versus myosin filaments resulting, therefore, in changes in muscle length without alteration of actin-myosin overlap (Knapp and Mill 1971). Since the development of force depends on an optimal overlap between actin and myosin, obliquely striated muscles are, therefore, particularly adapted for development of force over a wide range of muscle lengths (Toida et al. 1975).

The Internal Membrane System. This is differently organized in muscle fibres of annelids and nematodes (cf. Sect. 2.1.2). In the latter, a well-developed, transverse tubular system invaginates from the cell membrane and penetrates along the Z-line into the interior of the fibre where it forms diadic contacts with cisternal elements of the sarcoplasmic reticulum (Rosenbluth 1968, 1972). In annelid muscles, on the other hand, the SR forms calcium-storing sarcotubules (Heumann and Zebe 1967) which extend from the fibre interior along the Z-line to the cell membrane with which they form diadic structures (Mill and Knapp 1970).

Fast and Slow Fibres. In many annelid muscle fibres the contractile elements are situated near the cell membrane of fibres which may be flat or may contain a non-contractile fibre core. Thus, fairly rapid contractions may be induced if calcium is released from the subsarcolemmal vesicles or enters the cell through the sarcolemma during the action potential. Such phasic contractions may be superimposed on slow maintained tonic contractions, thus raising the question whether the two kinds of contractile responses may be due to two fibre populations. The body wall muscles of the polychaete *Mercierella* has indeed been found to contain, in addition to slow circular muscle cells, two types of longitudinal muscle fibres which may be either thick and fast or thin and slow. Only the fast fibres are excited by calcium-dependent, all-or-none action potentials (Skaer 1974). In muscles of *Glycera* the slow fibres apparently contain longer thick filaments and less sarcoplasmic reticulum than the fast fibres (Rosenbluth 1968), whereas in the earthworm fast and slow fibre types may be distinguished histochemically (D'Haese and Carlhoff 1987). In these muscles, the myosin of the fast type has higher ATPase activity than slow myosin and its light chains have molecular weights of 18 kDa and 25 kDa. In slow fibres, on the other hand, the molecular weights of the light chains are 18 kDa and 28 kDa respectively, while the myosin-ATPase activity is as low as in crustacean and vertebrate slow muscle fibres. A multitude of myosin isozymes has also been recognized in the muscles of nematodes (Epstein et al. 1976b).

The Regulatory Proteins. The myosin light chains of molecular weight 25 kDa or 29 kDa appear to be crucial in the regulation of annelid muscle contraction, since they may confer calcium sensitivity to a hybrid contractile protein formed from annelid myosin and skeletal muscle actin in vitro (D'Haese 1980). The regulatory role of light chains, however, seems to be complemented by the proteins troponin and tropomyosin, which constitute the calcium switch of the thin actin filaments (Donahue et al. 1985). Like many types of arthropod muscle, therefore, obliquely striated muscles of nematodes and annelids are dually controlled in the sense that both myosin and actin filaments are calcium-regulated (Lehman and Szent-Györgyi 1975). These muscles are, therefore, quite distinct from vertebrate striated muscle which appears to be exclusively regulated by troponin and tropomyosin, and from molluscan muscles which seem to be entirely myosin-regulated, as discussed below (Chap. 6).

5.6 Generalizations and Conclusions

Investigations of striated muscle of vertebrates, arthropods, annelids and nematodes have distinguished types of fast and slow fibres which differ in speed and holding economy. These differences may be due partly to modification of the sliding-filament contractile mechanism and its contractile proteins, but they may also be due to differences in excitation-contraction coupling. Large differences in shortening speed are caused by differences in the number of contractile linkages between myofilaments acting in series or in parallel, or by the different turnover number of these linkages, as reflected by the different ATPase activity. In addition, the design properties of the calcium delivery and reuptake systems may also be adapted in a coordinated manner to the different functional needs (symmorphosis). In fast muscle, the T-tubular system is extensive, and the junctional surface between the sarcoplasmic reticulum and the T-system appears to be correlated with greater speed of contraction. The diffusion distance from the calcium-release site at the level of terminal cisternae to the myofibrils and from there to calcium-uptake sites in the sarcoplasmic reticulum tends to be as small as possible. Furthermore, the calcium affinity of the regulatory proteins is comparatively low, so that the contractile machinery may be switched off rapidly when the Ca^{2+} concentration is lowered by the extensive sarcoplasmic reticulum. In general, the sarcoplasmic reticulum is more abundant in fast than slow muscles, but astonishingly, it is scarce in rapidly oscillating fibrillar muscles of insects. The latter are unusual in that oscillatory contractions are not triggered by calcium release from the sarcoplasmic reticulum, but by the mechanical strain of the myofilaments caused by the vibrations of the thorax during flight. In this case, the actomyosin contractile system appears to be activated directly by stretch rather than by an increase in myoplasmic Ca^{2+} concentration in phase with the oscillations. Thus, the energy required for calcium handling must be assumed to be negligibly small in these powerful muscles.

Chapter 6

Myosin-Linked Regulation of Molluscan Muscle

In all muscles, contraction is triggered by an increase in intracellular free calcium. The precise nature of the response to elevated sarcoplasmic Ca^{2+} concentration, however, depends on the kind of regulatory proteins that sense and respond to intracellular concentrations of Ca^{2+}. These proteins are diverse. In vertebrate striated muscles, as we have seen in Chap. 4, the calcium sensor is troponin-C, located in the thin actin filaments. In such a thin filament, or actin-linked regulatory system, contractile activity is turned off by the tropomyosin-troponin system when it does not have calcium bound to it. In molluscan muscle, on the other hand, the calcium switch is incorporated into the myosin molecule. Regulation is thus myosin-linked. In contrast to vertebrate muscle, in molluscan muscle the same protein molecule serves both a contractile and regulatory function. In the relaxed state, the actin-activated myosin ATPase is repressed by an inhibitor, the so-called regulatory light chain of molluscan myosin (Sect. 6.1). A different regulatory mechanism is found in the slow adductor smooth muscle of bivalves showing the "catch phenomenon" (Sect. 6.2). Because of the ease with which the regulatory light chain can be reversibly removed from myosin, the striated adductor of the scallop (*Pecten sp.*) proved to be unusually suitable for studies of the calcium-regulated mechanisms (Chantler and Szent-Györgyi 1980; Szent-Györgyi and Chantler 1986).

6.1 Calcium Regulation in the Striated Adductor of the Scallop

By rhythmical twitch contractions of its cross-striated adductor, *Pecten* is capable of swimming over a short distance at a speed of nearly half a meter per second, using a simple mechanism: the sudden expulsion of a stream of water, produced by the rapid adduction of the shells, causes jet propulsion (Buddenbrock 1911; Thompson et al. 1980). In vitro, electrical stimulation of the isolated adductor evokes propagated, calcium-dependent action potentials which induce twitch contractions. The time course and the amount of force and heat produced during a twitch is quite similar to that in twitching sartorius: the latter, however, is more resistant to fatigue (Rall 1981). Although both types of muscle show the same band pattern of cross-striation in their fibres, details of structure and excitation-contraction coupling are quite different. The fibres of adductors are very flat, and since the innermost myofilaments are not more than 1 μm from the cell membrane, a rapid inward conducting membrane system is not required for rapid activation. The T-system is entirely absent, but there is an abundance of subsarcolemmal sarcoplasmic reticulum (Sanger 1971), whence activator calcium is released to the myofibrils where it is bound to myosin at the onset of contraction.

6.1.1 Recognition of Myosin-Linked Regulation

For many years it has been known that the myofibrillar ATPase of the adductor muscle of *Pecten* is strongly activated by trace calcium. This calcium sensitivity was, however, a most puzzling phenomenon, since the thin filaments were not found to contain a functional troponin. Clearly, therefore, the "key" for calcium sensitivity had to be sought outside the thin filament.

The answer to this riddle came in 1970 when Kendrick-Jones and colleagues in Szent-Györgyi's laboratory in Boston reported an important discovery. When they mixed myosin extracted from the adductor of *Pecten* with purified actin from rabbit skeletal muscle, a "synthetic" hybrid actomyosin was formed which, in the presence of calcium ions, was as active as a synthetic actomyosin composed of skeletal muscle myosin and actin. But whereas the skeletal muscle actomyosin remained fully active after reducing the free Ca^{2+} concentration with EGTA, the hybrid actomyosin ATPase was inhibited by over 80%. Since the skeletal muscle actin was not contaminated by troponin, these results could only be taken to mean that, unlike skeletal muscle myosin, its molluscan counterpart conferred calcium sensitivity to the enzyme preparation; it obviously contained a calcium switch which sensed and responded to Ca^{2+} in the range of 10^{-8} to 10^{-5}M.

The Boston researchers went on to carry out the converse experiment. They prepared the thin (actin-containing) filaments from the molluscan muscle and mixed them with myosin from rabbit skeletal muscle. In this way, they obtained a hybrid actomyosin which was not calcium-sensitive; its Mg-dependent ATPase was fully active even in the virtual absence of Ca^{2+}. Unlike their skeletal muscle counterparts, the thin filaments of molluscan muscle were obviously incapable of conferring calcium sensitivity to the contractile protein system. This result was not surprising, for as mentioned already, the thin filaments did not contain a functional troponin, but it meant that molluscan muscles did not possess an actin- or thin-filament-linked regulation, in addition to myosin-linked regulation.

These findings of Kendrick-Jones and colleagues also suggested a simple probe for myosin-linked regulation in an unknown sample of actomyosin. If the addition of purified skeletal muscle actin activated the myosin ATPase in the absence of Ca^{2+}, the myosin present was obviously not regulated, but if it did not, it was. Then, actin activation of myosin ATPase would be inhibited in the absence, but not in the presence, of Ca^{2+}. Using this simple actin-competition test, Lehman and Szent-Györgyi (1975) screened the animal kingdom for myosin-linked regulation which was found to be widely distributed among different phyla. The muscles of all molluscs and of the phyla Brachiopoda and Nemertina as well as of Echinodermata were found to be myosin-regulated. The skeletal muscles of vertebrates, protochordata and certain crustaceans, on the other hand, appeared to be entirely actin-regulated, whereas regulation in most other phyla

Table 6.1. Some invertebrates with thin- and/or thick-filament-linked regulation[a]

Phylum	Species	Regulation[b]
Arthropoda	*Gryllus domesticus*	A+M
	Lethocerus cordofanus	A+M
	Schistocerca gregaria	A+M
	Limulus polyphemus	A+M
	Balanus nubilus	A+M
	Homarus americanus (crusher claw)	A+M
Annelida	*Lumbricus terrestris, glycera sp.*	A+M
Mollusca	*Pecten sp.*	M
Nematoda	*Ascaris lumbricoides*	A+M
Sipunculida	*Golfingia gouldi*	A+M
Echinodermata	*Thyone briareus*	M

[a] Synonymous to actin-(A)- and myosin-(M)-linked regulation. From Lehman and Szent-Györgyi (1975).
[b] Key: A, actin- or thin-filament-linked regulation; M, myosin- or thick-filament-linked regulation.

seemed to be both myosin- and actin-linked (Table 6.1). *Pecten,* however, remained the preparation of choice for studies on the mechanisms of regulation as its actomyosin could be readily desensitized simply by removing the regulatory light chains.

6.1.2 The Role of Myosin Light Chains

While molluscan myosin from the scallop differs from skeletal muscle myosin immunologically (Wallimann and Szent-Györgyi 1981 a, b), its gross structure appears to be quite similar. The hexameric myosin molecule consists of six peptides, namely two heavy chains forming two oblong globular heads each bearing two light peptide chains, the essential light chain and a regulatory light chain. If the "tail end" of the myosin molecule (called light meromyosin or LMM) is removed by controlled proteolysis, a heavy meromyosin fraction (HMM) which retains myosin's capacity to hydrolyze ATP and to interact with actin is obtained. The acto-HMM thus formed hydrolyzes Mg-ATP in a calcium-dependent manner. In fact, many studies on calcium regulation of ATPase were carried out using the highly soluble acto-HMM rather than actomyosin which is insoluble at low ionic strength.

When myofibrils from *Pecten* were washed with a solution containing the chelating agent EDTA (ethylene-diamine tetraacetic acid) to remove bound calcium and magnesium from myosin, one type of light chain having a molecular weight of 17 kDa was extracted and the ATPase became calcium-insensitive (Szent-Györgyi and Szentkiralyi 1973; Chantler and Szent-Györgyi 1980). From these washed myofibrils Szent-Györgyi and colleagues prepared myosin or HMM that lacked the light chains, was unable to bind the Ca^{2+} with high affinity and was, therefore, no longer calcium-regulated. It was desensitized, i.e. its actin-activated

ATPase activity was as high as that of skeletal muscle actomyosin regardless of the Ca^{2+} concentration. One conclusion, therefore, seemed obvious. The light chain that could be removed by EDTA treatment must have had a regulatory function in as much as it inhibited actin-activated ATPase in the absence of calcium, but not in its presence, and it seemed to be responsible for calcium binding to myosin. This proposed role of the regulatory or EDTA light chain, as it was called, was demonstrated more directly by reconstitution experiments.

6.1.2.1 A Molluscan Regulatory Light Chain Confers Calcium Sensitivity

After incubating "desensitized" scallop myosin with isolated molluscan regulatory light chain, Szent-Györgyi and Szentkiralyi (1973) could fully restore calcium sensitivity. In the absence of Ca^{2+}, the actin-activated myosin ATPase was suppressed, whereas added calcium ions were bound to myosin and derepressed the ATPase. According to Chantler and Szent-Györgyi (1980), these effects can

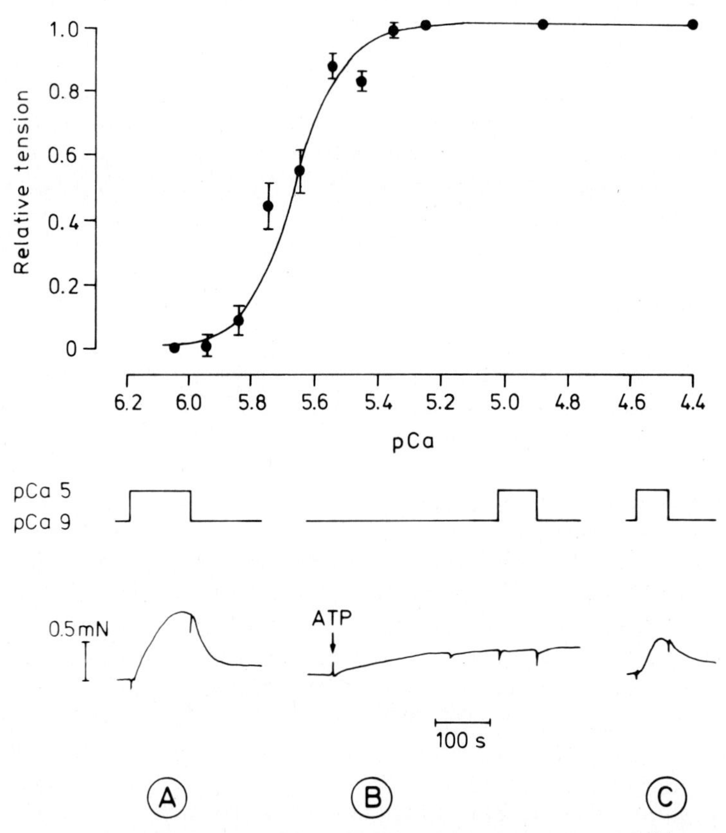

Fig. 6.1. Above: Dependence of force development on the neg. log of Ca^{2+} concentration in skinned fibres of the striated adductor of a scallop. The bathing solution contained 6 mM Mg-ATP, 1 mM Mg^{2+}, 5 mM CaEGTA buffer, imidazole pH 7.1. (Simmons and Szent-Györgyi 1985). *Below:* Calcium sensitivity of skinned fibres depends on the myosin regulatory light chain. *A* Response before and *B* after extraction of regulatory light chain. *C* Note that the capacity of relaxation at low ionized calcium is restored after readdition of the light chain. (Simmons and Szent-Györgyi 1985)

be correlated with the binding of the regulatory light chain to the actomyosin heavy chain, suggesting that the light chain acts as an inhibitor and confers calcium sensitivity to the molluscan contractile system. The latter can be studied in its native, structurally intact state after functional isolation. Simmons and Szent-Györgyi (1985) removed the cell membranes from adductor muscle fibres by skinning with detergents. As shown in Fig. 6.1, these skinned fibres contracted in a calcium-dependent manner when they were suspended in a salt solution containing ATP as an energy source. However, after extraction of the regulatory light chain with EDTA, calcium sensitivity was lost and the fibres contracted maximally even in the absence of Ca^{2+}. These contracted fibres, relaxed after addition

Table 6.2. Role of light chains in regulation of molluscan myofibrils

A. Loss and reconstitution of calcium regulation

Pecten myofibrils	ATPase ratio $-Ca^{2+}/+Ca^{2+}$	Ca binding (%)
Native	0.02	100
Desensitized[a]	1.2	4
Hybrid *(Aequipecten)*	0.06	90
Hybrid *(V. mercenaria)*	0.11	92
Hybrid *(Limulus)*	0.11	12
Hybrid (Cricket)	0.08	27
Hybrid (Chicken gizzard)	0.05	–

B. Effect of skeletal muscle light chains and light chain mutants on desensitized scallop myofibrils[b]

Desensitized myofibrils	ATPase (units)		Ca binding (%)
	$-Ca^{2+}$	$+Ca^{2+}$	
Control	0.16	0.15	(4)[e]
+Rabbit LC	0.05	0.08	(11)[e]
+Chicken LC	0.06	0.1	15
+*E. coli* LC[c]	0.05	0.07	14
+*E. coli* mutant LC[d]	0.25	0.23	4

[a] Myofibrils were desensitized by extraction of the regulatory light chains and reconstituted by replacing the native light chain with foreign regulatory light chain from *Aequipecten* adductor, *Venus mercenaria* adductor, *Limulus polyphemus* muscle, from cricket *(Gryllus domesticus)* or from chicken gizzard. Data from Sellers et al. (1980).

[b] Myosin heavy chain-light chain hybrids obtained by replacing extracted *Pecten* regulatory light chain with light chain from skeletal muscle of the rabbit *(Oryctolagus cuniculus)*, of chicken *(Gallus domesticus)* or with chicken light chains synthesized in *Escherichia coli* by genetic engineering techniques.

[c] Wild type with Asp. residue 47.

[d] Mutant type with Ala residue in position 47 (obtained by site-directed mutagenesis). Data from Reinach et al. (1986).

[e] Numbers in ( ) in % of Ca-binding of native scallop light chains (1 mole/mole LC). From Sellers et al. (1980).

of regulatory light chains which restored calcium sensitivity. These experiments show that in adductor muscle fibres calcium regulation is entirely myosin-linked and dependent on the EDTA light chain.

Since the addition of light chains restored the capacity of myosin to bind Ca^{2+} with high affinity (Table 6.2), the question arose as to whether the regulatory light chain was the calcium receptor. This possibility seemed particularly intriguing since the light chain is capable of binding both Ca^{2+} and Mg^{2+} with high affinity at a site which, as pointed out by Bagshaw (1980), has structural resemblance to the calcium-binding sites of troponin and parvalbumin. Though the calcium-magnesium-binding site seems to be important for its attachment to the heavy chain, it is obviously not directly involved in the regulation of contraction. For at physiological concentrations of Mg^{2+}, activator calcium ions cannot be bound to the light chain unless they displace the bound magnesium by a comparatively slow reaction. Thus, it is not the EDTA light chain, but the myosin heavy chain, that carries the functionally important selective calcium-binding site. For reasons that are not yet fully understood, however, the calcium affinity to this site is low unless the heavy chain interacts with the regulatory myosin light chain (Bagshaw and Kendrick-Jones 1979).

6.1.2.2 Myosin Heavy Chain – Light Chain Hybrids

The sequence homology of the regulatory light chain of molluscan muscle with myosin light chains from other types of muscle is most striking. All of these structurally related light chains may bind to molluscan myosin deprived of its own regulatory light chain; all of them bind calcium and magnesium and may, at least to a certain extent, replace the molluscan regulatory light chains functionally, as shown in Table 6.2. Indeed, when only one regulatory light chain was removed from molluscan myosin by treatment with ice-cold EDTA solution, calcium sensitivity was lost, but could be fully restored by replacing the lost molluscan light chain with the regulatory light chain of any other muscles (Szent-Györgyi and Szentkiralyi 1973). When, however, both regulatory light chains were removed from molluscan muscle by extraction with EDTA at 30 °C (Chantler and Szent-Györgyi 1980), restoration of calcium sensitivity depended on the type of the regulatory light chain used for substitution (Sellers et al. 1980).

Diversity of Regulatory Light Chains. Light chains from the muscles of molluscs or of brachiopods, called class-1 light chains, restored calcium sensitivity fully. Class-2 regulatory light chains are derived from non-molluscan muscle showing myosin-linked regulation, such as *Limulus* muscle, insect muscle or vertebrate smooth muscle. These light chains, as well as light chains from motile cells prepared from the thymus glands or from blood platelets, restored calcium regulation of molluscan myosin, too. However, the Ca^{2+} concentration required for 50% activation is much higher in the myosin light chain-heavy chain hybrids than in the parent native molluscan myosin. Class-3 light chains stem from muscles without myosin-linked regulation, such as lobster tail muscle and all vertebrate striated muscles. These peptides are unable to restore calcium sensitivity but they bind to scallop myosin and inhibit its ATPase activity, as shown in Table 6.2.

A further distinction of light chains concerns covalent modification by phosphorylation. The regulatory light chains of class-2 may be phosphorylated on serine residue 19 by the calcium-calmodulin-dependent myosin light chain kinase. This has a clear effect on biological activity as these light chains inhibit actin-activated ATPase of myosin only when they are not phosphorylated (cf. Chap. 8). Even in hybrids of molluscan myosin heavy chains and class-2 regulatory light chains, inhibition of actin-activated ATPase can be partly overcome by light chain phosphorylation, in much the same way as by raising the Ca^{2+} concentration (Kendrick-Jones and Scholey 1981). The regulatory light chains (class-1) of fast adductor of *Pecten*, on the other hand, are somewhat shorter than class-2 light chains. They lack the particular peptide sequence at the N-terminal end which recognizes and binds the myosin light chain kinase and cannot, therefore, be phosphorylated by this enzyme. Phosphorylatable regulatory light chains have, however, been isolated from slow adductor muscle of *Pecten,* which is a catch muscle (Sohma et al. 1985).

6.1.2.3 The Essential Light Chains

After removing the regulatory light chain by EDTA, another type of light chain can be removed by SDS and appears to be the molluscan counterpart of the essential light chain or alkali light chain of vertebrate skeletal muscle. Although structurally quite similar to the EDTA extractable light chains, the isolated essential light chains are unable to bind calcium and magnesium. Nonetheless they seem to be also essential for calcium regulation in molluscan muscles, which can be abolished by monoclonal antibodies directed against these peptides (Wallimann and Szent-Györgyi 1981 b). It seems quite possible, therefore, that both the essential and the regulatory light chains are involved in coordinated binding of Ca^{2+} to the myosin heavy chain and in the regulation of contraction.

6.1.3 Light-Chain-Dependent Calcium Binding and Contraction: Cooperativity

The regulatory light chains convert low-affinity calcium-binding sites on the heavy chain (or on the alkali light chain) to a specific high-affinity calcium-binding site (Bagshaw and Kendrick-Jones 1979). Both myosin heads, however, must be occupied by a regulatory light chain, suggesting cooperativity between heads. Indeed, the removal of only one regulatory light chain is already sufficient to desensitize molluscan myosin to calcium. Thus, if both regulatory light chains are removed from the myosin molecule, calcium sensitivity cannot be restored by replacing only one of them. Moreover, isolated subfragment-1 is not calcium-sensitive and finally, the steep dependence of the acto-HMM ATPase on the Ca^{2+} concentration suggests cooperativity between the two heads, although calcium binding per se is not cooperative (Chantler and Szent-Györgyi 1980).

Two moles calcium are bound per mole molluscan myosin with an apparent binding constant of about $10^6 M^{-1}$. While calcium binding occurs over a wide concentration range of approximately 10^{-8} to 10^{-6}M, the ATPase activity of acto-HMM is barely activated below 10^{-7}M, but exhibits a steep calcium dependence in the range between 10^{-7}M and 10^{-6}M free calcium. This suggests that acto-HMM is activated only when both myosin heads (S-1) are occupied with calcium. Note also that in myofibrils and skinned fibres the calcium dependence of the actin-activated ATPase and force development is even much steeper, suggesting an even greater cooperativity. This appears to be conveyed by the molluscan regulatory light chain, for in myofibrils in which the molluscan regulatory light chain had been replaced by the corresponding muscle light chain from *Limulus*, the relation between ATPase activity and free Ca^{2+} concentration is less steep and similar to the calcium dependence of calcium binding (Chantler 1983).

It is tempting to speculate on the possible biological significance of the extremely steep calcium-force dependence illustrated in Fig. 6.1, which is unique to scallop muscle. Is it of advantage for regulation of muscle showing fast all-or-none twitch responses? Consider that after the release of calcium into the myoplasm at the onset of activation a fast and nearly maximal contraction would be triggered suddenly as soon as a critical threshold of free calcium, about 1 μM, is reached. For relaxation only one, rather than two, calcium ions would have to be removed from the myosin molecule by the pump of the sarcoplasmic reticulum. In scallop adductor these processes are facilitated by a soluble calcium-binding protein (Collins et al. 1983) which serves as a calcium trap as well as a calcium shuttle between the calcium-binding sites at the myofilaments and the sarcoplasmic reticulum in much the same way as parvalbumin does in skeletal muscle.

6.1.4 Light-Chain Location and Movement

As pointed out already, regulation of contraction not only involves the regulatory light chain, but also the essential light chains, since antibodies directed against either type of peptide abolish calcium sensitivity of molluscan myofibrils (Wallimann and Szent-Györgyi 1981 a, b). Indeed, both light chains appear to be located in close proximity near the neck region of myosin which links the globular head with subfragment-2, thus giving rise to a barbed appearance in electron micrographs (Craig et al. 1980). Wallimann et al. (1982) demonstrated the closeness of the regulatory and the essential light chain more directly by using photosensitive heterobifunctional crosslinkers. In a typical experiment with scallop myofibrils they replaced the intrinsic regulatory light chain by the light chain of the clam *Venus mercenaria* to which they previously attached the photolabile crosslinker para-azido-phenylacylbromide. This reagent labelled the only cysteine residue of the light chain, which is 50 amino-acid residues away from the N-terminal end. As long as the preparation was kept in the dark, the substituted light chain carrying the label remained only loosely bound to the myosin head. After a flash of light, however, the reagent reacted with the neighbouring essential light chain, thus forming a molecular bridge of only 0.8 nm, corresponding to the length of the bifunctional crosslinker, between the regulatory and the essential light chain. From these studies Wallimann and colleagues concluded that the two kinds of light chain must indeed lie in close proximity on the myosin head. They

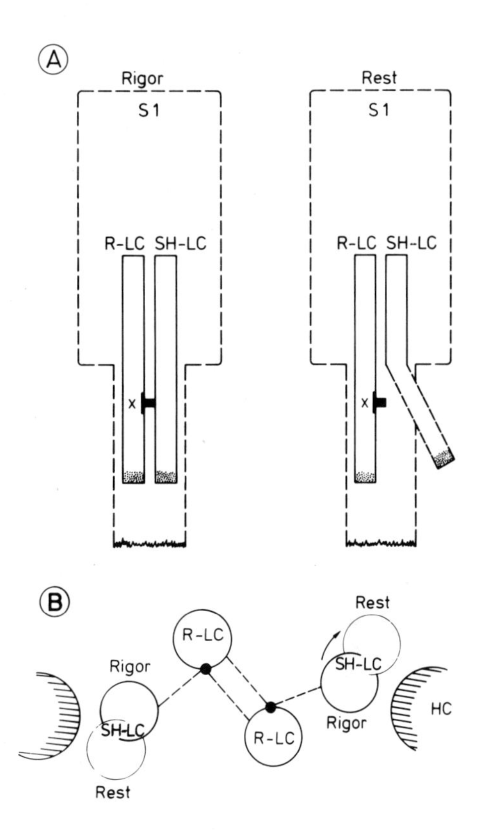

Fig. 6.2 A, B. Model of regulation in molluscan muscle. *A* Alteration of the distance between regulatory myosin light chain (*R-LC*) and essential light chain (*SH-LC*), at the level of amino acid residue 50. Note that a crosslinker (×) on residue 50 of a substituting gizzard light chain is able to form crosslinks with the SH light chain in the rigor state (or activated state), but not in the relaxed state. S_1: subfragment-1 of myosin. (Based on Hardwicke et al. 1983; but modified according to Hardwicke and Szent-Györgyi 1985). *B* Schematic diagram of cross-section through HMM (*HC*, heavy chains; *LC*, light chains) on the level of amino acid residue 50 of *R-LC*. Note movement of SH light chain (*arrow*) relative to regulatory light chain when muscle in rigor-contraction is relaxed with ATP at low Ca^{2+} concentration. (Hardwicke and Szent-Györgyi 1985)

are apparently bound to a small peptide fragment near the C-terminal end of the myosin subfragment S-1 (Szentkiralyi 1984; Bennet et al. 1984).

The schematic diagram of Fig. 6.2 summarizes crosslinking results obtained both with relaxed myofibrils and myofibrils in a state of rigor contraction. Note that successful crosslinking between the amino acid residue 50 and the essential light chain was only possible when the myofibrils were in rigor, but not when they were in the relaxed state occurring in the presence of ATP and at very low Ca^{2+} concentrations. According to Hardwicke et al. (1983), the 0.8-nm-long crosslinker was then obviously unable to bridge the gap between the light chains, suggesting that the peptides must have moved apart during relaxation. Under these conditions, the myofibrillar ATPase was inhibited. But in rigor or when the myofibrils were activated by Ca^{2+}, the N-terminal end moved away from the "off-position" in the direction of the essential light chain to which it could be crosslinked. Interestingly, the actomyosin ATPase of molluscan myosin containing crosslinked light chains was high, irrespective of the presence or absence of calcium, suggesting that the crosslinker "freezes" the regulatory light chain in the activated position. It would seem, therefore, that a tiny molecular movement of the regulatory light chain relative to the essential light chain may be an important step in switching the contractile machinery on and off. As shown in the diagram of Fig. 6.2, all movement is probably restricted to the N-terminal part of the light chains, since crosslinkers between groups more distant from the N-terminus could easily be formed between light chains even in the relaxed state. More recent experiments by Hardwicke and Szent-Györgyi (1985), however, showed that the change of the relative distance between two light chains is due to a motion of the essential light chain rather than of the regulatory light chain.

The light chain movements just described are, in many ways, reminiscent of the movement of tropomyosin from an inhibitory position on the periphery of thin filaments to a non-inhibitory position when muscle is activated (cf. also Chap. 4). However, the changes occur in the myosin neck region, a considerable distance from the actin-myosin interaction site. Nevertheless, because of its great length (Stafford and Szent-Györgyi 1978), the essential light chains could cover a large part of the interface between the myosin head and actin, thus preventing by a steric hindrance the formation of strong contractile linkages. However, as will be discussed below, steric hindrance of bond formation between actin and myosin is unlikely to be the mechanism of ATPase inhibition during relaxation. One should also keep in mind that apart from changes in light chain position other structural alterations take place during activation as well. For instance, as found by Vibert and Craig (1983), crossbridges are highly ordered in relaxed scallop muscle, but they may become disordered during activation. Moreover, tropomyosin molecules located on thin filaments move during activation as in vertebrate skeletal muscle. They slip into the groove between the actin strands of the thin filaments, perhaps as a consequence rather than as a cause of crossbridge attachment (Vibert et al. 1972). Certainly, calcium regulation may involve more than structural changes in the myosin heads alone. How else might one account for the fact that the ATPase of single-headed myosin molecules (Stafford et al. 1979), but not that of isolated heads (subfragment-1), may be activated by actin in a calcium-dependent manner?

6.1.5 Mechanism of ATPase Activation by Calcium

In the relaxed state, i.e. in the absence of calcium, the ATPase of scallop contractile proteins may be almost 600 times less active than in the contracted state, according to Wells and Bagshaw (1984). As discussed above and proposed by Szent-Györgyi and Szentkiralyi (1973), this inhibition of actin-activated ATPase may be due to the action of the regulatory and essential light chains which may effectively block the interaction of actin and myosin when they are in the "off-position" in the absence of calcium. Subsequent studies, however, showed that, as in the case of vertebrate striated muscle (cf. Chap. 4), it is not the interaction of actin and myosin per se that is blocked, but rather a subsequent step in the enzymic mechanism of ATP splitting which is required for forming force-generating bonds between actin and myosin. The reasons for adopting this view are manifold.

First, doubts arise as to whether light chains attached to the neck region of the myosin could really sterically block the attachment of myosin heads to actin. Second, the affinity between actin and the myosin subfragment HMM increases very little when the actomyosin ATPase is activated by Ca^{2+} (Chalovich et al. 1983). Third, the ATP-splitting rate may be accelerated by trace calcium to some

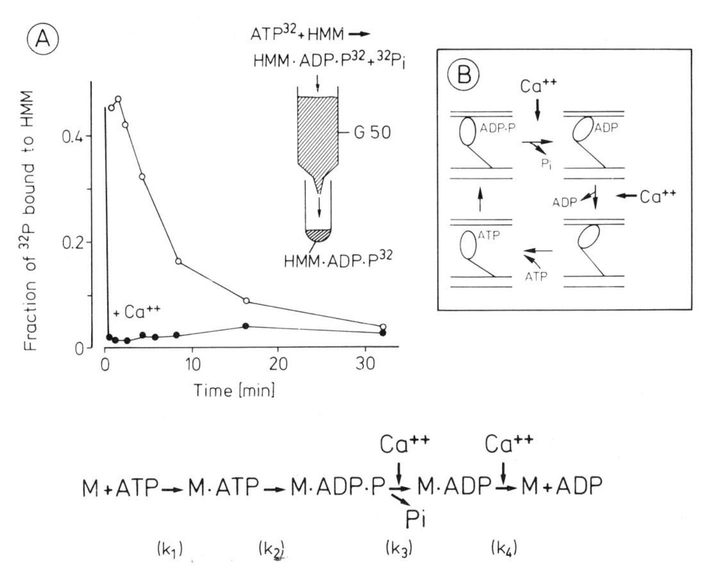

Fig. 6.3 A, B. The mechanism of calcium regulation of myosin ATPase in molluscan muscle. *A* Time dependence of P_i release from scallop HMM following the hydrolysis of the "energy-rich" bond of ATP in the presence (*solid circles*) and in the absence of Ca^{2+} (*open circles*). The *ordinate* is the fraction of HMM to which the products of ATP hydrolysis, P_i and ADP, are still bound. Note that in the absence of calcium, the decomposition of the enzyme-product complex requires several minutes. *Below:* Sequence of reactions occurring when ATP is hydrolyzed by myosin. As the product release is accelerated by calcium ions, the rate of continued ATP splitting is increased, too. k_1, k_2, k_3, and k_4 are rate constants. *Inset:* Method of determining the rate of P_i release: HMM and γ-^{32}P-ATP were rapidly mixed and incubated. After the time given on the *abscissa* the reaction products HMM·ADP·^{32}P, HMM·ADP and $^{32}P_i$ were rapidly separated on a Sephadex G-50 column (column centrifugation technique), see text. (Wells and Bagshaw 1985). *B* Hypothetical crossbridge cycle in a molluscan muscle. Both product-release steps may be accelerated by Ca^{2+}. When the Ca^{2+} concentration in activated muscle is suddenly lowered, the muscle relaxes presumably because crossbridges detach by reversing the attachment pathway rather than by way of the path involving dissociation of the products ADP and P_i (since this pathway is inhibited at low Ca^{2+} concentration)

extent even in the absence of actin, as found by Wells and Bagshaw (1984) and Ashiba et al. (1980).

According to Wells and Bagshaw (1984, 1985), the ATP-splitting mechanism of molluscan myosin involves several steps, including the rapid formation of an enzyme-substrate complex (M-ATP), the hydrolytic step forming the myosin-product complex (M-ADP-P) and the release of bound phosphate which is eventually followed by ADP release. Figure 6.3 showing this reaction scheme also illustrates the point that crossbridges in the myosin-ADP state, but not in the myosin-ADP-P_i state form strong force-generating crosslinkages with actin. In the absence of Ca^{2+} the formation of a myosin-ADP state is inhibited; the rate constant k_3 is extremely low. Consequently, there is an accumulation of crossbridges in the myosin-ADP-P state which form only weak bonds with actin. An increase in the Ca^{2+} concentration increases the rate constant k_3 by a factor of 1000 leading to the rapid formation of crossbridges in the myosin-ADP state which react with actin and generate tension. At the same time, the rate of continued ATP splitting is increased as calcium activates the rate-limiting step of the ATPase mechanism and the release of ADP and P_i (by the rate constants k_3 and k_4). Remember that a new cycle of ATP splitting cannot commence unless the enzyme first releases the bound products.

In principle, an elevated Ca^{2+} concentration would activate the myosin ATPase even under conditions in which crossbridges cannot interact with actin and cause contraction, for instance, when sarcomeres are stretched to the extent that actin and myosin no longer overlap. In this respect, Rall's observation (1981) is of note that adductor muscles of *Pecten* may produce a similar amount of heat during a twitch as frog sartorius, but unlike sartorius, a third of this heat is produced during relaxation when crossbridges are actually detaching. In vertebrate skeletal muscle, in contrast, the mechanical performance and the ATP-splitting rate are usually tightly coupled in the sense that the amount of work done determines the extent of the chemical reaction (cf. "Fenn effect," discussed in Chap. 1). A. F. Huxley (1980) attributed this feature to the two-way coupling between mechanical and enzymic steps in the crossbridge cycle. Crossbridges perform a power stroke during the transition between the weakly attached myosin-ADP-P state and the strongly actin-bound myosin-ADP state, but unless the crossbridge is allowed to rotate and perform the power stroke, this enzymic transition cannot take place. Then the ATPase is inhibited. In other words, head rotation and ATP splitting are interdependent; non-cycling crossbridges cannot hydrolyze ATP. The relaxed state of the troponin-tropomyosin system may prevent head rotation and hence, inhibit ATP splitting. Intuitively, therefore, one can visualize a purely actin-linked regulation as being more efficient in coordinating the rate of energy expenditure and mechanical activity than myosin-linked regulation. Only for an actin-linked regulation is there a built-in guarantee that crossbridges do not hydrolyze ATP unless they attach to actin, generate force and perform work.

6.1.6 Comparison of Myosin- and Actin-Linked Regulation

Whenever the formation of strong, force-generating bonds between actin and myosin is inhibited, contractile activity is turned off and the muscle relaxes. Formation of the strongly bound state of myosin (cf. AM-ADP state in Fig. 6.3) may be prevented by blocking attachment sites at the actin filaments or by inhibiting the enzymic mechanism of the myosin ATPase, in particular the transition between the different nucleotide-bound states. The first mechanism is probably realized in actin-linked regulation, the second may be more typical for myosin-linked regulation. In the latter case, inhibition is mediated by the regulatory and essential light chains and the contractile mechanism is disinhibited when calcium ions bind to myosin (molluscan muscle, Sect. 6.1.2) or when the light chain is phosphorylated (vertebrate smooth muscle, cf. Chap. 8). In myosin-linked regulation of molluscs the on-off switch is part of the contractile molecule myosin itself, a mechanism that may intuitively appear to be more primitive than actin-linked regulation which requires an additional regulatory protein system (Szent-Györgyi 1975). This idea was originally supported by comparative studies, showing that myosin-linked regulation was found predominantly in the lower invertebrates, while actin-linked regulation seemed to be confined to vertebrate striated muscle (Lehman et al. 1973). However, this concept was no longer supportable when Lehman and Szent-Györgyi (1975) discovered both actin- and myosin-linked regulation in such lowly creatures as the nematodes; most of their obliquely striated muscles are dually regulated. Dual regulation is also found in insect flight muscles (Lehman et al. 1974) and is widespread among invertebrate phyla (Table 6.1). This perhaps is an expression of "safety in redundance," a principle employed by engineers in the design of safe control systems. Double control based on diverse regulatory mechanisms may be a safeguard in the case that one of them fails.

During evolution, the myosin of vertebrate muscle became immunologically quite distinct from invertebrate myosin (Wallimann and Szent-Györgyi 1981 a) and it seems to have lost its capacity to bind calcium selectively with high affinity, so that regulation of vertebrate muscle apparently relies entirely on troponin. We should note, however, that at least under certain special experimental conditions, it has been possible to obtain a myosin from vertebrate skeletal muscle which formed a calcium-sensitive actomyosin with troponin-free actin (Pulliam et al. 1983), but the physiological significance of these findings remains open. Molluscan muscle, too, may be dually regulated since, after all, it is capable of synthesizing troponin, according to Lehman et al. (1980). Though controversial (according to Lehman 1983), it is, however, generally held that the scallop fast adductor does not make use of this regulatory protein, but relies on myosin control (Simmons and Szent-Györgyi 1980, 1985). Myosin-linked regulation is also exhibited by molluscan smooth muscles and, as remarked by Lehman and Szent-Györgyi (1975), it may be particularly suited for regulating catch.

6.2 Catch Muscles

In contrast to the fast and rapidly fatiguing striated adductor of the scallop, the smooth slow adductor can stay contracted and keep the shells closed for many days, apparently without fatigue and with little oxygen consumption. Thus, it is extremely difficult to open a clam. To account for this phenomenon, physiologists of the late 19th century and early 20th century proposed a catch or latch mechanism (von Uexküll's "Sperrtonus" 1912) which, though unable to develop force actively, would nevertheless be capable of "locking" muscle in the contracted state. It would resist stretch without using energy, in the words of Coutance (1878) much like a knot „incapable de se faire soi-même mais capable de résister à la traction". Isn't there a striking analogy of this proposed catch mechanism to the "latch mechanism" of vertebrate smooth muscle, which will be discussed in Sects. 6.2.5 and 8.2? As will be seen, the "latch state" may be ascribed to actin-attached crossbridges that are either slowly cycling or even not cycling at all.

After years of controversy and mainly due to the work of Jewell (1959), the catch concept became generally accepted by physiologists. Still, the molecular mechanism of catch and its regulation remained puzzling, as it was proposed that catch linkages might be different from contractile crosslinkages between actin and myosin (cf. Rüegg 1971). Is it possible to solve the problem by attempting a detailed structural, physiological and biochemical comparison of catch muscle with more familiar muscles?

6.2.1 Structural Features of Catch Muscle

To understand catch contraction and its regulation, it is necessary to know its structural basis. Do muscles that are capable of catch, such as the slow adductor muscle of many bivalves and also the byssus retractor muscle of *Mytilus,* differ structurally from molluscan phasic fast muscle? As reviewed by Twarog (1976), molluscan catch muscles are unstriated. They consist of mononucleated, spindle-shaped cells which may be up to 2 mm long, but only 5 μm wide, and these contain interdigitating thick and thin filaments in an apparently disorderly arrangement (Twarog 1967). In contrast to more familiar smooth muscles, however, catch muscles contain thick filaments of an extraordinary structure (Cohen 1982), consisting of a core of paramyosin covered by a monolayer of myosin molecules. These filaments seem to be excellently adapted to the requirement of supporting high forces. They are very thick, extremely long, 70–100 μm are not exceptional (Elliott and Bennet 1982), with crossbridges that may be formed with up to 17 to 20 surrounding actin filaments. Thus, the actin-myosin overlap zone of each filament is unusually large, so that many crossbridges of one thick filament will combine forces as they pull in parallel on the same "rope" of actin, much like a ship's crew that hauls in a rope hand over hand. Because of these structural features, it is not astonishing that molluscan catch muscle holds a record in force production of well over 150 Ncm^{-2} cross-section, despite the fact that the myosin content of these molluscan muscles is many times smaller than that of fast striated muscle which develops only 30 Ncm^{-2}. Let us now consider the structural and biochemical peculiarities in more detail.

6.2.1.1 Paramyosin Filaments

As the paramyosin- and myosin-containing thick filaments of catch muscle accommodate an unusually large number of force-generating crossbridges, they must be capable of bearing much greater tensile strength than thick filaments of fast adductor of *Pecten* and vertebrate striated muscle (Szent-Györgyi and Chantler 1986). This extra strength is provided by paramyosin, which contributes to the great thickness of the filaments, in particular in the middle region where tension is greatest, while the filament ends may be tapered. Accordingly, the filament diameter, as observed in cross-sections, is quite variable, reaching extreme values from 20 to well over 100 nm. As mentioned already, the surface of the thick filaments is covered only by a monolayer of myosin (Szent-Györgyi et al. 1971; Cohen 1982). Indeed, it would have been a poor design, had the myosin enzyme been buried within a 100-nm-thick filament where substrate is unavailable.

Paramyosin, which makes up the core of the thick filament, is an enzymically inert protein, which may be even more abundant than myosin in catch muscle, and was first isolated and crystallized by Bailey (1957). It forms rod-like molecules of 220 kDa molecular weight, 130 nm length and 2 nm diameter, consisting of two entirely α-helical peptide chains (Kendrick-Jones et al. 1969). Paramyosin is widespread in the animal kingdom; it is found in muscles of all invertebrate phyla (Winkelman 1976), but in the words of Chantler (1983) "its presence does not appear to have progressed further up along the evolutionary tree than members of the protochordata." In molluscan smooth muscles abundance of paramyosin seems to be correlated with the diameter of thick filaments (Margulis et al. 1979), to a certain extent with their length (Levine et al. 1976) and with the catch property. This finding, in particular, prompted the suggestion that paramyosin might somehow be implicated in the catch.

As reviewed by Rüegg (1971), two mechanisms have been suggested to account for catch. According to one proposition, contractile force is developed by contractile actin-myosin crosslinkages, but maintained in catch due to paramyosin-paramyosin interactions. Indeed, paramyosin filaments were reported to interact and aggregate in catch, a conclusion derived from electron micrographs of cross-sections of muscle fibres fixed with glutaraldehyde in the catch state, whereas in the relaxed state the thick paramyosin filaments always appeared to be separated and surrounded by thin filaments (Rüegg 1968a). However, further investigations are required to determine whether these findings are artefacts arising during fixation or whether they represent a structural basis of a catch mechanism (cf. Bennett and Elliott 1989); alternatively catch may be caused by the "setting" of actin-myosin crosslinkages after termination of contraction (Lowy et al. 1964; Lowy and Poulsen 1982). These, however, might be modified by paramyosin (Cohen 1982). In this context, the following observations may be of note: (1) the tensile strength of artificial paramyosin threads and their colloidal state (paramyosin solubility) varies greatly in the physiological range of pH 6.5 to 7.2 (Rüegg 1961, 1964). (2) In aerated seawater the pH of catch muscle is about 7, but may reach 6.5 during catch in anaerobiosis (Ellington 1983). (3) The actin-activated myosin ATPase of catch muscle is of much lower activity than that of molluscan fast muscle (e.g. Morita and Kondo 1982), but its activity can be even further reduced by disorderly aggregated "amorphous" paramyosin (Szent-Györgyi et al. 1971). (4) This inhibitory effect of paramyosin depends on its colloidal state: when actomyosin was mixed with the orderly aggregated paramyosin core of thick filaments, its ATPase activity was not inhibited (Epstein et al. 1976a). In conjunction then, these findings might suggest that a phase change within the paramyosin filaments might not only represent an alteration in paramyosin-paramyosin interaction, but it would also influence paramyosin-myosin interaction, perhaps the actomyosin-ATPase activity and possibly crossbridge movement, as discussed by Chantler (1983) and Cohen (1982). Such a phase change of paramyosin might perhaps be induced by a change in pH or by phosphorylation of those parts of the paramyosin core which are not entirely covered by myosin (as indicated by the capacity to react with paramyosin antibodies). As found by Achazi (1979a, b) and Cooley et al. (1979), paramyosin may be phosphorylated in vitro by a cAMP-dependent protein kinase as well as by a specific paramyosin kinase obtained from molluscs.

In addition to paramyosin, the regulatory light chain of molluscan catch muscle might also be implicated in the regulation of the rate of crossbridge cycling, as pointed out by Morita and Kondo (1982). According to these authors, the regulatory light chains of catch muscles differ from those of molluscan fast muscles in that they can be phosphorylated by myosin light-chain kinase (Sohma et al. 1985). Perhaps it is the type of phosphorylated regulatory light chain attached to myosin that determines the activity of the actomyosin ATPase and possibly the rate of crossbridge cycling (but cf. also Castellani and Cohen 1987).

6.2.1.2 Thin Filaments

During contraction, and probably during catch, crossbridges projecting from thick filaments may be attached to surrounding thin filaments. As reviewed by Chantler (1983), the structure of these filaments is quite similar to thin filaments from other muscles. However, whereas all molluscan thin filaments contain actin and tropomyosin, only some of them contain troponin-C and troponin-I in small quantities. Molluscan actin belongs to the β-type; it is slightly less acidic than the α-actin from vertebrate skeletal muscle (DeCouet et al. 1980). As in all muscle, there is only one tropomyosin molecule for every seven actin monomers. All tropomyosin molecules consist of two α-helical peptide chains which form a coiled structure, 2 nm wide and 41 nm long. According to Bailey (1957), the amino acid composition is similar to that of paramyosin which was, therefore, formerly classified as a tropomyosin-like protein denoted as tropomyosin-A. Since molluscan muscle regulation is purely myosin-linked, the function of tropomyosin does not seem clear. Perhaps it has a role in consolidating the structure of very long thin filaments which must bear considerable tension in molluscan catch muscle. This tension will be transmitted either to attachment plaques of the cell membrane or to dense bodies. The latter may be homologous to Z-discs of cross-striated muscles, they may even subdivide small myofilament bundles into sarcomere-like structures (Sobieszek 1973), similar perhaps to the "mini-sarcomeres" proposed for vertebrate smooth muscle (cf. Chap. 8).

The contractile force produced by the interaction of sliding thick and thin filaments is regulated on the myosin molecule by calcium ions which either enter the cell from the extracellular space through the cell membrane or are released from internal calcium stores (Muneoka and Twarog 1983).

6.2.1.3 Calcium Stores

Most of the intracellular calcium is stored within subsarcolemmal vesicles, the cisternae, which are part of the superficial sarcoplasmic reticulum (Atsumi and Sugi 1976). In paramyosin catch muscle and other molluscan smooth muscle, the cisternae appear to be attached to the inner surface of the cell membrane by electron-dense particles, thus giving rise to a diadic structure (Nunzi and Franzini-Armstrong 1981). In some molluscan muscles, such as snail buccal retractor, which belong to the type of classical smooth muscle, the cell membrane adjoining the cisternae is invaginated, thus forming a rudimentary T-system (Dorsett and Roberts 1980; cf. also Richardot and Wautier 1971). A true T-system, however, is absent in all molluscan muscle.

When smooth muscle is stimulated, the membrane depolarization causes the release of calcium from the subsarcolemmal stores, either directly or indirectly by means of the calcium-induced calcium release mechanism triggered perhaps by calcium entering the cell through the membrane (Muneoka and Twarog 1983). According to Sugi and Suzuki (1978), application of acetylcholine causes calcium release from internal stores, which may be blocked by procaine, while the K^+-induced contracture of the byssus retractor is not blocked under these conditions (Sugi and Yamaguchi 1976). A K-contracture is, however, blocked by lanthanum (Muneoka et al. 1977) or by calcium-channel blockers (Wabnitz and von Wachtendonk 1976) and may, therefore, be ascribed to the influx of external calcium. Thus, there is a remarkable diversity in the different modes of calcium regulation of molluscan smooth muscle. Relaxation may be accomplished by sequestering the released calcium activator in the cisternae (Stössel and Zebe 1968; Heumann 1969; Atsumi and Sugi 1976) or by calcium extrusion through the cell membrane (Bloomquist and Curtis 1975a).

6.2.2 Phasic and Tonic Contraction of the Anterior Byssus Retractor Muscle of *Mytilus* (ABRM)

Because of the parallel arrangement of fibres and the ease with which it can be dissected, the anterior byssus retractor muscle of *Mytilus edulis* proved ideal for studies on the mechanics and regulation of catch (Winton 1937). When stimulated with alternating current, this muscle responds with a tetanic contraction which gives way to a rapid relaxation after cessation of stimulation. The same muscle, however, contracts in a tonic manner after stimulation by acetylcholine or direct current. These types of stimulation cause a membrane depolarization and release of calcium into the myoplasm (Bloomquist and Curtis 1975b). After cessation of stimulation the membrane repolarizes, the intracellular Ca^{2+} concentration is lowered and yet the muscle fibres do not relax quickly or completely. Instead, they enter a "catch state" which is characterized as follows. The rate of tension decay is so low that it may take more than 15 min until the force developed during contraction drops to its half-value. Under these conditions, tension is maintained very economically; to maintain a given tension, at least ten times less energy is required than in the contraction preceding the catch (Baguet and Gillis 1968). Moreover, tension does not seem to be supported by active contractile processes; rather, it is due to a rigor-like stretch-resistant state which, however, may be unlocked by serotonin (Twarog 1954) or by stimulation of inhibitory nerves which, upon stimulation, release serotonin as transmitter.

Jewell (1959) demonstrated the passive nature of catch tension in a now classical experiment (cf. Fig. 6.4). During stimulation with acetylcholine, both ends of the isolated ABRM were rigidly fixed so that the muscle developed an isometric contraction. Force developed as the contractile elements shortened, thereby stretching the elastic elements in series. When, at maximum contraction, one end of the muscle was suddenly released by 5% of its length (quick release), the force dropped immediately to baseline since the stretched elastic elements became discharged. Soon after, however, tension recovered as the activated contractile elements shortened, thereby restretching the series elastic components and redeveloping force to a new value determined by the "active state" of the preparation

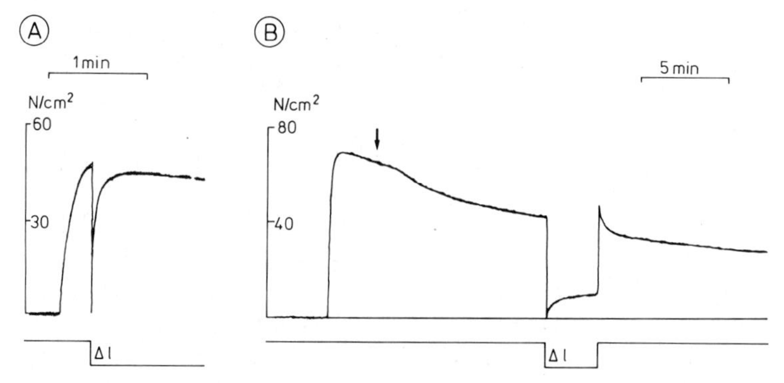

Fig. 6.4 A, B. Active state (*A*) and catch state (*B*) in a molluscan muscle. *A* Isometric contraction of isolated anterior byssus retractor muscle of the bivalve *Mytilus edulis* (ABRM) after addition of acetylcholine (10 μM) to the bathing medium (seawater). A quick release of the muscle by 5% of its length causes an immediate drop in tension (due to the discharge of series elastic elements) followed by the active redevelopment of tension by the contractile elements. *B* Isometric contraction (induced by 10 μM acetylcholine) of ABRM is terminated after washing the drug out (at *arrow*). During the very slow relaxation that follows, the ABRM is in catch. Tension is maintained passively, i.e. the "active state" is negligibly small, as indicated by the smallness of the active tension recovery following the quick release (by 5% L_0). Force can, however, be restored after restretching the fibres passively. (Rüegg 1965)

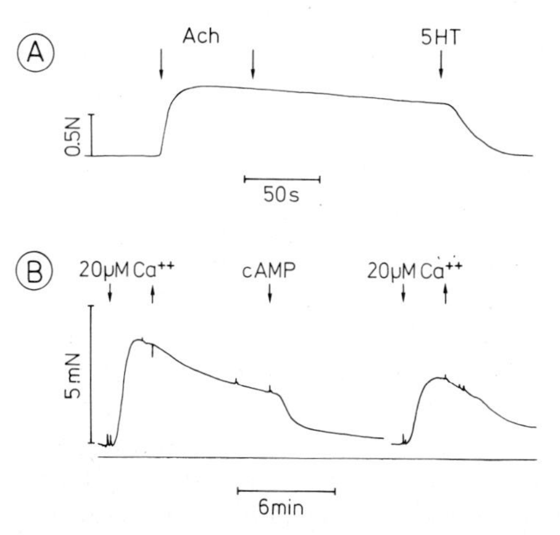

Fig. 6.5 A, B. Abolition of catch by serotonin may be mediated by cyclic AMP. *A* Catch contraction of the isolated anterior byssus retractor of *Mytilus edulis* induced by acetylcholine (added at *arrow*) and removed at *second arrow*. The slow relaxation which follows is accelerated by serotonin (5-hydroxytryptamine, 5 HT, 10^{-5}M added at *arrow*) which abolishes the catch phenomenon without inhibiting active contraction. (Rüegg 1965). *B* Catch-like contraction of skinned (demembranated) byssus retractor induced by Ca^{2+} (20 μM) which was removed (with 5 mM EGTA) at *second arrow*. The slow relaxation, which follows, is accelerated by addition of cyclic AMP (10–100 μM added at *arrow*). A subsequent calcium-induced active contraction is only slightly inhibited by cAMP. All solutions contained 5 mM Mg-ATP and imidazole 20 mM, pH 6.7 (Pfitzer and Rüegg 1982). Note that c-AMP-dependent myosin kinase may induce the phosphorylation of regulatory light chains (Sohma et al. 1988). This may abolish the catch by promoting the decomposition of the actomyosin-ADP crossbridge state [cf. Eq. (6.1)] according to M. Takahashi et al. (1988)

(cf. Ritchie 1954). When this quick release technique was performed during the prolonged maintenance of tension in the catch state, something quite different happened. The tension drop during the quick release was not followed by a quick recovery of tension; but the initial force could be restored by restretching the muscle fibre to the length it had before the release. This showed that stiffness was high, though muscle was not in an "active state." The muscle behaved as if it were in rigor; but unlike rigor mortis the ATP level remained high, of the order of 1 μmol g^{-1} muscle (Rüegg and Strassner 1963; Nauss and Davies 1966), and the catch was reversible. As already mentioned (cf. Fig. 6.5 A), it could be abolished by serotonin, an effect which is associated with a slight increase in pH (Ishii et al. 1991; Zange et al. 1990). The mechanism by which this catecholamine induces relaxation is still not fully understood (Twarog 1976). Serotonin does not alter the membrane potential (Muneoka et al. 1977), instead it stimulates the adenylate cyclase associated with the cell membrane and causes a rise of the intracellular level of cAMP which acts as a messenger between the receptors of the cell membrane and intracellular targets, such as the sarcoplasmic reticulum or the contractile apparatus (Achazi et al. 1974). Thus, cAMP may cause relaxation either by enhancing calcium sequestration or by influencing the contractile mechanism. The latter possibility seems much more likely as intracellular calcium levels are already low in catch (Atsumi and Sugi 1976; Muneoka and Twarog 1983; Ishii et al. 1989 a) which can be abolished by serotonin without any change in myoplasmic free calcium (Ishii et al. 1989 b) and also because of results obtained with skinned fibres of ABRM.

6.2.3 Analysis of Catch Regulation in Skinned Fibres – Role of Calcium and cAMP

A catch-like state could also be produced in skinned fibres from ABRM which are devoid of a functional cell membrane and sarcoplasmic reticulum and in which the composition of the interfilament space can be controlled (Rüegg and Weber 1963; Baguet 1973). Such fibres offer an excellent opportunity to determine whether, following a calcium-induced contraction, the catch-like state is indeed induced by lowering the free Ca^{2+} concentration and whether this state can be abolished by cAMP-dependent processes without altering Ca^{2+} concentration. Figure 6.5 shows the force development of skinned ABRM in ATP salt solution when the Ca^{2+} concentration is increased to 1 μM. Under these conditions, tension production and ATP-splitting rate are as high as in intact muscle, about 50 N cm^{-2} and 1 s^{-1}, respectively (Schumacher 1972). After lowering the Ca^{2+} concentration to levels of 0.1 μM or even lower, the ATPase activity is inhibited at once (Güth et al. 1984), although force declines slowly; the mechanical state now exhibits the characteristics of a catch state. When the bundle is suddenly released by 5% of the resting level, the force barely recovers, but the initial tension can be restored by simply restretching the fibre to the original length. As shown in Fig. 6.5, this catch-like state can be rapidly abolished either by the addition of cAMP (Cornelius 1982) or catalytic subunit of cAMP-dependent protein kinase (Pfitzer and Rüegg 1982). This relaxing effect of cAMP cannot be attributed to a lowering of the Ca^{2+} concentration which is already kept at a low and constant level by EGTA. The effect of cAMP cannot even be attributed to a decrease in calcium sensitivity of the contractile mechanism (Cornelius 1982); rather, it may be associated with and perhaps due to phosphorylation of contractile or regulatory proteins, such as paramyosin (Achazi 1979 a, b), the myosin heavy chain, or the regulatory myosin light chain (Sohma et al. 1985). Calcium-induced contraction is not inhibited by cAMP. These findings demonstrate that catch and active contractile state are regulated differently. Contraction is induced by increasing the free Ca^{2+} concentration, whereas catch occurs following a calcium-induced contraction when the Ca^{2+} concentration is lowered; catch, but not contraction, may be modulated by cAMP-dependent protein kinase, and may be due to a "locked" crossbridge state (AM-ADP state, cf. M. Takahashi et al. 1988) as described below.

6.2.4 A Biochemical Catch Mechanism

Crossbridge cycling during calcium-induced contraction may be associated with the formation of intermediates of the ATP hydrolysis reaction as follows:

$$\mathrm{M \cdot ATP} \rightarrow \mathrm{M \cdot ADP \cdot P} \underset{}{\overset{+\mathrm{A}}{\rightleftharpoons}} \mathrm{AM \cdot ADP \cdot P} \xrightarrow{\nearrow \mathrm{P_i}} \mathrm{AM \cdot ADP} \xrightarrow{\nearrow \mathrm{ADP}} \mathrm{AM} \xrightarrow{\downarrow \mathrm{ATP}} \mathrm{A + M \cdot ATP}. \tag{6.1}$$

In this scheme the actin-myosin-ADP complex is the force-generating state which forms when phosphate is released from the actin-myosin-ADP·P complex (AM·ADP·P) and which is itself decomposed when ADP is released (Geeves et al. 1984). Crossbridges detach as the released ADP is replaced by ATP. When the Ca^{2+} concentration is lowered at the onset of catch, the ATP-splitting mechanism just described is drastically inhibited by slowing both the release of inorganic phosphate from the myosin product complex as well as the subsequent dissociation of the myosin-ADP complex (Wells and Bagshaw 1984, 1985). This has two consequences: (1) new crossbridges will not be formed, since the reaction pathway leading to the actomyosin-ADP state (the force-generating state) is blocked. Therefore, the muscle is unable to develop force actively and to shorten. It is not in the active state. (2) Already existing linkages of actin- and myosin-ADP, on the other hand, will detach very slowly since the release of ADP is also inhibited at low Ca^{2+} concentration. Thus, relaxation of precontracted muscle is very slow when the Ca^{2+} concentration is lowered. It is tempting to speculate then that this reaction may be modulated by the interaction of either paramyosin or perhaps myosin light chains with the myosin heavy chain, as already briefly discussed before. Is it possible, for instance, that a cAMP-dependent kinase-induced covalent protein phosphorylation unlocks catch by accelerating the release of ADP? As we have seen, ADP release per se does not cause crossbridge detachment, but it leads to the formation of rigor linkages (nucleotide-free myosin-actin complex) which then rapidly detach in the presence of ATP. Still, many questions remain unanswered: Which proteins, for instance, become phosphorylated in situ by cAMP-dependent protein kinase? In which way does protein phosphorylation affect the ATP-splitting mechanism and, hence, the rate of crossbridge cycling? In this context it is interesting to note that because myosin "heads" and paramyosin molecules are in direct contact, there is a structural match between them at the surface of the thick filament, according to Cohen (1982).

Since in the absence of the regulatory light chains the reactions of ATP hydrolysis are fast (Chantler 1983), the special regulatory light chains found in catch muscles might be responsible for the extremely low myosin ATPase and for the slow detachment of crossbridges. These light-chain effects may well be modulated by cAMP-dependent phosphorylation (Sohma et al. 1985, 1988) that may cause the rapid breakdown of the Am-ADP-crossbridge state according to M. Takahashi et al. (1988).

6.2.5 Comparison of Catch and Latch in Smooth Muscle

Catch-like states have been found in muscles of many phyla, e.g. in Arthropoda (Wilson and Larimer 1968), Annelida and Nematomorpha (Swanson 1971). Many vertebrate smooth muscles in particular are also capable of exhibiting a catch-like state, which has recently been named "latch" (Murphy et al. 1983). Previous work on smooth muscle catch or "viscous tone," as it was also called, has been reported by many groups, including Winton (1930), Rüegg (1963), and A. P. Somlyo (1967), and briefly reviewed by Rüegg (1971). There seems to be general agreement now that the catch state is characterized by slow relaxation after a tonic contraction and after the cessation of an "active state." Tension is maintained with little or even no energy expenditure (Siegman et al. 1980, 1985). Naturally, the question arises whether in these vertebrate smooth muscles the catch mechanism is similar to that in molluscs. Here, as in the case of molluscan muscle, "latch" may follow an active contraction when the intracellular free calcium levels decline (Remboldt and Murphy 1986; cf. Fig. 8.8). In skinned (demembranated) smooth muscle fibres, too, the catch-like state may be reproduced by lowering the Ca^{2+} concentration.

Skinned guinea pig taenia coli muscle contracts after raising the Ca^{2+} concentration, but relaxes very slowly when it is lowered. Under these conditions, tension is maintained passively, but can be rapidly abolished by inorganic phosphate in concentrations (5 mM) that may naturally occur in smooth muscle (Schneider et al. 1981). The reduction in Ca^{2+} concentration leading to a catch-like state of skinned fibres also causes the inhibition of ATPase activity and apparently lowers the detachment rate of crossbridges (Güth and Junge 1982). It is possible that this effect of calcium reduction is mediated via the phosphorylation state of the regulatory light chains in vertebrate smooth muscle which are phosphorylated during contraction, but dephosphorylated during catch. As discussed in more detail in Chap. 8, the regulatory light chain is phosphorylated by calcium- and calmodulin-dependent protein kinase during contraction, and when calcium is removed myosin is dephosphorylated by a calcium-independent protein phosphatase fairly rapidly, while tension declines much more slowly. It seems possible, therefore, that it is not the lowering of Ca^{2+} concentration per se, but dephosphorylation of the regulatory light chains which may block the release of ADP from myosin and crossbridge detachment. Indeed, the non-phosphorylated regulatory light chain of vertebrate smooth muscle may have a similar inhibitory effect on the crossbridge-cycling mechanism as the regulatory light chain of molluscan muscle. Its phosphorylation may, according to Murphy and colleagues, be instrumental in maintaining a moderately high frequency of cycling of crossbridges in smooth muscles. Thus, when myosin becomes dephosphorylated, non-cycling or latch crossbridges may be formed (Dillon et al. 1981; Murphy et al. 1983; Aksoy et al. 1983) which, though unable to develop tension actively, are still capable of bearing tension passively. As suggested by the findings of Siegman et al. (1985) and Remboldt and Murphy (1986, presented in Fig. 8.8), the lowering of the myoplasmic Ca^{2+} concentration might cause the "latching" of crossbridges. By analogy with molluscan muscle (Sect. 6.2.4), it seems possible that at low calcium levels and following dephosphorylation of attached crossbridges, the latter are in a "locked state" (latch), because the release of the ADP product from the enzyme is inhibited (cf. Butler et al. 1990) under these conditions. However, it is not clear whether latch bridges can form only from previously phosphorylated, attached bridges by rapid dephosphorylation as proposed by Hai and Murphy (1988), or whether cooperative reattachment of dephosphorylated bridegs is also possible (cf. Somlyo and Himpens 1989). Future work will have to show whether the latch of vertebrate smooth muscle is due to the persistence of a locked crossbridge state (actomyosin-ADP state, cf. Butler et al. 1990) as in the case of catch in molluscan smooth muscle (cf. M. Takahashi et al. 1988).

6.3 Summary

In contrast to vertebrate skeletal muscle, regulation of molluscan muscle is myosin-linked. The on-off calcium switch regulating actin-myosin interaction and contraction is located on the myosin filaments. When the Ca^{2+} concentration in the interfilament space is low, the contractile machinery is kept in the off-state by the regulatory light chains which are attached to the myosin heads and inhibit actin-activated myosin ATPase. In concentrations of 1–10 μM the activator calcium is bound to the myosin heavy chain, while the unspecific metal-binding sites of regulatory light chains remain occupied by Mg. Subsequently, the light chains move; they are then no longer inhibitory and the contractile mechanism is switched on, as demonstrated by a high actin-activated ATPase activity associated with the continuous cycling of crossbridges, as well as force development or shortening of muscle fibres in the active state. When the Ca^{2+} concentration is reduced again by the calcium pump of the sarcoplasmic reticulum, the contractile mechanisms of fast phasic muscles, such as the striated adductor of the scallop or the tinted (translucent) adductor of oysters, relax as the crossbridges detach rapidly. In the slow tonic adductor, on the other hand, termination of the on-state does not immediately give way to relaxation. Here, the lowering of the Ca^{2+} concentration in the interfilament space often induces a catch state in which linkages formed during contraction break extremely slowly.

Chapter 7

The Vertebrate Heart: Modulation of Calcium Control

Throughout a lifetime the myocardium contracts and relaxes rhythmically and without fatigue, while constantly adapting its power output to the variable hemodynamic demands. In the beating heart, the contractile mechanism is switched on and off in a most regular manner by the rise and fall of the intracellular free calcium. To understand the control of the myocardium, therefore, we must first consider how the myoplasmic Ca^{2+} concentration is kept low in the "off-state" or diastole (Sect. 7.1), and how then the "on-state" or systole is initiated by calcium release from the sarcoplasmic reticulum and/or by a transsarcolemmal calcium influx (Sect. 7.2). The contractile machinery of the heart is never maximally stimulated so that the heart always has reserves; in fact, the calcium activator released into the myoplasm causes, at most, half-activation of the contractile proteins. Thus, there is enough "play" for up- as well as for down-regulation according to the variable demands of circulating blood. For instance, force or "contractility" increases when more calcium is released into the myoplasm so that a higher fraction of the calcium-binding sites on the troponin molecule become occupied (Sect. 7.3). However, the decisive parameter determining contractility is obviously not the Ca^{2+} concentration per se, but the calcium occupancy of troponin. Thus, contractility may also be altered by changing the calcium sensitivity of the regulatory protein and, therefore, the responsiveness of the myofilaments to calcium ions (Sect. 7.4). By all of these mechanisms, the activity of the myocardium may be adapted over a wide range.

7.1 Calcium-Transport Mechanisms

The contractile machinery consisting of cross-striated myofibrils is turned off when the free Ca^{2+} concentration of the myoplasm around the myofilaments is less than about 0.2 μM because, under these conditions, the actomyosin-ATPase activity and the forceful interaction of actin and myosin are inhibited by the regulatory protein system troponin and tropomyosin (Winegrad 1979). Even in the relaxed state, however, some crossbridges remain attached, as judged by X-ray diffraction analysis. These are probably non-cycling (Matsubara and Millmann 1974), but may be responsible for diastolic tone. The passive tension observed after stretching cardiac muscle must, however, be mainly ascribed to structures in parallel to the actomyosin-contractile system, such as intracellular protein filaments (e.g. connectin or titin; cf. Wang 1984, Itoh et al. 1986) or the sarcolemma and connective tissue around the myocytes.

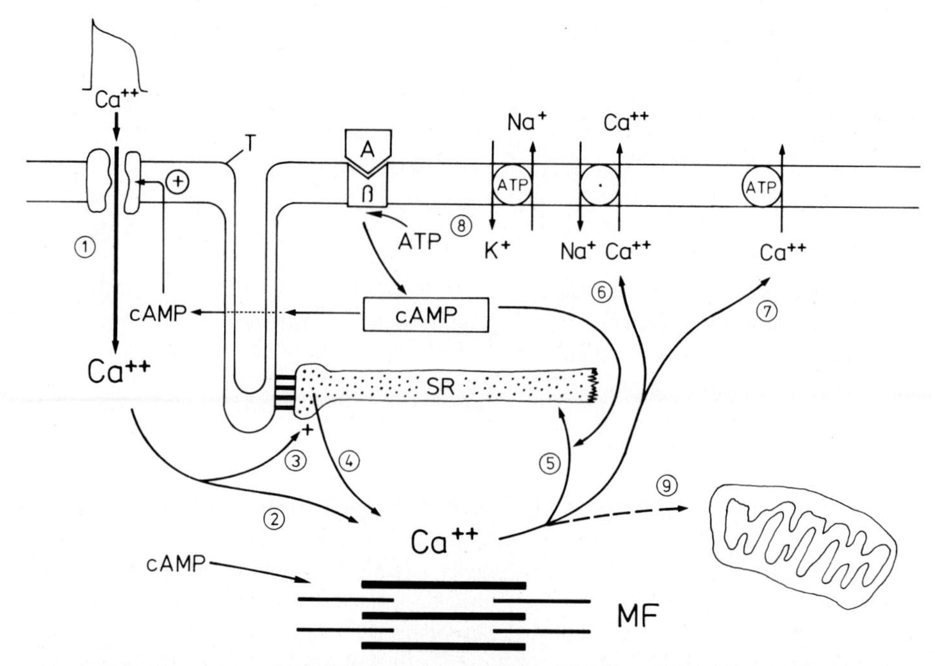

Fig. 7.1. Calcium movements in mammalian cardiac muscle. *1* Calcium influx through calcium channel; *2* direct calcium action on myofilaments; *3* calcium-trigger effect on sarcoplasmic reticulum causing calcium-induced Ca^{2+} release (*4*); *4* calcium release from sarcoplasmic reticulum (*SR*); *5* calcium reuptake into SR; *6* calcium extrusion by Na-Ca exchange; *7* calcium extrusion by active transport (membrane calcium pump); *8* Na-K pump; *9* mitochondrial calcium uptake at very high $[Ca^{2+}]$. Note that cAMP activates calcium channels and the sarcoplasmic reticulum calcium pump. *MF*=Myofilaments; *SR*= sarcoplasmic reticulum; *T*=T-tubules; *A*=agonist reaction with β-receptor

The down-regulation of the free Ca^{2+} concentration to a low level may be due to different homeostatic mechanisms of the cell (cf. Fig. 7.1) in lower and higher vertebrates. In fish and frog heart, for instance, which contains very little sarcoplasmic reticulum and mitochondria, intracellular free calcium is almost entirely controlled by the cell membrane, whereas in the heart of warm-blooded animals the sarcoplasmic reticulum may be more important.

7.1.1 Calcium Sequestration by the Sarcoplasmic Reticulum and the Role of Mitochondria

The free Ca^{2+} concentration of the myoplasm in relaxed heart muscles is approximately 0.1–0.2 μM (Marban et al. 1980), whereas the total calcium amounts up to 1 mmol kg^{-1} tissue. Much of this calcium is bound to the sarcoplasmic reticulum which is well developed in the avian heart (Sommer and Johnson 1969) and in mammalian heart (Fawcett and McNutt 1969).

Internal Membrane System in Mammalian Ventricles. As in the case of skeletal muscle, this consists of two components, the T-tubular system and the sarcoplasmic reticulum sui generis. But it is not until a few weeks after birth that T-tubules invaginate from the cell membrane at the level of the Z-lines and penetrate transversely into the interior of the muscle fibre. In contrast to their skeletal muscle counterpart, cardiac *T-tubules* are rather wide and may also extend in the axial direction (Forssmann and Girardier 1970). In the avian heart and in the atria of mammals, T-tubules are absent. Comparative studies show that the speed of contraction is not, however, correlated with abundance of the T-tubular system (Sommer and Johnson 1979), and in the mammalian ventricle its function is not yet clearly defined. Moreover, cardiac muscle cannot be activated locally by stimulation of the Z-line region of the muscle fibres containing the T-tubules. Attempted detubulation with glycerol has been unsuccessful (Strosberg et al. 1972). As reviewed by Sommer and Johnson (1979), the *sarcoplasmic reticulum* network may be located near the Z-lines, and beneath the membrane (subsarcolemmal sarcoplasmic reticulum) or within the cell. A part of this reticulum is designated as junctional since it forms contacts with sarcolemmal invaginations of the T-tubules and with the cell surface membrane. These junctions are said to be diadic since one T-tubular element and one sarcoplasmic reticulum element are linked by the electron-dense "feet" to form a twin structure or diad which may be analogous to the triads of skeletal muscle. The junctional sarcoplasmic reticulum consists of large cisternae which may correspond to the calcium-containing terminal cisternae in skeletal muscle but, in addition, cardiac muscle also contains calcium-filled cisternae that are not in contact with either T-tubules or the surface membrane. This extended junctional sarcoplasmic reticulum, as these cisternae are called, is especially abundant in avian cardiac muscle lacking T-tubules, and it is probably also involved in the calcium-release mechanism. In addition, cardiac myocytes also contain a longitudinal system of free sarcoplasmic reticulum which is responsible for the reuptake of Ca^{2+} from the myoplasm.

Calcium Storage. In cardiac muscle the sarcoplasmic reticulum is much less abundant than in skeletal muscle (Bossen et al. 1978; cf. Table 2.1). Correspondingly, the amount of calcium that can be taken up and released by the sarcoplasmic reticulum is quite small, approximately 70 nmol ml^{-1} cell volume in rat ventricle and even smaller in hearts of other species. This quantity of calcium, which is much smaller than the total calcium within the cell, is just sufficient to induce one twitch contraction after which most of it can be reaccumulated by the reticulum with a rate corresponding to the speed of relaxation. As shown by Levitsky et al. (1981), vesicles of the sarcoplasmic reticulum that are isolated from 1 g cardiac tissue can accumulate 70 nmol calcium within 200 ms.

The level of free intracellular calcium is self-regulated since any increase in Ca^{2+} concentration activates the calcium pump of the sarcoplasmic reticulum which then lowers the free calcium. The receptor for calcium is calmodulin which, when occupied by four calcium ions, activates a protein kinase in the sarcoplasmic reticulum causing the phosphorylation of the regulatory protein phospholamban (Pifl et al. 1984). A phosphorylation of phospholamban causing a stimulation of the calcium pump of sarcoplasmic reticulum may also be elicited by the calcium- and lipid-dependent protein kinase (Movsesian et al. 1984) as well as by the cAMP-dependent protein kinase which is activated by increased intracellular cAMP levels (Tada et al. 1974a, b). This self-regulating mechanism decompensates, however, when the myoplasm is flooded by entrance of calcium from the extracellular space, for instance under ischemic conditions or in other pathological states, such as hereditary cardiomyopathy of hamsters (Wrogemann and Nylen 1978). Under these conditions, the increased intracellular Ca^{2+} concentration may even inhibit calcium uptake by the sarcoplasmic reticulum which cannot cope further with the calcium overload. In this extreme situation, calcium is taken up by the mitochondria (Carafoli 1982), which may be calcified and damaged, while the energy metabolism is impaired. Ultimately, the cells die, digested by their own proteinases (Fleckenstein 1983). These examples emphasize the importance of sarcoplasmic reticulum and mitochondria in calcium homeostasis and the role of the cell membrane as a barrier that protects the myoplasm from the high "toxic" concentrations of extracellular calcium.

7.1.2 Calcium Movements Across the Cell Membrane

In the myocardium the net calcium transport across the cell membrane depends on the balance of calcium influx and outflow. Calcium ions enter the cell primarily via calcium channels, but also by means of the Na-Ca exchanger; they leave the cell again by means of the exchanger or by an active transport mechanism (Fig. 7.1).

7.1.2.1 Calcium Channels: Control of Transmembrane Influx of Calcium

The plasmalemma separates the extracellular fluid containing approximately 1 mM Ca^{2+} from the myoplasm where the free calcium is around 0.1–0.2 μM. Hence, there is, in addition to the electrical gradient, a large concentration gradient favouring the entry of calcium into the myoplasm. Whether or not calcium ions enter the cell will then depend on the functional state of calcium-channel proteins which are inserted into the lipid phase of the membrane. There are approximately 3–5 calcium channels per μm^2 membrane area (McDonald et al. 1986). Under resting conditions when the membrane is polarized, these channels are mostly closed so that calcium ions are practically prevented from diffusing into the cell. The channels open, however, when the membrane depolarizes during excitation or also under anoxic or ischemic conditions: due to the lack of energy supply, the Na-K pump of the cell membrane is inhibited so that intracellular Na^+ accumulates and K^+ is lost, and the membrane potential becomes less negative (but cf. also Allen and Orchard 1987).

Studies on Myocytes. The importance of extracellular calcium for the structural integrity of the cell membrane is demonstrated by the calcium paradox phenomenon. When isolated cardiac cells (myocytes) are prepared in saline solution with very low Ca^{2+} concentration, the cell membrane may become fairly calcium-permeable, so that on readmission of calcium the cells die because of calcium overload. If, however, the occurrence of this "calcium paradox" can be prevented, myocytes become calcium-tolerant (e.g. Powell 1985) and have been found useful for metabolic studies as well as for electrophysiological investigations using the patch clamp technique developed by Neher and Sakmann and colleagues (Hamill et al. 1981).

The "patch clamp technique" made it possible to record the potential-dependent opening events of a single calcium channel within a very small membrane area, as shown in Fig. 7.2 (Trautwein and Pelzer 1985; McDonald et al. 1986). When a constant (clamped) voltage is applied across such a membrane patch, current is flowing in an all-or-none-pulsed manner depending on whether the channel is open or closed. The open-probability of the calcium channels is a function of the membrane potential. The gates of the channel open only when the potential becomes less negative than about −30 mV, but during prolonged depolarization, channels become inactive and close again. On a millisecond time scale single channel recordings may then show closely spaced bursts of brief current pulses corresponding to intermittent channel openings and closings. During longer periods of approximately 0.1–1 s, several bursts of channel openings may occur in succession or they may be followed by periods of rest due to channel inactivation. Because of the stochastic nature of opening and closing of channels, the calcium current flowing into the cell depends on the ensemble of the average of the single

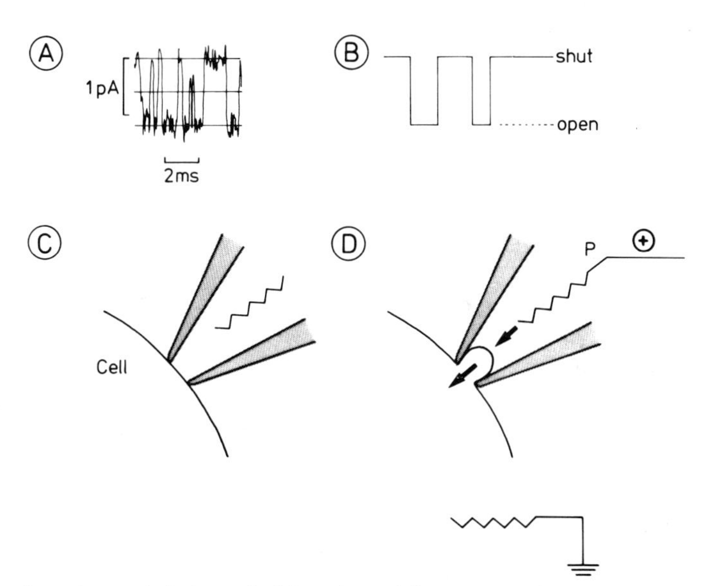

Fig. 7.2 A–D. Recording of openings and closings of calcium channels in myocytes of guinea pig ventricle. *A* Cell-attached recording of elementary currents through a single calcium channel. *B* Interpretation of recordings in terms of channel openings and closings. There are two closed states, one of them being refractory. The net calcium current is NpI, where N is the total number of active channels, p is the open-probability and I is the elementary current. Upon membrane depolarization the open-probability greatly increases. *C* Method of patch clamp technique: Pipette (electrode filled with ion carrier, e.g. $BaCl_2$, and connected to current voltage converter) in contact with clean surface of cell. *D* Upon suction a tight "Giga-ohm" seal is formed between membrane patch and the pipette (*P*) which may be used for cell-attached recordings. If a voltage is applied through the pipette across the membrane patch, a membrane current will flow from the pipette into the cell, if and only if, a (single) ion channel is in an open state. (Trautwein and Pelzer 1985)

calcium currents, i.e. on the number of channels, the probability of their being open, as well as on the calcium-driving potential. The last is the difference between the membrane potential and the calcium equilibrium potential (E_{Ca}), as calculated from the Nernst equation, $E_{Ca} = (RT/2F) \cdot (\ln[Ca]_o/[Ca]_i)$.

Certain dihydropyridines and other organic compounds listed in Table 7.1 (p. 187) are called *calcium antagonists,* because they block, figuratively speaking, the calcium channels. They bind with high affinity at the inside of the cell membrane to the channel proteins, thereby reducing the mean open lifetime of the channel. With some calcium antagonists, like verapamil, this effect is particularly large when the membrane is depolarized (Reuter 1984). In this way, calcium antagonists reduce the calcium influx into the cell even under hypoxic conditions, such as coronary occlusion, and they reduce intracellular necrosis caused by calcium overload of the cell, in particular by the calcium activation of intracellular proteases (Fleckenstein 1983). There are at least two types of calcium channels, one of them being gated at a membrane potential of -50 mV and the other requiring greater depolarization, but deactivating more slowly (Nilius et al. 1985; Bean 1985).

7.1.2.2 Calcium Extrusion by the Na-Ca Exchanger and the Sarcolemmal Calcium Pump

As the cell membrane is somewhat leaky to calcium, these ions must be continuously expelled from the cell either by an ATP-driven calcium pump of the sarcolemma (Caroni et al. 1980) or, and perhaps more importantly, by the Na-Ca exchange mechanism (cf. Sect. 2.3.5.2) described by Reuter and Seitz (1968).

The Na-Ca Exchanger. This system uses the free energy released by Na^+ when they enter the cell in exchange for Ca^{2+}. The overall driving force on Na^+ then equals:

$$E_{Na} - E_M ,$$

where E_{Na} is the sodium equilibrium potential and E_M is the membrane potential. This force is used to overcome the electromotive force acting on Ca^{2+}, which is

$$2(E_{Ca} - E_M) ,$$

since Ca is divalent. Only if the electromotive force on Na^+ is greater than the imposing force on calcium, can the latter be expelled by a rotary movement of the exchanger, coupling the inflow of three sodium ions to the outflow of one calcium ion. No net movement of Ca^{2+} and Na^+ occurs under equilibrium conditions when the electromotive forces on Ca^{2+} and Na^+ are exactly balanced.

Under these conditions,

$$n(E_{Na} - E_M) = 2(E_{Ca} - E_M) , \qquad (7.1)$$

where n, the number of Na^+ ions exchanging for Ca^{2+} ions, is 2 to 3 (Pitts 1979). The equilibrium potentials E_{Ca} and E_{Na} depend, of course, on the ratios $\frac{[Ca]_o}{[Ca]_i}$ and $\frac{[Na]_o}{[Na]_i}$ and may be calculated using the Nernst equation. Substituting E_{Ca} and E_{Na}

in Eq. (7.1), Blaustein (1974) calculated the intracellular free calcium under equilibrium conditions:

$$[Ca]_i = [Ca]_0 \cdot \frac{[Na]_i^n}{[Na]_0^n} \exp[(n-2) E_M F/RT], \tag{7.2}$$

where $[Ca]_i$, $[Na]_i$, $[Ca]_0$, and $[Na]_0$ are the intracellular and extracellular concentrations of sodium and calcium ions respectively, E_M is the membrane potential and F, R, and T have their usual meaning.

Assuming $n=3$, the Blaustein equation predicts that at the known physiological concentrations of Ca^{2+} and Na^+ outside and inside the cell, the exchange mechanism is in equilibrium at a membrane potential of about -40 mV. At more positive potentials, calcium enters the cell in exchange for sodium leaving the cell and intracellular calcium increases, but at more negative potentials calcium ions are extruded in exchange for extracellular sodium. Since three positively charged sodium ions are probably exchanged for two positive charges carried by a calcium ion, there is a net inward current in relaxed or repolarizing heart due to this electrogenic Na-Ca exchange (Mullins 1979; Reeves and Sutko 1980). As this calcium movement is completely dependent on the energy supply by the electrochemical gradient of Na^+, its direction may be reversed by lowering the extracellular or increasing the intracellular Na^+ concentration; then the intracellular free calcium tends to rise. In cardiac muscle of mammals, the sarcoplasmic reticulum and mitochondria, as well as the calcium pump of the sarcolemma, might, however, cope with the influx by the Na-Ca exchange mechanism. In frog or fish heart, on the other hand, where the sarcoplasmic reticulum is sparse, the intracellular free calcium would increase dramatically, thus causing a contracture.

Studies with Isolated Sarcolemmal Vesicles. These investigations have allowed the biochemical characterization of the calcium pump (Caroni et al. 1980, cf. Sect. 2.3.5.2) and of the Na-Ca exchange mechanism (Reeves and Sutko 1979). Vesicles may form spontaneously from isolated sarcolemmal fragments and may accumulate up to 20 nmol calcium s^{-1} and mg^{-1} protein in exchange for 60 nmol sodium under optimal conditions (Reinlib et al. 1981). At a Ca^{2+} concentration of about 1 μM outside the vesicles and at a physiological concentration of Na^+, the sarcolemmal, ATP-driven calcium pump exhibited only 3% of the maximal calcium uptake activity of the Na-Ca exchange mechanism. However, at resting values of free calcium (about 0.2 μM or 0.1 μM) the exchange mechanism is only 5% activated, whereas the calcium pump is still 50% activated. Chapman (1983) calculated, therefore, that in the resting myocardium about two-thirds of calcium extrusion is due to the exchange mechanism and one-third may be due to the sarcolemmal calcium pump. At a high internal Ca^{2+} concentration, therefore, as early in relaxation, calcium extrusion must be due predominantly to the exchange mechanism rather than the calcium pump.

Because of their calcium dependence, both the calcium pump as well as the Na-Ca exchange mechanism are part of a feedback mechanism. Any rise in the intracellular free Ca^{2+} concentration would automatically stimulate calcium extrusion, and any fall below a critical level might shut off the pump as well as the exchange mechanism. By controlling the permeability of Ca^{2+} as well as the activity of the calcium-extrusion mechanism, the cell membrane may, therefore, play an important regulatory role in calcium homeostasis. Additionally, the external side of the cell membrane also binds calcium ions in the vicinity of the calcium channels so that they are rapidly available for transmembrane calcium influx when the muscle fibre is excited (Philipson et al. 1980; Langer et al. 1982).

7.2 Calcium Movements as the Link Between Excitation and Contraction

In contrast to the heart of many invertebrates, the vertebrate heart beat is not elicited by nervous impulses, but by pacemaker cells of the sinoatrial node. The action potentials generated by these cells then spread along the myocardium of the atrium to the atrioventricular node and then along the bundle of His and the conducting system of Purkinje fibres to the working myocardium of the ventricles. The action potentials propagate from cell to cell via low resistance pathways, the nexus or gap junctions of the intercalated discs. In isolated trabecular muscles of the ventricle or papillary muscle, action potentitals followed by twitches may be elicited by "pacing" the preparation with brief electrical shocks.

As cardiac fibres are electrotonically coupled by gap junctions, the heart muscle contracts, in response to a just supra-threshold stimulation of even a few cells, always with all of its fibres or it does not respond at all if the threshold is not reached. This is the "all-or-none response" of the heart. As the action potential and, hence, the "refractory period" may last as long as the contraction phase, cardiac muscle cannot usually be tetanized by repetitive stimulation except at very high extracellular Ca^{2+} concentrations. The importance of calcium for heart muscle contraction has been recognized since Ringer's discovery (1883) that the isolated heart ceases to beat in the absence of calcium in the bathing or perfusion solution. Since then it has become increasingly clear that calcium ions do indeed play a key role in cardiac excitation and excitation-contraction coupling.

7.2.1 Action Potential and Calcium Entry

Several current components generate at least three phases of the action potential of the working myocardium. The first phase rises within a millisecond from the resting membrane potential of about -80 mV to an overshoot at approximately $+20$ to $+40$ mV. This peak is followed by rapid repolarization to a plateau (phase 2) around $+20$ mV which is then maintained for a fraction of a second. Finally, the initial membrane potential is restored during repolarization (phase 3). As in the case of skeletal muscle, the rapid reversal of the membrane potential during the first millisecond may be ascribed to a fast inward current of sodium ions initiated when the membrane potential becomes less negative than the "threshold potential" of about −60 mV. Though this Na current is inactivated within a few milliseconds, this short time is sufficient to gate potential-dependent calcium channels through which a stream of Ca^{2+} enters the cell driven by the electrochemical gradient of calcium ions that carry the *"slow" inward current* denoted as SI (Beeler and Reuter 1970a; cf. also Rougier et al. 1969). More recently, it has been possible to dissect this calcium current into a faster and slower phase (cf. Noble 1984). With time, K conductance increases, too, thus giving rise to a K outward current that balances the calcium inward currents after the peak of the action potential. Throughout this state of balance membrane depolarization is maintained at approximately 20 mV (the plateau phase of the action potential) until the calcium current also becomes inactivated. Then the membrane is rapidly repolarized by a (cAMP-dependent) chloride current and by the potassium outward current. In Purkinje fibres, the latter may be activated by the increased intracellular free Ca^{2+} concentration at the end of the plateau phase (Isenberg 1977). The inward "plateau current" was identified as a calcium current by the finding that, unlike the fast current, it did not depend on the extracellular Na concentration (cf. Isenberg 1982) and could not be blocked by tetrodotoxin, but was a function of the extracellular Ca^{2+} concentration (Beeler and Reuter 1970 a). These discoveries were a milestone in the understanding of the pathways leading from excitation to contraction, as they seemingly explained the calcium dependence of heart muscle contraction. Later it was shown, however, that the role of slow inward current was quite different in cardiac muscle of various classes of vertebrates. In fish and amphibian hearts, the calcium current seems to be responsible mainly for causing a maintained membrane depolarization that would then allow the Na-Ca exchanger to drive the Ca^{2+} into the cell in an amount sufficient to elicit contraction. In mammalian and avian hearts, on the other hand, the inflowing calcium ions, with a few exceptions discussed below, do not seem to activate the contractile machinery directly, but exert a trigger function in as much as they induce calcium release from the sarcoplasmic reticulum. Additionally, inflowing calcium serves to replenish the intracellular calcium stores, thus facilitating calcium release in subsequent contractions. These aspects will now be considered in more detail.

7.2.2 Activation of Myocardial Myofilaments of Lower Vertebrates by Transmembrane Sarcolemmal Calcium Influx

In contrast to hearts of higher vertebrates, frog ventricles contain a sparse sarcoplasmic reticulum which is mostly situated beneath the cell membrane. Thus, activation has been ascribed to the entrance of Ca^{2+} through the cell membrane rather than to calcium release from the sarcoplasmic reticulum. Since the distance between the cell membrane and the centre of the fibre is not more than 5 μm, diffusing calcium ions would reach their target in time to elicit the delayed and slow tonic contraction characteristic of the frog heart (Morad and Goldman 1973). Much of the calcium entry actually occurs during the action potential which may last up to 1 s and which, as pointed out above, is characterized by a long plateau phase supported by the inward calcium current (Rougier et al. 1969).

A relationship between calcium inward current and contraction has been demonstrated by many investigators. Thus, application of stimulating currents during the plateau phase or application of noradrenaline prolongs the plateau, and enhances the duration of the inward current, thereby causing an increase in the duration and extent of force development which is elicited by the action potential. In contrast, acetylcholine abbreviates the action potential and reduces contractile force, as reviewed by Morad and Goldman (1973).

7.2.2.1 Role of Calcium Inward Current

To find out whether the calcium currents rather than the membrane potential per se determine the force development, Morad and Orkand (1971) studied the relationships among membrane potential, membrane current and force in a more systematic manner. Using voltage-clamp techniques, these authors suddenly altered the membrane potential from its resting value of about −80 mV to a potential between −50 mV and +80 mV. A potential jump to −10 mV, for instance, opened calcium channels, caused a slow inward current, and, after a delay of about 80 ms, a tonic rise in tension which was maintained throughout the depolarization. When the external Ca^{2+} concentration exceeded the physiological value of about 1 mM, the tonic response was preceded by a transient phasic component that commenced soon after the stimulus and which, in contrast to the tonic contraction, was not maintained. Larger depolarizing steps increased the calcium conductivity of the membrane, the inward current and force development. The voltage dependence of the calcium current was "bell-shaped" (Pizarro et al. 1985). When the membrane potential reached values more positive than 10 mV, the intensity of the calcium current and the extent of phasic contraction declined, reaching zero values at membrane potentials between +60 mV and +80 mV. Assuming that calcium current and phasic contraction were coupled, this decline was not unexpected, for the calcium current, of course, depends on the calcium conductance as well as on the driving potential for Ca^{2+} as follows:

$$I_{Ca} = g_{Ca}(E_M - E_{Ca}) . \qquad (7.3)$$

Assuming a free Ca^{2+} concentration of 0.2 mM outside and 0.5 μM inside the cell, the calcium equilibrium potential E_{Ca} may then be calculated from the Nernst equation to be +75 mV. At this potential then electromotive force $(E_M - E_{Ca})$ driving the Ca^{2+} across the cell membrane should be zero. But surprisingly, tonic force production was maximal. This experiment clearly suggested that unlike phasic contraction, development of tonic force cannot be induced by Ca^{2+} carried by the slow inward current, but must be caused by some other effects of the change in membrane potential. This conclusion was strengthened by the finding that calcium-channel blockers, such as dihydropyridins, inhibit the phasic contractile response, but not the tonic response of frog heart. The search for the source of activator calcium, under these conditions, has led to a controversy. On the one hand, Ca^{2+} may be released from intracellular calcium stores, such as the subsarcolemmal vesicles, since force production does not seem to be paralleled by a corresponding calcium influx (Pizarro et al. 1985; cf. Fig. 7.3). Additionally, or perhaps alternatively, a trans-sarcolemmal calcium influx may be brought about

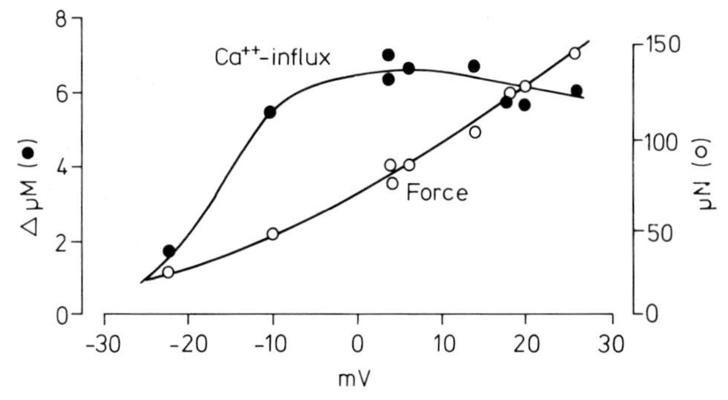

Fig. 7.3. Dependence of calcium influx (●) into frog heart ventricle cell and contractile force (○) on the membrane potential in voltage clamp experiments (*abscissa*). The calcium inward movement leads to a measurable lowering of extracellular calcium levels (in the vicinity of the cells, cf. left *ordinate*), which can be estimated with calcium probes. Note that at low negative, but not at positive, membrane potentials force is correlated with calcium inward movement. (Replotted from Pizarro et al. 1985)

by the Na-Ca exchange mechanism already described in Sect. 7.1 (cf. Chapman 1983).

7.2.2.2 Role of Na-Ca Exchange in Contraction and Relaxation

Though still controversial, this exchange mechanism may be electrogenic since one Ca^{2+} ion is probably exchanged for three Na^+ ions (Mullins 1979). Consequently, an outward Na current coupled with an influx of calcium must be facilitated by the action potential when the outside of the cell becomes negative with respect to the cell interior. This exchange will then continue until equilibrium is reached and under these conditions, the Ca^{2+} concentration may be calculated from the Blaustein equation [Eq. (7.2)]. Other factors being equal, intracellular free calcium should depend on the ratio $[Ca]_0/[Na]_0^n$, where n is between 2 and 3 and $[Ca]_0$ and $[Na]_0$ are the extracellular concentrations of calcium and sodium. This prediction is in agreement with studies showing that in frog heart the calcium-dependent force is constant as long as the ratio $[Ca]_0/[Na]_0^n$ is also kept constant (Lüttgau and Niedergerke 1958), while withdrawal of extracellular sodium causes a contracture. This suggested that the intracellular free calcium may indeed be related to the ratio $[Ca]_0/[Na]_0$ (Benninger et al. 1976). As predicted [Eq. (7.2)], intracellular free calcium and force development can, of course, be enhanced by increasing the extracellular calcium levels.

After the end of the action potential plateau the cell membrane repolarizes when the inward calcium current ceases. Then an outward current, due partly to potassium efflux and partly to the Na-Ca exchange mechanism, may contribute to a rapid repolarization of the cell membrane. In frog heart, this potential change coincides with relaxation since the Na-Ca exchange reverses as soon as the membrane potential becomes more negative than -40 mV. Under these conditions then, the inward movement of sodium ions carrying the inward current is favoured which, in turn, causes a net calcium efflux followed by muscle relaxation. The energy required for the process stems, as discussed already in Sect. 7.1.2, from the electrochemical gradient of sodium which is maintained as long as sodium ions are pumped out of the cell by the ATP-driven electrogenic sodium pump. If the pump is inhibited, for instance by vanadate or cardiac glycosides, the frog heart may fail to relax. As predicted by Eq. (7.2), intracellular free calcium and, hence, force also increases when the extracellular sodium concentration is lowered.

Changes in the osmolarity and, hence, in the sodium concentration of the extracellular medium occur naturally in fish that inhabit estuarine waters where they are likely to be exposed to rapid changes in salinity. When fish move from seawater to fresh water the osmolarity of the plasma decreases from 350 mOsmol l^{-1} to 250 mOsmol l^{-1} mainly as a result of a lower sodium chloride concentration (Vislie and Fugelli 1975). At this low salinity, the rate of calcium influx may be increased, while the rate of calcium efflux is decreased causing an intracellular rise in free calcium. Under these conditions, the resting tension as well as the systolic force of fish heart increases considerably according to Gesser and Mangor-Jensen (1984). But the pumping function of the heart is impaired because of incomplete relaxation and ventricular filling.

In conclusion, the early phasic part of force production in cardiac ventricular muscle of lower vertebrates appears to be elicited by the inward calcium current during the action potential, whereas the later tonic response seems to be due either to calcium release from intracellular stores or to calcium entering the cell via a Na-Ca exchange mechanism which, when operating in reverse, is also responsible for relaxation. With regards to the heart of higher vertebrates, it seems unlikely that the Na-Ca exchanger supplies the activator calcium for the myofilaments, as proposed by Langer and colleagues (1982). Rather, it may have a more indirect effect by influencing the filling of the sarcoplasmic reticulum with calcium at least under conditions where the ratio $[Na]_i/[Na]_0$ is increased (cf. Sect. 7.3.2).

7.2.3 Calcium Release from the Sarcoplasmic Reticulum During Contraction of Mammalian Hearts

A direct activation of myofilaments by calcium ions entering the myocardium cell through the sarcolemma has been proposed by Langer et al. (1982). Alternatively, and even more importantly, activator calcium may be released from the sarcoplasmic reticulum. The relative importance of these pathways, illustrated in Fig. 7.1, may, however, differ widely in the hearts of different mammals (Bers 1985).

As proposed by New and Trautwein (1972), it may be the inward calcium current per se that causes the release of calcium from the sarcoplasmic reticulum in a quantity sufficient to activate the myofilaments. Such a release mechanism would be particularly important in Purkinje fibres, avian heart and mammalian atria which lack a functional T-tubular system. Trautwein's hypothesis, which has been greatly elaborated and extended by Fabiato (1983), is now supported by a wealth of data showing that (1) there is a close relationship between calcium inward current and contraction, and (2) the myoplasmic free Ca^{2+} concentration expected as a result of calcium entry through the cell membrane may be sufficient to induce calcium release from the sarcoplasmic reticulum, but insufficient to activate the myofilaments directly.

7.2.3.1 Relation Between Calcium Current and Contraction

The calcium current can be enhanced by increasing the extracellular Ca^{2+} concentration, a manoeuvre which also enhances within a few seconds the contractile force in superfused thin bundles from ventricular trabeculae paced by stimulation at 20 min^{-1} (Kitazawa 1984). The quantitative relationship between calcium current, extent of contraction and membrane potential has been studied in isolated myocytes (Isenberg et al. 1985) as well as in multicellular preparations using the voltage-clamp technique (Beeler and Reuter 1970b; Maylie and Morad 1984). Both calcium current and contraction, which require a similar threshold depolarization, depend on the membrane potential in a similar bell-shaped manner, maxi-

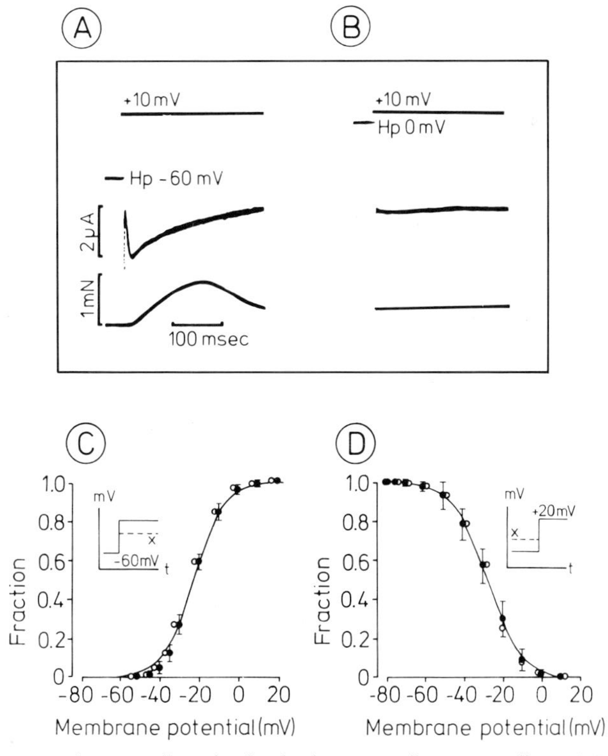

Fig. 7.4 A–D. Inward calcium current and contractile activation in the mammalian myocardium. *A, B* Effect of depolarizing pulse (to +10 mV membrane potential) on calcium inward current and contraction. The holding potential before the depolarizing pulse was −60 mV in *A* or 0 mV in *B*. (After Trautwein et al. 1975). *C* Dependence of relative force (o) and calcium conductance (●) of cell membrane on the extent of membrane depolarization or membrane potential (*x*, given on *abscissa*). Membrane potential prior to depolarization = −60 mV. *D* Steady state inactivation. Dependence of relative force (o) and calcium conductance (●) on membrane potential, (*x*, given on *abscissa*) prior to the contraction inducing depolarization (to +20 mV). *Inset:* Sequence of membrane potential changes in *C* and *D*: −60 mV to *x* or *x* to +20 mV. (After Trautwein et al. 1975)

mal contraction and maximal calcium current being elicited when the membrane is depolarized to +10 mV (New and Trautwein 1972). But as the inside of the cell is clamped to even more positive potentials, force and calcium current decrease in parallel, suggesting that the latter determines the contractile strength with two exceptions. These exceptions are the fast contractile response of single myocytes with calcium-loaded sarcoplasmic reticulum (Isenberg et al. 1985) and the so-called tonic contraction, sometimes obtained after prolonged depolarization in cardiac muscle. Both contractile responses may occur at potentials around +100 mV when the inward calcium current is not flowing. Direct depolarization-induced calcium release from the sarcoplasmic reticulum (mediated by couplings between the T-system and the sarcoplasmic reticulum) or calcium entry by the Na-Ca exchange mechanism may be involved in these cases.

To illustrate further the parallelism between contractility and calcium current, Fig. 7.4 shows the results of voltage-clamp experiments of Trautwein et al. (1975) in which the membrane is suddenly depolarized from −60 mV to +10 mV (Fig. 7.4 A) or alternatively from 0 mV to a pulse potential of +10 mV (Fig. 7.4 B). Though the final membrane potential is the same in both cases, the responses are quite different. In the first case, a pronounced calcium current, shown as a transient downward deflection on the oscilloscope tracing, and a prominent contractile response may be observed. In the second case (Fig. 7.4 B), both phenomena are missing because the calcium-conducting system had been inactivated during the prolonged depolarization to 0 mV. However, a partial or even complete reactivation occurs if the membrane is repolarized for a brief moment, e.g. about 0.5 s, prior to the application of the +10-mV pulse voltage. The extent of recovery or rather the extent of remaining inactivation then is a function of the membrane potential prior to the application of the +10 mV-pulse potential, as shown by the steady-state inactivation curve in Fig. 7.4 D. This figure also shows that the steady state inactivation of the calcium current and calcium conductance, on the one hand, and of the contractile capacity, on the other hand, are related. Obviously, the "parallelism" between calcium conductance and contractile capability (Fig. 7.4 C, D) suggests a possibly causal relationship between calcium entry and contraction.

7.2.3.2 Calcium-Induced Calcium Release

After demonstrating a quantitative relationship between inward calcium current and contraction, the question arises as to whether the amount of calcium entering the cells with the current is sufficient to activate either the myofilaments directly or to release calcium from the sarcoplasmic reticulum in a quantity sufficient for activation. According to Isenberg (1982), the calcium entry following depolarization increases the total cellular calcium concentration by as much as 25 μM within 100 ms. Under certain conditions, e.g. after a period of rest, this calcium may be sufficient to activate the myofilaments directly (cf. Reiter et al. 1984). Normally, the situation may be more complex.

Fabiato (1983) calculated that the total calcium in the cell would rise from 7 to 14 μmol l^{-1} cell water during the action potential. Since much of this calcium would be bound to internal membranes or proteins, the free calcium would, ac-

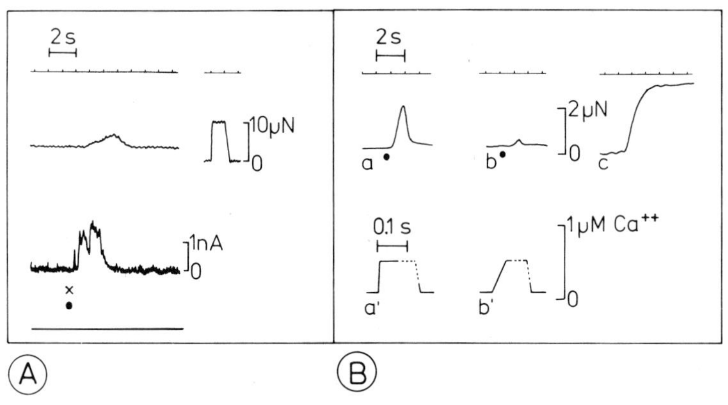

Fig. 7.5 A, B. Evidence of calcium-induced calcium release in cardiac muscle. *A* Method of investigating the effect of calcium ions on calcium release from the sarcoplasmic reticulum of a mechanically skinned cardiac cell (20-μm-long dog Purkinje cell) perfused first with relaxing solution pCa 7 and then for exactly 0.1 s with a solution of pCa 6.3 (x). This "calcium pulse" evokes a calcium release causing a large calcium transient detected by aequorin luminescence (noisy tracing) as well as a transient increase in force. The fibre relaxes as the calcium is taken up by the sarcoplasmic reticulum. Calibration *bar* for force: 10 μN and for the photomultiplier-current measuring luminescence: 1 nA. (Fabiato 1985a). *B* A fast (*a'*), but not a slow rise in Ca^{2+} in the perfusion fluid (*b'*) elicits a calcium release (not shown) causing a transient contractile response (cf. *a* and *b*). At (*c*) the skinned fibre was directly and maximally stimulated by perfusion with saline containing high Ca^{2+} (pCa 4.2). All perfusion solutions contained ATP and Mg^{2+} ions (pMg-ATP 2.5, pMg 2.5, pH 7.1, 0.17 ionic strength). (After Fabiato et al. 1985a, b)

cording to Fabiato, only rise from 0.1 to 0.2 μM. This concentration would be below the threshold required for activating the myofilaments; however, it would be sufficient to release calcium from the sarcoplasmic reticulum, as demonstrated most elegantly by Fabiato (1982b, 1983, 1985a, b, c).

As an experimental model, Fabiato used single avian or mammalian myocytes from which the outer cell membrane had been removed by microdissection, but in which the contractile myofibrils, as well as the sarcoplasmic reticulum, had been left intact. These mechanically skinned fibres were then incubated in a medium resembling the intracellular milieu containing magnesium, ATP and various salts in the appropriate concentrations as well as 0.1 mM EGTA to keep the Ca^{2+} concentration below 0.1 μM. Figure 7.5 shows that a sudden increase of the Ca^{2+} concentration to 0.2–0.5 μM Ca^{2+} elicited a transient contraction, provided that the sarcoplasmic reticulum was intact; but when it was destroyed by detergents, no contraction occurred. To explain the phenomena, Fabiato suggested that the rise in free calcium was so small that it would not activate the myofilaments directly, but would nevertheless release stored calcium from the sarcoplasmic reticulum. This calcium would then cause a contraction followed by relaxation as the released calcium was again recuperated by the sarcoplasmic reticulum. The size of the postulated, brief calcium transient could actually be estimated from the luminescence of the photoprotein aequorin which had been added to the incubation medium in order to monitor the free Ca^{2+} concentration from moment to moment. As shown in Fig. 7.5, a fraction of a second following the sudden exposure of the skinned fibre to 0.5 μM Ca^{2+} the free calcium rose transiently to 1 μM, a concentration similar to the free Ca^{2+} concentration measured by Allen and Kurihara (1980) in a living cardiac cell during a twitch contraction. Moreover, the contraction elicited in the skinned fibre was similar in size to the twitch contraction elicited by a current pulse in the same preparation before the skinning-procedure. Fabiato's experiments, therefore, showed that the amount of calcium entering the cell during the action potential and causing a rise in intracellular free Ca^{2+} concentration to 0.2–0.5 μM could induce the release of sufficient calcium from the sarcoplasmic reticulum to cause a twitch contraction of physiological strength and duration. It is important to note, however, that during these contractions the contractile machinery is only half-maximally activated, since skinned preparations may develop twice as much force if the Ca^{2+} concentration surrounding the myofilaments is increased to 10 μM (Fabiato 1981; cf. also Fig. 7.5). These experiments clearly demonstrate that the beating heart may have a large contractile reserve.

Fabiato (1985 b) proposed that the calcium channels of the sarcoplasmic reticulum were cycling among three states, a closed state, an open state and a refractory state, which are dependent on calcium in different manners. He noted that only a rapid increase of the Ca^{2+} concentration released calcium from the sarcoplasmic reticulum (Fig. 7.5 B). A slow increase in Ca^{2+} concentration, however, had little effect since the calcium-carrying system would become refractory within less than 1 s after the onset of the calcium rise, and it would not recover unless the Ca^{2+} concentration were again lowered considerably. The extent of recovery or, for that matter, the degree of remaining "steady-state inactivation," therefore, depend on the Ca^{2+} concentration to which the release sites of the sarcoplasmic reticulum are exposed prior to the application of the calcium pulse.

7.2.3.3 Replenishment of Calcium Stores

In Fabiato's experiments with skinned fibres, a slow rise in free calcium not only failed to induce a calcium release, but actually stimulated calcium uptake and thus served for the refilling of the calcium stores in the sarcoplasmic reticulum. The loading of the stores with calcium then depends on the time during which they are exposed to the slightly elevated Ca^{2+} concentration, and there is evidence that the calcium taken up by the longitudinal or free sarcoplasmic reticulum reaches the release sites in the terminal cisternae or in the junctional sarcoplasmic reticulum only after a delay so that it is not immediately available as activator calcium. When the sarcoplasmic reticulum is filled to a greater extent, a subsequent calcium pulse will then induce a larger release of calcium and cause a greater contraction than under normal conditions of filling (Fabiato 1985 c). If, on the other hand, the sarcoplasmic reticulum is kept at a low Ca^{2+} concentration for a prolonged time, it loses much of its calcium by leakage so that a subsequent calcium pulse will only induce a small contraction. These in vitro observations may be relevant for our understanding of time-dependent activation in cardiac muscle, to be discussed in Sect. 7.2.4.

As mentioned already, the inward calcium current could be dissected into an earlier, fast-inactivating component coinciding with the very early phase of the action potential plateau and a later, slow-inactivating calcium current flowing during the action potential plateau (cf. Noble 1984). According to Fabiato, the fast component may be responsible for calcium-induced calcium release, whereas the slow component may supply the calcium required for replenishment of the leaky calcium stores. Thus, a prolongation of the action potential plateau should improve the calcium filling of the sarcoplasmic reticulum by the slow component of the inward calcium current so that a subsequent action potential might cause a greater calcium release and a larger contraction.

7.2.4 Delayed Effects of Excitation on Contraction

Duration of the Action Potential. As mentioned in the preceding section, a filling of the calcium stores by the calcium inward current and perhaps also by the Na-Ca-exchange mechanism may provide a mechanism for influencing contractility in an indirect manner following a prolongation of the action potential. Whereas a change in duration of the action potential has an immediate positive inotropic effect on the contractile force of frog heart, its effect on contractility of the mammalian myocardium is delayed: a considerable time elapses between the change in the action potential duration and its visible effect on the contractile mechanism (Wohlfart and Noble 1982). For instance, if papillary muscles are stimulated at regular intervals, e.g. 1 s^{-1}, the action potential and the accompanying contractions are quite similar and reproducible. A prolongation of an action potential by injection of currents, however, does not lead to an immediate increase in the strength of the accompanying twitch, but influences the next following contraction; the increase in force is even observed if the action potential associated with that contraction is again normal (Antoni et al. 1969; Wood et al. 1969). Obviously, the heart remembers, so to speak, the previous beat. This result is, of course, just what one would expect from the hypothesis, elucidated in Fig. 7.6, that the calcium entering during the plateau phase serves to increase the time-dependent filling of the calcium stores so that in the next following contraction a greater amount of calcium is available for release. The reversed effect is observed when in a regularly stimulated papillary muscle one action potential is artificially shortened; the contraction elicited by this shortened action potential is not decreased, but the next following one is, since now the store is less filled. Obviously, calcium is continuously leaking out of the stores and escaping the cell so that calcium replenishment by the slow inward current is essential for maintaining contractility.

Positive Staircase. When stimulation is interrupted for a prolonged time, the calcium stores become empty, as shown by electron probe analysis (Wendt-Gallitelli and Jacob 1982) as well as by measuring the calcium loading of the heart using the ^{45}Ca isotope (Lewartowski et al. 1984). The first contraction following a longer period of rest (rested state contraction, cf. Koch-Weser and Blinks 1963) is, therefore, smaller and delayed by up to 100 ms, for it is obviously not due to calcium release from the empty sarcoplasmic reticulum, but to calcium entering the cell during the action potential via the membrane calcium channels (Reiter et al. 1984). When stimulation continues, stores are gradually filled and contraction

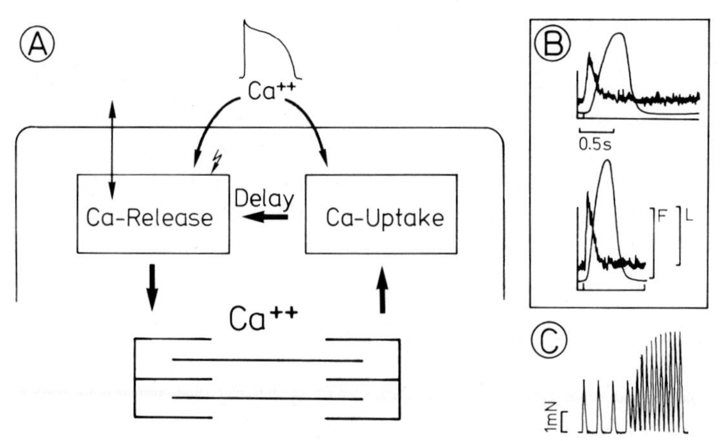

Fig. 7.6 A–C. Force-frequency relation in the heart. *A* Model of cardiac excitation-contraction coupling explaining the stimulus force-frequency relationship. During the plateau phase of the action potential, an early calcium current may be involved in calcium-induced calcium release, whereas the "slow" component of the Ca^{2+} current seems to replenish the Ca-release sites of the sarcoplasmic reticulum with a delay (cf. Fabiato 1985c). It is assumed that the filling state of the SR determines the amount of Ca^{2+} which can be released from the SR by subsequent action potentials and, hence, the size of subsequent contractions. (After Wohlfart and Noble 1982). *B* Twitch contraction and calcium transient (noisy tracing, aequorin luminescence) in cat papillary muscle at low stimulation frequency (0.1 Hz top) and at higher frequency (1 Hz bottom). (Morgan and Blinks 1982). Force calibration $F = 1.3\ N\,cm^{-2}$ luminescence, $L = 86$ counts s^{-1}. *C* Delayed increase in twitch force (positive staircase) when the stimulation frequency is abruptly increased from 0.5 Hz to 1.7 Hz. (After Wohlfart and Noble 1982, cf. also Schouten et al. 1987)

increases with each beat in a positive "staircase." The increase in contractility is associated with and possibly due to a greater uptake and storage of calcium (Lewartowski et al. 1984) which appears to be located in the sarcoplasmic reticulum near the Z-line (Wendt-Gallitelli and Wolburg 1984).

Force-Stimulus Interval Relationship. A change in the interval between stimulation or in the frequency of the action potentials also changes force (Koch-Weser and Blinks 1962) since it alters the amount of calcium entering the cell per unit time, the loading of the sarcoplasmic reticulum and, hence, the amount of calcium that can be released during the beat (Wohlfart and Noble 1982). Although the action potential is shorter at higher frequency, it occupies the contractile cycle for a relatively longer period than at low frequency. Consequently, an increase in stimulation frequency may enhance the amount of calcium entering the cell by inward calcium current during the plateau phase. This would result in a greater filling of the calcium stores and a larger amount of calcium that would be available for release in each contraction. In this way, a sudden increase in frequency of stimulation increases the contractile force with a delay, as illustrated in Fig. 7.6. However, a second effect of the rise in intracellular calcium is also to be noted. As first shown for Purkinje fibres, calcium ions also activate the potassium channels, thus accelerating repolarization of the membrane by enhancing the K^+ efflux (Isenberg 1977). Consequently, the plateau phase of the action potential is shortened, thus causing a secondary decrease in the duration of the calcium current and a reduction in the intracellular Ca^{2+} concentration. This mechanism, therefore, constitutes a negative feedback damping the primary, positive inotropic effect of an increase in stimulation frequency.

Comparative Aspects. The frequency-dependent phenomena differ among various mammalian species (Bers 1985; Rumberger and Reichel 1972). For instance, rat ventricles and atria of many species contract less strongly when the frequency of stimulation is increased. Unlike non-hibernating chipmunks, the hearts of hibernating chipmunks contract maximally even when the frequency of stimulation is lowered. This feature may reflect a special adaptation to allow the myocardium of hibernating chipmunks to contract forcefully at low temperatures when the heart is beating only a few times per minute. Even when these hearts are stimulated after a long period of rest they contract forcefully and, in contrast to the heart of non-hibernating chipmunks, the contractile strength does not increase further if rhythmical stimulation is resumed. These experiments suggested to Kondo and Shibata (1984) that in rats and hibernating chipmunks the sarcoplasmic reticulum is apparently not leaky and that its filling is independent of the slow inward calcium current. Indeed, in the hearts of rats and hibernating chipmunks the action potentials exhibit only a very brief (50 ms) plateau phase so that the extent of inward calcium flow is small. In non-hibernating chipmunks, in contrast, the action potentital is prolonged and contractile strength depends on the replenishment of calcium stores by the slow inward currents. This is shown by the strong dependence of the contractility on stimulation frequency and on the extracellular calcium supply in these muscles.

In conclusion, there seem to be large differences in the calcium-handling mechanisms of various species, frog ventricle being most and rat ventricle least dependent on calcium influx from the extracellular space (Bers 1985). The slow inward current-dependent, calcium-replenishing mechanism of the sarcoplasmic reticulum may, therefore, be of variable importance, but in most mammalian hearts, it may appear to constitute a short-time memory of past stimulation history. In addition, stimulation may also mediate perhaps via an increase in the time-averaged, free Ca^{2+} concentration a number of biochemical alterations. These may include increased phosphorylation of the regulatory myosin light chains (Bárány and Bárány 1981; Sayers and Bárány 1983; Silver et al. 1986) or an altered expression of myosin genes responsible for the synthesis of myosin isozymes (e.g. Rupp 1982; Srihari et al. 1982; Cummins 1983; Swynghedauw et al. 1982). These changes may constitute a biochemical long-term memory of the past average of free Ca^{2+} concentration and muscle activity in the heart, but detailed discussion thereof is certainly beyond the scope of this brief consideration.

7.3 Myoplasmic Free Calcium, a Major Determinant of Contractility

As already described, heart muscle has a large contractile reserve, so that its performance can be increased by a variety of manoeuvres; these include a rise in the extracellular calcium or intracellular sodium concentration, the increase in frequency of stimulation and the application of compounds with a positive inotropic action. In most of these cases, the increase in force or positive inotropy appears to be due ultimately to an increase of intracellular free Ca^{2+} concentration reached during contraction.

In 1978, Allen and Blinks succeeded in determining the change in the intracellular free Ca^{2+} concentration of frog atrial cells which had been injected with the calcium-sensitive, luminescent protein aequorin. Following an electric stimulation, calcium entered the cell and diffused into the fibre interior so that the Ca^{2+} concentration rose slowly, reaching a maximum after 200 ms (Fig. 3.2). In mammalian ventricles, however, maximum values were reached within less than 25 ms (Allen and Kurihara 1980; Morgan and Blinks 1982) since activator calcium is predominantly released from intracellular stores. Figure 7.8 A shows the time course of action potential, calcium transient and twitch of ferret ventricular muscle measured after a single electric shock. If the frequency of stimulation is increased, the filling of the sarcoplasmic reticulum is improved, as already discussed, so that each stimulus induces a greater release of calcium from the sarcoplasmic reticulum and a correspondingly larger calcium transient, as shown in Fig. 7.6. Due to the higher Ca^{2+} concentration reached during systole, the contractile force also increases, but takes a longer time to develop fully. The rate of relaxation, however, is barely affected, although the calcium transient is decaying faster at high frequency of stimulation.

7.3.1 The Dependence of Force and Intracellular Calcium Transients on Extracellular Calcium Concentration

As described more than a hundred years ago by Ringer, the increase in extracellular calcium has a positive inotropic effect on frog heart, which fails to contract in calcium-free saline (cf. Sect. 1.2.2). Very thin fibre bundles of mammalian cardiac trabecula respond to a sudden increase in external Ca^{2+} concentration with an increase in force nearly without a delay (Kitazawa 1984). According to Allen and Kurihara (1980), the intracellular calcium transient is also increased. As the electromotive force for calcium is increased at high external Ca^{2+} concentration, the influx of calcium during the action potential is enhanced. It is not yet clear, however, how much the inflowing calcium contributes directly to the increase in intracellular free calcium and to what extent it acts in a more indirect manner by triggering the release of more calcium from the sarcoplasmic reticulum or by improving the filling of the stores with calcium that can be released in subsequent contractions. The latter mechanism, however, seems to be rather unlikely, according to Kitazawa (1984), since thin cardiac muscle fibres respond almost immediately to high external calcium.

The variation of force with alterations in extracellular Ca^{2+} concentration may be of particular importance in lower vertebrates in which extracellular Ca^{2+} concentration appears to vary greatly. In diving turtles, for instance (Jackson and Heisler 1982), the extracellular Ca^{2+} concentration may increase dramatically as a result of severe exercice or during hypoxia, thus producing a positive inotropic effect on the heart.

7.3.1.1 Relationship Between Force and Intracellular Calcium Ion Concentration

It is interesting to relate the peak force development to the intracellular Ca^{2+} concentration reached during force development (Allen and Kurihara 1980). Figure 7.12 shows that this relationship is the same regardless of whether force has been changed by altering the frequency of stimulation or by altering the extracellular Ca^{2+} concentration. But even during a positive inotropic intervention, such as increasing the extracellular free calcium, the force development is far from maximal. The full potential of force development can only be revealed after skinning the fibres and exposing the membrane-free contractile structures suspended in ATP salt solution to a Ca^{2+} concentration that is many times larger than that reached in vivo during a maximal positive inotropic effect. As pointed out already, the contractions elicited in the skinned fibres at maximal activation with 10^{-5}M Ca^{2+} are about twice as large as those of the same fibre before skinning (Fabiato 1981). This suggests that even under optimal conditions, living cardiac muscle is never more than about half-maximally activated so that force can easily be varied by increasing or decreasing the free Ca^{2+} concentration. On the contrary, skeletal muscle is probably maximally activated by an intracellular Ca^{2+} concentration reaching about 10^{-5}M during twitch so that its force development is quite independent of small variations of internal or external Ca^{2+} concentration.

The relationship between free calcium in the vicinity of the myofilaments and force development of the contractile machinery has also been studied after removing the external cell membrane. In such skinned fibres, the threshold calcium concentration for eliciting a contraction is of the order of 1 μM as in the case of living fibres. Compare Figs. 7.14, 7.7, and 7.12 and note that the relationship between Ca^{2+} concentration and contraction is similar in living cardiac muscle and in isolated contractile structures where the free Ca^{2+} concentration has been adjusted by calcium buffers. Interestingly, the calcium-force relationship of cardiac muscle is much less steep than that of skeletal muscle. This peculiarity of cardiac muscle may reflect an adaptation that allows the fine adjustment and regulation of force by changes in intracellular Ca^{2+} concentration. In skeletal muscle, in contrast, a similar precise regulation of force would, of course, be difficult since a small increase in the free Ca^{2+} concentration would transform a barely threshold response into a maximal contraction. A different steepness of the calcium-force relationship in skeletal and cardiac muscle is, therefore, of great functional significance and may be accounted for by a difference in the calcium-troponin interaction.

7.3.1.2 Interaction of Calcium and Cardiac Troponin

In vertebrate cross-striated muscle, calcium activation is intracellularly mediated by the calcium receptor troponin and the calcium occupancy of its specific binding sites determines the contractile force. Like its skeletal muscle counterpart, cardiac troponin consists of a calcium-binding subunit (TnC), an inhibitory subunit (TnI) and a tropomyosin-binding peptide (TnT). Troponin-C differs, however, from skeletal muscle TnC by having only one instead of two calcium-specific sites, whereas the two unspecific, high-affinity calcium-binding sites are identical (Holroyde et al. 1980). This difference might be partly responsible for the lack of steepness in the relationship between Ca^{2+} concentration and force in cardiac muscle. Thus, when troponin-C from fast skeletal muscle was replaced by cardiac troponin-C, the relationship between Ca^{2+} concentration and force development of skinned fibres became more shallow, now resembling the less cooperative calcium-force relationship in cardiac muscle (Moss et al. 1986, cf. also Pan and Solaro 1987).

Troponin-I of skeletal muscle and cardiac muscle are also distinct. The latter is prolonged at its N-terminal end by a segment of 26 amino acids. These include a serine residue in position 20 that may be phosphorylated by cAMP-dependent protein kinase, whereby the calcium affinity of troponin-C is reduced (Solaro et al. 1981; cf. also Buss and Stull 1977 and Sect. 7.4.1).

7.3.2 Toxins and Drugs Influencing Force and Intracellular Free Calcium

Many species depend on the defense by toxins which are poisonous to possible predators. Some of these toxins poison the heart by interfering with excitation-contraction coupling, but in low doses they might also be of therapeutic use in the treatment of congestive heart failure.

Cardiac Glycosides. Because of its positive inotropic action, the poison from the foxglove (*Digitalis purpurea*) has been used for many decades in the management of cardiac failure. Application of digitalis, ouabain or other cardiac glycosides to isolated papillary muscle or Purkinje fibres increases the intracellular calcium transient and force (Fig. 7.9; cf. Morgan and Blinks 1982; Wier and Hess 1984). The cellular mechanisms that enhance intracellular Ca^{2+} concentration are still not fully understood; their mode of action may be complex (Scholz 1984). Although an intracellular action must also be considered (Isenberg 1984), cardiac glycosides seem to act mainly by inhibiting the Na- and K-dependent ATPase of the cell membrane and, hence, the Na extrusion, thereby slightly increasing the intracellular Na concentration. As a consequence, less calcium is extruded by the Na-Ca exchange mechanism and the myoplasmic free calcium increases until a new equilibrium is reached. Under these conditions, the relationship between $[Ca]_i$ and $[Na]_i$ is described by the Blaustein equation [Sect. 7.1.2, Eq. (7.2)] showing that other factors being equal, $[Ca]_i$ is proportional to $[Na]_i^n$, where n equals 2 to 3 and corresponds to the number of Na ions exchanged for each Ca^{2+} ion. As shown by Wasserstrom et al. (1983) and Grupp et al. (1985), the positive inotropic effect of cardiac glycosides indeed appears to be quantitatively related to the increase of intracellular Na concentration, an increment of 1 mM being required to double the twitch force in electrically paced papillary muscle. After application of toxic concentrations of cardiac glycosides, the intracellular free calcium increases not only during systole, but also during diastole so that eventually the heart relaxes only incompletely: a contracture develops. An endogenous digitalis-like substance produced by the adrenal glands may play a role in the regulation of contractility and vascular tone (Hamlyn et al. 1991).

A similar kind of inotropy followed by contracture develops in hearts of fish which are poisoned by sea anemone toxins that interact with the sodium channel. These toxins as well as the veratrum alkaloids retard the inactivation of the sodium current with the result that the myocyte becomes flooded by sodium during excitation and the intracellular Ca^{2+} concentration rises due to Na-Ca exchange (Scholz 1984). In contrast, tetrodotoxin, the poison of the Japanese puffer fish, inhibits the excitatory sodium currents completely so that Na-dependent action potentials can then no longer be elicited.

Calcium Antagonists. As mentioned above (Sect. 7.1.2.1), certain dihydropyridine compounds and other substances listed in Table 7.1 block calcium channels and, hence, calcium entry. Figure 7.7 shows that both the calcium transient and force are diminished by the calcium antagonist nitrendipine, but the relation between intracellular free calcium and force remains unaltered (Wier and Yue 1985). Significantly, the drug concentrations required for half-maximal inhibition of calcium transients, calcium entry and force generation are almost identical and usually in the nanomolar range. This suggests that calcium antagonists act by decreasing the calcium influx through the cell membrane during the action potential rather than by inhibiting calcium release from the sarcoplasmic reticulum. Their efficacy is often potential- and "use-dependent." In the case of verapamil and dil-

Table 7.1. Calcium modulators

a) *Calcium channel blockers* (calcium antagonists: calcium entry blockers)	
Dihydropyridines:	*Other compounds:*
Nifedipine	Verapamil
Nitrendipine	Gallopamil (D-600)
Felodipine	Diltiazem
	Fendiline
b) *Calmodulin antagonists*	
Phenothiazines (trifluoperazine, chlorpromazine); naphthalene derivatives (W-7); calmidazolium (R 24571)	
c) *Calcium agonists*	
Dihydropyridines (BAY K 8644, CGP 28392)	
d) *Calcium sensitizers* [a]	
Imidazole derivatives:	Sulmazole (AR-L 115 BS),
	Pimobendan (UD-CG 115 BS),
	BM 14.478 [b]
	Carnosine [c]
	N-acetylhistidine [c]
Others:	EMD 53998, see Sect. 10.2.3.4

[a] See also Sect. 10.2.3.4.
[b] Müller-Beckmann et al. (1986).
[c] Harrison et al. (1985).

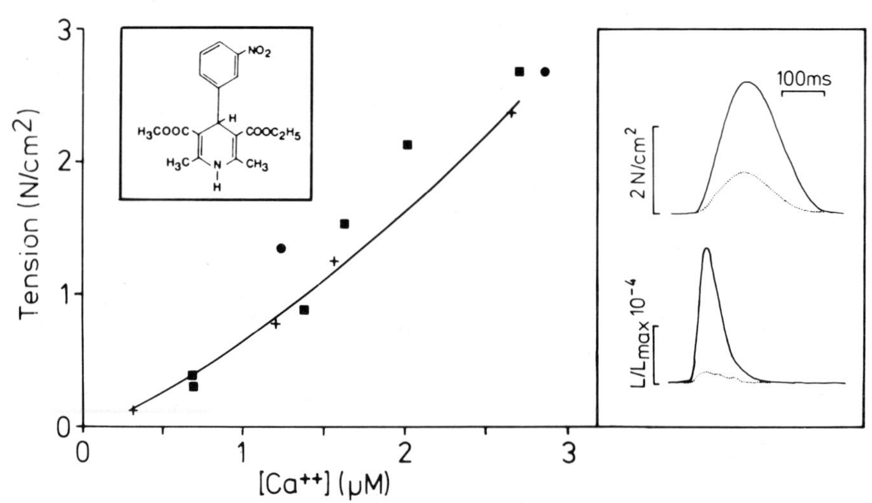

Fig. 7.7. Calcium antagonists (Ca-channel blockers) inhibit contractility by reducing the intracellular calcium transient. *Left* The relationship between intracellular Ca^{2+} concentration and twitch force at various extracellular Ca^{2+} concentrations in the absence of nitrendipine (+) and in its presence (■ 200 nM, ● 1 μM). *Inset* Structure of nitrendipine. *Right* Nitrendipine (0.4 μM, dotted line) decreases twitch contraction and intracellular calcium transient detected by aequorin luminescence (L/L_{max}) in surviving ferret papillary muscle. (Wier and Yue 1985)

tiazem, inhibition of calcium channels is enhanced when the membranes are depolarized or excited at increasing frequency. In this way, calcium antagonists exert a beneficial, energy-saving effect on the working myocardium as they lower the intracellular free calcium and, hence, the myofibrillar ATPase activity. As was pointed out by Fleckenstein (1983), calcium antagonists are, therefore, particularly useful in the management of coronary heart disease when the oxygen supply to the myocardium is limited.

Miscellaneous. In contrast to the channel blockers, ryanodine and dantrolene block the release of calcium from intracellular stores, whereby decreasing the calcium transient (Fig. 7.8; cf. Wier et al. 1985). Caffeine and theophylline, on the other hand, increase and prolong the intracellular calcium transient as well as the twitch contraction by stimulating the release from and inhibiting calcium reuptake by the sarcoplasmic reticulum (Konishi et al. 1984; P. Hess and Wier 1984). In addition to having an effect on the sarcoplasmic reticulum, caffeine also inhibits the enzyme phosphodiesterase (PDE) which is engaged in the degradation of cyclic adenosine monophosphate (cAMP) within the cell. Under the influence of caffeine and other PDE inhibitors, therefore, the intracellular cAMP levels rise. As discussed below (Sect. 7.3.3), this cyclic nucleotide increases the calcium influx through membrane channels during the action potential, thereby facilitating calcium-induced calcium release or the filling of the internal stores with calcium which can be released into the myopolasm in subsequent contractions. In this way, PDE-inhibiting cardiotonic compounds, such as piroximone, amrinone and milrinone, may increase intracellular free calcium during systole and exert a positive inotropic, i.e. force-enhancing, effect so that these compounds might be a welcome stimulant for a "weak" or insufficient heart (Endoh et al. 1986; Morgan

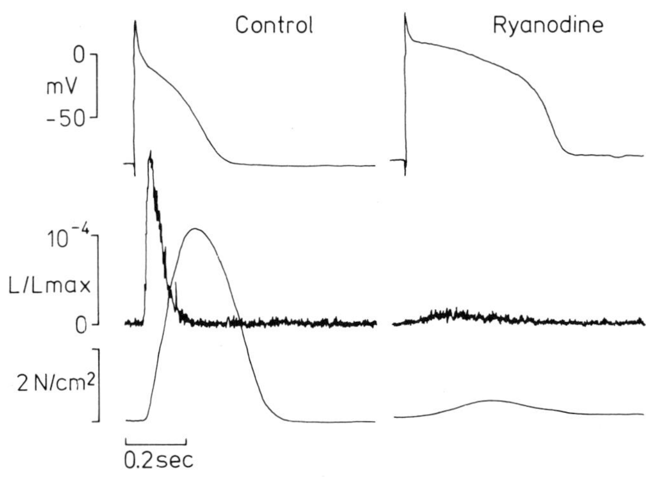

Fig. 7.8 A, B. Simultaneous recording of action potential, calcium transient and twitch in ferret papillary muscle. *A* In the absence and *B* in the presence of 5 µM ryanodine. Intracellular free calcium was estimated by measuring aequorin luminescence (L/L_{max}, cf. Fig. 3.7 B). Note that ryanodine does not inhibit the action potential and the calcium current, but the release of calcium from the sarcoplasmic reticulum, the calcium transient and the contractile response: in ferret ventricles activator calcium stems from the sarcoplasmic reticulum rather than from the calcium influx through the membrane. (Wier et al. 1985)

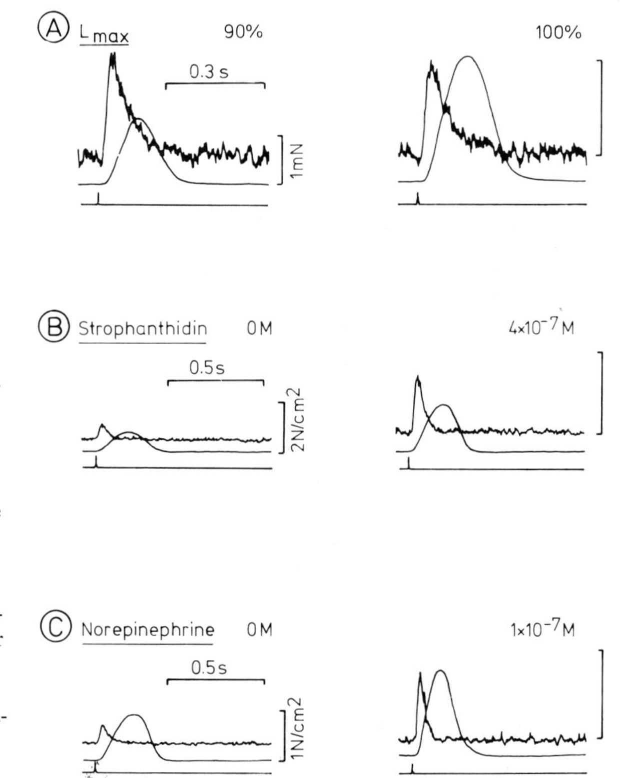

Fig. 7.9 A–C. Relationship between intracellular free calcium and contraction. *Continuous lines:* Force. *Noisy tracings:* aequorin luminescence of electrically stimulated cat papillary muscle. Calibration bars: force (N) or photomultiplier counts s^{-1}; right bars: A, 86 c s^{-1}; B, 600 c s^{-1}; c, 1000 c s^{-1}. Increase in force by stretching (*A*) is not associated with increase in free Ca^{2+}: calcium sensitivity modulation. Increase in force by application of cardiac glycoside (*B*) or noradrenalin (*C*) is due to an increase in intracellular free calcium: calcium amplitude modulation. (After Morgan and Blinks 1982)

et al. 1986). At the same time, however, the myofibrillar ATPase is also enhanced due to the increased intracellular Ca^{2+} concentration so that the oxygen requirement increases greatly.

Forskolin is a toxin extracted from Indian plants of the genus *Forskolea*, which activates adenylate cyclase causing thereby dramatic increases in cellular cAMP levels, intracellular free calcium and contractile force (Endoh et al. 1986). Other positive inotropic compounds stimulating adenylate cyclase are the catecholamines, e.g. the hormones noradrenaline, adrenaline and serotonin or the drug dobutamine. Serotonin enhances force in the heart of the mollusc *Dolabella* when it is released from stimulatory serotoninergic neurons (Higgins and Greenberg 1974), while noradrenaline and adrenaline are released from sympathetic nerves and from the medulla of the suprarenal glands in vertebrates (see Nilsson 1983 for review).

7.3.3 How Noradrenaline Increases Contractility

It is well known that the sympathetic nerves innervating the heart of vertebrates and their transmitter noradrenaline increase heart rate, speed of atrio-ventricular conduction as well as the myocardial force. Reuter and Scholz (1977) ascribed the latter positive inotropic effect to an enhancement of the inward calcium current which increases the calcium transient (Allen and Kurihara 1980) and, thus, the development of force (Fig. 7.9). The mechanisms by which catecholamines enhance the calcium currents have been analyzed in detail (Reuter 1983, 1984) and have been shown to be cAMP-dependent.

Role of cAMP. In a first step, the catecholamines occupy specific receptors situated at the cell membranes, predominantly β-receptors, thereby stimulating the membrane-bound enzyme adenylate cyclase which catalyzes the formation of cAMP (Robison et al. 1971, cf. Fig. 2.11 A). In this way, the message carried by the neurotransmitter is handed over to a second messenger, cAMP, which transmits the stimulatory signal inside the cell to specific protein targets, the cAMP-dependent protein kinases. When activated by cAMP, these kinases transfer the terminal phosphate group of ATP to certain cellular proteins that are recognized by the enzyme. There is now much evidence that such a biochemical pathway is involved in the signal transference to the calcium channels. For instance, Brum et al. (1983) mimicked all actions produced by noradrenaline or adrenaline by injecting cAMP or active cAMP-dependent protein kinase into cardiac cells. Figure 7.10 shows that following injection, the duration of the action potential plateau and, hence, the slow inward current are increased. Injected myocytes contract more forcefully, but relax faster. All kinases act by phosphorylating proteins, and in fact, various proteins including troponin-I (Solaro et al. 1976) as well as various membrane proteins (Bárány and Bárány 1981) have been found to be phosphorylated in catecholamine-treated hearts.

Alteration of the Calcium Channels. Such alterations after administration of noradrenaline or other catecholamines have been studied by electrophysiological techniques. Reuter (1983) succeeded in measuring the calcium conductance of a

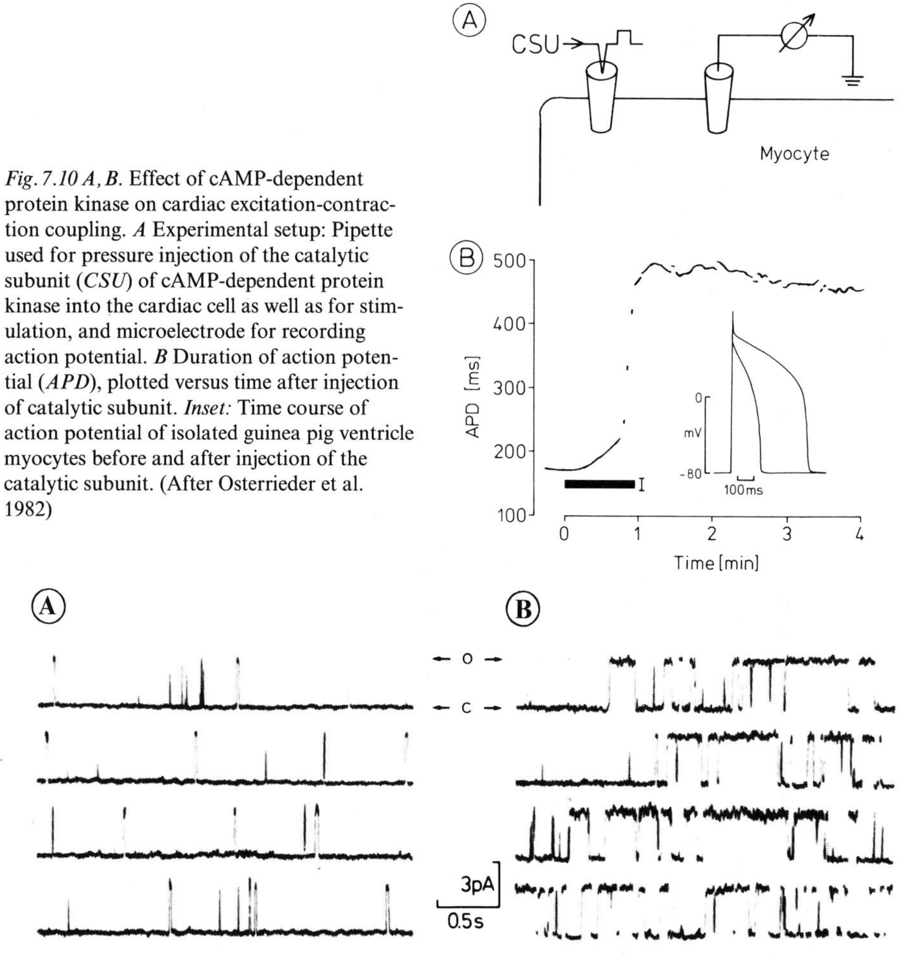

Fig. 7.10 A, B. Effect of cAMP-dependent protein kinase on cardiac excitation-contraction coupling. *A* Experimental setup: Pipette used for pressure injection of the catalytic subunit (*CSU*) of cAMP-dependent protein kinase into the cardiac cell as well as for stimulation, and microelectrode for recording action potential. *B* Duration of action potential (*APD*), plotted versus time after injection of catalytic subunit. *Inset:* Time course of action potential of isolated guinea pig ventricle myocytes before and after injection of the catalytic subunit. (After Osterrieder et al. 1982)

Fig. 7.11 A, B. Effect of cAMP-dependent phosphorylation on conductivity of purified, isolated calcium channels incorporated into artificial membrane. Current pulses (single channel opening and closing events) were recorded with the patch clamp technique (cf. Fig. 7.2) before (*A*) and after (*B*) phosphorylation of the channel protein by 1.25 μM catalytic subunit of cAMP-dependent protein kinase in presence of 1 mM ATP-γ-S as phosphate donor and 0.3 mM $MgCl_2$. Note that phosphorylation prolongs the channel opening times, shortens the closed intervals between the openings and increases the open-probability from 0.04 to 0.5 at a given membrane potential (+165 mV in the electrode pipette). Method: Calcium channels were isolated from muscle, purified on a nitrendipine affinity column and incorporated into a membrane bilayer covering the orifice of a micropipette (patch electrode) filled with 100 mM $BaCl_2$. (From Flockerzi et al. 1986, with permission; for structure of calcium Channels cf. Tanabe et al. 1987)

tiny patch of cell membrane which was so small that it contained only one or two calcium channels. When a potential was applied across this membrane patch and held constant (clamped) by means of a special feedback system it was possible to measure a current which was directly proportional to the calcium conductance. In such "patch clamp experiments" current does not flow continuously, but in discrete quanta. Obviously, the conductance changed in an all-or-none manner from moment to moment, suggesting that the single channel of the patch was either open or closed. Reuter (1983) showed that noradrenaline increases the open probability of the channel rather than the number of the channels that may be open. This effect of noradrenaline could also be mimicked by the application of cAMP-dependent protein kinase which increased the open probability even of single, isolated calcium channels incorporated into an artificial membrane (see Fig. 7.11; cf. Flockerzi et al. 1986) and Kameyama et al. (1986) showed that phosphorylation of calcium channels by cAMP-dependent protein kinase also increases their open probability in myocytes.

Enhancement of Force, Stiffness and Shortening Velocity. By increasing the calcium influx during the slow inward current catecholamines cause an increased release of calcium from the sarcoplasmic reticulum, thus enhancing the intracellular transient of calcium and force development during the twitch (Allen and Kurihara 1980; Morgan and Blinks 1982). Force is paralleled by a similar increase in immediate stiffness which, according to the theory proposed by A. F. Huxley and Simmons (1973), may be proportional to the number of crossbridges attached to actin at any one moment. Herzig (1978) concluded, therefore, that adrenaline and other catecholamines may increase the force of cardiac contraction by increasing the number of crossbridges pulling on the actin filaments. But, in addition, the crossbridge cycling rate and, hence, the shortening velocity and rate of relaxation may also be enhanced. These kinetic effects are of great physiological importance since they allow the heart to beat at a higher frequency.

The increases in force, shortening velocity and stiffness may all be attributed to the increased levels of myoplasmic free calcium for, as shown by Herzig and Rüegg (1980), an increased level of ionized calcium increases the shortening speed as well as the stiffness of isolated contractile structures (skinned fibres) from cardiac muscle suspended in an ATP salt solution.

Acceleration of Relaxation by Noradrenaline. In the presence of noradrenaline or adrenaline, the calcium transient, following stimulation of cardiac muscle, is shorter than in its absence, suggesting that the released calcium is taken up more rapidly by the sarcoplasmic reticulum (Allen and Kurihara 1980). Under these conditions, the calcium pump is stimulated by the phosphorylation of a special regulatory protein of the sarcoplasmic reticulum called phospholamban. This protein can be phosphorylated at high Ca^{2+} concentration by a calcium-calmodulin-dependent protein kinase (Pifl et al. 1984) and by a phospholipid-dependent protein kinase (protein kinase-C; Movsesian et al. 1984); most importantly however, cAMP-dependent protein kinase phosphorylates phospholamban, thereby activating the calcium uptake mechanism of the sarcoplasmic reticulum (Tada et al. 1974 b). It is possible, however, that the rate of relaxation of cardiac muscle may depend not only on the speed at which the intracellular Ca^{2+} concentration is lowered after the end of contraction, but also on the rate at which calcium comes off troponin. Indeed, the calcium off-rate may be increased under the influence of cAMP. As we mentioned already, the increase in cAMP levels following the administration of adrenaline causes phosphorylation of troponin-I by cAMP-dependent protein kinase which, in turn, increases the calcium off-rate of troponin-C (Solaro et al. 1981). Since the contractile force is dependent on the calcium occupancy of troponin-C, an increased calcium off-rate should then cause an increased rate of relaxation. Whereas the off-rate is increased, the on-rate is calcium-diffusion-limited and, therefore, unlikely to be affected by cAMP. Consequently, the calcium affinity of troponin is decreased at high levels of cAMP since it is the ratio of on-rate and off-rate. Calcium occupancy and force will, therefore, decrease at any given intermediate level of free calcium in the presence of cAMP, i.e. the relationship between force and calcium will be altered (Allen and Kurihara 1980). These findings suggest that cardiac function may not only be controlled by the intracellular free Ca^{2+} concentration, but also by changes in calcium sensitivity of myofilaments, as will be discussed next.

7.4 Alteration of Contractility by Changes in Calcium Responsiveness of Myofilaments

In the past it had been assumed that the force of cardiac muscle is uniquely determined by the free Ca^{2+} concentration within the cell. As recently shown, however, the calcium-force relationship may vary over a wide range, mainly because of alterations of the calcium sensitivity of the myofilaments. The factors known to modulate the calcium responsiveness are the intracellular pH, the concentration of cyclic nucleotides, and the sarcomere length as well as some novel cardiotonic drugs, the "calcium sensitizers" (see Table 7.2).

Table 7.2. Modulation of calcium sensitivity in cardiac muscle

Modulator[a]	Effect[a]	Ref.
Increase in sarcomere length	+	Hibberd and Jewell (1982)* Allen and Kurihara (1982)**
Phosphorylation of regulatory myosin light chain	+	Morano et al. (1985)*
Phosphorylation of troponin I	−	Herzig et al. (1981b)*
Hypoxia:	−	Allen et al. (1982)**
Acidosis		Schädler (1967)*
Inorganic phosphate	−	Herzig and Rüegg (1977)* Kentish (1986)*
New cardiotonic drugs (calcium sensitizers):		
Sulmazole	+	Herzig et al. (1981a)* Solaro and Rüegg (1982)*
Pimobendane	+	Rüegg et al. (1984b)*
BM 14.478	+	Müller-Beckmann et al. (1986)*

[a] Increase (+) or decrease (−) in calcium sensitivity of cardiac myofilaments as determined in skinned cardiac muscle preparations* or in vivo**. Ca^{2+} sensitivity as used here is defined as the pCa (neg log of Ca^{2+} concentration) inducing half-maximal activation of demembranated (skinned) fibre preparations in saline containing Mg-ATP.

7.4.1 Calcium Desensitization of Myofilaments by Cyclic Nucleotides

Morgan and Blinks (1982) as well as Allen and Kurihara (1980) discovered that catecholamines altered the calcium-force relationship in the living heart, as shown in Fig. 7.12 A. They stimulated cardiac muscle with current pulses in the presence or absence of noradrenaline and determined force as well as the intracellular Ca^{2+} concentration using aequorin as a probe. When they increased the catecholamine concentration in steps, intracellular free calcium, cAMP and force rose at first concomitantly. But when the catecholamine concentration was increased even more, force remained constant, while the Ca^{2+} concentration increased still further, raising the question whether the responsiveness of the contractile apparatus to calcium had decreased. Such a desensitizing effect would have to be ascribed to cAMP since, as already mentioned, the latter also desensitizes isolated contractile proteins. A desensitizing effect can be most readily demonstrated using skinned fibres in which the filaments are directly accessible to calcium ions and cAMP (Herzig et al. 1981 b). In these fibres an addition of cAMP or of the catalytic subunit of cAMP-dependent protein kinase alters the relationship between calcium and force in the sense that a higher concentration of Ca^{2+} is required to produce 50% activation (Fig. 7.12 B). This decrease in calcium sensitivity was paralleled by an increase in phosphorylation of troponin-I by cAMP-dependent protein kinase (cf. Buss and Stull 1977). Ray and England (1976) prepared synthetic actomyosin from myosin, actin, tropomyosin and troponin which

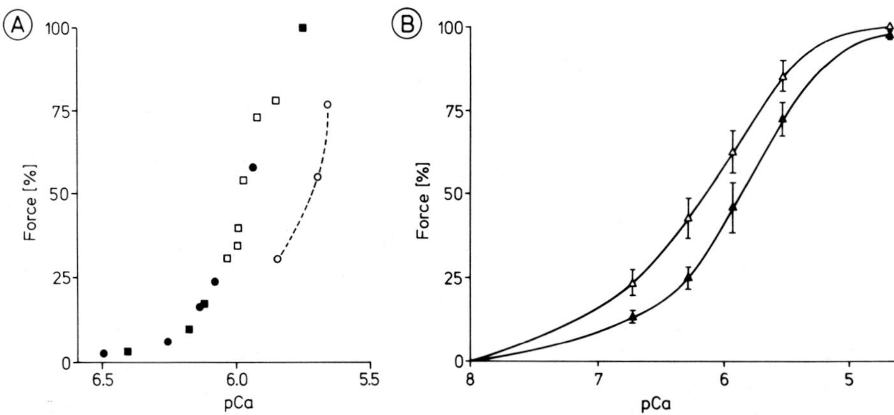

Fig. 7.12 A, B. Relationship between Ca^{2+} concentration and force in cardiac muscle. *A* Intact rat papillary muscle. Twitch force of a "paced" papillary muscle (*ordinate*) plotted versus pCa of the myoplasm determined by aequorin luminescence at peak of Ca transient. Note a similar relationship when force is increased by various positive inotropic interventions, except for noradrenaline (*open circles*) which causes a shift to the right of the Ca-force relation. Key to symbols: ■ = stimulation at different frequencies; □ = paired pulse stimulation; ● = change of the extracellular calcium; ○ = addition of noradrenaline. (Allen and Kurihara 1980). *B* Experiment suggesting that Ca desensitization following noradrenaline (*A*) may be mediated by cAMP. *Ordinate:* Relative force of chemically skinned pig ventricle trabecula as a function of Ca^{2+} in the Mg-ATP-containing bathing medium; *open symbols:* in the absence of cAMP; *closed symbols:* in the presence of cAMP. (Herzig et al. 1981 b)

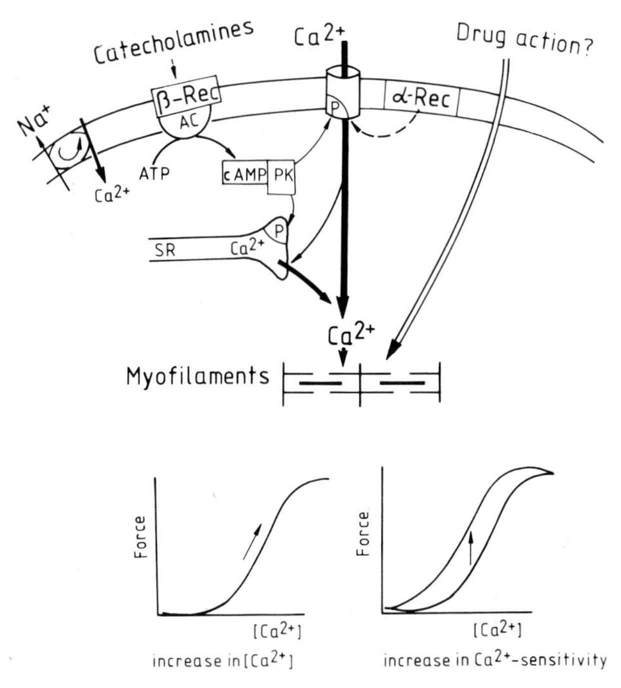

Fig. 7.13. Inotropic mechanisms in cardiac muscle. Force may be increased by increasing the myoplasmic free Ca^{2+} concentration (action of catecholamines and inhibitors of phosphodiesterases) or by increasing the Ca^{2+} sensitivity of the myofilaments. The shift to the left of the calcium-force relationship may be induced by an increase in sarcomere length or by calcium-sensitizing cardiotonic drugs. AC = adenylate cyclase, PK = cAMP-dependent protein kinase phosphorylating the calcium channel, the sarcoplasmic reticulum (SR) and Troponin-I

either was or was not phosphorylated by cAMP-dependent protein kinase (Buss and Stull 1977). In the former case, a much higher concentration of calcium was required for half-maximal activation of the contractile actomyosin ATPase.

Based on these results, we may envisage the following reaction cascade taking place when the β-receptors of the myocardium are stimulated with catecholamines. Due to the stimulation of adenylate cyclase, cAMP is formed which has a complex effect: on the one hand, it causes an increase in the inward calcium current inducing a rise of intracellular free Ca^{2+} concentration and, therefore, a stimulation of force and ATPase activity (Fig. 7.13). On the other hand, cAMP accelerates relaxation by stimulating the calcium pump of the sarcoplasmic reticulum and decreases the calcium responsiveness of the myofilaments by activating cAMP-dependent protein kinase to phosphorylate troponin-I. As the Ca^{2+} concentration within the myocardium is not sufficient to saturate troponin, the reduction of calcium affinity of troponin reduces the calcium occupancy of troponin-C and, hence, myofibrillar ATPase activity and force development (Solaro et al. 1981). Surely such a mechanism would be cardioprotective since, at high concentration of adrenaline or noradrenaline, the intracellular Ca^{2+} concentration rises so dramatically that, given a normal calcium sensitivity, the myofibrillar

ATPase would be overstimulated. The rate of ATP hydrolysis would then be faster than the rate of ATP synthesis, thus causing a loss of high-energy phosphate compounds, in particular a decrease in the level of creatine phosphate and ATP. This situation may arise under severe stress, and as it leads to cardiac necrosis it may be one of the reasons for a sudden death (Fleckenstein 1983). Nature's own way of preventing these harmful effects of adrenaline and noradrenaline is to reduce the responsiveness of the actomyosin ATPase to calcium ions.

7.4.2 Hypoxic Insufficiency

Besides cAMP-induced desensitization to calcium ions there are other cardioprotective mechanisms in the cell when there is an imbalance between synthesis and breakdown of ATP. Figure 7.14 shows an experiment of Allen and Orchard (1983 b), who reported that the calcium transients following the application of single shocks to papillary muscle are nearly the same under normoxic and hypoxic conditions, although the force development is much reduced in the hypoxic heart, as also found by Allen and Orchard (1983 a), under conditions of intracellular acidosis. H-ions reduce the calcium affinity of troponin (Blanchard et al. 1984) and decrease the calcium sensitivity of the myofilaments, as shown in skinned-fibre studies (Schädler 1967). Hypoxia also causes rapid hydrolysis of myoplasmic creatine phosphate and the accumulation of inorganic phosphate and creatine (Allen et al. 1985). An increase in inorganic phosphate to 10 mM reduces the force de-

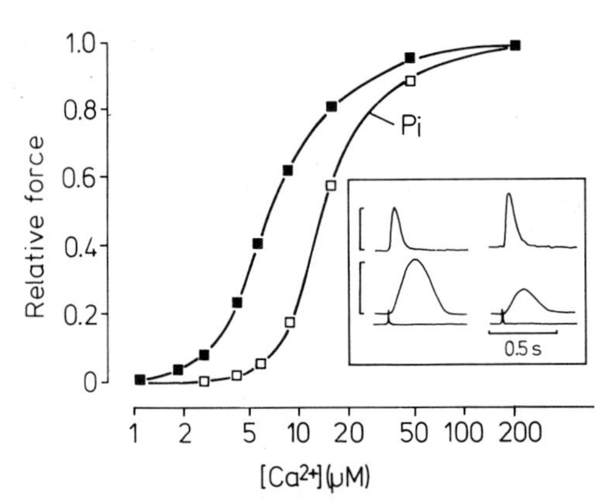

Fig. 7.14. Ca^{2+} sensitivity modulation in rat ventricle by inorganic phosphate and hypoxia. Relationship between force and Ca^{2+} concentration in the presence (*open symbols*) and in the absence of 10 mM inorganic phosphate (*closed symbols*) in chemically skinned fibres from rat heart ventricle. Note that in the presence of phosphate Ca^{2+} sensitivity is decreased, while cooperativity is increased. All solutions contained ATP (total 5 mM), 3 mM Mg^{2+}, ionic strength 200 mM, pH 7.0, at 20 °C. Ca^{2+} was clamped with CaEGTA buffers to give free Ca^{2+} between 0.2 and 200 μM. (Kentish 1986). *Inset* Transients of intracellular [Ca^{2+}] (aequorin luminescence) and twitch force in rat ventricle under normal conditions (*left*) and hypoxic conditions (*right*). Calibration *bar*: 25 μAmp (*above*) and 1 N cm^{-2} (*below*). (Allen and Orchard 1983 b). Under hypoxic conditions creatine phosphate decomposes, while intracellular levels of inorganic phosphate exceed 10 mM, as shown by nuclear magnetic resonance studies. (cf. Allen et al. 1985)

veloped by skinned fibres at a given Ca^{2+} concentration (Herzig and Rüegg 1977). As shown in Fig. 7.14, phosphate shifts the calcium-force relationship to the right, indicating a decrease in calcium sensitivity of the myofilaments (Kentish 1986). In fish, the inhibition of contractile performance of the heart under anoxic or acidotic conditions also is extremely pronounced, but cardiac insufficiency can be overcome by the positive inotropic actions of calcium ions that accumulate in the blood under hypoxic conditions (Nielsen and Gesser 1984). In the mammalian heart similar compensating effects are caused by the increased filling of the ventricle.

7.4.3 Frank-Starling Mechanism

When the calcium responsiveness of myofilaments is decreased during cardiac insufficiency in the hypoxic heart, the function of the heart as a pump is, of course, also impaired. One of the consequences of this down-regulation is an increased central venous pressure that improves the filling of the heart and may cause dilation of the ventricles. For a long time, it has been known that with increasing sarcomere length development of cardiac force and performance in general increase, a phenomenon called "law of the heart" or Frank-Starling mechanism. Until recently, it was generally held that the increase in force with increasing sarcomere length might be due entirely to a change in overlap of thin and thick filaments as in the case of vertebrate skeletal muscle. Here, as we recall (Chap. 1), force development and crossbridge activity are reduced at short sarcomere length because of the impairment of crossbridge activity by the double overlap of interdigitating filaments. In cardiac muscle, however, the force-length relationship is much steeper than in skeletal muscle, suggesting that the geometry of interaction between thick and thin filaments cannot be the ultimate cause for the alteration of force by changes in sarcomere length in these muscles (Winegrad 1979). The most important factor responsible for the increase in contractile force by stretching appears to be an increase in calcium sensitivity of the myofilaments (Allen and Kentish 1985). As discovered by Hibberd and Jewell (1982), the calcium-force relationship is shifted to the left when skinned fibres are stretched, indicating an increase in the calcium sensitivity of the myofibrils. In mammalian intact papillary muscle, stretching caused a dramatic increase in force, while the amplitude of the calcium transient was barely affected (Fig. 7.9, cf. also Allen and Kurihara 1982). A stretch-induced increase in free Ca^{2+} concentration is, however, likely to occur in the molluscan heart where stretching prolongs the action potential (Nomura 1963).

The haemodynamic improvement of cardiac function by increasing the sarcomere length is, however, not as large as suggested by the increase in force, since in the enlarged heart the muscles of the ventricular wall must exert a greater force to develop a given systolic pressure than in the case of a small heart, for according to the "Laplace relation," the wall tension is related to the product of pressure exerted and the radius of the hollow organ. As the rate of energy expenditure is mainly related to the wall tension, and the heart work depends on the change in volume and pressure, the efficiency is low in a dilated heart. For this reason, from an energetic point of view, it would be preferable to increase calcium sensitivity and cardiac force by positive inotropic drugs rather than by stretching.

7.4.4 Positive Inotropic Drugs as Calcium Sensitizers

Cardiotonic drugs may increase force at a given sarcomere length either by increasing the free Ca^{2+} concentration or by increasing the Ca^{2+} sensitivity of the myofilaments (Fig. 7.13). An increase in force produced at a given intracellular Ca^{2+} concentration has been observed in living beating papillary muscle following the application of phenylephrine, a stimulator of α-receptors, as well as following the administration of the novel cardiotonic drug sulmazole (Blinks and Endoh 1984, 1986). This compound stimulates the calcium binding to troponin and the myofibrillar ATPase at a given submaximal Ca^{2+} concentration (Solaro and Rüegg 1982) and it decreases the free Ca^{2+} concentration required for a 50% activation of the skinned fibres (Herzig et al. 1981 a).

Since in the hypoxic heart and perhaps also in the failing heart, the calcium sensitivity may be depressed, it seems to be more logical to cure the failing heart by increasing calcium sensitivity than by applying drugs that raise the intracellular calcium levels with a risk of causing a calcium overload in the heart (Herzig 1984; Rüegg 1986). Drugs listed in Table 7.1, which increase the calcium responsiveness of the myofilaments, may be called calcium sensitizers, in contrast to the classical positive inotropic agents, such as cardiac glycosides, phosphodiesterase inhibitors or catecholamines, which ultimately act by raising the intracellular free Ca^{2+} concentration (cf. Colucci et al. 1986 a, b).

7.4.5 Ischemia, Necrosis and Stunned Myocardium

The cardioprotective calcium-desensitizing effects of H-ions and phosphate during hypoxia (cf. Fig. 7.14) are beneficial since they inhibit the utilization of ATP by the actomyosin ATPase and, therefore, prevent a rapid decay of the ATP level. When, however, after prolonged ischemia, e.g. occurring as a consequence of coronary thrombosis, the intracellular levels of ATP fall below a critical limit, a completely different situation arises: at Mg-ATP concentrations below 10 μM the actomyosin ATPase and the crossbridge force generators are no longer inhibited at low resting levels of intracellular free calcium. On the contrary, actomyosin ATPase and force generation may be potentiated (Winegrad 1979). As a consequence, force increases without increases in intracellular free calcium in living cardiac fibres in which the energy metabolism has been poisoned (Allen et al. 1984). Force development in the myocardial cell and ATPase activity may then be twice as high as at the physiological free Ca^{2+} concentration (about 1 μM) which exists in the beating myocardium during systole (cf. also Bowers et al. 1992).

In the absence of oxidative regeneration of ATP the greatly activated ATPase activity causes a rapid exhaustion of intracellular ATP and the development of rigor, while the larger forces developed may cause microlesions of the cell membrane, in particular near the attachment plaques of the actin filaments and perhaps near the gap junctions in the intercalated discs. Because of the mechanical damage, the membrane may also become leaky so that intracellular enzymes leave the myocytes and extracellular calcium ions invade the cell. Thus, the intracellular free Ca^{2+} concentration rises as a consequence of the development of rigor contraction (Cobbold et al. 1984; Allshire et al. 1987). Finally, the calcium ions accumulate at such a high concentration that they are taken up by the mitochondria where they inhibit the energy metabolism (Nayler et al. 1979; Tsokos et al. 1977), so that the cell becomes necrotic. However, it could be shown that cytosolic calcium rises considerably even in less severe ischemia (Lee et al. 1987; Steenbergen et al. 1987; Allen and Lee 1989) both during systole and diastole (cf. also Lee and Allen 1991 b, for review) when contractile force is reduced while resting tension is rising.

Reperfusion injury and stunned myocardium. If ischemia is brief, e.g. less than 1 h, the heart may be reperfused following thrombolysis. Thereafter, the metabolic state is improved; intracellular pH and levels of inorganic phosphate and creatine phosphate are normalized within a few minutes (Kusuoka et al. 1987, 1990). Despite these improvements, the myocardium is often left in a stunned condition, in which contractility is considerably depressed. Remarkably, this loss of contractile function is not due to energetic deprivation, as systolic pressure can be normalized by applying catecholamines or high extracellular calcium and intracellular calcium transients are normal or even above normal. Thus, the diastolic levels of intracellular calcium determined by F19-nuclear magnetic resonance (NMR) of reperfused ferret hearts loaded with the calcium indicator 5-F19-bapta were normal, while the peak calcium levels in systole were even above normal. These results of Kusuoka et al. (1990) indicate that the calcium responsiveness of the myofilaments is decreased in the stunned myocardium as the developed pressure decreased by 20%, while intracellular calcium ion levels doubled, reaching a peak value of nearly 1 μM. Independently, and using a different method, Allen et al. (1989) arrived at similar conclusions.

The mechanism of calcium densensitization ist not yet clear. The calcium sensitivity of myofibrillar ATPase isolated from stunned myocardium is within the normal range (Krause 1988). It seems possible that calcium responsiveness had been altered in vivo by the accumulation of metabolites which had been washed out from myofibrils. Indeed, it is known that the glutathione status is altered in the stunned myocardium following reperfusion, in that the level of reduced glutathione (GSH) decreases, while that of oxidized glutathione (GSSG) greatly increases. As shown by Bauer et al. (1989), the myofibrillar calcium sensitivity, as assayed in a skinned fibre system, is high in the presence of physiological concentrations of GSH (about 4 mM), but may be greatly reduced, if the ratio GSH/GSSG declines as is the case during reperfusion and oxidative stress. During reperfusion, GSH is oxidized by the glutathione-peroxidase reaction that protects the heart from the oxygen free radicals and H_2O_2 that are formed during reperfusion. In fact, if the ischemic heart is treated with scavengers of oxygen free radicals or glutathione just prior to reperfusion (Bolli et al. 1989; Bauer et al. 1991), "stunning" could be avoided in post-ischemic, reperfused, open chest dog hearts. These results also indicate that myocardial stunning may be caused by oxygen free radicals and may be mediated perhaps by an alteration in the glutathione status that causes desensitization to calcium.

As the formation of oxygen free radicals may be a consequence of calcium overload during ischemia (Kusuoka et al. 1987), a calcium theory of stunning has been proposed (cf. Opie 1991). Thus, calcium overload has been implicated both as a cause of reflow injury and in the genesis of arrhythmias. The mechanisms of intracellular calcium elevation occurring in reperfusion (cf. Kusuoka et al. 1990) are, however, not yet clear. Most probably, it is due to a hyperactivity of the sodium-calcium exchanger causing a vigorous inflow of calcium at high intracellular levels of sodium. The latter may be due to an inhibition of the sodium-potassium ATPase in the energy deprived ischemic heart, but also to an activation of the sodium-hydrogen exchanger when the intracellular H-ion concentration rises in the ischemic heart (cf. Opie 1989). As a consequence of calcium overload, more calcium is taken up by the sarcoplasmic reticulum and released again during systole, thereby increasing the free calcium transients. The rise in Ca^{2+} levels may be seen as an adaptive mechanism that counteracts the decrease in calcium responsiveness. On the other hand, it may also be a maladaptation inasmuch as additional energy has to be spent to cycle the extra calcium that has been released. The rate of energy spent for calcium cycling and the oxygen consumption in general are therefore much greater in the stunned myocardium than in the normal heart (Laxson et al. 1989). Manoeuvres to offset the decrease in force of the stunned myocardium should therefore include glutathione or pharmacological agents that increase myofibrillar calcium responsiveness rather than positive inotropic drugs that increase intracellular free calcium.

Chapter 8

Vertebrate Smooth Muscle

Muscles of inner organs of vertebrates, such as gut, uterus, bronchial tree and blood vessels, are termed "smooth muscles" as they lack ordinary striations. Smooth muscles may be spontaneously active, or they may contract in response to nervous or hormonal stimulation. For example, think of the constriction of blood vessels that may be influenced by sympathetic nerves releasing noradrenaline as a transmitter. Hormones and neurotransmitters exert their action on the smooth muscle cell by reacting with specific targets on the cell membrane, the receptors, but they may or may not alter the membrane potential. The question of how these membrane events are coupled to the contractile responses has been one of the most intriguing problems in smooth muscle research in recent years.

As in skeletal and cardiac muscle, excitation-contraction coupling involves the release of calcium ions into the myoplasm, but the mechanism by which calcium activates the contractile machinery is, however, quite different. Thus, the intracellular calcium switch has been identified with calmodulin rather than with troponin. Calmodulin is a ubiquitous calcium-binding protein which, in conjunction with a specific protein kinase, activates the contractile machinery via phosphorylation of smooth muscle myosin. Perhaps it is not so surprising then, that there is no unique relationship between the intracellular free Ca^{2+} concentration and contraction. Thus, following a brief and transient increase in free Ca^{2+} concentration during smooth muscle stimulation, contraction may be maintained even though intracellular calcium levels may have declined to low values (Morgan and Morgan 1984a). Is this capacity to maintain tension at low Ca^{2+} concentration due to a modification of the contractile mechanism or rather to peculiarities of the calcium regulatory system? In search for an answer we shall, therefore, first consider the special features of the smooth muscle contractile machinery (Sect. 8.1). Then we address the problem of how the contractile structures are regulated by calcium ions (Sect. 8.2), how the Ca^{2+} concentration itself is regulated at a cellular level (Sect. 8.3) and how intracellular calcium handling and calcium effects are modulated by cyclic nucleotides (Sect. 8.4).

8.1 Contractile Mechanism

Basically, the contractile apparatus of smooth muscle operates, as in the case of skeletal muscle, by means of a sliding filament mechanism. But unlike in striated muscle, the actin and myosin filaments are not in register, but are arranged in a less orderly manner so that no striation pattern is visible under the light microscope.

8.1.1 Organization of the Contractile Structure

Smooth muscle cells are spindle-shaped, about 5 to 50 μm wide and up to 0.5 mm long. They contain twice as much actin and tropomyosin as skeletal muscle, but four to five times less myosin (Murphy et al. 1974). This difference has a structural correlate. Within the smooth muscle cells actin and myosin filaments are organized in bundles in such a way that each myosin filament is surrounded by up to 10 or 15 actin filaments. The latter may be several μm long and 8 nm thick (Small 1974). Myofilament bundles are obliquely oriented within the spindle-shaped smooth muscle cells as they run diagonally from an attachment point of the cell membrane through the cell and insert at an attachment patch on the opposite cell side (Fig. 8.1). Quite often, attachment patches of neighbouring cells match. Since the narrow gap between these patches is bridged by electron-dense material, the cells become mechanically coupled. In this way, smooth muscle cells form a contractile network to which collagen fibres may be attached (Gabella 1984).

Myofilament bundles contain several "dense bodies" which are anchored by a network of 10-nm-thick, so-called intermediate filaments, forming a cytoskeleton. Dense bodies appear to be analogous to Z-bands in as much as they form the border lines of "mini-sarcomeres" (cf. Small 1974; Fay et al. 1982). Each of these contractile units contains several myosin filaments overlapping with actin filaments. The latter insert on dense bodies or attachment patches of the cell membrane. Myosin filaments of smooth muscle may be structurally different from those of skeletal muscle. Rather than forming a bipolar filament, the myosin molecules arrange themselves in such a way that all molecules on one side of the filament point with their "heads" in the same direction. On the other side, they point in the opposite direction so that the structure appears as "side polar" (Small and Squire 1972; cf. Squire 1981).

A side polar or "face polar" filament structure has an interesting functional consequence. All myosin heads forming crossbridges on one side of the filament would pull an adjacent actin filament into the same direction, thus allowing it to slide over and past the myosin filament (Fig. 8.1). On the other side of such a filament, the actin filament would slide into an opposite direction. As can be seen,

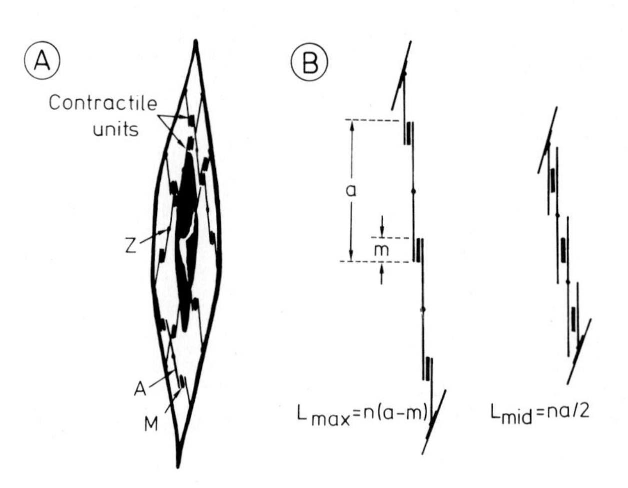

Fig. 8.1. A Schematic diagram showing sarcomere-like, diagonally arranged contractile units of actin (*A*) and myosin (*M*) within a smooth muscle cell; Z: dense bodies, homologous to Z-lines in striated muscle. *B* Overlap of sliding actin and myosin filaments in contractile units in shortened and extended fibres. (After Small and Squire 1972)

such a mechanism would allow considerable shortening of the "mini-sarcomere," possibly up to 80% of the initial length. According to the model proposed by Small et al. (1986), muscle shortening as well as tone, however, may also be affected by the non-contractile, cytoskeletal intermediate filament domain of the smooth muscle cell. This domain contains filamin and actin filaments rather than actin and myosin filaments. Smooth muscle cells are able to develop about twice as much force per cross-sectional area as skeletal muscle. This is puzzling since smooth muscle, as we recall, contains rather little myosin whereas the actin content is high (Murphy et al. 1974). But note that the comparatively long sarcomere-like structures of smooth muscle would generate the same force as several short sarcomeres in series (as they occur in skeletal muscle), despite the fact that the latter contains much more myosin. In other words, a given force can be maintained with much fewer ATP-splitting myosin crossbridges and, therefore, with a lower "energy cost" than in the case of skeletal muscle. This "advantage" is, however, offset by a much lower shortening speed due to the longer, sarcomere-like structure (cf. also Chap. 5, Fig. 5.3). Unfortunately, the length of the smooth muscle sarcomere-like contractile units is rather controversial. It may measure several tens of microns (Squire 1981) or be as short as 3 μm (Fay et al. 1982; A.V. Somlyo et al. 1977 a); also myosin filaments may be bipolar as in skeletal muscle or face polar. In any case, however, the sarcomere length is not large enough to account for the difference of several hundred times in shortening speed and holding economy between fast striated and smooth muscle. Other factors, such as low "cycling frequency" (cf. Chap. 5) of the crossbridges, must also play a role.

8.1.2 The Crossbridge Cycle

In analogy with skeletal muscle, let us assume that force is generated and maintained by crossbridges undergoing a cyclic process of attachment to actin, rotation and detachment. Even during isometrically maintained contraction, crossbridges perform internal work when they rotate and stretch elastic elements within the crossbridge. This work is "paid for" by the hydrolysis of ATP. Thus, the less the crossbridges cycle in a given time, the less energy is required for maintaining tension. In other words, the hydrolysis of one molecule of ATP by one crossbridge serves to maintain tension for a comparatively longer time than in the case of skeletal muscle, as after the hydrolytic step, the crossbridge occupied with the reaction product ADP remains, for a comparatively long time, in an attached force-generating state. Hence, the quotient tension × time/mole ATP split (the holding economy) is rather high; thence smooth muscles with slowly cycling crossbridges are indeed very economical. They are, however, very slow since slowly moving crossbridges cannot pull the actin filaments past the myosin filaments at high speed. As discussed in more detail in Chap. 5, the velocities of shortening and holding economy are inversely correlated in various types of muscle.

The crossbridge cycling frequency may be determined from the oxygen consumption during the maintenance of maximal force in a vascular smooth muscle, since one molecule of O_2 may be consumed in synthesis of six molecules of ATP which, in the steady state, is consumed by the myosin ATPase. From an oxygen consumption rate of about 0.3 μmol O_2 min^{-1} g^{-1} tissue, an ATP production and consumption of roughly 2 μmol min^{-1} may be derived; about two-thirds serve to supply the actomyosin contractile system with ATP. The division of the amount of ATP split per second by the amount of myosin (about 25 nmol g^{-1} cell) yields the molecular turnover number of myosin. Assuming that a myosin crossbridge splits one molecule of ATP in each crossbridge cycle, we obtain a crossbridge cycling frequency of approximately 0.5 s^{-1} to 1 s^{-1}. This is about 30 to 50 times smaller than the crossbridge cycling frequency of fast skeletal muscle at the same temperature (Paul and Rüegg 1976). Since, compared with skeletal muscle, smooth muscle requires about five to six times fewer myosin crossbridges for maintaining a given force and since the cycling frequency is so low, smooth muscle requires approximately 300 times less energy for maintaining tension. Conversely, maximal shortening speed is about 300–500 times smaller than that of fast skeletal muscle.

It is noteworthy that the low rate of energy expenditure by smooth muscle, as reflected by a low rate of oxygen consumption, may be largely accounted for by the low actomyosin-ATPase activity. Arterial smooth muscle contains about 13 mg myosin g^{-1} muscle which has a specific activity of about 90 nmol min^{-1} mg^{-1} protein at body temperature. Obviously, the slowness of smooth muscle as well as its high holding economy may be largely due to the low ATPase activity of the myosin molecule.

The mechanism of ATP-hydrolysis by the actin-activated myosin ATPase involves several steps as in the case of skeletal muscle actomyosin (Sellers 1985). First, ATP is bound to actomyosin. Subsequently, it causes a very rapid dissociation of actin and myosin which is followed by the hydrolysis of ATP. Thereafter, the myosin-product complex formed (myosin-ADP-P) reattaches to actin. This step presumably corresponds to the reattachment of the crossbridges in the crossbridge cycle. In smooth muscle the attached state must presumably be especially long-lived, since the dissociation of the actomyosin-ADP complex is by far the slowest, i.e. the rate-limiting step of the enzymic cycle, lasting about 0.5 to 1 s (Marston 1983). Since the attached, force-generating state of crossbridges persists for so long after splitting one molecule of ATP, the quotient tension × time per ATP is especially large, i.e. the holding economy is very high.

During prolonged contraction, the crossbridge cycling rate becomes considerably reduced and part of the attached crossbridges may even stop cycling completely. Such slowly-cycling or non-cycling crossbridges ("latch-bridges") may account for the ability to maintain tension with very little energy consumption. Evidence for a reduction in the crossbridge cycling rate during prolonged contraction has been obtained by Butler and Siegman (1983) and Krisanda and Paul (1984). They found that during prolonged contraction, the rate of oxygen consumption or ATP consumption may be progressively reduced, although tension is maintained. At the same time, the muscle looses its capacity to shorten quickly. In summary, the lifetime of attached crossbridges and the rate at which the actomyosin-ADP complex decomposes may be extremely variable and may be regulated in smooth muscle as discussed in Sect. 6.2.5. It may well be that the specific features of the smooth muscle crossbridge cycle are essentially due to special properties of the smooth muscle contractile proteins.

8.1.3 The Proteins Associated with Contraction

As reviewed by Hartshorne and Gorecka (1980), the smooth muscle contractile proteins actin and myosin are in many respects quite similar to their skeletal muscle counterparts. Actin and myosin interact with each other to form an actomyosin complex which is contractile in the presence of certain activating factors: it shrinks or "superprecipitates" slowly after addition of ATP (Filo et al. 1963; Schirmer 1965). On the other hand, there are also noteworthy differences: these proteins can be extracted from the minced tissue by salt solutions of very low ionic strength and the actin-activated, magnesium-dependent myosin ATPase has an extremely low activity unless activating proteins are added (see Sect. 8.2.1).

Actin. With a molecular weight of 42 kDa, actin of smooth muscles resembles its skeletal muscle counterpart in amino acid composition and sequence (Zechel and Weber 1978), its ability to polymerize from the globular form into the fibrillar form, its ability to interact with myosin and to activate the magnesium-dependent myosin ATPase of skeletal muscle as well as in the capacity to bind to smooth muscle *tropomyosin*. The latter is almost entirely α-helical and consists of two subunits (peptide chains of molecular weight 36 and 39 kDa respectively) which form two parallel strands (cf. also Sect. 8.2.4).

Smooth Muscle Myosin. This protein, which has a molecular weight of 470 kDa, is immunologically different from its skeletal muscle counterpart (Gröschel-Stewart 1980). It consists of two heavy chains to which light peptide chains are attached. Using the proteases papain or chymotrypsin as selective proteolytic "knives," the myosin molecule can be cut into two fragments: (1) the light meromyosin (LMM) forming the tail portion of the molecule which is embedded in the myosin filament and (2) heavy meromyosin (HMM) which consists of two heavy peptide chains that form the "neck" (called subfragment S-2, HMM-S2) bearing two globular heads (called subfragment S-1, HMM-S1). As in skeletal muscle, the myosin heads contain the catalytic site that binds nucleotides and splits ATP, as well as the site involved in the interaction with actin (for review cf. Morano 1992).

The Myosin Light Chains. Two kinds are attached to myosin heads. One has a molecular weight of 17 kDa and is essential for ATPase activity. The other light chain of molecular weight 20 kDa has a regulatory function; it binds 1 mol calcium mol^{-1} light chain and may be phosphorylated. Unlike the 20 kDa light chain of skeletal muscle to which it is structurally homologous, this light chain cannot be removed by DTNB [5,5′-dithiobis(2-nitrobenzoic acid)]. The regulatory light chain of smooth muscle is also structurally homologous to molluscan regulatory light chain, in fact so much that it can replace it functionally: hybrid myosins formed from molluscan myosin heavy chain and regulatory light chain of vertebrate smooth muscle exhibit an actin-activated ATPase which is as calcium-sensitive as natural molluscan actin-activated myosin ATPase (Kendrick-Jones et al. 1983). But whereas light chains of scallop fast adductor cannot be phosphorylated, the phosphorylation of the regulatory light chain of vertebrate smooth muscle by myosin light-chain kinase may have a profound effect on both the structure and enzymic activity of myosin. According to Ikebe et al. (1983), the phosphorylation of one serine residue promotes the interconversion from a low-activity configuration to a high-activity state (cf. also Suzuki et al. 1985).

Conformational Changes of Myosin. That smooth muscle myosin can exist in two functional states had been suspected for many years (Schirmer 1965). The phosphorylated form is characterized by an extended molecular shape, a slow rate of sedimentation in the ultracentrifuge (with 6 Swedberg-units) and a comparatively high viscosity at physiological salt concentration. Because of their low solubility at low ionic strength, 6 S myosin molecules readily aggregate into myosin filaments. The second (dephosphorylated) form sediments with a high rate (10–12 S) and has a comparatively low viscosity because of the flexed configuration of the molecule (Onishi and Wakabayashi 1982; Craig et al. 1983). Unlike skeletal muscle myosin, 10 S smooth muscle myosin is very soluble in the presence of ATP even at low physiological salt concentrations. In contrast to skeletal muscle myosin, smooth muscle contractile protein can, therefore, be extracted at low ionic strength from minced muscle. To distinguish it from conventional myosin, its discoverers Laszt and Hamoir (1961) gave it a special name: "tonomyosin," but this notation has now been abandoned.

Besides the phosphorylation of the myosin light chains there are, however, several other possibilities for transformation of 10 S myosin into "conventional" 6 S myosin which is less soluble and has a high actin-activated, magnesium-dependent ATPase activity (Ikebe et al. 1983), i.e. by increasing Mg^{2+} concentration, by lowering free ATP concentration or simply by raising the ionic strength. The interconversion of the kinked myosin configuration (10 S myosin) into its elongated 6 S form by phosphorylation processes greatly activates the myosin ATPase and may, therefore, constitute an important step in the activation of smooth muscle contraction, as will be discussed next.

8.2 Calcium Activation of the Contractile Apparatus

It is now generally agreed that smooth muscle contraction is initiated by an increase in the myoplasmic free Ca^{2+} concentration to about 1–10 μM, but the mechanism by which calcium ions activate the contractile machinery remains puzzling. How is the Ca^{2+} concentration sensed by the contractile apparatus, and in which way does the calcium signal influence the contractile machinery? In smooth muscle research this is one of the most burning problems, that both biochemists and physiologists have sought to solve. In biochemical experiments actin-activated myosin ATPase is quite generally taken to represent the "biochemical correlate" of contraction. Micromolar concentrations of calcium were found to activate smooth muscle actomyosin ATPase several times, and its dependency on the Ca^{2+} concentration was remarkably similar to that of skeletal muscle actomyosin (Sparrow et al. 1970; cf. also Schirmer 1965), but the calcium regulatory mechanism appeared to be quite different. For instance, attempts to isolate troponin as a regulatory component were quite unsuccessful (Hartshorne and Gorecka 1980). As we recall, the calcium-dependent inhibitory function of troponin was first discovered when actomyosin precipitates of skeletal muscle were repeatedly "washed" with dilute salt solutions. The resulting purified actomyosin was active and, unlike the original native actomyosin, could not be inhibited when the Ca^{2+} concentration was lowered. Calcium sensitivity, however, was restored when troponin and tropomyosin were added. When the same washing procedure was applied to smooth muscle actomyosin, it also became calcium-insensitive (Sparrow and van Bockxmeer 1972), but there was a difference: now the activity was completely inhibited even at elevated free Ca^{2+} concentration as if an activator had been removed (cf. Ebashi 1980). Subsequently, the nature of the activating factor could be, at least partly, elucidated. Of prime importance are calmodulin and myosin light-chain kinase, the enzyme catalyzing the phosphorylation of myosin. However, additional protein factors may also be involved in smooth muscle activation (Sect. 8.2.4; cf. Marston 1982).

8.2.1 Calmodulin and Myosin Light-Chain Kinase Activate Muscle Contraction

Myosin Light-Chain Kinase Phosphorylates Myosin. When the thoroughly washed, inactive smooth muscle actomyosin was mixed with a fraction isolated from the washing solution, it regained calcium sensitivity. Then, addition of calcium ions activated the actomyosin ATPase which, at the same time, became phosphorylated. Sobieszek (1977) identified the site of phosphorylation with the regulatory light chain of myosin where, according to Maita et al. (1981), a serine residue in position 19 is phosphorylated. Hence, the activating principle conferring calcium sensitivity was probably an enzyme that phosphorylated and activated actomyosin in a calcium-dependent manner. Since ATPase activity and phosphorylation of myosin rose in parallel when the Ca^{2+} concentration was increased in a stepwise manner, a causal relationship seemed likely (Aksoy et al. 1976). The hypothesis was, therefore, advanced that calcium activates smooth muscle via phosphorylation of myosin light chains by the myosin light-chain kinase, and that, unlike skeletal muscle, smooth muscle regulation was "myosin-linked" rather than thin filament-linked.

The experiments just described, however, did not exclude the possibility that the activation and phosphorylation, although correlated, were due to different activating proteins contained in the same crude protein fraction. Further progress was, therefore, crucially dependent on the isolation and purification of the calcium-dependent enzyme which phosphorylated myosin, termed myosin light-chain kinase (cf. Adelstein and Klee 1981). Could this protein with a molecular weight of 130 kDa alone play the part of the activator and restore calcium sensitivity? Surprisingly, the kinase lost its activity with increasing purification, suggesting that an important cofactor might have been lost. Indeed, one of the fractions separated by column purification reactivated the kinase fraction and made it calcium-sensitive. According to Dabrowska et al. (1977), the active principle was identical with, and could be replaced by the calcium-dependent regulator (CDR) of cyclic nucleotide diesterase, discovered by Cheung (1970) and later termed "calmodulin." The addition of calmodulin as cofactor was, however, only required when myosin light chain kinase was intact. After its partial digestion with certain proteinases it was transformed into a highly active calcium-insensitive enzyme that required neither calcium nor calmodulin for its activity (Walsh et al. 1982b).

Calmodulin Activation of Myosin Light-Chain Kinase. Calmodulin, as we recall (Chap. 4.3.2), is a highly coiled globular molecule with a MW of 17 kDa containing 148 amino acid residues organized in four calcium-binding domains (Babu et al. 1985). Each of these contains a binding site with a moderately high affinity for calcium (K_{Ca} about10^{-6}M, cf. Dedman et al. 1977). Thus, calmodulin is capable of sensing the free Ca^{2+} concentration in the physiologically important range of 10^{-7}M to 10^{-5}M. When occupied with calcium ions, calmodulin changes its shape just slightly. The α-helical content increases somewhat and hydrophobic sites from the interior of the molecule become exposed to the outside. Consequently, the protein becomes more hydrophobic and the affinity for calmodulin target proteins, such as myosin light-chain kinase, increases (cf. Adelstein and Klee 1981). The complex, formed from myosin light-chain kinase and calmodulin (CaM) occupied with four calcium ions, has an apparent dissociation constant (K_{CaM}) of only 10^{-9}M.

By increasing its affinity for myosin light-chain kinase in a calcium-dependent manner, calmodulin acts both as a calcium sensor and calcium switch for the regulation of the enzymic phosphorylation of smooth muscle myosin. For it is the ternary complex of calcium, calmodulin and myosin light-chain kinase that represents the active enzymic moiety. Its concentration, of course, will not only depend on the free Ca^{2+} concentration, but also on the concentration of free calmodulin. Thus, it is clear that the enzymic activity of myosin light-chain kinase will also depend on the concentration of both calcium and calmodulin. According to Blumenthal and Stull (1980), the Ca^{2+} concentration required for 50% enzymic activation (the calcium sensitivity; EC_{50}) may then be described by the relation:

$$\text{calcium sensitivity} = K_{Ca}\,(1 + K_{CaM}/[CaM])\,. \tag{8.1}$$

For instance, with the values (10^{-6}M and 10^{-9}M) given above for the dissociation constants, a Ca^{2+} concentration of 2 μM would be required for half-maximal activation at a calmodulin concentration of 10^{-9}M. This Ca^{2+} requirement would, however, be increased if the free calmodulin concentration or the affinity between myosin light-chain kinase and calmodulin were lowered. In this way, calcium activation of myosin light-chain kinase is inhibited when this enzyme is phosphorylated at two sites by cAMP-dependent protein kinase (Conti and Adelstein 1981; cf. Sect. 8.4.2).

As the calcium sensitivity of myosin light-chain kinase depends on the calmodulin concentration, it would be important to know this value in smooth muscle.

Most of the total calmodulin contained in smooth muscle in an apparent concentration of 30–50 μM is presumably bound to calmodulin-binding proteins (Grand and Perry 1979) so that only a very small fraction of calmodulin is available and responsible for smooth muscle activation (Rüegg et al. 1984a). In this context it should be remembered that calmodulin has innumerable functions in eucaryotic cells (Means et al. 1982).

Phosphorylation Hypothesis. It states that myosin light-chain kinase and calmodulin activate the actomyosin contractile system via phosphorylation of the regulatory myosin light chain. As reviewed by Hartshorne and Mrwa (1982), phosphorylation of myosin is obviously a key event in a cascade of reactions leading to smooth muscle activation (Fig. 8.2). When the Ca^{2+} concentration increases to 10^{-5}M, calmodulin binds four calcium ions and then reacts with myosin light-chain kinase to form an active ternary enzyme complex which in turn catalyzes the phosphorylation of myosin. Actin then interacts with phosphorylated myosin and thus increases its ATPase activity. In this way, the calcium signal is greatly amplified, since a complex formed of myosin light chain kinase and calmodulin occupied with four calcium ions is capable of phosphorylating and activating many myosin molecules. The price for such amplification, however, is slowness of response, so that there is a comparatively long delay of 0.2–1 s or even longer between an electric stimulus and the onset of smooth muscle contraction. According to Marston (1982), 1 kg gizzard smooth muscle contains 1.6 μmol myosin light-chain kinase with a molecular turnover number of 20 s^{-1} at 25° C. Thus, the myosin contained in 1 kg muscle (25 μmol) could easily be phosphorylated within the 5 s which a smooth muscle may take to contract, even if intact myosin is less readily phosphorylated than isolated light chains (cf. also Table 8.1).

Table 8.1. Activity of phosphorylating and dephosphorylating enzymes in vertebrate smooth muscle. Is it large enough to account for phosphorylation-induced contraction and relaxation?

Myosin light chains (content)	73 μM
Calmodulin (content)	30 μM
MLCK, Myosin light chain kinase (content)	0.36 μM
MLCK activity (K catal: 30 °C)	20 μmol P min^{-1} mg^{-1} protein
MLCK activity total (38 °C)	31 μmol P s^{-1} l^{-1} cell water
Phosphatase activity	0.26 s^{-1}
Phosphatase activity total	1.9 μmol P_i s^{-1} l^{-1} cell water
Extent of phosphorylation	0.65 mol P_i mol^{-1} P-light chain
Pseudo-ATPase activity[a]	14 μmol ATP s^{-1} l^{-1}
ATP consumption rate in vivo (38 °C)	145 μmol ATP s^{-1} l^{-1} cell water

[a] It may be calculated from these activities in bovine trachealis muscle that 65% of the myosin may be phosphorylated in 2 s at maximal activity of myosin light chain kinase and phosphatase, and the rate constant for dephosphorylation is 0.25 s^{-1}. These calculated rates are even higher than the experimentally determined rate of phosphorylation and dephosphorylation in vivo and the observed rates of contraction and relaxation. Note that pseudo-ATPase activity (originating from continued phosphorylation and dephosphorylation of myosin during steady state phosphorylation) amounts to approximately 10% of the total ATPase during the early phase of tension maintenance. During prolonged tension maintenance (in the "latch state"), the extent of phosphorylation, pseudo-ATPase and actomyosin ATPase all decrease to about 20% of the initial values. (From Kamm and Stull 1985a, b).

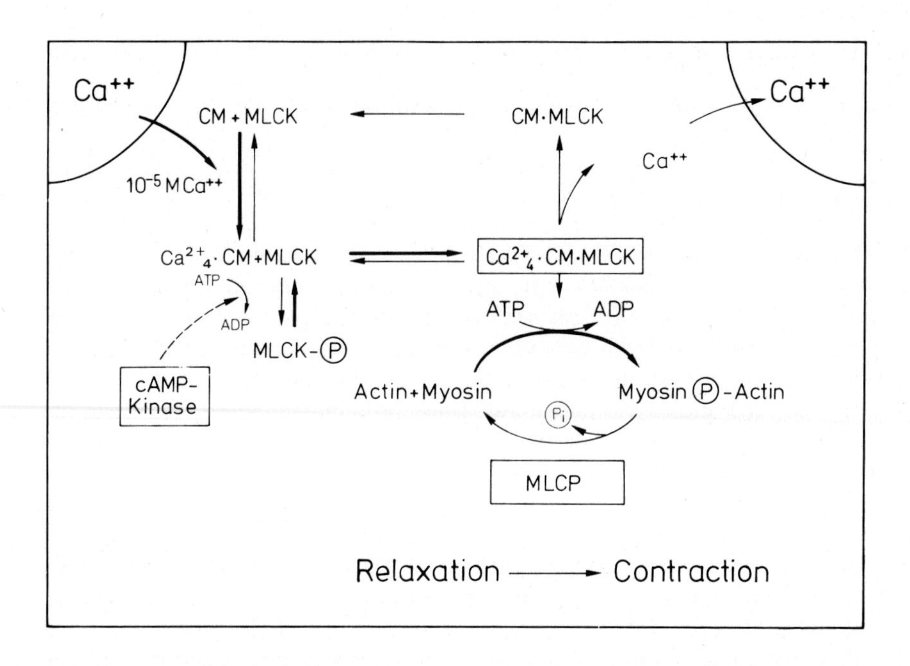

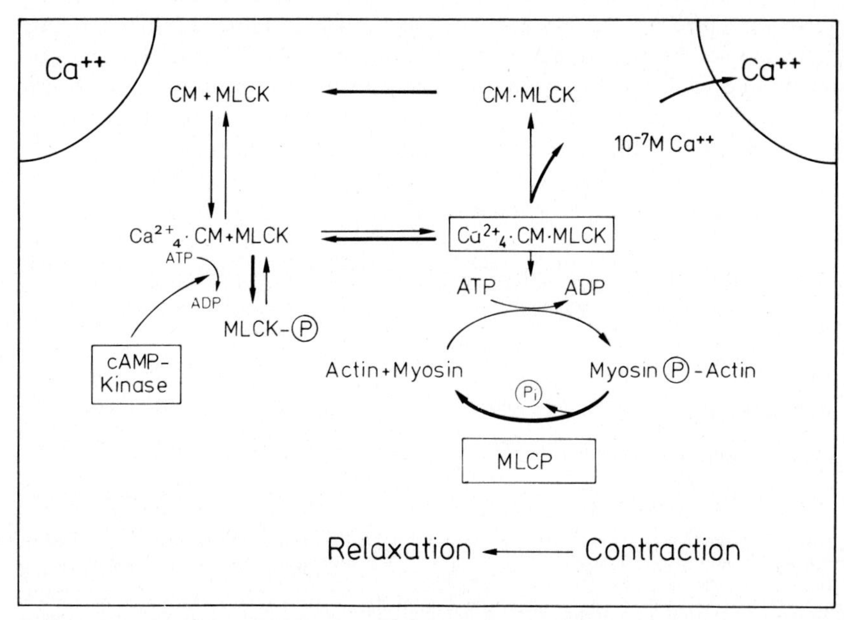

Fig. 8.2. Myosin phosphorylation regulates smooth muscle contraction. Activation is initiated by Ca^{2+}, *MLCK* (myosin light chain kinase) and *CM* (calmodulin) forming the active ternary complex which catalyzes the phosphorylation of the myosin light chain at serine residue 19 (formation of myosin-P). When Ca^{2+} is lowered to 10^{-7}M, the CM-MLCK complex decomposes and myosin is dephosphorylated by myosin-MLCP (myosin light chain phosphatase). Only phosphorylated myosin interacts with actin to form "cycling," ATP-hydrolyzing contractile crosslinkages. MLCK may be also inactivated by cAMP-dependent protein kinase forming MLCK-P (myosin light chain kinase phosphorylated at two sites). Adapted from Kamm and Stull 1985a

8.2.2 Myosin Phosphatase

We have seen that calcium activation of smooth muscle actomyosin ATPase involves phosphorylation of the regulatory myosin light chain. When the Ca^{2+} concentration is reduced to less than 10^{-7} M, the myosin light-chain kinase is repressed and myosin is dephosphorylated resulting in an inhibition of actomyosin ATPase. The enzyme catalyzing dephosphorylation is a myosin phosphatase which is spontaneously active, i.e. it does not require cations for its activation (DiSalvo et al. 1983). Its activity is the same in the presence and absence of calcium ions. Thus, during calcium activation of the actomyosin ATPase, the degree of myosin phosphorylation and, hence, ATPase activity should depend on the balance of kinase and phosphatase activity, as illustrated in Fig. 8.2. At high Ca^{2+} concentration the kinase activity would predominate over the phosphatase activity, thus causing a net phosphorylation until a steady state is reached when the rate of phosphorylation matches that of dephosphorylation. The latter, of course, depends not only on the phosphatase activity, but also on the concentration of myosin, that is already phosphorylated. At low free Ca^{2+} concentration, on the other hand, the phosphatase activity predominates, hence causing dephosphorylation of the light chains, inhibition of the actomyosin ATPase and relaxation of the contractile mechanism (cf. Haeberle 1985; Takai et al. 1987; Bialojan et al. 1988). Dephosphorylated myosin is less able to interact with actin and even if it does, its ATPase activity is rather low. Under these conditions, the rate-limiting step of the ATP-splitting mechanism, presumably the dissociation of the actomyosin-ADP state, appears to be inhibited (Marston 1982; Sellers 1985). It is possible, therefore, that the dephosphorylated light chain is an inhibitor of actin-myosin interaction in smooth muscle. Indeed, myosin subfragment S-1 is fully enzymically active in the presence of actin when the regulatory light chain has been removed by careful digestion with papain (Mrwa and Rüegg 1977). Besides, molluscan myosin lacking a regulatory light chain is fully activated by actin even in the absence of calcium, but can be inhibited by the addition of regulatory light chains of vertebrate smooth muscle (Kendrick-Jones and Scholey 1981, cf. Sect. 6.1.2.2).

8.2.3 A Futile Cycle of Myosin Phosphorylation and Dephosphorylation May Regulate Smooth Muscle Contraction

According to the phosphorylation hypothesis, the contractile state depends on the extent of myosin phosphorylation. Now suppose that during a maintained weak contraction at slightly elevated Ca^{2+} concentration both force and myosin phosphorylation are low, but in a steady state which results from a balance of phosphorylation and dephosphorylation processes. In this case, both reactions form a futile substrate cycle as defined by Newsholme and Crabtree (1976), involving continuous phosphorylation and dephosphorylation of myosin and, hence, a net ATP hydrolysis without, however, causing a net change in myosin phosphorylation. In a sense, such futile ATP expenditure by this pseudo-ATPase may be considered as a waste of energy. On the other hand, futile substrate cycling is also one of nature's ways of gaining a particularly sensitive control over a metabolic pathway. To illustrate this principle, let us consider how a small change in myosin light-chain kinase activity might be amplified by the substrate cycle to produce a large change in ATPase activity or force of the actomyosin-contractile system. Suppose, for instance, that phosphorylation and dephosphorylation occur concomitantly at a rate of 100 units s^{-1}. A small calcium-induced increase of myosin light-chain kinase activity by 10% would then result in a net myosin phosphorylation of 10 units s^{-1}, but a further increase by another 10% only would increase the net rate of phosphorylation to 20 units s^{-1} and, hence, presumably double the rate of contraction. In the absence of futile cycling, on the other hand, activation of myosin light-chain kinase would have led to a proportional increase in the rate of phosphorylation and contraction. The futile cycle also provides a means for rapid down-regulation of contractile activity. Consider again the case in which phosphorylation and dephosphorylation processes balance. A reduction of myosin light-chain kinase activity, e.g. by 10%, because of a tiny reduction in free calcium, would then cause net dephosphorylation and a slow relaxation. But both rate processes would be doubled if myosin light-chain kinase were further inhibited by an additional 10%, indeed, a remarkable amplification!

From an energetic point of view, futile cycling is, of course, a high price for keeping smooth muscle in a state of readiness to respond to a signal with either an increase or decrease in force. This is because it involves, as pointed out above, the continuous hydrolysis of ATP which is rate-limited by the activity of the myosin phosphatase and which may represent over 10% of the total energy turnover during contraction (cf. Table 8.1).

8.2.4 Alternative Mechanisms of Smooth Muscle Activation

The phosphorylation hypothesis just described is now widely accepted. It seems clear that in many instances phosphorylation of myosin is a prerequisite for smooth muscle contraction. The question arises, however, whether it is also per se sufficient for activation. In addition to calmodulin-dependent phosphorylation of myosin, at least three different mechanisms have been proposed for smooth muscle activation in the absence of additional phosphorylation of myosin: (1) Calcium binding to myosin; (2) activation by leiotonin; and (3) regulation by the thin-filament proteins tropomyosin, caldesmon and calponin.

8.2.4.1 Activation by Direct Calcium Binding to Myosin

According to Chacko and Rosenfeld (1982), calcium ions may activate the actomyosin ATPase even without further increasing the extent of myosin phosphorylation when they are bound to the regulatory light chains. In their experiments Chacko and Rosenfeld first phosphorylated smooth muscle myosin by myosin light-chain kinase and purified it from contaminating kinases and phosphatases by column chromatography. In this way, they obtained a stable-phosphorylated myosin which could be activated by actin even in the absence of calcium ions, suggesting that myosin phosphorylation was necessary and sufficient for activation. Surprisingly, however, this basal activation could then be further increased by the addition of micromolar calcium which was bound to the light chains, but if binding was inhibited by magnesium ions (5–10 mM), no extra activation was observed (cf. Kaminski and Chacko 1984). The calcium-dependent activation could therefore only be observed at very low Mg^{2+} concentrations (below 2 mM). However, high levels of intracellular free calcium may well affect shortening velocity as suggested by Barsotti et al. (1987; cf. also Siegmann et al. 1984).

8.2.4.2 Leiotonin

According to Ebashi (1980), leiotonin is a calcium-dependent activator of smooth muscle contraction; it binds calcium, requires tropomyosin as cofactor and is said to increase actomyosin-ATPase activity without any increase in myosin phosphorylation. However, confirmatory reports from other laboratories are lacking. Thus, it remains difficult to assess the importance of this regulatory system as well as that of other protein factors which have been found to activate smooth muscle actomyosin without increasing the extent of myosin phosphorylation (Persechini et al. 1981; Cole et al. 1983).

8.2.4.3 Thin-Filament Regulatory Proteins: Caldesmon, Calponin and Tropomyosin

In addition to myosin-linked regulation, vertebrate smooth muscles appear to have a calcium switch located on the thin filaments. Thus, Marston and Smith (1984) obtained from vascular smooth muscle a preparation of actin filaments that activated the magnesium-dependent myosin ATPase of skeletal muscle in the presence, but not in the absence, of trace calcium. Since purified actin is known to activate myosin even in the absence of trace calcium, Marston's experiments clearly show that the thin filament proteins must contain regulatory proteins conferring calcium sensitivity to the contractile system. The nature of these proteins, however, was puzzling. All attempts to prove the existence of troponin in these filaments have failed; rather than troponin, smooth muscle thin filaments may contain, in addition to actin and tropomyosin, another protein system, caldesmon and calmodulin, which may perhaps play a role analogous to that assigned to troponin-I and troponin-C in skeletal muscle.

Caldesmon. This is a major calmodulin- and actin-binding protein of MW 130 kDa that is quite abundant in smooth muscle; it is probably the calcium regulatory component of the thin filaments (Marston and Lehman 1985). In the absence of calcium ions caldesmon binds actin rather than calmodulin, thereby inhibiting actin-myosin interaction and, hence, contraction. When, however, the Ca^{2+} concentration increases to 1–10 μM, caldesmon (CaD) binds to calmodulin (CaM) instead of actin, so that its inhibitory action on actin-myosin interaction is relieved (Sobue et al. 1982; cf. Sobue and Sellers 1991).

$$\text{CaD}-\text{Actin}+\text{CaM} \xrightarrow{+\text{Ca}^{2+}} \text{CaD-CaM}+\text{actin}\,. \tag{8.2}$$

As reviewed by Lehman (1991), caldesmon binds to the N-terminal region of actin (S. Adams et al. 1990) and may block crossbridge attachment and hence contraction (Brenner et al. 1991). This blockage may be relieved by calmodulin when it binds to caldesmon in a calcium-dependent manner and, also, perhaps by caldesmon phosphorylation. Indeed, the inhibition of actomyosin-ATPase activity by caldesmon may be derepressed by its covalent phosphorylation (Sutherland and Walsh 1989).

Calponin, another thin filament protein, which is structurally related to troponin-T also inhibits actomyosin ATPase; again, this inhibition is derepressed by protein phosphorylation (T. Takahashi 1988). Calponin, as caldesmon, also binds to the N-terminal region of actin, and it may therefore compete with caldesmon as discussed by Lehman (1991). Since caldesmon inhibits contraction without affecting the state of myosin phosphorylation (Pfitzer et al. 1992), it seems therefore plausible that both caldesmon and calponin may modulate the relation between contraction and myosin phosphorylation or, for that matter, intracellular free calcium.

Tropomyosin. Phosphorylation of a thin-filament protein component by means of a calmodulin-dependent protein kinase increases calcium sensitivity of myofilaments so that calcium activation is possible even in the range below 1 μM. Under these conditions, a thin-filament calcium switch may, according to Marston (1982), be more important for regulation than activation via phosphorylation of smooth muscle myosin. One of the protein factors involved in increasing calcium responsiveness is presumably tropomyosin which is located in thin filaments (Marston and Smith 1984). According to Merkel et al. (1984), it also increases the calcium sensitivity of a hybrid actomyosin made up from skeletal muscle actin and smooth muscle myosin and supplemented with myosin light-chain kinase and calmodulin. At an intermediate Ca^{2+} concentration, myosin is partly phosphorylated and its ATPase partly activated, whereas at the same Ca^{2+} concentration addition of tropomyosin increases ATPase activity without increasing the extent of phosporylation.

In conclusion, therefore, tropomyosin apparently improves phosphorylation-contraction coupling and thus allows the contractile system to operate at a lesser degree of myosin phosphorylation.

8.2.5 Phosphorylation-Contraction Coupling in Actomyosin Systems and Intact Smooth Muscle

Because of the discovery of additional activating factors, the question arises whether or not actomyosin and, hence, contraction may be activated exclusively via the pathway of myosin phosphorylation.

In Vitro Studies. It is possible to work with enzyme systems that are seemingly devoid of "unwanted" extra factors. An interesting model is an "opened cell" of the alga *Nitella* that contains long strands of actin cables. When these were over-

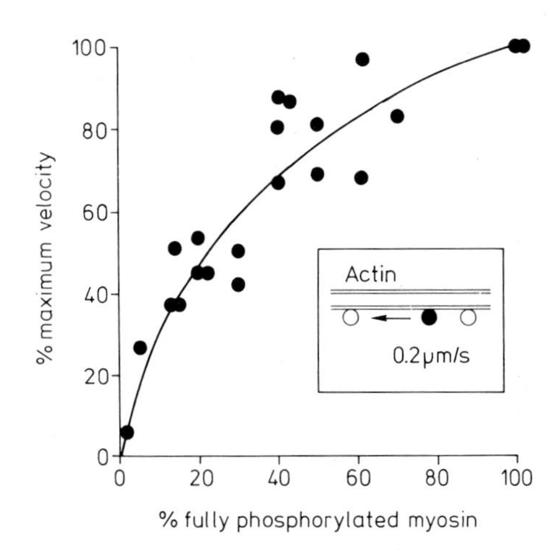

Fig. 8.3. Myosin phosphorylation increases the crossbridge cycling rate. Relation between phosphorylation of myosin and the velocity of migration of myosin-coated beads attached to actin strands of the alga *Nitella*. Fully phosphorylated and non-phosphorylated beads were mixed. Non-phosphorylated beads did not move, while fully phosphorylated myosin beads moved with a speed depending on the mixing ratio given on the *abscissa*. *Inset:* Coated fluorescent beads attached to actin strands in a medium containing 1 mM ATP, 4 mM $MgCl_2$, 10 mM KCl, 2 mM EGTA, pH 7. (Sellers et al. 1985)

laid with microspheres coated with phosphorylated smooth muscle myosin, the beads started to slide along the actin filaments in a manner analogous perhaps to the sliding of thin versus thick filaments in muscle contraction. As shown in Fig. 8.3, this movement is inhibited by the presence of non-phosphorylated myosin beads, which by themselves do not move at all, according to Sellers et al. (1985). These experiments clearly suggest that myosin phosphorylation alone may be essential for crossbridge cycling in smooth muscle, a conclusion that is also reached by biochemical studies on ATP hydrolysis by smooth muscle myosin. According to Chacko (1981), for instance, smooth muscle myosin-ATPase, is activated by actin alone, even in the absence of added calcium ions, provided myosin has been previously phosporylated.

However, if myosin phosphorylation were both necessary *and* sufficient for smooth muscle activation, it should be strictly correlated with contraction as well as with its biochemical correlate, the actomyosin-ATPase activity. As mentioned already, such correlations have been amply described (cf. Sobieszek and Small 1976). According to Persechini and Hartshorne (1981), however, the relation between the extent of phosphorylation and activation may be rather curvilinear: the relationship is steep at high degrees of phosphorylation, but there is very little activity if less than 50% of the light chains are phosphorylated, suggesting perhaps a cooperative action of the two phosphorylated myosin heads. Obviously, the light chains of both myosin heads must be phosphorylated to induce activation.

In Vivo Studies. Phosphorylation of light chains seems sufficient for activation as intracellular injection of a Ca^{2+}-insensitive, constitutively active myosin light-chain kinase causes surviving smooth muscle cells to contract (Itoh et al. 1989). In contrast to isolated, tropomyosin-free actomyosin preparations, intact smooth muscle cells are fully activated even if less than 50% of the myosin light chains are phosphorylated. Moreover, myosin phosphorylation precedes tension development in isometric contraction of smooth muscle, but subsequently, the extent of myosin phosphorylation may decrease dramatically, though force is maintained for prolonged periods (Aksoy et al. 1983); during relaxation myosin is rapidly dephosphorylated, while force declines slowly (Gerthoffer et al. 1984).

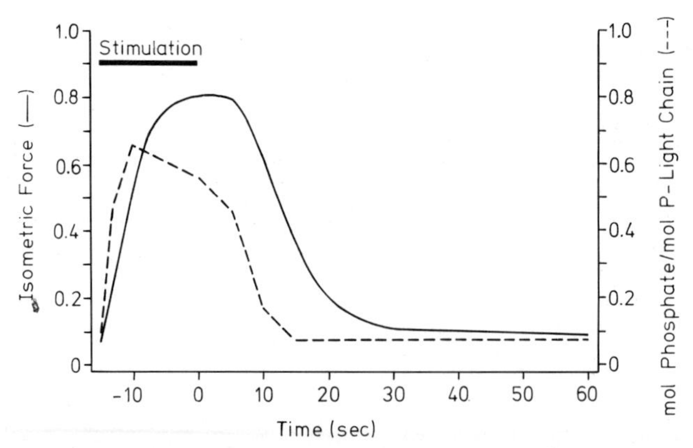

Fig. 8.4. Myosin phosphorylation (---) during contraction and relaxation (———) of bovine trachealis smooth muscle. (After Kamm und Stull 1985 b). Contraction of isolated trachealis muscle was induced by neural stimulation. At various stages a muscle was frozen and the amount of light chain phosphorylation was determined by isoelectric focussing techniques

In many instances contraction was not at all associated with myosin phosphorylation (Murray and England 1980). Very thorough studies of the relationship between myosin phosphorylation and force were carried out by Miller et al. (1983) and by Kamm and Stull (1985 b). To estimate and quantitate myosin phosphorylation, muscles were frozen in different states of contraction and after extraction of the proteins the phosphorylated and non-phosphorylated light chains were separated by isoelectric focussing techniques according to the difference in protein charge. When trachealis muscle was neurally stimulated, phosphorylation values of myosin and fibre stiffness increased within seconds from basal levels to 0.6 mol phosphate mol^{-1} regulatory light chain, while contractile tension rose more slowly (Kamm and Stull 1985 b, 1986). This temporal relationship of force and phosphorylation is illustrated in Fig. 8.4 which also shows a rapid dephosphorylation and relaxation after cessation of stimulation.

"Latch State." During prolonged neuronal or hormonal stimulation, the situation was, however, quite different (Fig. 8.5). While the extent of myosin phosphorylation increased in the first minutes of contraction, it decreased again during prolonged maintenance of tension (latch state). Relaxation could then be induced by the addition of isoprenaline without any change in the extent of myosin phosphorylation (Kamm and Stull 1985 a). Would this experiment not suggest that force production may be associated with phosphorylation, but that for continued contraction phosphorylation of myosin may not be necessary? Was there a non-phosphorylation-dependent activating system that could be inhibited by isoprenaline? It is interesting to note here, that while force is maintained, the rate of ATP splitting declined, indicating perhaps a slowing of crossbridge cycling (Butler and Siegman 1983). These findings have been taken to mean that there may be different regulatory mechanisms for force maintenance and for maintaining a high cycling frequency of crossbridges, as originally suggested by Dillon et al. (1981). While the latter may be phosphorylation-dependent, force may be

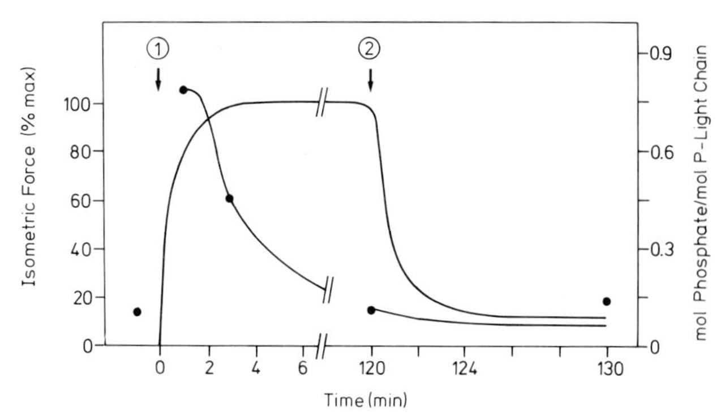

Fig. 8.5. Prolonged tonic contraction (———) of bovine trachealis smooth muscle elicited by acetylcholine (at *1*) and terminated by isoprenaline, a β-agonist (at *2*). Note that contraction is initiated by a transient phosphorylation of the myosin regulatory light chain (–●–), but is maintained at low levels of phosphorylation. The initial contraction phase is associated with relatively rapid crossbridge cycling (e.g. $1\ s^{-1}$), whereas prolonged contraction is due to slowly cycling dephosphorylated bridges (latch-bridges). (After Kamm and Stull 1985 a)

maintained by slowly cycling or non-cycling, dephosphorylated crossbridges (cf. Hai and Murphy 1988; Fischer and Pfitzer 1989; Paul 1990; cf. also Sect. 6.2.5). In contrast to actomyosin ATPase, unfortunately, the force-maintaining system cannot be studied with isolated contractile proteins in vitro, and there is, therefore, a definite need for a simplified system allowing the investigation of force regulation at a biochemical level. As will be discussed next, demembranated or skinned smooth muscle fibres provide an excellent model for such investigations and for testing the various hypotheses of calcium activation in smooth muscle.

8.2.6 Regulation of Contraction in Skinned Fibres

The nature of the intracellular regulatory mechanism controlling development and maintenance of force may be analyzed after isolation of the structurally intact contractile system from other cell components, such as the cell membrane, the sarcoplasmic reticulum, the mitochondria and the myoplasm. This functional isolation of the contractile machinery may be achieved by a skinning procedure with the detergent triton X-100 by which the membranes are destroyed, and much of the myoplasm is extracted (Gordon 1978). In this way, the contractile structures become accessible to and may be influenced at will by exogenously added regulatory proteins and enzymes, while the desired Ca^{2+} concentration can be precisely maintained by the addition of CaEGTA buffers. Since extracted fibres rapidly lose the low molecular weight components of the myoplasm, the bathing medium must, of course, be complemented by ATP, magnesium ions and other constituents, such as pH-buffers and reducing agents. Skinned fibres are relaxed when the free Ca^{2+} concentration is kept below 10^{-8}M. Increasing the Ca^{2+} concentration in the range of 1 to 10 μM increases force in a graded manner (Filo et al. 1965). But if the preincubation in relaxing solutions lasts for several hours, calmodulin becomes extracted and the capacity to respond to calcium is lost. It

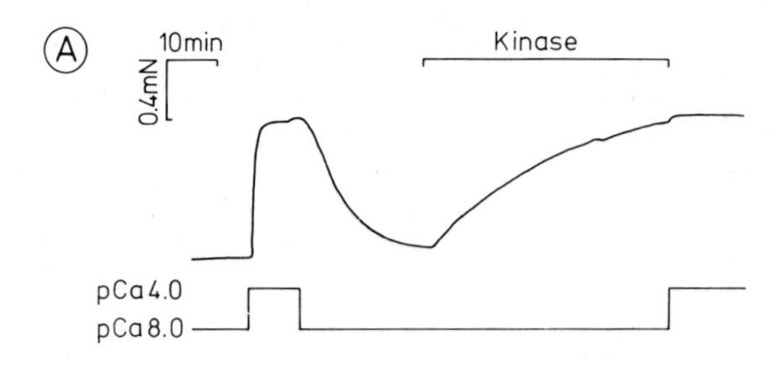

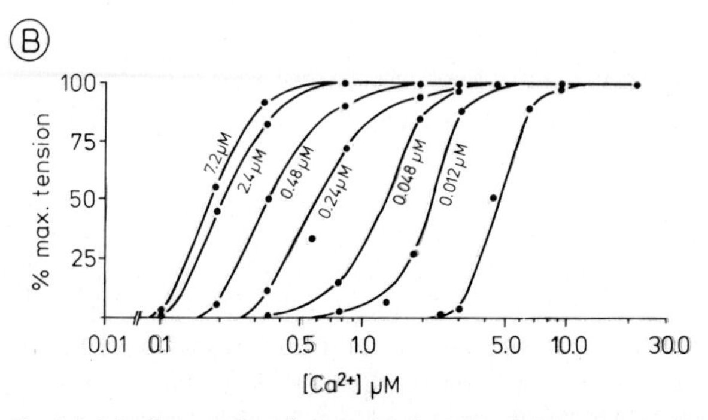

Fig. 8.6 A, B. Contraction via myosin phosphorylation. *A* Following a cycle of contraction and relaxation, the addition of proteolyzed (calcium-independent) myosin light chain kinase to the bathing solution causes a slow contraction of skinned chicken gizzard muscle fibres even at very low Ca^{2+} (pCa 8). The force development is associated with myosin phosphorylation and equal to Ca-induced force before (control) or after addition of calcium-insensitive MLCK. The bathing solution contained 70 mM KCl, 2 mM Mg-ATP, 1 mM Mg^{2+} and Ca^{2+} as indicated. (Walsh et al. 1982a). *B* Dependence of force on Ca^{2+} concentration in skinned fibres of guinea pig taenia coli suspended in ATP salt solution at various concentrations of calmodulin (as indicated on the curves) in the bathing solution. Note that calmodulin increases calcium sensitivity. (Sparrow et al. 1981)

may, however, be restored by the addition of calmodulin in less than micromolar concentrations (Sparrow et al. 1981). As the calmodulin concentration is increased, the Ca^{2+} concentration required for 50% activation is progressively reduced (Fig. 8.6). This suggests the calcium-calmodulin complex as the regulatory moiety (Fig. 8.6B). During contraction, furthermore, myosin light-chain phosphorylation increases from about 0.1 mol phosphate mol^{-1} regulatory light chain to about 0.6 mol. If the Ca^{2+} concentration was lowered again, the light chains were dephosphorylated and the skinned fibres relaxed. A similar kind of relaxation was obtained after addition of trifluoperazine and other "calmodulin antagonists" (listed in Table 7.1), which are bound to calmodulin in a calcium-dependent manner and inhibit myosin light-chain kinase activity (Hidaka et al. 1979). In conjunction then, all of these findings implicate a calmodulin-dependent regulation of force development via phosphorylation of myosin light chains. However, we are left with a problem!

How can it be safely excluded that phosphorylation, although necessary, may per se not be sufficient for inducing contraction. For instance, calcium binding to myosin may be another necessary step involved in activation. To solve this problem it would be desirable to phosphorylate the light chains of skinned fibres without simultaneously raising the free Ca^{2+} concentration. These crucial experiments became possible after Walsh et al. (1982b) succeeded in rendering the myosin light chain kinase calcium insensitive by controlled proteolysis. This enzyme phosphorylates light chains even in the absence of calcium ions, thereby activating the actomyosin ATPase fully. Furthermore, after addition of calcium-insensitive kinase, relaxed skinned fibres rapidly developed maximal force, and the regulatory light chains became phosphorylated, although the free Ca^{2+} concentration remained low (Fig. 8.6; cf. Walsh et al. 1982a). Smooth muscle myocytes also contracted after intracellular injection of Ca^{2+}-insensitive myosin light-chain kinase (Itoh et al. 1988). Clearly, therefore, myosin phosphorylation alone was both necessary and sufficient for turning on the contractile machinery and for inducing maximal activation of ATP splitting and force production. Equally important, of course, was the converse experiment, the experimental demonstration that inhibition of actomyosin ATPase and relaxation of skinned fibres by calcium removal were entirely due to the concomitant dephosphorylation of myosin rather than to the lowering of the free Ca^{2+} concentration per se. Indeed, addition of the highly purified myosin phosphatase to actomyosin reduced its ATPase activity even though the Ca^{2+} concentration was kept constant; similarly, force development of skinned fibres was inhibited by myosin phosphatase (Bialojan et al. 1985). The importance of myosin phosphorylation for smooth muscle contraction was further demonstrated by using the ATP analogue ATP-γ-S as a substrate. With this compound light chains became irreversibly thiophosphorylated, and thiophosphorylated skinned fibres contracted even in calcium-free solution (Kerrick et al. 1980).

There are, however, skinned-fibre experiments that do not readily fit into the framework of the phosphorylation scheme. For instance, the speed at which skinned fibres can shorten without load also depends on the Ca^{2+} concentration, but the calcium dependence seems to be different from that of force activation. Force development requires less calcium for half-maximal activation and saturates at lower free calcium levels than unloaded shortening velocity (Paul et al. 1983) and activation of the ATP-splitting rate of fibres (Arner and Hellstrand 1983). Conversely, force production requires a higher free Ca^{2+} concentration and a higher degree of myosin phosphorylation than force maintenance (Chatterjee and Murphy 1983). These studies in skinned fibres may be helpful in the analysis of related phenomena in intact smooth muscle. We have already mentioned that force can be maintained for a prolonged period even though the calcium transient is brief, and intracellular free calcium (Morgan and Morgan 1984a) as well as the extent of myosin phosphorylation (Miller et al. 1983; Kamm and Stull 1985a) may return to low values soon after stimulation. The results seem to favour the hypothesis that ATPase and force development may be regulated by phosphorylation, while maintenance of force may perhaps be dominated by other calcium-dependent regulatory systems. The variable relationship between phosphorylation and contraction must indicate, of course, that phosphorylation cannot be the ultimate determinant of contractile force. Additional, yet unknown, mechanisms must be postulated to couple myosin phosphorylation and contraction. In addition, there is also a variable relationship between force maintenance and Ca^{2+} concentration which will be considered in more detail in the following section.

8.3 Regulation of the Intracellular Calcium Ion Concentration

The intracellular calcium ion concentration is the principal determinant of smooth muscle tone; we also have seen that calcium ions activate the contractile mechanisms mainly via phosphorylation of myosin. Now we address the questions of how much calcium must be released into the myoplasm to activate smooth muscle contraction, how the release process is itself regulated and how the intracellular Ca^{2+} concentration is lowered again during relaxation.

8.3.1 Intracellular Free Calcium

In resting smooth muscle the intracellular Ca^{2+} concentration determined with Fura-2 (Williams et al. 1985; Pritchard and Ashley 1986, cf. Fig. 8.7A) and by calcium-selective electrodes is approximately 0.1–0.2 μM. Stimulation by depolarizing electric currents causes an increase in intracellular free Ca^{2+} concentration that reaches a maximum value and then declines to low resting values within seconds, while the time course of contraction is much slower. In the experiments of Morgan and Morgan (1982), application of the vasoconstrictor angiotensin caused a more sustained contraction associated with a biphasic change in free calcium. The latter increased at first transiently, but this phase was followed by a much smaller and more sustained transient. Since force remained high, it seemed clear that tension maintenance required a lower Ca^{2+} concentration than force production. This conclusion was strengthened by studies on the phasic vascular smooth muscle of ferret portal vein (Morgan and Morgan 1984a). Here again, application of phenylephrine resulted in a large, but brief, calcium transient while force was maintained. Potassium-induced contractures were equally large, but the concomitant calcium transients were sustained (Fig. 8.7B, C).

Figure 8.8 shows an experiment of Remboldt and Murphy (1986) in which they stimulated a tonic smooth muscle, the hog carotid artery, with high K^+ saline and then lowered the K^+ concentration. At first the myoplasmic Ca^{2+} concentration rose (to 1 μM), thus causing an increase in force. After partial reduction of the K^+ concentration, however, force remained high, though the free Ca^{2+} concentration returned to nearly basal levels. Under these conditions, force seems to be maintained by extremely slow cycling crossbridges (latch state!), as also suggested by the experiments of Siegman et al. (1985) and by the previously mentioned experiments of Kamm and Stull (1985a). In conclusion, the relationship between intracellular free Ca^{2+} concentration and force is not unique, but depends on the stimulation pattern and the duration of contraction (Bradley and Morgan 1987; Somlyo and Himpens 1989; Suematsu et al. 1991; cf. Fig. 8.7).

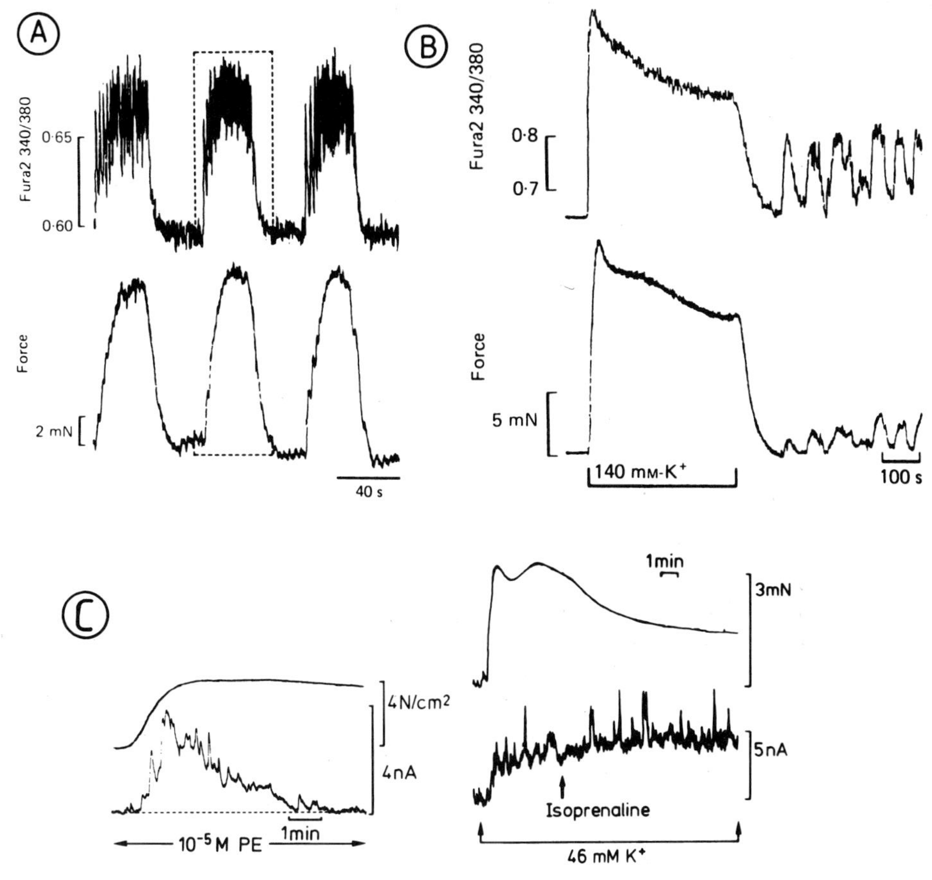

Fig. 8.7. A Spontaneous rhythmic contractions of isolated guinea-pig ileum triggered by calcium transients resembling calcium action potentials superimposed on slow waves of membrane potential (Himpens and Somlyo 1988). *Upper trace*: calcium transients (determined by fura-2 fluorescence ratio signal at 340 and 380 nm). Tension fluctuations are associated with and probably due to spontaneous oscillation of the membrane potential (not shown) on which calcium action potentials are superimposed (cf. Liu et al. 1969; Connor et al. 1977). *B* Depolarization with 140 mM K^+ induces an increase in intracelluar [Ca^{2+}] and contraction. Note correlation of force and contraction (Himpens and Somlyo 1988). *C* Force and calcium transient (noisy tracing, photomultiplier output of aequorin luminescence) in a vertebrate smooth muscle (ferret portal vein) after addition of 10 μM phenylephrine (*PE*) or following membrane depolarization with 46 mM KCl. Note that following addition of *PE* force is maintained, though the calcium transient is brief. K-depolarized muscle relaxes with isoprenaline while aequorin luminescence persists. (From Morgan and Morgan 1984 a, b, with permission)

The level of intracellular Ca^{2+} concentration of smooth muscle depends on a multitude of regulatory processes which are schematically shown in Fig. 8.9 (cf. A.P. Somlyo 1985 b). These involve calcium influx through calcium channels at the cell membrane, calcium release and reuptake by internal calcium stores, such as the sarcoplasmic reticulum or subsarcolemmal binding sites at the membrane, and lastly, calcium extrusion through the cell membrane. Before discussing these

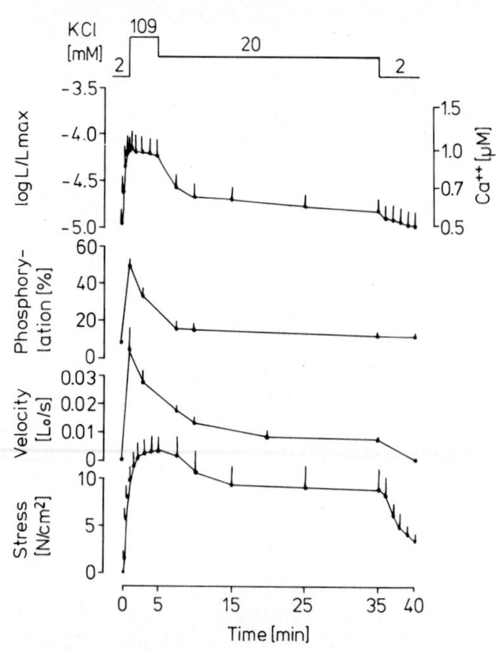

Fig. 8.8. Relationship between intracellular Ca^{2+} concentration (detected by aequorin luminescence L/L_{max}) and force during and after a K-induced membrane depolarization. Also shown is the extent of myosin light chain phosphorylation and velocity of unloaded shortening measured by isotonic releases. Note that a partial reduction of external K^+ and membrane repolarization barely reduces force, though the internal calcium levels and the extent of phosphorylation as well as the capacity to shorten quickly and the rate of energy expenditure (not shown) are drastically reduced: latch state! Preparation: hog carotid artery at 37° C. (After Remboldt and Murphy 1986, with permission of the authors and the American Heart Association)

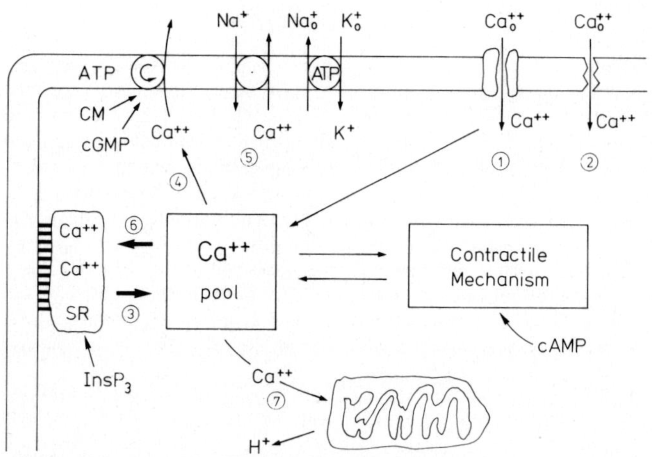

Fig. 8.9. Summary of calcium homeostasis in smooth muscle. Intracellular free calcium levels depend on the balance of calcium release and uptake mechanisms. Calcium release into myoplasm occurs via voltage- and receptor-operated membrane channels (*1, 2*) and by release from the sarcoplasmic reticulum (*SR*) (*3*). Calcium removal from the myoplasm is effected by the calcium pump of the sarcoplasmic reticulum (*6*) and of the sarcolemma (*4*) or by the Na-Ca exchanger (*5*), while mitochondrial calcium binding is unimportant under physiological conditions (*7*). The membrane calcium pump is Ca^{2+}-calmodulin (*CM*)-dependent and stimulated by cGMP-dependent protein kinase

various calcium movements let us first consider the excitation phenomena at the level of the cell membrane which precede and are responsible for increased calcium influx or calcium release from the sarcoplasmic reticulum.

8.3.2. The Relative Importance of Membrane Depolarization for Activation

Electromechanical Coupling. In many smooth muscles contraction is mediated by action potentials which are dependent on the extracellular free Ca^{2+} concentration and must, therefore, be due to an inward current of calcium ions (Johansson and Somlyo 1980). Continued maintenance of force may then be ascribed to bursts of action potentials causing a tetanic contraction which may differ from the analogous phenomena of skeletal muscle only by the slowness and the low fusion frequency of the constituent twitches; force and frequency of action potentials are then often correlated (Bülbring 1955). In spontaneously active smooth muscle, such as intestinal smooth muscle, uterus, the ureter or the portal vein, action potentials are myogenic, that is they originate in pacemaker cells whence they spread to other smooth muscle cells (the follower cells) via gap junctions between neighbouring membranes. Stretch, hormonal or neurogenic influences may increase the contractile force by increasing the action potential frequency, while spontaneous rhythmical fluctuations of pacemaker firing frequency may cause rhythmical contractile activity, such as the movements of the gut.

In contrast to myogenic tone, the neurogenic tone of less spontaneously active smooth muscles, such as in the vas deferens and certain blood vessels, is dependent on the firing of innervating sympathetic fibres, which by releasing noradrenaline as neurotransmitter, may elicit excitatory postsynaptic potentials which then summate and trigger action potentials (Hirst 1977; Hirst and Neild 1980a, b); these action potentials can, therefore, only be elicited at increased stimulation frequency. For instance, by increasing the rate of firing, the sympathetic nerves cause vasoconstriction of resistance vessels and, hence, a rise in blood pressure.

In smooth muscles of larger arteries application of noradrenaline induces a prolonged contraction which is elicited by long-lasting membrane depolarization rather than by bursts of action potentials (A.P. Somlyo and Somlyo 1968). In these tonic smooth muscles calcium-dependent, spike-like action potentials can only be elicited if the potassium channels of the membrane are blocked pharmacologically (Harder et al. 1979). Without this intervention electric depolarizing stimuli would not only open calcium channels of the membrane, thereby causing a calcium influx, but they would also activate potassium channels. As a result, there is an increased efflux of potassium ions supporting a current which compensates the inward calcium current and, therefore, prevents the generation of a regenerative calcium action potential. It is worth noting here that the activation of potassium channels and its damping effect on excitation provides a feedback loop in excitation-contraction coupling, since it is itself due to increased intracellular Ca^{2+} concentration. Electromechanical coupling, i.e. the relationship between membrane potential and contraction, has been studied extensively. In coronary smooth muscle, for instance, the resting potential has a value of -55 mV, but when the potential is increased to -45 mV, a mechanical threshold is reached (Ito et al. 1979) above which the internal Ca^{2+} concentration of smooth muscle increases (cf. Neering and Morgan 1980) so that the muscle develops force. Half-maximal and maximal contractile activation are obtained at -30 mV and -10 mV, respectively.

Pharmacomechanical Coupling. In some vascular smooth muscle, however, membrane depolarization does not appear to be a prerequisite for contraction (A. P. Somlyo and Somlyo 1968). For instance, noradrenaline may elicit contraction of rabbit ear arteries without any change in membrane potential (Droogmans et al. 1977), and in coronary arteries acetylcholine may elicit a contraction without causing depolarization (Ito et al. 1979) provided the smooth muscle is not covered by the endothelium. If the endothelium is present, however, stimulation by acetylcholine or anoxia causes relaxation without alteration of the membrane potential (Furchgott 1983). These examples of non-electric or pharmacomechanical coupling (A. V. Somlyo and Somlyo 1968) between the cell membrane and the contractile structures demonstrate that membrane depolarization is not always essential in smooth muscle activation. The crucial step in activation is the release of calcium ions into the myoplasm regardless of whether this is caused by electromechanical or pharmacomechanical coupling.

8.3.3 Calcium Channels and Calcium Influx

Whereas "tonic" smooth muscles, e.g. in veins and arteries, are often capable of maintaining contractile tension for prolonged periods even in the absence of calcium ions in the external fluid, this is not the case in so-called phasic smooth muscles, such as in portal vein or guinea pig taenia coli. Following a stimulus these muscles respond with a transient (phasic) contraction, suggesting that continued activity of the contractile machinery may depend on calcium supply from the outside of the muscle cell. Calcium ions may cross the cell membrane through different types of calcium channels which may be distinguished by their dependence on the membrane potential and their different susceptibility to calcium channel-blocking drugs, such as the "calcium antagonists" verapamil, nifedipine and diltiazem (cf. Fleckenstein 1983; Sperelakis and Caulfield 1984).

Calcium Channels and Channel Blockers. The channels that are most readily blocked by calcium antagonists are potential-sensitive in as much as they open when the membrane potential becomes more positive and close at more negative potentials. It is these voltage-dependent channels which allow the influx of activator calcium during potassium-induced contraction and they may also be engaged in the generation of smooth muscle action potentials. Figure 8.10 shows an experiment of Klöckner and Isenberg (1985) in which they measured the dependence of the "inward calcium current" in isolated smooth muscle cells on the membrane potential, using the patch clamp technique developed by Neher and Sakmann (Hamill et al. 1981). For recording, a small membrane patch containing only one or two calcium channels is slightly sucked into a micropipette, as previously described in Chap. 7, Fig. 7.2. After applying a constant voltage between

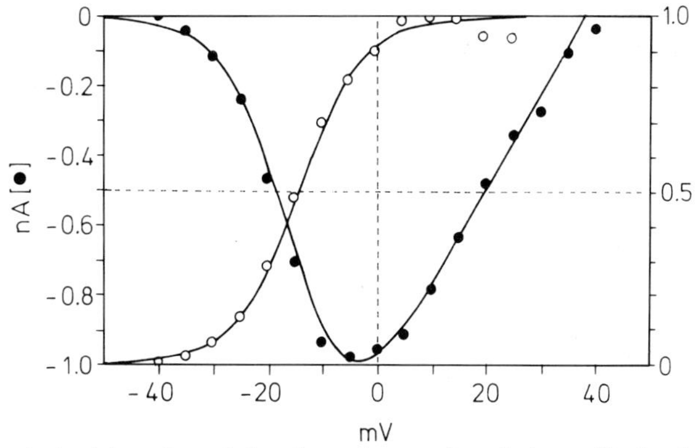

Fig. 8.10. Evidence of voltage-dependent calcium channels in cell membranes of patch clamped isolated smooth muscle cells from the urinary bladder of guinea pigs. Influence of the membrane potential (voltage clamped) on the peak calcium inward current (*closed circles*) and on the peak calcium conductance of the membrane (*open circles*). *Abscissa:* Clamped potential reached after a depolarizing step. *Ordinate:* calcium current (*left*) or relative conductance (*right*). (Klöckner and Isenberg 1985)

the electrode within the pipette and the bathing solution, a measurable current will flow across the ionic channels of the membrane only, since the seal between membrane and pipette is absolutely tight. Under these conditions, the current flows in an "all-or-none" pulsed manner depending on whether the channels are open or closed. The opening probability of calcium channels and, hence, the calcium conductivity increases when the membrane depolarizes, but following prolonged depolarization, the channels become inactivated. These mechanisms are of greatest physiological and pathophysiological importance. Let us think, for instance, of the resistance vessels, the small arterioles regulating our blood pressure. In these, potential-dependent calcium channels of the smooth muscle cell membrane are probably gated when noradrenaline is released from the sympathetic nerve endings, thereby reacting with the so-called γ-receptors rather than with the α_1-receptors (Hirst and Neild 1980 a). Indeed, noradrenaline modulates potential-dependent L-type calcium channels of small ear arteries even after blockage of α_1-receptors in that it increases the rate of calcium influx caused by membrane depolarization (Benham and Tsien 1988). Rather than recruiting new channels, noradrenaline seems to increase the open probability of voltage-dependent calcium channels of arterial smooth muscle, in fact, so much that they open already at physiological membrane potentials (Nelson et al. 1988). Thus, a small depolarization in the presence of noradrenaline, might cause a comparatively large calcium influx that might be prohibited by an equally small membrane hyperpolarization which occurs e.g. after activating potassium channels. It will also be stopped by Ca-channel blockers (Nelson and Worley 1989). Noradrenaline modulation of voltage-dependent L-type calcium channels will therefore be important in promoting sympathetic vasoconstriction in resistance vessels that are very sensitive to organic calcium antagonists. In fact, the blockage of these channels by calcium antagonists has been found to be extremely useful in the management of hypertension (cf. Fleckenstein 1983, for review of the early work).

Other types of calcium channels seem to be less readily blocked by calcium channel blockers (Meisheri et al. 1981; Benham and Tsien 1987). These channels open independently of the membrane potential when a receptor is occupied by its agonists. They have therefore been called receptor-operated channels. For instance, by combining with α_2-receptors, noradrenaline is said to cause a calcium influx and elicit a contraction without any change in the membrane potential (pharmaco-mechanical coupling) of vascular smooth muscle (Hondeghem and Katzung 1986), but the extent to which α_2-receptors open calcium channels that are less sensitive to channel blockers is questionable (Cavero et al. 1983). According to Weiss (1981), calcium may also flow into the cell through leakage channels which cannot be modified by the membrane potential, or calcium antagonists, but they can be blocked by lanthanum.

Calcium Influx Measurement. The calcium entering the myoplasm can be estimated from the size of the "calcium membrane current" or by determining the time-dependent uptake of $^{45}Ca^{2+}$ by vascular smooth muscle (van Breemen et al. 1973). The maximal influx of calcium during a noradrenaline- or potassium-induced contraction may be as large as 200 nmol g^{-1} tissue. These values have been confirmed by Bond et al. (1983), who used electron-probe techniques for measuring the total intracellular mypolasmic calcium; they suggested that the calcium released into the myoplasm would be more than sufficient to saturate the four binding sites of the calcium-binding protein calmodulin occurring in an intracellular concentration of about 30 μM (Grand and Perry 1979). Since depolarization- and noradrenaline-induced calcium influx as well as contraction can be inhibited by calcium channel blockers (van Breemen et al. 1980), it would appear that activator calcium predominantly stems from the extracellular space, at least in the case of phasic smooth muscles, in which the sarcoplasmic reticulum is less important (Sugi et al. 1982). This view has, however, been challenged by Bond et al. (1984), who showed the importance of intracellular calcium release for smooth muscle contraction.

8.3.4 Calcium Release from Intracellular Stores and Phosphoinositide Metabolism

As mentioned already, larger arteries and veins contain smooth muscles that are capable of sustained tonic contraction even when calcium is removed from the extracellular space. In such tonic smooth muscles the sarcoplasmic reticulum (SR) is much more abundant than in phasic smooth muscles (Devine et al. 1972); it forms an internal calcium store and may play a major role in the regulation of contractility.

Evidence for Intracellular Stores. Concentrations of calcium antagonists sufficient to block the calcium entry through potential-dependent channels and, therefore, suppress potassium-induced contraction do not inhibit the contracture induced by noradrenaline. Even when the membrane is made completely impermeable to calcium ions by the application of lanthanum, noradrenaline is still able to evoke *one* contraction. This response, the "noradrenaline-release" contraction, is mediated through α_1-receptors; it must obviously be due to the release of calcium from internal stores which then become emptied so that a second application of noradrenaline is usually no longer effective (van Breemen et al. 1973, 1980). These stores may, however, be refilled either by reuptake of calcium from the myoplasm or perhaps through receptor-operated calcium channels during a prolonged exposure to noradrenaline (Saida and van Breemen 1984).

In coronary smooth muscles, the existence and intracellular localization of calcium stores could be demonstrated by Itoh et al. (1982a), who found that both acetylcholine and caffeine evoked a contraction after incubation in calcium-free saline. Both contractions seem to be caused by calcium release from the same intracellular store as in the case of guinea pig taenia coli (cf. Endo et al. 1980; Casteels and Raeymaekers 1979). This caffeine-labile store is functional in muscle fibres even if the outer membrane has been made permeable by chemical skinning procedures involving saponin: skinned fibres are relaxed when kept in an ATP salt solution containing EGTA (0.1 mM) to keep the free Ca^{2+} concentration below 10^{-7}M, but they contract following the addition of caffeine, suggesting that calcium has been released from an internal store. These internal stores involve diadic junctions with the cell membrane, and the "deep" sarcoplasmic reticulum which is not in contact with the cell membrane.

What is the physiological signal that causes a release of calcium from the deep stores when the α_1-receptors of the cell membrane are occupied by noradrenaline? Since there is no T-system in vertebrate smooth muscle, the signal transmission from membrane to stores must be chemical. Indeed, calcium itself has been proposed as the second messenger, as calcium released from subsarcolemmal calcium pools may diffuse to the internal calcium stores where it causes calcium-induced calcium release (Saida and van Breemen 1983). However, other intracellular messengers may be primarily implicated.

Mobilization of Calcium by Phosphoinositide Metabolites. When noradrenaline activates α_1-receptors, it causes a cascade of reactions involving the breakdown of phosphoinositol lipids of the cell membrane by phospholipases, the phosphatidylinositol response (Fig. 8.11). Michell et al. (1981) suggested that this reaction sequence may be essential for the activation by the receptor and the mobilization of intracellular calcium within the cell.

Inositol Trisphosphate (Fig. 2.11). As one of the breakdown products of phospholipids which is formed within seconds after receptor occupation, inositol trisphosphate has been shown to release calcium from internal stores of secretory cells of the pancreas (Streb et al. 1983; Berridge and Irvine 1984) and from stores of the smooth muscle following activation of α_1-receptor by noradrenaline (Suematsu et al. 1984; A.V. Somlyo et al. 1985; Hashimoto et al. 1986; cf. Fig. 8.12). Alpha-1-mediated smooth muscle activation is, therefore, brought about by a mechanism very different from noradrenaline activation induced by α_2-receptor (and γ-receptor) activation which may cause an increased calcium influx into the cell. Both effects, however, will be counteracted by β-receptor activation by catecholamines which causes smooth muscle relaxation, as discussed in more detail in Sect. 8.4.

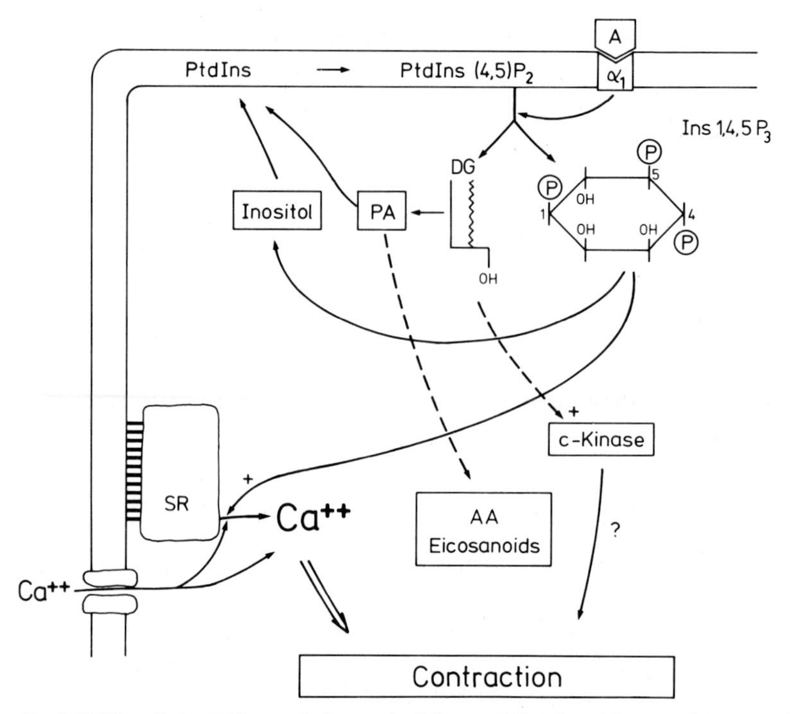

Fig. 8.11. Phosphoinositide metabolites and calcium mobilization. After agonist occupation of a membrane receptor phosphatidylinositol bisphosphate [*PtdIns (4,5) P_2*] is broken down to diacylglycerol (DG) and inositol triphosphate (*Ins 1,4,5 P_3*). Both compounds are recycled to form phosphoinositides (*PtdIns*). Diacylglycerol may be rephosphorylated to form phosphatidic acid (*PA*), which is a substrate for the formation of arachidonic acid (*AA*) and the eicosanoids leukotrienes and prostaglandins. *Ins 1,4,5 P_3* is a second messenger which induces the release of calcium from the sarcoplasmic reticulum (which is also induced by calcium entering the cell through voltage-dependent and receptor-operated calcium channels). *DG* is a second messenger activating the lipid-dependent protein kinase-C, which may cause a sustained contractile response. (After Berridge 1985, cf. also Berridge 1987)

Diacylglycerol (Fig. 2.11). As another second messenger, diacylglycerol is formed together with inositol triphosphate by the action of phospholipase C (a phosphodiesterase) on membrane-bound polyphosphoinositides, e. g. phosphatidylinositol 4,5 bisphosphate (Berridge and Irvine 1984; cf. Fig. 8.11). Whereas the calcium-mobilizing inositol trisphosphate is rapidly degraded, the effect of diacylglycerol is longer-lasting, since it activates a lipid-dependent protein kinase (kinase C) causing the phosphorylation of specific target proteins. Protein kinase C is a polypeptide of MW 77 kDa which may be activated by high concentrations (10^{-5}M) of calcium. In the presence of diacylglycerol, certain phospholipids or the tumour-promoting phorbolester (TPA), the calcium requirement of the protein kinase C is markedly reduced and because of this sensitivity modulation, the enzyme is now active even at resting levels of intracellular calcium. Protein kinase C has also been implicated in the activation of smooth muscle contraction since it is capable of phosphorylating a threonin residue of the regulatory light chains (Endo et al. 1982). Furthermore, TPA, activating kinase C may induce a slow and pro-

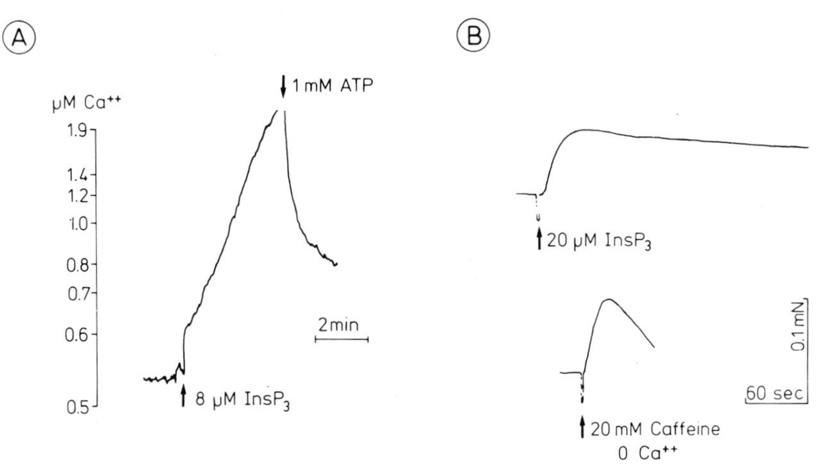

Fig. 8.12 A, B. Evidence of inositol trisphosphate (*InsP*$_3$)-induced calcium release from the sarcoplasmic reticulum. (A. V. Somlyo et al. 1985). *A* Detection of calcium release with a calcium electrode. Addition of 8 μM InsP$_3$ causes an increase in free Ca^{2+} concentration in the bathing solution containing saponin skinned (permeabilized) strips of rabbit main pulmonary artery smooth muscle. *B* InsP$_3$-induced calcium release (*above*) and caffeine-induced calcium release (*below*) cause similar increases in force of skinned pulmonary artery smooth muscle. The calcium stores were loaded by preincubating the fibre in ATP salt solution containing CaEGTA buffer. Subsequently, fibres were immersed in relaxing solution containing low levels (0.2 mM) of EGTA in order to facilitate the detection of released calcium (cf. also Walker et al. 1987).

longed contraction of smooth muscle. The question arises, therefore, whether diacylglycerol-mediated activation may be capable of eliciting a contraction even at resting levels of intracellular free calcium (Rasmussen and Barrett 1984). The research in this field is, however, rather controversial since phosphorylation of the threonin residue of regulatory light chains in smooth muscle has been reported to inhibit rather than activate the actomyosin ATPase (Nishikawa et al. 1984b). Furthermore, diacylglycerols serve as precursors for the formation of arachidonic acid and its derivatives, the prostaglandins and the leukotrienes, which are local hormones for intercellular communication and regulation of smooth muscle contraction.

8.3.5 Calcium Reuptake by the Sarcoplasmic Reticulum and Calcium Extrusion Through the Cell Membrane

Whereas potassium-induced contraction of smooth muscle is supported by calcium influx, relaxation due to reduction of potassium concentration is associated with calcium efflux. Calcium extrusion may be due either to active ATP-dependent transport across the cell membrane (van Breemen et al. 1973; Casteels and van Breemen 1975) or it may be due to a Na-Ca-exchange mechanism in which intracellular calcium ions are traded for extracellular sodium, as proposed by Reuter et al. (1973) and Blaustein (1977) and discussed by Pritchard and Ashley (1986). According to Droogmans et al. (1985), the exchange mechanism may, however, be of much less importance than the sarcolemmal calcium pump which has recently been isolated (Wuytack et al. 1985) and shown to be stimulated by a calcium- and calmodulin-dependent protein kinase. Interestingly, this ion pump is also activated by cyclic guanosine monophosphate (cGMP), and inhibited by the hormone ocytocin which induces uterine smooth muscle contraction (Popescu et al. 1985a, b).

While the transport processes through the cell membrane appear to be important for the calcium homeostasis of the cell and also for controlling the level of intracellular calcium during contraction, they may, however, not be directly responsible for smooth muscle relaxation. The reason is that the latter always precedes calcium efflux unless intracellular calcium sequestration is inhibited by caffeine. Moreover, after noradrenaline-induced contraction vascular smooth muscle relaxes even when the calcium efflux is completely blocked by lanthanum application (van Breemen 1976). Consequently, relaxation must be ascribed to calcium sequestration by an intracellular calcium store. As suggested by A.P. Somlyo et al. (1982), the sarcoplasmic reticulum is predominantly involved in calcium reuptake, while mitochondria play a minor role, if any, at least under physiological conditions. The sarcoplasmic reticulum may amount to 3% or even 5% of the cell volume and bind calcium as demonstrated directly by electron probe analysis of muscles in situ. In vitro fragmented sarcoplasmic reticulum recuperates calcium actively (Raeymaekers and Hasselbach 1981); the low activity of this calcium pump is probably one of the causes of the slowness of smooth muscle relaxation, and may also be affected by the SR-associated protein phospolamban (cf. Eggermont et al. 1988).

8.4 Modulation of Calcium Activation by Cyclic Nucleotides and G-Proteins

Both intracellular Ca^{2+} concentration as well as the calcium responsiveness of the contractile structures in smooth muscle may be under the control of cyclic nucleotides. It is generally held that in many instances smooth muscle relaxation may be mediated by cyclic adenosine 3',5'-monophosphate (cAMP) (Kramer and Hardman 1980). Indeed, cAMP is one of the most important intracellular messengers: when certain receptor sites of the cell membrane are occupied by a primary messenger, such as a hormone, the message it conveys may be signalled by a "second messenger," cAMP, to intracellular target proteins (Robison et al. 1971). In all known cases, these have been identified with cAMP-dependent protein kinases which transfer the terminal phosphate group of ATP to specific enzymes, thereby changing their activity (Nimmo and Cohen 1977). cAMP itself is formed from ATP in a reaction sequence catalyzed by the membrane-bound enzyme adenylate cyclase which may be activated by an associated hormone receptor (Fig. 2.11). The cAMP formed, however, is continuously broken down to AMP by other enzymes, the cyclic nucleotide phosphodiesterases. As the intracellular concentration of cAMP is primarily regulated by the rate of synthesis and degradation, it is crucially dependent on the relative activities of adenylate cyclase and cyclic nucleotide phosphodiesterases. When these enzymic activities are high, but balanced, the cAMP concentration is in a steady state; but a small change in either activity may cause a fast and dramatic alteration in the cAMP level. Significantly, the adenylate cyclase as well as one of the phosphodiesterases require calcium and calmodulin as activators.

The importance of cAMP for smooth muscle relaxation has been demonstrated by showing that phosphodiesterase inhibitors, such as papaverine, may also increase intracellular cAMP levels and may cause smooth muscle relaxation. The same result, namely an increase in cAMP levels and relaxation, is also achieved by stimulation of adenylate cyclase with forskolin or by noradrenaline or adrenaline. While forskolin has a direct activating effect on adenylate cyclase, noradrenaline and adrenaline stimulate more indirectly through activation of the β-adrenergic receptors of the cell membrane which are closely linked with adenylate cyclase. It is, therefore, generally believed that cAMP mediates β-adrenergic actions of adrenaline and noradrenaline, such as stimulation of glycogen metabolism and β-adrenergic relaxation.

8.4.1 Mediation of Beta-Adrenergic Relaxation of Vascular Smooth Muscle by cAMP

The Discovery of β-Action of Adrenaline. For many decades it has been known that adrenaline levels in the blood increase during "stress" and that this may have diverse effects on smooth muscles in different vascular beds. In running dogs, for instance, the spleen and its blood vessels contract, while the coronary arteries of the heart dilate. This dual action has, of course, a synergistic effect in an emergency situation, such as fight or flight: blood reservoirs in the spleen and elsewhere supply more blood to the systemic circulation, while the perfusion of the heart and, hence, the blood supply to the heart muscles improve. The different reactivity of the smooth muscle of coronary arteries and spleen to the same catecholamine transmitter substance thus allows a functional coordination of the ergotrophic sympathetic effect (W. R. Hess 1948). This diversity, as we recall, can be understood in terms of two different receptor populations, α- and β-receptors, which are involved in mediating the action of the hormone (Ahlquist 1948, 1962): when noradrenaline occupies α-receptors of blood vessels, the membrane calcium channels open or calcium may be released from intracellular stores and cause contraction, which is mediated by inositol trisphosphate (cf. Sect. 8.3.4). When combined with β-receptors, the same transmitter causes relaxation which appears to be mediated by cAMP (Shepherd et al. 1973).

Evidence for the Involvement of cAMP. A possible role of cAMP in β-adrenergic relaxation was first suspected when it was shown that adrenaline or noradrenaline causes an increase of cyclic nucleotide levels before the onset of relaxation and that there was a dose-dependent relation between cAMP levels and the relaxing effect. Even so, however, the temporal and quantitative correlation between relaxing response and cAMP level does, of course, not necessarily imply a causal relationship. More convincing are experiments in which cAMP is injected into smooth muscle cells which then subsequently relax (Scheid et al. 1979). Instead of injecting cAMP, it is also possible to apply a cAMP analogue (dibutyryl-cAMP), which unlike cAMP itself, crosses the cell membrane and induces relaxation. Last but not least, cAMP may be applied after stripping the muscle cell from the impermeable cell membrane by using a chemical-skinning procedure. In this way, the contractile structures become freely accessible to and may be influenced by added cAMP. These skinned fibres, which contract when supplied with ATP in the presence of the appropriate concentration of free calcium, may relax following the addition of cAMP in concentrations of 2 μM only (Meisheri and Rüegg 1983; Rüegg and Pfitzer 1985). Since such cAMP concentrations occur naturally during β-adrenergic stimulation it seems possible that these increased cAMP levels may indeed be responsible for the observed relaxing effect of β-agonists.

8.4.2 The Mechanisms of cAMP-Mediated Relaxation

cAMP is known to activate cAMP-dependent protein kinase within the cell by dissociating the enzyme into its regulatory subunits and its catalytic subunit (Nimmo and Cohen 1977). The liberated catalytic subunit is enzymically active. Following the application of β-agonists to coronary arteries, cAMP levels rise and the activity levels of cAMP-dependent protein kinases increase (Silver et al. 1982). The target proteins that become phosphorylated by cAMP-dependent protein kinases are, however, manifold. For instance, the enzyme myosin light-chain kinase (MLCK) of smooth muscle may be phosphorylated and inhibited (Adelstein and Hathaway 1979; DeLanerolle et al. 1984), as mentioned already in Sect. 8.2.1. In addition, the proteins of the cell membrane and the protein phospholamban of the sarcoplasmic reticulum may also be phosphorylated. In this way, cAMP may cause a reduction of intracellular free calcium and, hence, relaxation by one of the following mechanisms: (1) calcium influx through calcium channels may be inhibited (Meisheri and van Breemen 1982); (2) extrusion through the cell membrane may be promoted (Bülbring and den Hertog 1980), though perhaps indirectly, via stimulation of the Na-K ATPase and Na-Ca exchange (Scheid et al. 1979): as the intracellular Na^+ concentration is reduced following stimulation of the Na-K ATPase, the exchange of extracellular sodium for calcium is facilitated; (3) cAMP may, however, also enhance the uptake of calcium into the sarcoplasmic reticulum of smooth muscle by causing phosphorylation of phospholamban and stimulating the calcium pump of the sarcoplasmic reticulum (Mueller and van Breemen 1979).

cAMP Effects on Calcium Homeostasis. Smooth muscle fibres skinned by the detergent saponin are very suitable for study of the role of the sarcoplasmic reticulum. Itoh et al. (1982b) found that cAMP and cAMP-dependent protein kinase promoted calcium uptake into the intracellular calcium stores of skinned fibres when they were exposed to ATP salt solutions containing calcium ions in a concentration of 1 μM. Increased filling of the calcium stores under the influence of cAMP was indicated by the observation that a subsequent addition of caffeine induced a greater release of calcium from the stores and, hence, a larger contraction of the skinned fibres than under control conditions without the cAMP treatment. While these experiments suggest that cAMP causes relaxation and inhibition of smooth muscle contraction by lowering the intracellular free Ca^{2+} concentration, there is also evidence that cAMP-mediated relaxation may not be associated with a decrease in intracellular free Ca^{2+} concentration. According to Morgan and Morgan (1984 b), preparations of ferret portal vein contract when the intracellular Ca^{2+} concentration is increased after raising the extracellular concentration of potassium. However, these fibres relax after application of isoprenaline, a β-agonist, although the intracellular Ca^{2+} concentration remains unaltered (Fig. 8.7C). These experiments indicate that the calcium sensitivity of the contractile system has been decreased, suggesting that cAMP may also have a more direct effect on the actomyosin system.

cAMP Effects on the Contractile Mechanism. When cAMP-dependent protein kinase phosphorylates the enzyme myosin light-chain kinase, its activity becomes inhibited (Adelstein and Hathaway 1979) so that the ratio of myosin light-chain kinase activity to phosphatase activity decreases. This, of course, results in a decrease in the extent of myosin phosphorylation in the actomyosin-contractile system and in a concomitant reduction of ATPase activity even if the Ca^{2+} concentration is held constant (Mrwa et al. 1979; Silver and DiSalvo 1979). Evidence for a similar cAMP-induced reaction in living smooth muscle has been obtained in experiments of DeLanerolle et al. (1984), who produced relaxation of tracheal smooth muscle by application of the adenylate cyclase stimulator forskolin. The intracellular cAMP levels rose to over 10 μM and the extent of myosin light-chain kinase phosphorylation increased, while myosin phosphorylation decreased. Without knowledge of the intracellular free Ca^{2+} concentration, DeLanerolle's experiments, however, do not exclude the possibility that additional mechanisms, such as stimulation of calcium uptake, may also be involved in cAMP-induced relaxation.

The extent to which the actomyosin-contractile system and its regulatory proteins may be directly influenced by cAMP cannot be easily determined in experiments with intact fibres. This question, however, can be studied, using skinned fibres which are devoid of the cell membrane and the internal membranes of calcium stores. In these preparations the actomyosin-contractile system and its regulatory proteins are directly accessible to calcium ions, calmodulin, cAMP and cAMP-dependent protein kinase. As shown in Fig. 8.13, a calcium-induced contraction of skinned guinea pig taenia coli may be partly reduced when the Ca^{2+} concentration is lowered again to 1.6 μM. Addition of cAMP, however, induces without change in Ca^{2+} concentration a more complete reduction in force which is paralleled by concomitant dephosphorylation of the myosin light chains from approximately 0.4 mol phosphate mol^{-1} light chain to approximately 0.27 mol (Rüegg and Pfitzer 1985). A more complete relaxation and further dephosphorylation of myosin was achieved after lowering the Ca^{2+} concentration to 10^{-8}M with EGTA. Since cAMP activates its protein kinase in these experiments by releasing the catalytic subunit, it is not surprising that the cAMP effect can be completely mimicked by addition of the catalytic subunit of cAMP-dependent protein kinase in micromolar concentrations (cf. Kerrick and Hoar 1981; Rüegg et al. 1981). As is also evident from Fig. 8.13, there is a steep relationship between

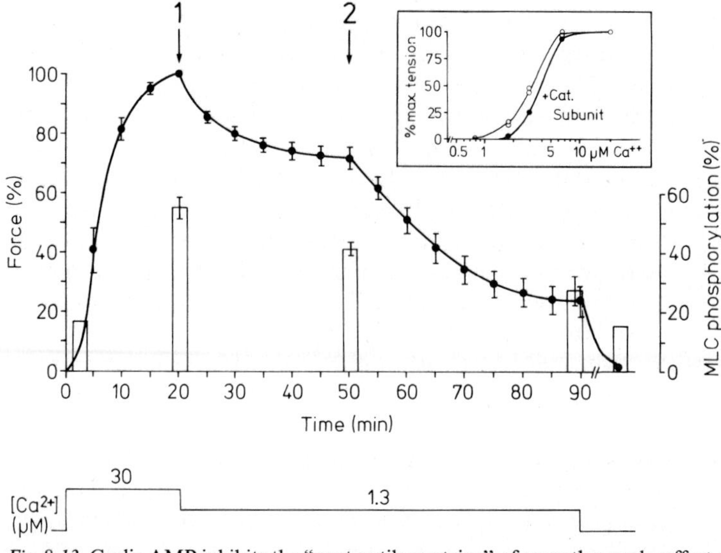

Fig. 8.13. Cyclic AMP inhibits the "contractile proteins" of smooth muscle: effect on force and myosin phosphorylation of skinned or demembranated muscle fibres of guinea pig taenia coli. Contraction induced by 30 μM Ca^{2+}. At *1*: Ca^{2+} lowered to 1.3 μM. At *2*: addition of cAMP. Note that cAMP causes relaxation of calcium-induced skinned fibre contraction without any change in Ca^{2+} concentration. Solutions, cf. Fig. 8.14. *Inset:* Relationship between force development of skinned fibres and Ca^{2+} concentration in the presence (–●–) and in the absence (–○–) of catalytic subunit of cAMP-dependent protein kinase. (After Rüegg and Pfitzer 1985)

myosin phosphorylation and contractile tension when the latter is altered by changing the concentration of cAMP or free calcium. This suggests that it is the change in myosin phosphorylation that determines the change in contractile tension regardless of whether this is brought about by a reduction in Ca^{2+} concentration or by cAMP-dependent protein kinase.

The inhibitory effects of cAMP-dependent protein kinase on myosin phosphorylation by myosin light chain kinase (MLCK) and on force development of skinned smooth muscle could only be observed at low but not at high concentrations of calcium ions or calmodulin (Meisheri and Rüegg 1983; Pfitzer et al. 1985). This is because MLCK, combined with calmodulin occupied with calcium (ternary complex), can be phosphorylated at one site only that is apparently not involved in the regulation of activity. In contrast, at intermediate Ca^{2+} concentration (e.g. 1 μM) the active ternary complex is in equilibrium with "free" MLCK that can be additionally phosphorlyated at a second (regulatory) site. The phosphorylation of MLCK at two sites then affects this equilibrium, because, as mentioned already (Sect. 8.2.1), doubly phosphorylated myosin light chain kinase has a much lower apparent affinity for calcium-calmodulin than the monophosphorylated or non-phosphorylated enzyme. This results in a decomposition of the active ternary complex and, therefore, in an inhibition of MLCK and contractile activity (Adelstein and Hathaway 1979; Conti and Adelstein 1981), as explained in Fig. 8.2. As force development of skinned fibres is inhibited by cAMP at low, but not at high, Ca^{2+} concentration, the calcium-force relationship is also altered. Thus the Ca^{2+} concentration required to produce 50% activation is increased, showing that the preparation becomes less calcium-sensitive or less responsive to calcium in much the same way as in intact smooth muscle cells after treatment with β-agonists (cf. Rüegg and Paul 1982; Morgan and Morgan 1984 b).

In conclusion, it would seem that relaxation mediated by cAMP may be due to its concerted action on the calcium-sequestering mechanisms of the sarcoplasmic reticulum and the calcium sensitivity of the contractile system: by stimulating calcium reuptake into the sarcoplasmic reticulum, the intracellular free Ca^{2+} concentration would be reduced, a condition necessary to allow an effect of cAMP on the actomyosin-contractile system. By inhibiting myosin light-chain kinase activity the apparent calcium sensitivity of the contractile system would be lowered. In conjunction, both effects should result in relaxation of precontracted smooth muscle and in a partial dephosphorylation of smooth muscle myosin as well as in a phosphorylation of myosin light chain kinase. This is the prediction, but so far experimental findings have been rather controversial (DeLanerolle et al. 1984; Nishikori et al. 1982; versus Hirata and Kuriyama 1980, Miller et al. 1983). Possibly, there is a great degree of diversity in the mechanisms of relaxation. With the advent of improved calcium-measuring techniques (Morgan and Morgan 1984 b), it has become possible, however, to determine the relative importance of changes in calcium sensitivity of contractile structures and changes in free Ca^{2+} concentration: experiments have shown that high levels of cAMP inhibit force production without reducing intracellular free calcium.

8.4.3 Cyclic Guanosine Monophosphate (cGMP)-Mediated Relaxation

The role of cGMP has been rather controversial, since increased cGMP levels were originally found both in contraction and relaxation of smooth muscle. However, several drugs or hormones that cause smooth muscle relaxation have now been found to cause a dramatic increase in intracellular cGMP levels which may rise to 0.2 or 0.5 μM (Kukovetz et al. 1979, Keith et al. 1982). cGMP may also be involved in endothelium-mediated relaxation of vascular smooth muscle (Holzmann 1982). Furchgott (1983) discovered that in intact coronary arteries acetylcholine application caused relaxation, but when the endothelium had been damaged by rinsing with distilled water, the response to acetylcholine was reversed. The nature of the coupling mechanism between endothelial cells and the contractile machinery is still mysterious. Apart from an "endothelium-derived relaxing factor," certain prostaglandins have also been implicated as local messengers between the endothelial cells and the vascular smooth muscle cells, in particular in the relaxation caused by anoxia. These factors then cause a rise in intracellular levels of cGMP which serves as an intracellular messenger causing relaxation (Rapoport and Murad 1983).

cGMP has also been implicated in drug-induced relaxation of vascular smooth muscle. Certain forms of heart attack are caused by spasms of coronary

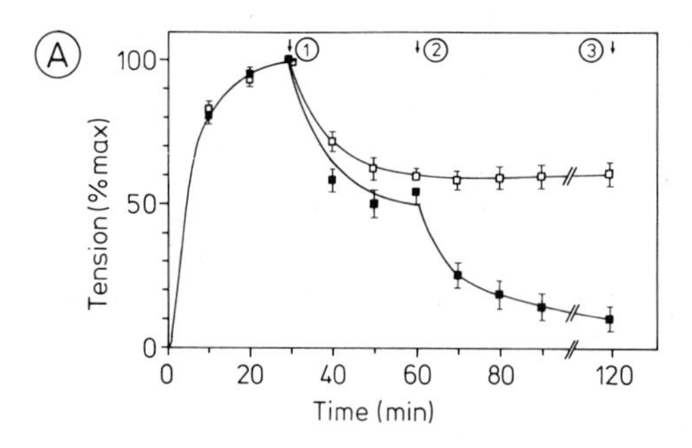

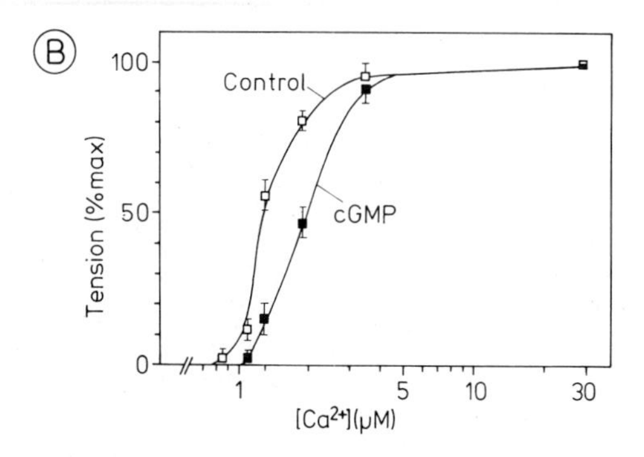

Fig. 8.14 A, B. Relaxing effect of cyclic GMP on skinned fibres from guinea pig taenia coli. *A* The chemically skinned fibres were contracted maximally by increasing Ca^{2+} to 30 μM in the presence of 0.2 μM calmodulin. At *arrow 1* partial relaxation was initiated by decreasing the Ca^{2+} concentration to 1.3 μM. At *arrow 2* cyclic GMP (5 μM) and cGMP-dependent protein kinase (0.1 μM dissolved in buffer) were added. Note marked relaxation (*filled symbols*), whereas no relaxation occured in controls where buffer was added without cGMP and kinase (*open symbols*). Means ±SE of 20 experiments. *B* Modulation of calcium sensitivity by cGMP. In the presence of cGMP and cGMP-dependent protein kinase the calcium-force relationship was shifted to the right. *Ordinate:* Tension in % of maximal tension at 30 μM Ca^{2+}. Maximal tension was not affected by cGMP and kinase. All solutions contained (mM): ATP total 7.5, $MgCl_2$ 10, KH_2PO_4 6, imidazole 20, creatine phosphate 10, creatine phosphokinase and 5 mM CaEGTA buffer, pH 6.7, 20° C. (Pfitzer et al. 1986)

vessels. For many years nitrogen oxide-containing compounds, such as nitroglycerine, were found beneficial in the treatment since they cause vasodilation of the coronaries as well as of the larger veins. The relaxation induced by these compounds has been found to be temporally and quantitatively paralleled by an increase in cGMP levels (Galvas and DiSalvo 1983), and all of these compounds also stimulate guanylate cyclase of coronary arteries (Gerzer et al. 1981). Moreover, the analogue 8-bromo-cGMP crosses the cell membrane, stimulates cGMP-dependent protein kinase and causes smooth muscle relaxation (Schultz et al. 1979). Other smooth muscle relaxants, such as the natriuretic factor produced by the atria of the heart, and the endothelium-derived relaxing factor (nitric oxide, cf. Palmer et al. 1987), also cause a rise in intracellular cGMP levels (Winquist et al. 1984; Schenk et al. 1985).

Mode of Action. cGMP exerts its effect by stimulating a cGMP-dependent protein kinase, thereby causing phosphorylation of target proteins. For instance, myosin light chain kinase may be phosphorylated and inhibited by cGMP-dependent protein kinase (Nishikawa et al. 1984a). This inhibition, in turn, would be expected to reduce the extent of myosin phosphorylation and contractility. Indeed, application of cGMP and cGMP-dependent protein kinase in micromolar quantities has been found to inhibit calcium-induced contractile responses of membrane-skinned preparations from pig coronary smooth muscle (Pfitzer et al. 1984), tracheal smooth muscle (Pfitzer et al. 1982) and guinea pig taenia coli (Fig. 8.14, cf. Pfitzer et al. 1986). In intact fibres, relaxation and increase in cGMP levels are paralleled by phosphorylation of diverse muscle proteins (Rapoport et al. 1983). It is possible that in living smooth muscle the direct inhibitory effect of cGMP on calcium responsiveness may well act in concert with other cGMP effects which reduce the amount of calcium stored in the sarcoplasmic reticulum (Itoh et al. 1983) and, therefore, diminish the contraction induced by calcium release from these stores. Alternatively, cGMP may reduce intracellular calcium levels (Morgan and Morgan 1984b) by stimulating the sarcolemmal calcium pump which extrudes calcium from the myoplasm (Popescu et al. 1985b).

8.4.4 Calcium Sensitization of Myofilaments Mediated by G-proteins

When smooth muscle is stimulated with α_1-adrenergic agonists such as phenylephrine or noradrenaline, the intracellular calcium ion concentration often rises very little or not at all, while force increases slowly and dramatically (Bradley and Morgan 1987; Rembold and Murphy 1988). This is in sharp contrast to stimulation with high potassium saline or other agents producing membrane depolarization, where the force transients are accompanied by large calcium transients. Obviously, the calcium responsiveness of the contractile structures may be quite different according to the conditions of stimulation, but it may also vary from one smooth muscle to another. For instance, it is high in tonic smooth muscle and low in phasic smooth muscle, which responds to stimulation with a short transient contraction (Somlyo and Himpens 1989). In many cases, stimulation can increase contractility of the smooth muscle by increasing the calcium responsiveness rather than by enhancing the intracellular free calcium ion concentration.

The nature of the mechanisms underlying calcium sensitization in smooth muscle is not clear at all, but evidence is accumulating that it may involve a small GTP-binding protein (p 21 G-protein) that may be related to the product of the rho-gene (Hirata et al. 1992). This new insight has been gained by investigating smooth muscle with a new technique called selective permeabilization. By using a bacterial toxin (*Staphylococcus*-α toxin) or the saponin-ester β-escine, the membrane can be perforated, so that it becomes permeable to small ions (or even small proteins in the case of β-escine), while retaining the larger proteins of the myoplasm. Since substrates and other small molecular components leak out, the preparation has to be bathed in a solution containing ATP as an energy source, but also pH buffers and calcium buffers to clamp the intracellular calcium ion concentration. The cell membrane, on the other hand, is completely depolarized and can no longer be voltage controlled. Interestingly, however, these preparations still respond to agonists such as phenylephrine and even noradrenaline, showing that receptor contraction coupling is intact and does not depend on a change in the state of membrane polarization or a change in the free calcium ion concentration which had been clamped (Nishimura et al. 1988; Kitazawa et al. 1989). Contraction is then ascribed to a sensitization to calcium and, interestingly, it can be mimicked by GTP or its analogue GTP-γ-S and may be antagonized by GDP-γ-S suggesting that G-proteins may be involved (Fujiwara 1989; Kitazawa 1991 a). A similar kind of sensitization can also be induced by activating protein kinase C with phorbol esters (Nishimura et al. 1990; cf. also Rembold and Murphy 1988 b), implying perhaps that receptor stimulation may cause the breakdown of phosphatidylinositolphosphates to give IP_3 and diacylglycerol, which may cause the activation of protein kinase C. Calcium sensitization may involve a G-protein-mediated inhibition of myosin light chain phosphatase, leading to enhanced Phosphorylation (Kitazawa et al. 1991 a, b) of light chains. Indeed, it has been shown previously that the inhibition of the myosin light chain phosphatase (with okadaic acid) can cause calcium sensitization (Bialojan et al. 1988). In addition, however, as already discussed (see Sect. 8.2.4), the proteins calponin and caldesmon might also be implicated in the altered calcium responsiveness.

In summary, then, the contractile responsiveness of smooth muscle is not solely dependent on the intracellular calcium ion concentration, but depends also on the calcium responsiveness of the contractile proteins that exhibits characteristic temporal and spatial patterns of variation within vascular beds. Thus, responsiveness is low in the muscles of portal vein, intermediate in large conduit arteries and high in the case of small arterioles (Boels et al. 1991), where it is subject to temporal alterations by G-proteins and cyclic nucleotides.

Chapter 9

Principles of Calcium Signalling in Muscle

Compared with magnesium, the calcium ion has a smaller charge density and hydration energy (Williams 1970), which enables it to move on and off protein-binding sites much more rapidly (Diebler et al. 1960; Eigen and Hammes 1963). These properties render calcium ions uniquely suitable for an intracellular messenger function. Indeed, their biological functions are many – ranging from the control of cell motility, secretion and membrane function to the modulation of DNA synthesis. In all muscles, as we have seen, calcium ions represent the trigger for contraction which is called into action by raising their intracellular concentration from about 0.1 μM to 10 μM. Thus, calcium ions are, with a few exceptions, the principal intracellular messengers carrying information from the excited cell membrane to the myofilaments in the interior of the fibres. The modes and mechanisms of this information transfer are, however, quite different in various types of muscle. In fast skeletal muscle, for instance, an electric signal, the action potential, travels along the membrane invaginations, the transverse tubules, and hits the triads where this extracellular message is transformed into an intracellular one: calcium ions are released from the terminal cisternae of the sarcoplasmic reticulum and diffuse into the myoplasm. This calcium signal is then recognized by troponin-C, the specific calcium-receptor protein, that turns on the contractile mechanism. In vertebrate smooth muscle, on the other hand, calcium ions may also be released from the cell membrane to affect a different calcium-sensing protein, calmodulin, which, in conjunction with myosin light-chain kinase, catalyzes the phosphorylation of the regulatory myosin light chain, thereby inducing contraction.

In discussing these different processes of intracellular information transfer in cybernetic terms, the calcium channel of the terminal cisternae or the calcium channels or binding sites at the cell membrane may represent the "sender" or "signal generator" that encodes and emits the calcium signal, a transient increase and decrease in myoplasmic free Ca^{2+} concentration. The myoplasm is the information transmission channel, whereas the specific calcium-binding proteins troponin-C, calmodulin, or myosin light chains are the receivers of the message that decode calcium signals for the user, the contractile elements (as explained in Fig. 9.1). In the following we shall briefly compare the various types of senders and consider some problems of encoding and transmission of signals as well as the various types of signal receiver and feedback control in different kinds of muscle.

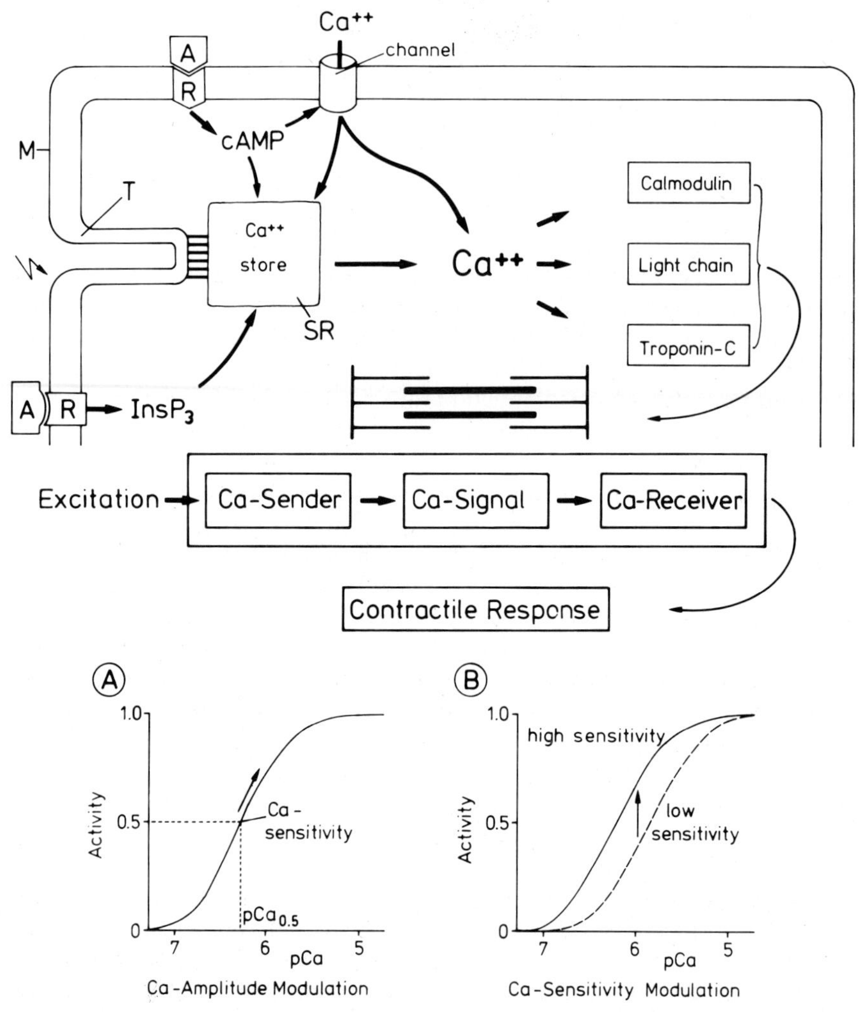

Fig. 9.1. Generalized scheme of excitation-contraction coupling in muscle, involving a calcium sender (sarcoplasmic reticulum, calcium channels of the cell membrane), encoding the calcium signal (= calcium transient), which is decoded by the signal receiver (troponin-C, calmodulin or myosin light chains). The excited cell membrane (*M*) or specific membrane receptors (*R*) communicate with the Ca^{2+} sender by means of the T-system (*T*) or by second messengers, such as cyclic AMP or inositol trisphosphate ($InsP_3$). The cellular response (contraction) depends on the amplitude of the signal (e.g. Ca^{2+} concentration, cf. *A*) or on the calcium sensitivity of the receiver or signal transducing system. (Sensitivity modulation, cf. *B*.) Important negative sensitivity modulators are cyclic nucleotides and certain metabolites (H^+, P_i), while certain drugs are positive sensitivity modulators (cf. Table 7.2)

9.1 Senders of Calcium Signals

Calcium Channels. These channels send out calcium signals to the myofilaments and are located either in the cell membrane (and the T-tubules) or in the part of the sarcoplasmic reticulum that forms diadic or triadic junctions with the membrane of the T-tubules. In both cases calcium channels connect a compartment of high Ca^{2+} concentration, approximately 0.1 to 1 mM, with the myoplasm in which the Ca^{2+} concentration is kept below 0.1 μM under resting conditions by means of calcium pumps of the sarcoplasmic reticulum or in the cell membrane. Upon depolarization of the cell membrane, these calcium channels open, thus permitting a flow of calcium ions along the concentration gradient. However, some differences between the calcium channels of the cell membrane and of the sarcoplasmic reticulum are noteworthy. In the former, the driving force of calcium ions is determined by an electrochemical gradient that depends on both the extracellular Ca^{2+} concentration and the membrane potential: it decreases when the cell depolarizes. In the sarcoplasmic reticulum, on the other hand, the calcium-driving force is largely protected from changes in membrane potential since the membrane of the cisternae is electrically isolated from the closely apposed T-tubular membrane. Depolarization of the muscle cell, therefore, does not alter the electrical gradient between cisternae and myoplasm, though it does increase calcium conductivity in the cisternal membrane. Here, the closing and opening of the calcium channel is obviously determined by the potential of a membrane different from that in which the channels reside. The nature of this coupling process between the T-tubular depolarization and the change in calcium conductivity in the sarcoplasmic reticulum, the so-called T-SR coupling, is however, still a mystery (Martonosi 1984; Somlyo 1985a; Ebashi 1991; cf. Sect. 10.1.2.5).

The amount of calcium released from the sarcoplasmic reticulum per unit time is dependent not only on the extent of depolarization, but also on the area of the calcium-permeable membranes in the junctional sarcoplasmic reticulum. This incidentally explains why in fast muscles containing a larger amount of junctional sarcoplasmic reticulum and more calcium channels, the rate of calcium release is higher and the calcium signals are faster than in slow muscles where the sarcoplasmic reticulum is sparse.

Second Messengers that Mobilize Calcium. As first pointed out by A.P. Somlyo and Somlyo (1968), membrane depolarization may not be the only way of triggering calcium release from the sarcoplasmic reticulum in smooth muscle. Other activation processes that precede and are responsible for the generation of a calcium transient may include calcium-induced calcium release as well as inositol trisphosphate-induced calcium release which may occur without membrane depolarization in smooth muscle. In cardiac muscle the gating of the membrane calcium channels is enhanced by cAMP. Calcium senders, therefore, may encode quite different kinds of intracellular messages into calcium signals which are then sent to the myofilaments.

In cardiac muscle, cAMP increases the open-probability of the calcium channels in the cell membrane, thereby increasing the calcium influx into the myoplasm which, in turn, causes the release of calcium from the sarcoplasmic reticulum by calcium-induced calcium release. As shown by Fabiato (1983), a small, but rapid increase in intracellular free Ca^{2+} concentration may act as a trigger releasing more calcium from the sarcoplasmic reticulum, particularly in vertebrate heart muscle, but also in many types of invertebrate striated muscle. The calcium transient thus generated is, however, comparatively brief. The reason for this is that as the level of free calcium rises in the myoplasm, it inhibits further calcium release with a delay, while the calcium already released may be taken up by the sarcoplasmic reticulum, bound by proteins or extruded through the cell membrane. Consequently, therefore, calcium entering the myoplasm via calcium channels during activation may act as a short-lived "mercurial" messenger (Rasmussen and Barrett 1984) between the excited cell membrane and the parts of the junctional sarcoplasmic reticulum that are not in direct contact with the cell membrane or its invaginations. As this "trigger-calcium" releases more calcium from the sarcoplasmic reticulum, the initial calcium signal is greatly amplified.

Inositol trisphosphate is a recently discovered intracellular messenger that transmits signals from membrane receptors of smooth muscle to the calcium-release sites of the sarcoplasmic reticulum (Berridge and Irvine 1984). For example, inositol trisphosphate is involved in the α-adrenergic receptor contraction coupling: when the α-receptors of the cell membrane are occupied with agonists, such as noradrenaline, the phospholipases of the cell membrane become activated and they degrade membrane lipids, in particular phosphoinositides to diacylglycerol and inositol trisphosphate. While the latter mobilizes calcium from the sarcoplasmic reticulum of smooth muscle, even after membrane skinning, the former bypasses the calcium signalling channel: diacylglycerol may activate protein kinase-C even at very low concentrations of myoplasmic free calcium, thus causing a phosphorylation of myosin light chains which, in turn, may be followed by the activation of the contractile mechanism. Diacylglycerol also serves as a substrate for the calcium-dependent formation of arachidonic acid which is the precursor of prostaglandins, leukotrienes and thromboxanes. The role of all these intercellular and possibly intracellular messengers in the regulation of motile and secretory activity of cells can hardly be overemphasized. Further research is urgently needed to understand their mode of action.

9.2 Transmission of Calcium Signals

The release of calcium from the sarcoplasmic reticulum causes a rapid increase in free calcium to 10 μM which is taken up again later on by the calcium pump of the sarcoplasmic reticulum (Inesi 1985), or it may be extruded by the cell membrane. The calcium transient thus generated is the signal transmitted to the myofilaments where it is received by the calcium-receptor proteins. Since the resting calcium level is so low, of the order of 0.1 μM, the signal–to–noise ratio is quite high, i.e. small fluctuations of calcium levels do not interfere with calcium-signal transmission.

Amplitude and Frequency Modulation. In most slow muscles calcium signals are amplitude-modulated. Thus, a greater depolarization of the T-tubular membrane may encode a greater calcium signal or calcium transient which, in turn, codes for a greater contractile force.

Amplitude-modulated signals, however, would not be suitable for the coupling of excitation and contraction in fast skeletal muscle as they might be readily disturbed by many factors, including muscle movement. Consider the case of sarcomere lengthening of frog twitch muscle which would increase the diffusion distance from calcium-emitting terminal cisternae near the Z-line to the force-generating actin-myosin overlap zone. As a consequence, calcium release from the sarcoplasmic reticulum would take longer to diffuse to the myofilaments, and it might be diluted to a greater extent by the myoplasm than in the case of shortened muscle fibres. Such a diminution of the signal would prima facie be expected to decrease force production were it not for two compensatory mechanisms: (1) sarcomere lengthening increases calcium sensitivity of the myofilaments. (2) The calcium signal is standardized by being supramaximal. This protects the code of calcium-signal transmission against "unwanted" mechanical or other disturbances. Following a nerve impulse when an all-or-none action potential reaches the triad, the calcium sender emits a supramaximal "packet" of calcium, probably enough to saturate the calcium-binding sites on the calcium-receptor proteins (Cannell and Allen 1984; Gillis 1985). Thus, the contractile elements respond to the signal always in the same stereotypic all-or-none manner regardless of whether the myoplasmic Ca^{2+} concentration is slightly higher or lower. The code, in other words, is now no longer amplitude-modulated, but "digital," transmitting only "yes or no" or "switch-on, switch-off" signals. As a result, signal transmission is protected from local influences, such as alterations in calcium sensitivity or mechanical disturbances, but it responds more reliably to the command impulses received from the central nervous system. Often, such a signal transmission is frequency-modulated. In fish muscle, for instance, tetanic force of fast or slow fibres can be considerably increased by simply increasing the frequency of muscle excitation (Flitney and Johnston 1979). Perhaps this is the way in which the muscle-activation mechanisms have adapted to the evolutionary trend of bringing muscle more and more under the control of the evolving central nervous system, at the same time freeing it from "unwanted" local influences. Clearly, such digital mechanisms would not be required or useful in species from lower phyla, in which peripheral control mechanisms may dominate, or in cardiac or visceral smooth muscles where amplitude modulation by peripheral metabolic factors may even be necessary to adapt performance optimally to the homeostatic needs of the organism. Think, for instance, of the vertebrate heart where the amplitude of the calcium signals and force of contraction may both be increased under the influence of the stress hormone, adrenaline.

Speed of Calcium-Signal Transmission. The rate of transmission in fast and slow muscle is quite different, for as already mentioned, the rate of increase in free Ca^{2+} concentration in the myoplasm depends on the rate of calcium release from the sarcoplasmic reticulum, as well as on the distance over which the calcium ions must reach the actin-myosin interaction sites. The calcium signal is transient because the released calcium is taken up again by the sarcoplasmic reticulum. However, contractile activity always outlasts the calcium transient (Gillis 1985) so that, for example, tetanic tension may be maintained during repetitive stimulation even though the free calcium levels may oscillate considerably (Cannell and Allen 1984). The phenomenon of persistent contractile activity after cessation of the calcium transient is observed in all striated muscle, but is particularly pronounced in vertebrate smooth muscles, for here the temporal calcium code is translated into a long-lasting structural code as myosin becomes phosphorylated by the calcium-calmodulin-dependent myosin light-chain kinase. In striated muscle the message of a brief calcium signal is also stored for some time when calcium is bound to the receptor protein troponin, thereby causing a conformational change in the response element. Here, too, the temporal code of the signal has been translated into a spacial or structural code which encodes the force-generating attachment of crossbridges to actin. Due to this attachment, moreover, the calcium affinity of troponin-binding sites increases so that the Ca^{2+} concentration may now be lowered without a loss in force (Brandt et al. 1982). Such hysteresis phenomena have also been studied in crustacean muscles (Ridgway et al. 1983; Ridgway and Gordon 1984), but they seem to be particularly pronounced in lamellibranch catch muscle. Here, contraction may be maintained after lowering calcium to basal levels, but may not even be due to the cyclic operation of crossbridges, but to some kind of locking or catch mechanism which does not require the continuous hydrolysis of ATP (cf. Rüegg 1971). In the following we shall consider in more detail the interaction of the calcium signal with its different receivers and the subsequent transformation of the temporal calcium code into a structural code.

9.3 Diversity of Calcium-Signal Receivers

The calcium message may be received by calcium-binding proteins, such as troponin-C, myosin light chains and calmodulin. However, all of these proteins contain similar calcium-binding domains having also a similar calcium affinity, and in all cases calcium binding results in the movement of a protein switch: in vertebrate and crustacean muscles that are troponin-regulated, tropomyosin slips into the groove between the actin-monomer strands, whereas in "myosin-regulated" scallop striated muscles, the two kinds of light chains attached to the myosin heavy chain move relative to each other when the switch is activated by calcium binding. In calmodulin-regulated vertebrate smooth muscle, on the other hand, calcium-activation mechanisms may be more complex. Calcium activation may induce a flip-flop movement of actin-bound caldesmon to a calmodulin-binding site on the thin filament, or myosin may undergo a conformational change when its regulatory light chain is phosphorylated by calmodulin-activated myosin light-chain kinase. This latter myosin phosphorylation-dependent regulatory mechanism is also the signal for coding an increased contractile activity in many other muscle and non-muscle cells (Kendrick-Jones and Scholey 1981; Walsh 1985). It is also a means of greatly amplifying the calcium signals, because only four calcium ions complexing with one calmodulin molecule may activate a molecule of myosin light chain kinase which catalyzes the phosphorylation of many myosin molecules. Calmodulin-dependent myosin phosphorylation may even play a role in the modulation of contractile activity in vertebrate skeletal and cardiac muscle since it increases the calcium sensitivity of the troponin system (Persechini et al. 1985; Morano et al. 1985). Obviously, actin-linked and myosin-linked activation pathways may influence each other.

Comparison of Different Calcium-Binding Proteins in Muscle. Myosin light chains, calmodulin and various forms of troponin-C evolved from a common ancestor, a primitive calcium-binding protein which must have existed already in the most primitive eucaryotes nearly a billion years ago. Since this time the structure of the calcium-binding domain, known as the "EF–Hand" (Kretsinger 1980), has been conserved because of the uniqueness of its properties. Like a hand, the binding domain grips and enwraps the calcium molecule for which it has a much higher affinity than for magnesium ions. Whereas magnesium-protein complexes form and dissociate very slowly, the calcium complex with the specific calcium-binding sites of the EF–hand domain may form and dissociate as rapidly as required for the messenger and trigger function of calcium (cf. also Eigen and Hammes 1963). An apparent exception is another protein with EF–hand structure, parvalbumin, which may function in the myoplasm as a temporary calcium sink and as one of the terminators of the calcium signal, for this protein does not bind calcium immediately after an increase in its myoplasmic concentration, but with a delay. The delay, however, is indirectly caused by magnesium that blocks the calcium sites of parvalbumin in relaxed muscle and dissociates only slowly within approximately a fraction of a second when the calcium signal is received (Gillis 1985).

Calmodulin is the universal multifunctional calcium-signal receiver of all eucaryotic cells, including protozoa and primitive fungi. Since it is a mobile protein in the myoplasm, it may react as a ligand with many different calmodulin-binding proteins or calmodulin receptors, thereby influencing innumerable cell functions at the same time. The specific effect induced by calmodulin then depends, of course, on the kind of receptor protein that is called into action as a "discriminator." In this way, calmodulin may activate, for instance, glycolysis as well as contraction of vertebrate smooth muscle in a coordinated fashion when the calcium level increases. The information flow mediated by calmodulin is comparatively slow because of the transit time required for calmodulin to reach the target proteins (protein kinases) which it activates and because of the slowness of the response involving phosphorylation of the regulatory proteins. Troponin-C, on the other hand, is then incorporated into thin filaments where it forms a structurally linked system with the other regulatory proteins troponin-I, troponin-T and tropomyosin. Thus, it will only react with its neighbouring regulatory protein, troponin-I, but it does so in a rapid and most direct manner. Since there is no diffusional delay, the transit time for information flow from one molecule to the next is very short indeed, suggesting that this calcium-binding protein evolved specifically to meet the need for rapid contractile activation in fast muscles. There are, however, differences in its mode of action in various muscles. In ascidian muscles, for instance, troponin and tropomyosin are calcium-dependent activators of the actomyosin-contractile system (Ebashi 1983), whereas in decapod crustacean and in vertebrate striated muscles troponin molecules are calcium-dependent inhibitors. Here, calcium binding to troponin-C derepresses the inhibition of the contractile mechanism induced by troponin-I in the absence of calcium.

In molluscan muscle calcium binding to myosin light chains rather than to troponin enables myosin heads to strongly interact with actin, thereby inducing contraction. Astonishingly, most invertebrate muscles are not solely myosin- or troponin-regulated, but dually controlled in as much as calcium switches are located both on the thick and thin filaments.

In summary, then, calcium binding proteins such as troponin-C, calmodulin and myosin light chains may serve to illustrate the principle of a common structural theme with variations in the molecular design allowing for adaptations to specific, functional requirements. Thus, the EF-hand structure comprising the calcium binding helix-loop-helix motif has been conserved throughout the evolution of Ca^{2+}-binding proteins, while other parts of the molecule evolved to meet specific functional demands. Calmodulin, for instance, acquired the capability to interact with a plethora of very different proteins such as actin, myosin light chain kinase and phosphorylase kinase.

9.4 Contractile Responsiveness to Calcium

As soon as the calcium signal is received by the calcium receptor protein troponin-C or calmodulin, it is transduced into a cellular response such as muscle shortening or force development. The latter not only depends on the size, shape and frequency of calcium transients as discussed above (p. 242), but, in addition, also on the calcium responsiveness of the myofilaments. Thus, the sigmoidally shaped relationship between the intracellular free calcium level (pCa) and force development might be affected in several ways. First, the *calcium efficacy* could be enhanced or diminished as indicated by an upward or downward displacement of the force/pCa curve, i.e. an increase in the maximally calcium-activated force. Secondly, there might be an increase in the *apparent calcium affinity* or *calcium sensitivity* as indicated by a displacement of the calcium force relation to the left, i.e. towards lower Ca^{2+} concentrations or higher pCa values as illustrated in Fig. 9.1. Conversely, muscle myofilaments could be desensitized to calcium ions. These modulations are particularly important in cardiac muscle, which is usually only about half-maximally activated by Ca^{2+} (cf. Chap. 7). Of course, alterations in calcium sensitivity and efficacy are often combined.

It is obvious that one cause for calcium sensitization could be seen in an increased calcium affinity of troponin-C. This then would lead to a larger calcium occupancy of troponin-C and force development at any intermediate level of calcium activation. On the other hand, calcium occupancy and force ought to be unaltered at maximal calcium activation, i.e. under conditions in which, presumably, the calcium binding sites of troponin-C are saturated. Clearly, therefore, an increased or decreased calcium efficacy cannot possibly be ascribed to a greater or smaller calcium affinity of troponin-C. Rather, it must be ascribed to mechanisms "downstream" of troponin, for instance, to alterations in crossbridge turnover kinetics.

As discussed in Chapter 1, steady state force depends on the apparent rate constants of crossbridge attachment and detachment, and these rate constants affect the calcium efficacy and the calcium sensitivity of the myofilaments, according to Brenner (1988). Slow muscles are characterized by having a low crossbridge detachment rate constant (p. 125) as well as a high apparent calcium sensitivity (see p. 68). Alterations in the attachment rate constants are also important in influencing calcium sensitivity. For example, phosphorylation of the regulatory myosin light chain increases the apparent attachment rate constant (Sweeney et al. 1990), thus causing an apparent increase in calcium responsiveness and calcium sensitivity (Persecchini et al. 1985). At any given Ca-level, therefore, force would be enhanced by myosin phosphorylation. However, since the latter ist also calcium-regulated, it is clear that calcium might affect contractile force both via the troponin-tropomyosin system (thin filament or actin-linked regulation) and the myosin light chain system (thick filment or myosin-linked regulation). Therefore, in muscle fibres containing a dual regulatory system involving both thin and thick filament-linked regulation, these two systems may then interact and influence or modulate each other in a complex manner, thereby enhancing the sensitivity of the response to the calcium messenger. In mammalian smooth muscle, regulation is even more complex since the relation between free calcium and phosphorylation of light chains is also variable. It might be regulated, for instance, by small G-proteins that increase Ca-sensitivity and by cyclic nucleotides that decrease it (cf. Sects. 8.4.2 and 8.4.4).

cAMP and the Modulation of Calcium Sensitivity. The different calcium regulatory systems for contractile mechanisms may also interact with other intracellular messenger systems, such as the cAMP system. Here, the signal generator is adenylate cyclase which catalyzes the formation of cAMP, a signal that is then received by different kinds of cAMP-dependent protein kinases acting as "discriminators" and directing the signal to specific intracellular targets. According to Rasmussen and Barrett (1984), cAMP and calcium may both act as "synarchic intracellular messengers" collaborating or antagonizing each other, as the case may be. For instance, cAMP-dependent protein kinase may phosphorylate and activate phosphorylase kinase in much the same way as calcium and calmodulin. Unlike the latter, however, it inhibits myosin light-chain kinase and, hence, the smooth muscle contractile mechanism. When myosin light-chain kinase is phosphorylated by cAMP-dependent protein kinase, its affinity for calcium-calmodulin decreases so that a higher concentration of calcium and/or calmodulin is required to activate the contractile system. This is an example for negative calcium sensitivity modulation occurring, however, only in smooth muscle. Modulation of calcium sensitivity is widespread not only in smooth muscle, but also in cardiac muscle. A well-known example is phosphorylation of cardiac troponin-I which is also mediated by cAMP-dependent protein kinase and which reduces the calcium affinity for troponin-C. Thus, force may be altered without necessarily altering the free Ca^{2+} concentration in the environment of the myofilaments.

9.5 Feedback Signals and Servoloops

Following the stages of the information flow from the calcium-signal sender to the calcium-signal receiver, we should be aware, however, that the direction of information flow may be reversed. For instance, while calcium occupancy of troponin causes crossbridge attachment and contraction, the reverse is also true. As mentioned above, crossbridge attachment may lead to increased calcium occupancy via increased calcium affinity of troponin-C (Brandt et al. 1982; Güth and Potter 1987). On the other hand, when calcium occupancy of troponin induces muscle shortening, the latter may lead to calcium desensitization and, hence, to a decrease in calcium occupancy of troponin and to a release of bound calcium into the myoplasm (Ridgway et al. 1983; Gordon et al. 1984). Because of these effects, shortening may deactivate the contractile mechanism (Edman 1975; Bozler 1972) in skeletal and cardiac muscle. Shortening may, however, also deactivate the contractile machinery directly, in particular in insect flight muscle where oscillatory phenomena depend on stretch activation and shortening deactivation (as reviewed by Tregear 1983). In autoregulatory vascular smooth muscle an isotonic contraction decreases, whereas stretching the muscle fibre increases the calcium permeability of the cell membrane and, thus, the intracellular free Ca^{2+} concentration (Bevan et al. 1985).

Following excitation of cardiac muscle, the intracellular Ca^{2+} concentration increases, thereby causing an increase of potassium conductivity of the cell membrane which, in turn, inhibits the excitation of the muscle cell (Isenberg 1977). All these examples of positive and negative feedback loops may suffice to demonstrate that excitation-contraction coupling in muscle may be a closed-loop rather than open-loop control sequence, as illustrated in the schematic diagram of Fig. 9.2.

Surely, this kind of consideration is particularly useful when looking at another aspect of muscle activation, the increase in metabolic rate as a result of increased intracellular free calcium. Thus, the calcium signal causes an increase in actomyosin-ATPase activity, which, in turn, is a prerequisite for an increased mechanical output by the contractile proteins. The latter, however, also affects the

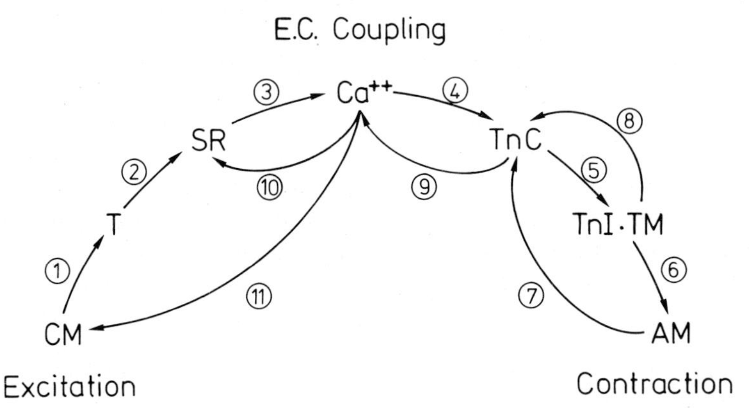

Fig. 9.2. Closed-loop control sequence in excitation-contraction coupling. *1* Membrane depolarization invades the T-system; *2* T-SR coupling; *3* Ca release from SR; *4* Ca binding to troponin-C (TnC); *5* alteration in the structure of thin filament proteins [troponin-I (TnI), tropomyosin (TM)]; *6* activation of actomyosin (AM) ATPase and contraction; *7* crossbridge attachment or muscle shortening alter Ca affinity of troponin-C, which is also affected by phosphorylation of troponin-I (*8*). The free Ca^{2+} concentration is affected by alterations in calcium affinity for troponin-C (*9*) and affects itself the sarcoplasmic reticulum (Ca-pump, Ca-induced Ca-release) (*10*) as well as the excitability of the cell membrane (CM) by activation of K-channels (*11*), see text

actomyosin ATPase, as discussed in more detail in Chap. 1. For example, the rate of energy consumption and the rate of ATP splitting is much larger when the muscle shortens under load than under merely isometric conditions, even if the extent of calcium activation is the same in both cases (Fenn effect, cf. Woledge et al. 1985). When the muscle fatigues, however, reaction products of ATP hydrolysis, such as phosphate and hydrogen ions, accumulate (Kushmerick and Meyer 1985) reducing the affinity of troponin-C for calcium, thereby inhibiting actomyosin ATPase and contraction. Further, while calcium activation alters crossbridge action, the reverse is also true inasmuch as calcium responsiveness may be influenced by the kinetics of crossbridge turnover as discussed in Sect. 9.4.

In summary, then, it is clear that calcium regulation in muscle activation is subject to feedback. With suitable delays in the control system, calcium levels will oscillate or calcium signalling might induce a calcium wave that may be propagated intracellularly, thus generating complex temporal and spatial patterns of free calcium within, say, a smooth muscle cell (cf. Tsien and Tsien 1990, for further discussion). At this time, we are only just beginning to appreciate the complexity of the cellular calcium regulatory system as a whole, which, in calcium homeostasis, has the capacity of self-regulation.

Chapter 10

Molecular Level Approaches to Excitation-Contraction Coupling in Heart and Skeletal Muscle

With the advent of molecular biology, our knowledge on the mechanisms of calcium signalling and signal transduction increased enormously. In this chapter, we highlight some of the spectacular advances and new approaches regarding excitation-contraction coupling in cardiac and skeletal muscle and the mechanisms by which the calcium signal elicits a contractile response and determines contractility.

10.1 Calcium Channels in T-System SR Coupling and Calcium Release

As already mentioned in Chapter 2, the control of the contractile activity is exerted by two membrane systems in skeletal muscle, the T-tubules and the membrane of the sarcoplasmic reticulum. During excitation-contraction coupling (EC-coupling), electrical depolarization of the T-tubular membranes induces the rapid release of calcium ions from the sarcoplasmic reticulum. However, it is not yet quite clear how the voltage is sensed in the T-tubules and transformed into a signal, which is then transmitted to the calcium release sites of the sarcoplasmic reticulum. Recent work has established that this unknown step involves essentially the communication between two types of calcium channels, one which resides in the T-tubular membrane and the other in the membrane of the sarcoplasmic reticulum (cf. Fig. 10.1).

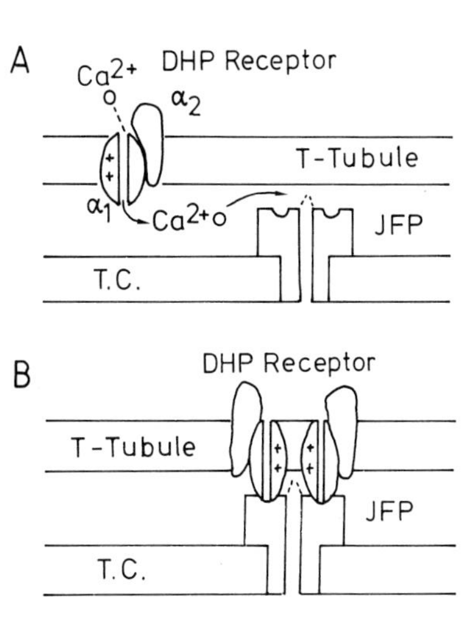

Fig. 10.1 A, B. Model of EC-coupling between T-system and the terminal cisternae (*T. C.*) of the sarcoplasmic reticulum (excitation-contraction coupling) in the triad. *A* Cardiac-type EC-coupling requiring calcium influx through fast-activated L-type calcium channels (DHP receptors) inducing calcium release from the junctional foot protein (*JFP*) or ryanodine receptor (calcium release channel). *B* Skeletal-type EC-coupling effective even in the absence of external Ca^{2+} because of (mechanical) interaction of the DHP receptor and calcium release channels (ryanodine receptor) (Adapted from Caswell and Brandt 1989)

10.1.1 The Calcium Release Channel of the Sarcoplasmic Reticulum

It has long been suspected that the calcium release channel of the sarcoplasmic reticulum may be located in the junctional feet that span the gap between the membrane system of the T-tubules and the sarcoplasmic reticulum (cf. p. 38). The proof came with the discovery that the calcium channels of the lateral cisternae could be isolated and shown to be identical in size and shape with the feet structure of the triads, as will be discussed below.

10.1.1.1 The Lateral Cisternae Contain Ryanodine-Sensitive Calcium Channels

Calcium is released from the lateral cisternae of the sarcoplasmic reticulum (cf. Sect. 2.2.2) at a high rate (Melzer et al. 1984, 1987) which is mediated via a highly conducting ion channel rather than by a carrier-coupled mechanism. Indeed, Meissner and his colleagues succeeded in reconstituting isolated vesicles of lateral cisternae containing the putative Ca^{2+} channels into a planar lipid bilayer separating two compartments of a plastic chamber filled with saline (Smith et al. 1985, 1986). By means of a current amplifier connected to two electrodes inserted into the compartments and suitably applied potentials, an ionic current could be measured across the bilayer due to the opening of pore proteins (cf. Fig. 10.3). The channels conducted preferentially Ca^{2+} and Ba^{2+} when these ions were present in the experimental chambers. Interestingly, the opening or "gating" of the channels could be modulated by various ligands including ATP and other adenine nucleotides, caffeine and calcium itself. The plant alkaloid ryanodine also modulated channel activity and turned out to be particularly useful, as it is tightly bound to the channel protein (Fleischer et al. 1985).

10.1.1.2 Identity of Isolated Calcium Release Channels and "Feet"

The capacity of the putative channels of the lateral cisternae to bind ryanodine greatly facilitated its isolation (Imagawa et al. 1987; Inui et al. 1987). Thus, during the purification procedure, the protein fractions containing the enriched ryanodine receptor could be identified simply by monitoring ryanodine binding. The isolated protein, having a molecular weight of approximately 500 kDa (Zorzato et al. 1990), was in its functional properties similar to the native sarcoplasmic reticulum channel (Imagawa et al. 1987; Lai et al. 1988). As shown by electron microscopy (cf. Saito et al. 1988; Wagenknecht et al. 1989), it may form a tetramer resembling in its size and shape the "feet" spanning the gap between T-tubular and sarcoplasmic reticulum membrane (see Fig. 10.2; Block et al. 1988; cf. also Ferguson et al. 1984). Thus, the feet, the ryanodine receptor and the calcium channel all seemed to be the same protein!

To study its *calcium conductance,* the ryanodine receptor protein was fused into liposomes which were then incorporated into an artificial lipid membrane separating the compartments of a double chamber, as described above. In this way, the ion-conductive properties of the reconstituted channel could be com-

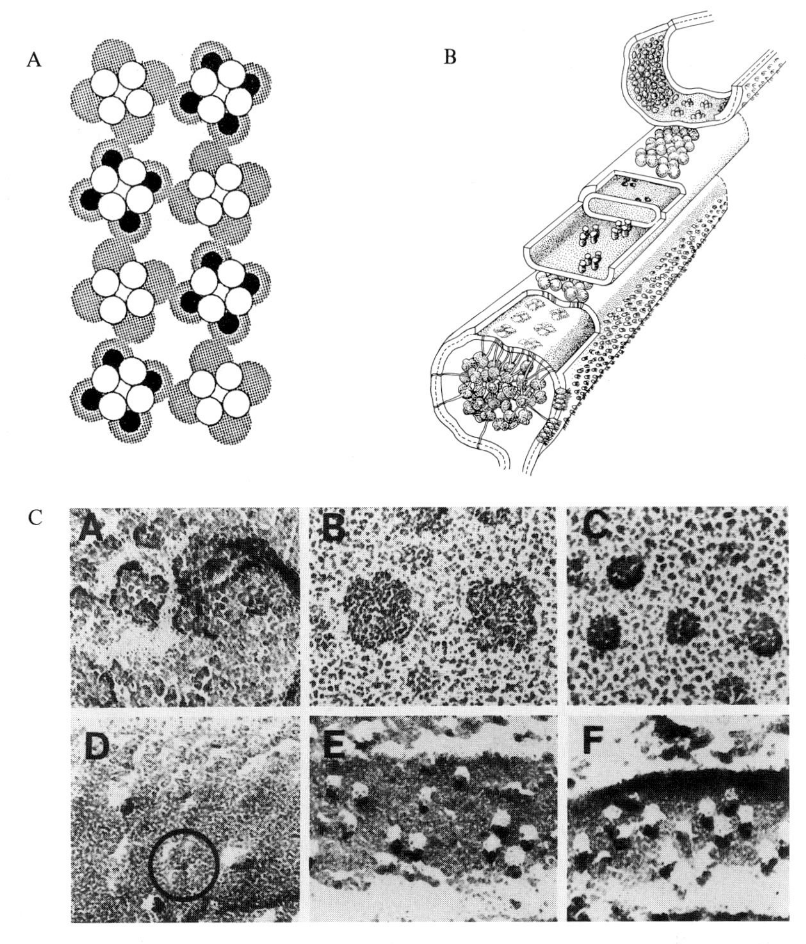

Fig. 10.2 A. The spatial relationship between T-tubular tetrads and foot protein (*shaded area* foot protein; *white circles* intramembrane parts of foot protein, *black discs* correspond to T-tubular tetrads) Note that every second tetrad matches a foot protein. *B* Model of three-dimensional reconstruction of the triad of fish muscle. Feet (four balls, i.e. four ryanodine receptor proteins forming a Ca^{2+} channel) occupy the gap between T-tubules and SR. *C* Fine structure of protein components of the triad: a) foot associated with SR membrane; b) isolated foot proteins (rotary-shadowed images); c) DHP receptors (tetrads), rotary-shadowed images (rabbit skeletal muscle); d) junctional SR "bumps" and (*E, F*) junctional T-tubular tetrads from triads of toad fish swimbladder muscle (electron microscope; freeze-fracture images). (Block et al. 1988; courtesy of Clara Franzini-Armstrong)

pared to that of the native sarcoplasmic reticulum membrane containing Ca-channels (Fig. 10.3; cf. also Williams 1992 for review).

Like native channels, the reconstituted Ca^{2+} release channel was preferentially selective for calcium ions (Lai et al. 1988, 1989; Smith et al. 1988). For K^+, for instance, it was about six times less permeable than for Ca^{2+}, when both ions competed for the channel. This could be shown by filling the two compartments of the double chamber with solutions containing Ca^{2+} ions on the "*cis* side" of the membrane and K^+ ions on the "*trans* side". It is clear that the currents of the

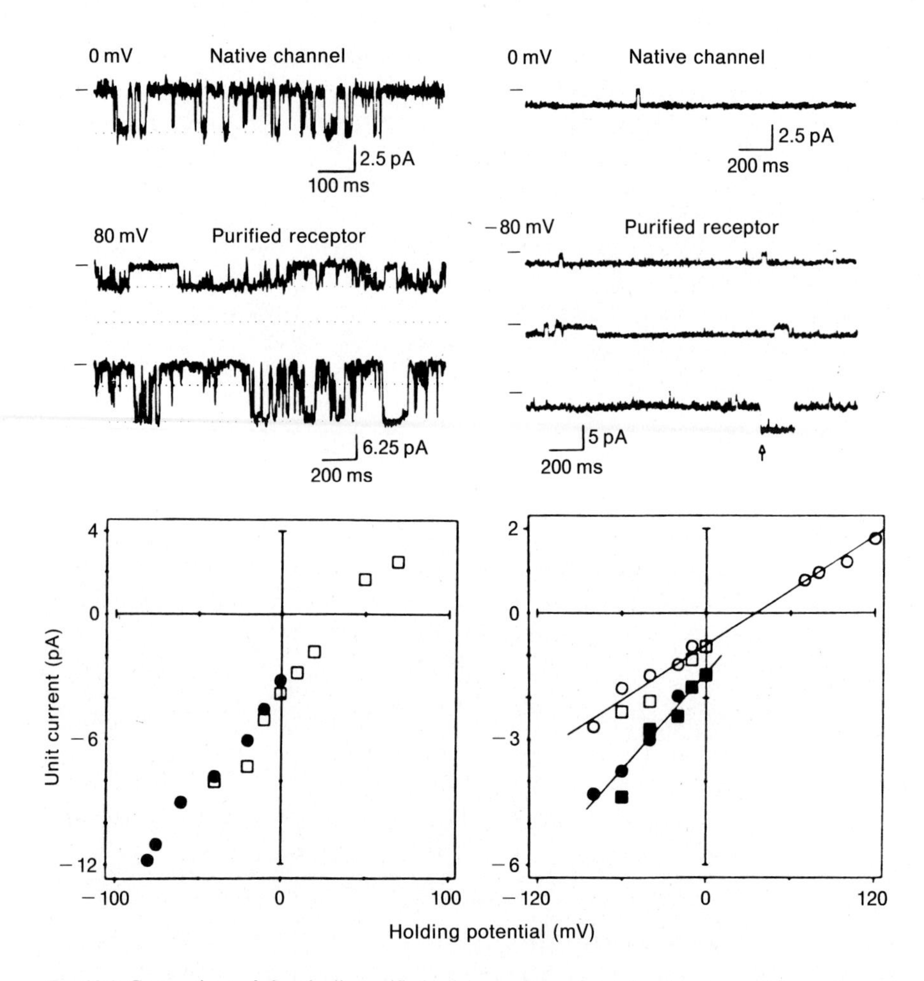

Fig. 10.3. Comparison of chemically purified calcium release channel with native calcium release channel before (*left*) and after blockage with ryanodine (*right*). LEFT PANEL *Top tracings* show Ca^{2+} currents across a channel incorportated into a lipid bilayer separating two chambers containing 1.2 μM free calcium (0.95 mM $CaCl_2$ and 1 mM EGTA in the cis-chamber) and 50 μM Ca^{2+} as a current carrier in the trans-chamber. The concentration of purified ryanodine receptor was 0.6 mg ml^{-1} in the cis-chamber. Due to the electrochemical gradient generated after imposing different voltages (0 or 80 mV), current pulses were generated when the channels in the lipid bilayer opened. The *upper recording* of the purified receptor shows a 50 pS conductance change, the *lower recording* a 110 pS conductance change. Intermittant channel openings cause downward deflections of the tracings (current spikes). *Below* Single channel current voltage relationships are shown for both native release channels (*squares;* SR vesicles incorporated into bilayers) and purified ryanodine receptor channels (*closed circles*) in the solutions described. Note that current reversal occurs at +38 mV. RIGHT PANEL Conditions as in (left panel), except that ryanodine (7 μM) has been added to the cis-site of the double chamber. *Top tracings* Ca-currents (*solid line* indicates baseline current). *Below* Single channel current versus voltage relationships shown for native channels (*squares*) and purified receptors (*circles*). *Open and closed symbols* are for small and large conductance states, respectively. Current reversal at +35 mV. From Smith et al. 1988, cf. Ashley et al. 1991

K^+ and Ca^{2+} ions would then flow in opposite directions as they depend on the ion concentrations and the conductivity of the channels for the respective ions. By applying suitable potentials across the membrane under voltage-clamp conditions, the net current flow along the electrochemical gradient may be opposed: the current between the compartments would become zero, at the so-called *reversal potential,* which is dependent on the respective ion concentrations and permeabilities as described by the Goldman equation (for its application to divalent and monovalent cations such as Ca^{2+} and K^+, cf. Baylor et al. 1984). In the special case, where one compartment of the double chamber contains K^+ and the other one Ca^{2+} at a given concentration, the permeability ratio of these two ions (P_{Ca}/P_K) could be readily determined from the reversal potential according to the equation

$$V_r = \frac{RT}{2F} \cdot \ln \frac{4P_{Ca}[Ca^{2+}]}{P_K[K^+]}, \qquad (10.1)$$

where V_r is the reversal potential, and R, T and F have their usual meaning. The permeability ratio for K^+ and Ca^{2+} ions estimated in this manner was remarkably similar to that already found for the calcium release channel of frog skeletal muscle sarcoplasmic reticulum. In these experiments, Stein and Palade (1988) were able to "patch" large sarcoplasmic reticulum vesicles (sarcoballs) and to investigate their calcium conductance by the patch-clamp technique described in a previous section (p. 169).

Interestingly, the values for (unit) Ca^{2+} conductivity of both native or reconstituted channels in planar bilayers were similar (ranging from about 100 to 240 pS depending on the conditions) to those found by Stein and Palade for calcium channels of the sarcoplasmic reticulum from frog skeletal muscle.

Furthermore, the open probability of the SR-calcium channels depended on the voltage across the sarcoplasmic reticulum membrane, or in the case of the purified ryanodine receptor, across the planar bilayer (cf. Ma et al. 1988). In both cases, the open probability of the channels increased with the free calcium ion concentration (Fill et al. 1990), suggesting that calcium itself was a modulator of the channel.

10.1.1.3 The Calcium Release Channel Could Be Cloned and Sequenced

The channel protein is extremely large with a molecular weight >500 kDa (cf. Zorzato et al. 1990). Thus, the establishment of the primary sequence of the approximately 5000 amino acid residues of the ryanodine receptor was a great achievement. It had been deduced by Takeshima et al. (1989) from the nucleotide sequence of the ryanodine receptor *complementary DNA* (cDNA), which corresponds to that part of the channel gene which codes for and is expressed as protein. It was cloned in *Escherichia coli* by standard procedures (cf. also Zorzato et al. 1990). Briefly, the total muscle poly-A-RNA containing all the different kinds of messenger RNA (mRNA) present in this tissue was extracted from muscle and transcribed into single-stranded complementary DNA (cDNA) using enzyme reverse transcriptase and then DNA polymerase to make double-stranded cDNA. In this way, a mixture or "library" of many different cDNA fragments was obtained which coded for the respective mRNAs. These fragments were then inserted into bacterial plasmids consisting of circular DNA. The constructs obtained in this way could then be used as vectors to transform bacteria (*Escherichia coli*) grown in culture. Statistically, only one plasmid containing one fragment of cDNA is usually taken up by any one specimen of *E. coli*. Given the very large number of different RNAs in the muscle extract, there was only a very small chance that the cDNA complementary to the ryanodine receptor mRNA was incorporated into a particular colony of *E. coli*. Thus, thousands of bacterial colonies had to be screened for those carrying the chimeric plasmid vector which contained the ryanodine receptor cDNA. In this procedure, the colonies were first transferred from agarose to nitrocellulose filters and then lysed. The DNA was then denatured, fixed on the filter and hybridized with a radioactively labelled oligonucleotide probe for ryanodine receptor cDNA. Successful hybridization was detected by autoradiography, of course, only if the sequence of the probe was complementary to and therefore matched a certain target sequence of the ryanodine receptor cDNA and if the latter was actually contained in a colony. To design a suitable oligonucleotide probe was not all that difficult, since

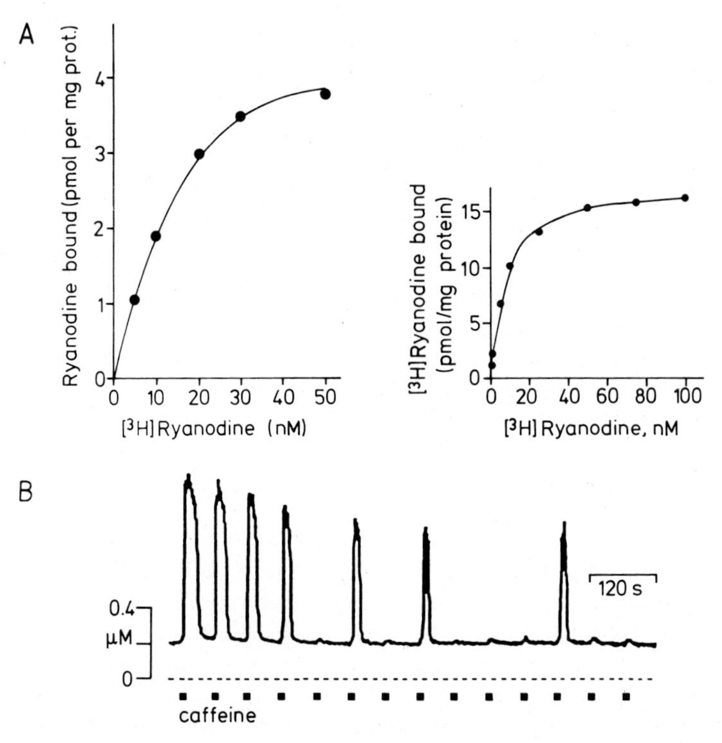

Fig. 10.4. The calcium release channel is a ryanodine receptor. *A* Ryanodine binding to isolated triads (*left*) as a function of ryanodine concentration (Imagawa et al. 1987) and ryanodine binding to isolated membrane preparations of CHO cells (*right*) that have been transfected with ryanodine receptor cDNA (cf. also Takeshima et al. 1989). *B* Calcium release of CHO cells that expressed calcium release channels after transfection with ryanodine receptor cDNA. Application of caffeine (1–50 mM) causes Ca-transients (cf. Penner et al. 1989)

knowledge of only a small part of the amino acid sequence of the ryanodine receptor was required. In fact, the sequence of the probe was chosen to correspond to the nucleic acid code for one of the known peptide fragments isolated from a proteolytic digest of purified ryanodine receptor.

Once bacterial colonies could be selected that carried the cDNA of ryanodine receptors, they were grown and propagated in culture to produce large quantities of "cloned" ryanodine receptor cDNA. The cDNA that was amplified in this way was then isolated and used either for sequencing or it was inserted into an "expression vector" (a plasmid) and injected, for instance, into the nuclei of hamster ovary cells in tissue culture. These transfected cells, after a few days, would then express cloned ryanodine receptor protein, which was incorporated into the intracellular membranes of the cells. Not surprisingly, it was identical to the chemically purified protein! In particular, the expressed recombinant protein reacted with antibodies directed against native calcium release channels and it bound ryanodine with the same high affinity as the native receptor (Fig. 10.4 A, B). Also, and most importantly, it was a functional calcium release channel (cf. Penner et al. 1989) that released Ca^{2+} upon stimulation of the cells as shown in Fig. 10.4 C.

The cDNA that was used to transfect the Chinese hamster ovary cells was of the "wild type". Alternatively, one might have used cDNA that was chemically altered by "site-directed mutagenesis". This method offers the prospect of very selective alteration of the expressed channel protein in order to study structure function relationships either in the transfected cells or after reconstituting the channel in lipid bilayers. Of course, the large quantities of cloned ryanodine receptors obtained by recombinant techniques also facilitated the elucidation of its structure as well as its reconstitution into lipid bilayers for functional studies.

10.1.1.4 Structure and Function Relationships of the Calcium Release Channel

The *ryanodine receptor* probably forms a homotetrameric complex, each monomer having a molecular weight of about 500 kDa and consisting of two portions (Fig. 10.5; cf. Takeshima et al. 1989), a large cytoplasmic domain and a much smaller membrane-spanning domain. The latter froms the C-terminal channel region that resembles the structure of the nicotinic acetylcholine receptor α-subunit. This structural resemblance makes it likely that the putative hydrophobic transmembrane segments in the C-terminal region of the ryanodine receptor of all four monomers will join together to surround a central pore, forming the channel. The bulk of the protein, however, constitutes the cytoplasmic domain which probably corresponds to the "junctional-foot" structure identified on electron micrographs of the triads. These morphological studies (Block et al. 1988; cf. Fig. 10.2) suggest a direct interaction of the foot protein with a protein component of the T-tubular membrane as depicted in Figs. 10.5 und 10.2.

The C-terminal domain probably also contains the binding sites for various *modulators of calcium channel activity*. These include ryanodine, caffeine, calmodulin, adenine nucleotides and the channel blocker ruthenium red, as well as an intracellular peptide described by Herrmann-Frank und Meissner (1989). All

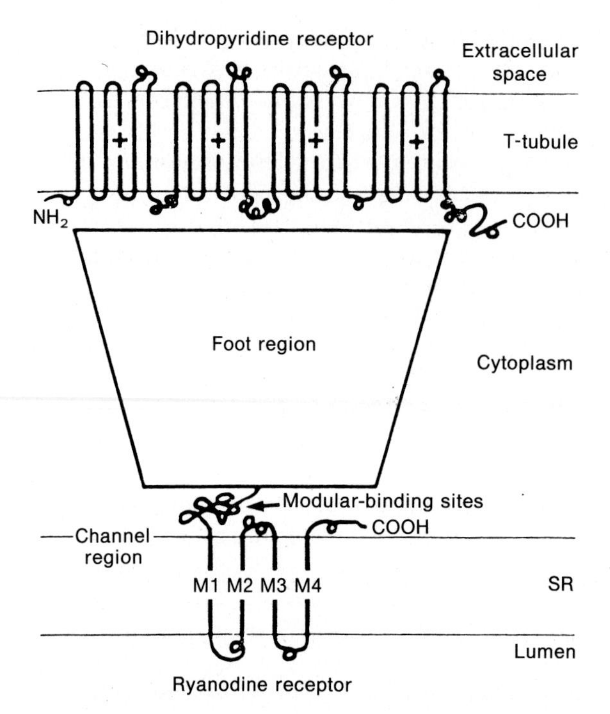

Fig. 10.5. Model of calcium release channel as coupled by the dihydropyridine receptor channel and ryanodine receptor-calcium channel by the foot region according to Takeshima et al. (1989). The transmembrane segments of the four units of repeat of the DHP receptor are shown and also the possible interaction with the foot region of the ryanodine receptor. The latter is anchored in the sarcoplasmic reticulum membrane by means of the four membrane-spanning segments *M1–M4*. Remember that four ryanodine receptor units form the tetrameric foot structure possibly corresponding to a channel. Tentatively, the sarcoplasmic effector loop region between units 2 and 3 of the dihydropyridine receptor may interact with the foot region, when the voltage sensors (possibly located) in the fourth helix of each unit of homology (indicated by +) sense the depolarization of the T-tubular membrane (Takeshima et al. 1989)

these compounds modulate the gating of the calcium channel reconstituted into planar bilayers. Furthermore, there are two second messengers that gate the open state of the calcium channel, reconstituted into planar bilayers. These are, as already mentioned, calcium as well as inositoltrisphosphate (cf. Hidalgo and Jaimovich 1989). The latter finding is particularly noteworthy, since there is a partial sequence homology between the ryanodine receptor of skeletal muscle and one of the calcium release channels of smooth muscle (cf. Mignery et al. 1989; Chadwick et al. 1990), which is physiologically gated by inositoltrisphosphate (cf. p. 227). Note, however, that smooth muscle also contains, in addition, a Ca^{2+}-gated, ryanodine-sensitive Ca^{2+} release channel (Herrmann-Frank et al. 1991).

As pointed out previously, the cloned channel reconstituted into a lipid bilayer can be gated by changing both the membrane potential or the calcium ion concentration. The latter effect is the basis for "calcium-induced calcium release", which is pathologically increased in muscles of patients suffering from *malignant hyperthermia* (Endo et al. 1983; cf. p. 45). This genetic disease is now known to be due to a point mutation (McLennan et al. 1990; Fujii et al. 1991; Otsu et al. 1991), causing a substitution of T for C at nucleotide 1843 in the ryanodine receptor gene, so that one amino acid residue (arginine 614) is substituted by cysteine (Gillard et al. 1991). As a consequence, the calcium sensitivity and conductivity of the calcium release channel are increased and its inactivation delayed (Fill et al. 1990). Calcium-induced calcium release, therefore, appears to be of pathophysiological importance in this disease. Whether or not, however, calcium is also a physiological mediator between the T-tubular membrane and the calcium release channel remains to be seen. If it were so, then, clearly, calcium channels of the T-tubular membrane would have to be primarily involved in excitation-contraction coupling!

10.1.2 The Calcium-Channel of T-Tubules in Cardiac and Skeletal Muscle

Molecular approaches have greatly contributed to our understanding of the control of the calcium release channel by the T-system. It has now become clear that the part of the T-system which senses the membrane potential and communicates with ryanodine receptors of the sarcoplasmic reticulum is also a kind of calcium channel. These channels, first characterized by Almers et al. (1981b) and Sanchez and Stefani (1983), bind calcium channel blockers such as dihydropyridines (Fosset et al. 1983; Varadi et al. 1991). Thus, the calcium channel of T-tubules is also known as the *dihydropyridine receptor (DHP receptor)*. In striated muscle this channel seems to be associated with the "tetrad structures" that face the "feet" structures containing the calcium release channel (Block et al. 1988; Leung et al. 1988). As is well known, cardiac and smooth muscle also contain dihydropyridine binding calcium channels in the cell membrane. However, their structure and function are somewhat different from that of the T-tubular calcium channel of skeletal muscle.

10.1.2.1 The Role of Calcium Inward Current in Skeletal and Cardiac Muscle

Cardiac muscle as well as smooth muscle contain two types of calcium channels, denoted as *T-channels* and *L-channels* respectively (Bean 1985; Nilius et al. 1985). The former does not bind dihydropyridines, it is gated at a membrane potential of about −50 mV and inactivates rapidly, whereas the latter requires greater depolarization but deactivates more slowly (L-type or long-lasting channel). In many invertebrate muscles as well as in heart muscle and smooth muscle, the L-type calcium channels of the surface membrane and of the T-tubular membrane presumably supply sufficient calcium for triggering *calcium-induced calcium release* in the sarcoplasmic reticulum. Evidence (reviewed by Ashley et al. 1991) for such a mechanism has been obtained by showing that a rapid rise in calcium surrounding the sarcoplasmic reticulum could induce the release of calcium. Thus, by flash photolysis of "caged calcium compounds" such as Nitr-5 injected into cells, a sudden increase in calcium could be induced experimentally within living cardiac muscle (Valdeolmillos et al. 1989) or in permeabilized barnacle muscle fibres (Lea and Ashley 1989); this led to the release of more calcium and to contraction. More importantly, the gating of cardiac calcium channels is sufficiently fast to account for the rapid rise in calcium level that is required for calcium-induced calcium release. As transmembrane calcium influx is required for cardiac EC-coupling, the latter is dependent on extracellular calcium, as already mentioned in Sect. 7.3.1.

In *vertebrate skeletal muscle,* excitation-contraction coupling differs in many ways from that of cardiac and invertebrate muscle. Though calcium channels are present in the T-tubular membrane, they probably constitute only part of the DHP receptors. A second fraction of the DHP receptors may not function as calcium channels (Schwartz et al. 1985; cf. also Lamb 1991). Gating of the T-tubular Ca-current is extremely slow (Sanchez and Stefani 1983), so that it may take over 100 ms to reach peak current (Feldmeyer et al. 1990). Thus, calcium influx occurs too late to account for the rapid onset of contraction by a calcium-induced calcium release mechanism. This is astounding, since the calcium release channel (ryanodine receptor) of skeletal muscle can in principle be gated by calcium to release more calcium. Apart from these considerations, there is also considerable experimental evidence that seems to rule out the necessity for calcium inflow to initiate calcium-induced release in skeletal muscle, at least under physiological conditions. For instance, depolarization of the surface membrane triggers calcium release from the sarcoplasmic reticulum of skeletal muscle even if intracellular calcium is heavily buffered with chelating agents such as Fura-2 (Baylor and Hollingworth 1988), or if the entire fibre is soaked in solutions containing the Ca^{2+} chelator EGTA (Armstrong et al. 1972; Lüttgau and Spiecker 1979). Furthermore, calcium currents and excitation-contraction coupling can be dissociated. For instance, perchlorate, a chaotropic agent, shifts the relationship between membrane depolarization and calcium release (Gomolla et al. 1983) without affecting the calcium inward current (Feldmeyer and Lüttgau 1988). Therefore, as mentioned above, in skeletal muscle, most of the DHP-binding calcium channel proteins are not functioning as typical calcium channels. They may act as a voltage sensor and signal transducer rather than as a true channel (Rios and Brum 1987).

10.1.2.2 The Dihydropyridine Receptor of Skeletal Muscle Is a Modified Calcium Channel

The DHP receptor consists of several subunits designated as α_1, α_2, β, γ and δ that form on oligomeric complex of 430 kDa (Campbell et al. 1988). The α_1-subunit has a molecular weight of 170 kDa. It has been isolated and cloned and found to be structurally and functionally similar to L-type calcium channels, in particular those of cardiac muscle, with which it has 60% sequence homology (cf. Tanabe et al. 1987; Mikami et al. 1989). Thus, the entire α_1-subunit contains four repeated units of homology, each of them having six α-helical membrane-spanning segments linked by hydrophilic loops (see Fig. 10.7). When the cDNA of the α-subunit, i.e. of the DHP receptor, was expressed in *Xenopus* oocytes, functional DHP-sensitive calcium channels were formed and incorporated into the cell membrane (Mikami et al. 1989; cf. also Perez-Reyes et al. 1989). If the other subunits were coexpressed as well, they modulated the channel properties, in particular the gating kinetics (Lacerda et al. 1991; Varadi et al. 1991).

The isolated α_1-subunit could be incorporated into *lipid bilayers* e.g. across the orifice of a patch micropipette filled with calcium chloride and containing an electrode. Obviously, it formed a pore; thus by applying a suitable voltage gradient, a current would flow through the membrane of the bilayer, provided the channels were open (Flockerzi et al. 1986; cf. Fig. 7.11.). Interestingly, the open probability of the channel could be increased by phosphorylation of the protein (at serine-687, with a c-AMP-dependent protein kinase; cf. Röhrkasten et al. 1988), but it was also a function of the voltage across the membrane. Indeed, the voltage dependence was found to be similar to that of the "native" calcium channel within intact muscle fibres (Rios et al. 1991). Obviously, the dihydropyridine binding protein of 170 kDa had both the properties of a calcium channel as well as that of a voltage sensor. As in other types of calcium channels, the open probability could be decreased by dihydropyridines (DHP) and other calcium antagonists that bound to the α-subunit of the channel (cf. Campbell et al. 1988).

Since calcium channel blockers also paralyzed contraction in muscle fibres (Eisenberg et al. 1983; cf. Feldmeyer et al. 1990a), the DHP-sensitive calcium channel or DHP receptor was implicated in excitation-contraction coupling. However, as mentioned already, the calcium inward current elicited by depolarization was too slow to account for a twitch in skeletal muscle, so that its role in excitation-contraction coupling remained a puzzle. What then was the function of the channel protein?

10.1.2.3 The DHP Receptor is Responsible for Excitation-Contraction Coupling

Direct evidence for an involvement of the DHP receptor in excitation-contraction coupling arose from *molecular genetic experiments* with dysgenic mice suffering from a lethal hereditary disease causing muscle paralysis (Beam et al. 1986). This defect is due to an uncoupling of excitation and contraction. The reason is that, obviously, an important component of the excitation-contraction coupling mechanisms is missing already in immature muscle cells called myotubes (Tanabe et al. 1988). Previous work, using antibodies, already showed that, unlike in myotubes of healthy mice, the T-tubular calcium channels or DHP receptors were probably not expressed. At least they were not readily detectable by monoclonal antibodies, and DHP binding was much reduced. "Tetrads" could not be identified in T-tubules by electron microscopy and all calcium inward currents were eliminated. Moreover, the gene coding for the DHP receptor was deleted or mutated, as shown by restriction endonuclease analysis and Southern blotting. Thus, specific oligonucleotide probes, which were complementary to and which hybridized with characteristic regions of fragments of the DHP gene, failed to detect the gene or its fragments in Southern blot analysis. Whereas in RNA extracts of healthy myotubes the probes would hybridize with messenger RNA coding for the calcium channel, this method (Northern blot

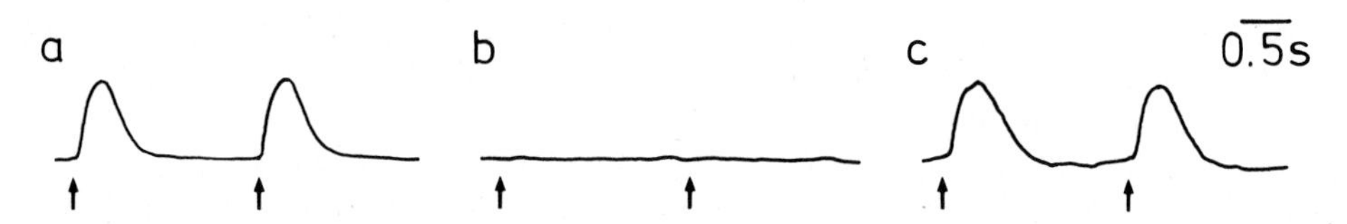

Fig. 10.6 a–c. Comparison of electrically evoked contractions in dysgenic myotubes from mouse skeletal muscle before and after its expression of transfected DHP-receptor cDNA. An expression plasmid containing the cDNA of skeletal or cardiac DHP receptor was injected into the nuclei of paralyzed myotubes from dysgenic mice. The contractile responses of these myotubes are compared with those of healthy myotubes. *a* Response of "healthy" myotubes expressing skeletal muscle DHP receptors. *b* Response of paralyzed dysgenic myotube (lacking DHP receptor). *c* Response of myotubes from dysgenic mice after "curing" paralysis by injection and expressing the cDNA of skeletal-type DHP receptor. The cDNA was inserted into an expression plasmid. For experimental conditions, cf. legend to Fig. 10.8 B. (Tanabe et al. 1988)

analysis) barely detected the specific messenger RNA in extracts of the diseased myotubes. Obviously, the DHP-mRNA was much reduced or even absent.

However, if messenger RNA of healthy myotubes was transcribed into complementary DNA (cDNA) and injected into the nuclei of diseased, immature muscle fibres (myotubes) lacking the gene product, excitation-contraction coupling could be restored (Tanabe et al. 1988). These *transfection experiments* were quite tedious, since the cDNA had to be packed first into an expression vector, a bacterial plasmid consisting of a circular DNA construct also containing promoters and reporter genes. When these constructs were injected into the nuclei of myotubes, functional DHP-receptor proteins were expressed after a short time and thereafter the myotubes were able to twitch again when electrically stimulated (Fig. 10.6). In this case, excitation-contraction coupling was also associated with movement of an electric charge (charge movement, see p. 40) within the T-tubular membrane, but, in addition, a slow calcium inward current could also be observed (B. A. Adams et al. 1990). The latter, however, did not seem to be essential for excitation-contraction coupling, since it occurred *after* the onset of contraction and since charge movement and contraction could also be elicited even in the absence of external calcium when no calcium current was flowing. Obviously, the expressed DHP binding protein of skeletal muscle was acting only as a voltage sensor in this case and not as a "true" calcium channel.

If, however, myotubes lacking DHP receptors were microinjected with a cDNA coding for cardiac DHP receptors, the myotubes were transformed (Tanabe et al. 1990a). They now showed the characteristic features of cardiac excitation-contraction coupling. Thus, unlike skeletal muscle but like cardiac muscle, they twitched only in the presence but not in the absence of external calcium (Fig. 10.8). Obviously, under these conditions, an influx of calcium ions through the "L-type" calcium channel *was* required to elicit contraction. Presumably, the inflowing calcium caused calcium-induced calcium release as described above.

These experiments, therefore, clearly indicate that the sarcoplasmic reticulum of skeletal muscle *is* in principle capable of calcium-induced calcium release, just as in the case of cardiac muscle. However, does this capability play a role physiologically in skeletal muscle? For instance, one might speculate that the released calcium could influence the release channels of the sarcoplasmic reticulum to cause further calcium release in the sense of a positive feedback. Given the fact that – at least in fish muscle – only half of the release channels face a voltage sensor (Block et al. 1988), such a mechanism might even be necessary to activate all channels. In any case, these experiments show that the release channels of skeletal muscle can be gated by calcium in much the same way as those of the heart, despite the fact that their primary structures are slightly different. Recall that in isolated calcium release channels incorporated into bilayers the open probability is also enhanced after increasing the free calcium ion concentration.

10.1.2.4 Structure Function Relationships of Cardiac and Skeletal Muscle DHP Receptors

We have just seen that in dysgenic mice the expression of the DHP receptors of cardiac muscle and skeletal muscle restores cardiac- and skeletal-type excitation-contraction coupling respectively. Naturally, the question arises as to what structural features are essential for these interesting functional differences. To approach this problem, Tanabe and colleagues dissected the complementary DNA of cardiac DHP receptors into several fragments using various restriction endonucleases, and subsequently replaced certain segments by the corresponding skeletal muscle counterparts (cf. Tanabe et al. 1990b). After annealing the fragments, chimeric cDNA constructs could be obtained which were then incorporated into "expression plasmids" that were then injected into the myotubes. In this way, *chimeric DHP-receptor* proteins were expressed containing features of both cardiac and skeletal muscle DHP receptors. In essence, the basic structural feature was that of the cardiac protein, while certain cytoplasmic loop regions were replaced by their skeletal muscle counterparts (Fig. 10.7). Interestingly, the replacement of the cytoplasmic loop region, linking the transmembrane segments II and III, altered the DHP protein from cardiac type to skele-

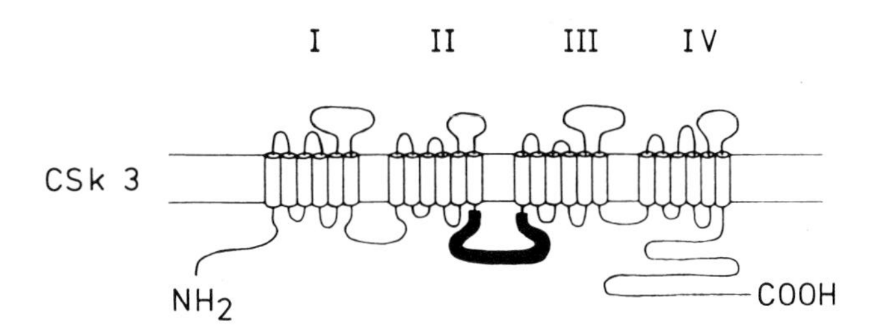

Fig. 10.7. Schematic representation of a chimeric DHP receptor (CSk 3) based on the sequence of the cardiac-type receptor. However, the cytoplasmic effector loop (between repeats 2 and 3) has been substituted by its skeletal muscle counterpart (shown in *black*). The cDNA encoding the chimeric receptor restored skeletal-type rather than cardiac-type EC-coupling in myotubes of dysgenic mice. These therefore contracted even in the absence of extracellular Ca^{2+}. *I* First unit of homology which is different in skeletal- and cardiac-type DHP receptor and responsible for gating properties. (Tanabe et al. 1990 b)

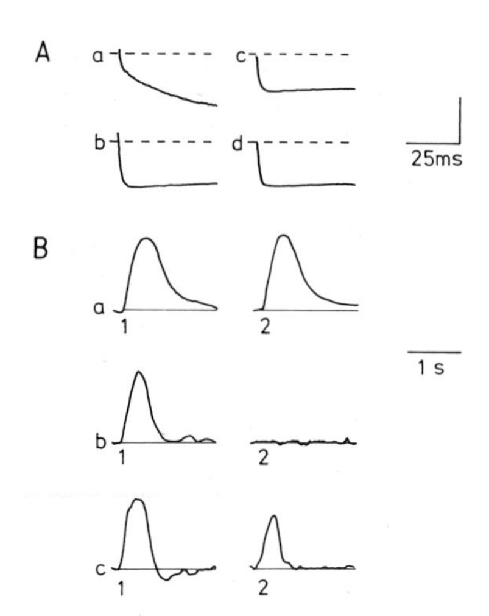

Fig. 10.8 A, B. Structure-function relationship in T-tubular calcium channels (DHP receptors). *A* Comparison of calcium currents in transfected myotubes from dysgenic mice expressing skeletal muscle DHP receptors (*a*), cardiac DHP receptors (*b*) or chimeric receptors (*c, d*), following transfection with expression plasmids containing the appropriate cDNA coding for cardiac DHP receptor with the cytoplasmic loop region, which links repeats II and III, being replaced by its skeletal muscle counterpart (mutant Sk 3, cf. Fig. 10.7) as shown in (*c*), or a chimeric receptor (*d*) in which one of the transmembrane helical segments has been replaced by the corresponding skeletal-type unit. The current was measured by a whole cell, patch-clamp recording with the patch pipette containing the following solution: 140 mM caesium aspartate, 5 mM $MgCl_2$, 10 mM caesium-EGTA, 10 mM HEPES buffer (pH 7.4 with CsOH). The bathing solution contained: 10 mM Ca^{2+}, 145 mM tetraethylammonium ion, 165 mM Cl^-, 0.003 mM tetrodotoxin, 10 mM HEPES buffer (pH 7.4 with CsOH). *B* Comparsion of electrically evoked contractions in dysgenic myotubes expressing *a* skeletal muscle DHP receptor; *b* cardiac-type receptor; and *c* a chimeric cardiac-type receptor, in which the cytoplasmic effector loop region linking repeats 2 and 3 has been replaced by the corresponding skeletal-type segment (construct denoted as Sk 3, cf. Fig. 10.7). Note that the cardiac-type DHP receptor restores EC-coupling only in the presence of external calcium (*1*), while the replacement of the effector loop region permits EC-coupling even in the absence of external calcium (*2*). Contractions were monitored on a videoscreen (arbitrary units) following an electrical stimulus of 5 ms duration in calcium containing bathing solution (*1*) or in calcium-free solution (*2*). Composition of the bathing solution: 146 mM NaCl, 5 mM KCl, 2 mM $CaCl_2$, 1 mM $MgCl_2$, 10 mM HEPES buffer (pH 7.4 with NaOH). In the Ca-free solution Ca^{2+} was substituted by Mg^{2+}. (After Tanabe et al. 1990 b)

tal type inasmuch as now excitation-contraction coupling could be restored in these myotubes even in the absence of external calcium. (Fig. 10.8 B). On the other hand, the fast gating properties characteristic of cardiac-type calcium channels were retained in this case, even if much of the cytoplasmic region was replaced by the corresponding segments of the skeletal muscle protein (Fig. 10.8 A). Obviously, the molecular structure responsible for the slow gating properties resides in other parts of the DHP receptor, most likely within the four repeated units of homology constituting the transmembrane regions.

Tanabe and colleagues (1991) then went on to characterize the structure-function relationship in even more detail by constructing a chimeric cDNA in which the first transmembrane-spanning unit of a repeat of the skeletal muscle DHP receptor had been replaced by its cardiac muscle counterpart. When this construct was transfected into the myotubes of dysgenic mice, the expressed DHP receptor showed the EC-coupling properties of the skeletal muscle DHP receptor, whereas the *gating properties* were typical for cardiac calcium channels. Perhaps, it is the molecular structure of the first unit of the repeat that determines whether a chimeric calcium channel shows fast cardiac-type or slow skeletal-type

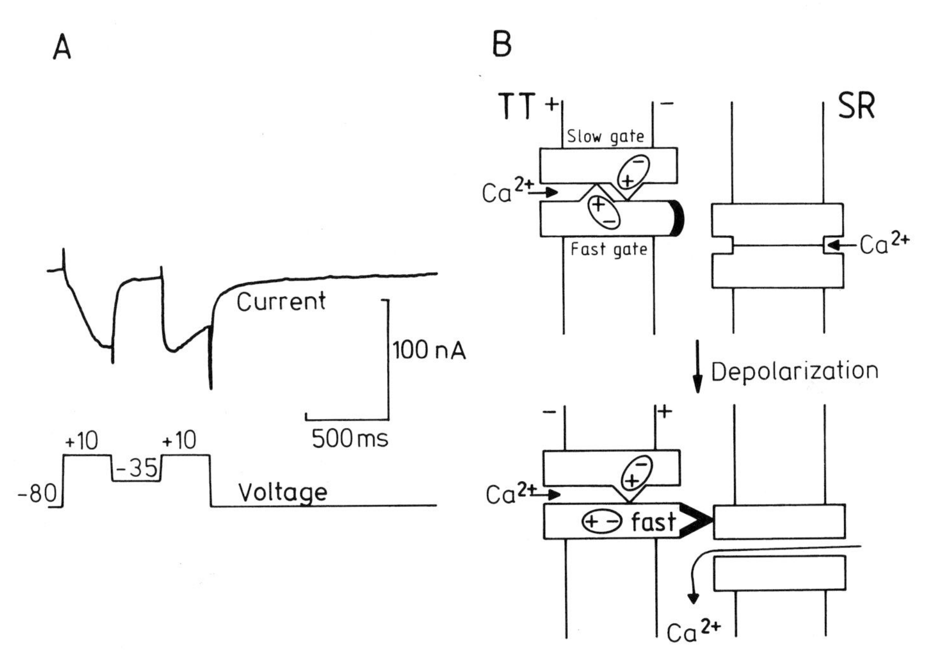

Fig. 10.9 A, B. Gating of T-tubular calcium channels (DHP receptors). Demonstration of a fast gating process in a dihydropyridine-binding calcium channel of T-tubules producing a slow calcium inward current in frog skeletal muscle (*A*), and a simple model showing how the fast gate could be coupled to the calcium release channel in the sarcoplasmic reticulum (*B*). *A* Two subsequent depolarizing pulses to the same membrane potential open the channel and induce a calcium current with a very different time course. This can be accounted for by assuming the presence of two different voltage-dependent gates. During the first pulse (conditioning pulse), the activation of the slow gate would be rate-limiting for channel opening, while at the second pulse (test pulse), the slow gate is still open, so that, now, the kinetics of the fast gate determines the speed of current activation. Note that conditioning pulse and test pulse are separated by a short interval at which the membrane is partly repolarized, so that the channels close (Feldmeyer et al. 1990 b). *B* Model of calcium channels gates as proposed by W. Melzer (unpubl.). The fast voltage-dependent gate controls the interaction of the effector-loop region of the DHP receptor with the calcium release channel site before the slow gate of the T-tubular calcium channel opens. In skeletal muscle, the slow gate may be associated with the first unit of repeat of the DHP receptor and may determine the normal slow opening of the channel, while the fast gate may be associated with the other three domains, which also have to interconvert into an activated form before the channels can open. A change in voltage causes a fast conformational movement (associated with charge movement), thus opening the fast gate. In the signal transmission to the calcium release channel, the crucial cytoplasmic effector loop between repeats II and III of the DHP receptor may be distorted by the rapid voltage- dependent conformational change in the neighbouring domain. Note that in cardiac muscle, the first unit of repeat of the DHP receptor has a different structure and may also be associated with a fast gate, so that the channel opens rapidly upon depolarization. (Courtesy of W. Melzer)

channel gating properties. It is tempting to speculate then that in skeletal muscle the first unit of the repeat may be associated with a slow channel gate which may be present in addition to a fast gate (cf. Fig. 10.9; Feldmeyer et al. 1990b). Alternatively, each of the four units of the repeat would have to be interconverted into an activated form for the channel to open, but in skeletal muscle the interconversion of the first unit may be slower than in cardiac muscle. Indeed, its primary structure greatly differs from its cardiac counterpart. Interestingly, the structurally most different region lies between the transmembrane-spanning helices S 5 and S 6 and may thus be close to the channel pore or it might even form part of the channel-lining. In future experiments, site-directed mutagenesis might help to identify the exact site responsible for the difference in the activation kinetics between a cardiac- and skeletal-type calcium channel. Recall, however, that the gating properties of the DHP receptor in vivo depend not only on the structure of the α-subunit, but as already mentioned above also on the interaction with the β-subunit as shown by co-expression of α- and β-subunits in Chinese hamster ovary cells (Varadi et al. 1991).

In summary, the experiments of Tanabe and colleagues just described have led to a molecular level understanding of the differences in cardiac and skeletal muscle EC-coupling and of the relative importance of calcium channel gating properties and voltage-dependent charge movement for the coupling process. It seems clear now that the calcium conductance of the channel and the gating properties are not per se essential for skeletal muscle-type excitation-contraction coupling. On the other hand, the voltage-sensing capacity may well be crucial. Thus, a sudden change in the membrane potential across the T-tubular membrane might give rise to a fast conformational change as indicated by the rapid *charge movement* in the T-tubular membrane (reviewed by Rios and Pizarro 1991) that may – at least partly – be associated with the movement of a fast gate suggested to be present in addition to the slow gate of the DHP receptor calcium channel (Feldmeyer et al. 1990b). Whether, however, these conformational processes associated with "gating" are also involved in the subsequent signal transmission to the junctional foot protein remains to be seen. More likely, perhaps, a small movement of (positively charged) residues within the cytoplasmic "effector" loop of the DHP receptor described above may be involved in the "mechanical coupling" of DHP receptor and ryanodine receptor protein that leads to calcium release from the sarcoplasmic reticulum (Rios and Pizarro 1991).

10.1.2.5 How Is the Calcium Release Channel Controlled by the DHP Receptor of the T-Tubule?

As we have seen, in the heart and in many types of invertebrate muscle, DHP receptors communicate with the calcium release channel by means of a chemical messenger, the calcium ion. Following calcium influx into the myoplasm, the calcium ions induce calcium release from the sarcoplasmic reticulum. As pointed out before, in the case of skeletal muscle, a similar mechanism has also been suggested, but seems unlikely, since excitation-contraction coupling is still possible when myoplasmic free calcium is buffered. One might speculate, however, that calcium may be tightly bound to the cytoplasmic face of the DHP receptor within the T-tubular membrane and may be released during stimulation to the target, the calcium release channel of the sarcoplasmic reticulum.

This hypothesis is, however, not supported by Heiny and Jong (1990), who showed, using potential-sensitive dyes, that the inner surface of the T-tubules becomes more positively charged during excitation-contraction coupling. Rather, this finding suggests that a negatively charged part of the voltage sensor may be moving away from the T-tubular surface in the direction of the calcium release channel. Conversely, a (positively charged) part of the above-mentioned cytoplasmic effector-loop region of the DHP receptor (cf. Fig. 10.7) may move in the opposite direction, thereby "pulling" as it were on the calcium release channel so that it becomes "unplugged" as suggested by Rios et al. (1991). Is it possible that the DHP receptor interacts in this way "mechanically" with the calcium release channel and in a most direct manner? Or is a "third party" (yet another protein or messenger) involved as suggested by Brandt et al. (1990)? These questions need to be answered to understand EC-copuling more fully and on a molecular level.

A powerful approach to these problems may be seen in the use of mechanically *skinned muscle fibres* in which triadic structures and depolarization-contraction coupling are still functional, while the composition of myoplasmic components in the bathing solution may be controlled at will. These preparations (Lamb and Stephenson 1989), in which the T-system had resealed spontaneously after mechanical skinning, could be depolarized simply by replacing potassium ions with sodium in the bathing solution. Interestingly, then, the depolarization-contraction coupling was found to depend on the concentration of *Mg-ions*. Thus, uncoupling occurred at high Mg^{2+} concentrations, since Mg-binding inhibits Ca-release. These findings suggested to Lamb and Stephenson (1990) that depolarization may stimulate calcium release via a reduction in Mg-affinity and Mg-binding to the calcium release channel. Clearly, this experimental model might also be useful for studies interfering in various other ways with the intracellular coupling process, for instance, by specific inhibitors and antibodies directed against functionally important epitopes. So far, however, the precise nature of the coupling between the DHP receptor and the calcium-release channel is still an open question, at least in the case of skeletal muscle.

10.2 Control of the Contractile Mechanism by Intracellular Free Calcium

In skeletal and cardiac muscle, the calcium signal generated by the calcium release channel is "picked up" and decoded by the calcium binding protein troponin-C and further processed by a set of thin filament proteins, notably troponin-I, troponin-T, actin and tropomyosin. These are the molecular mechanisms that call crossbridges into action. Ever since its first demonstration in 1964 by Portzehl, Caldwell and Rüegg, the dependence of contraction of intact musclefibres on micromolar levels of intracellular free calcium has been firmly established.

10.2.1 Molecular Properties and Role of Thin Filament Proteins

When calcium binds to troponin-C, its interaction with troponin-I is strengthened, whereas that of actin with troponin-I is weakened, thereby derepressing its inhibitory effect on actomyosin ATPase and contraction as already discussed in Chapter 4. Recombinant DNA technology offers a very promising approach to investigate structure-function relationships in these proteins by modifying the cDNA encoding the protein; the mutant polypeptide expressed in *Escherichia coli* can then be used in functional tests.

10.2.1.1 Troponin-C: Structure-Function Relationships Explored by Site-Directed Mutagenesis

The troponin-C (TnC) molecule has two independent *calcium-binding domains* which are located at the C-terminal and N-terminal, respectively, and are connected by a central helix (Satyshur et al. 1988). Recall (cf. P. 85) that each domain region contains a pair of calcium binding sites consisting of a helix-loop-helix motif (also known as EF-hand, see Fif. 4.1). Sites I and II are located in the N-terminal domain; they bind calcium specifically, and are thus involved in calcium regulation. In fast skeletal muscle troponin-C, both sites interact cooperatively and are necessary for full functional activity (Potter et al. 1991). However, site II seems to be particularly important as it has been conserved through evolution (Collins 1991). Thus, in slow skeletal muscle and cardiac muscle TnC, the only functional calcium-specific site is site II. Its paramount importance in muscle activation could be demonstrated by showing that a single point mutation of this site may abolish the specific calcium binding properties of troponin-C altogether (Putkey et al. 1989). In these experiments, cDNA of troponin-C from slow muscle or cardiac muscle was altered at specific sites by *site-directed mutagenesis*. One of the mutated cDNAs obtained coded for alanine rather than aspartic acid at residue 65, and was cloned in *Escherichia coli* to express mutated troponin-C. This protein no longer bound calcium specifically nor was it capable of activating the contractile system.

The functional consequences of substituting mutated TnC were demonstrated in a *skinned fibre assay* as follows. In permeabilized fibres from a slow muscle (rabbit soleus), troponin-C could be readily extracted by magnesium and calcium-free solution of low ionic strength. Such extracted fibres would then no longer contract in ATP salt solution after increasing the free calcium to 10^{-5} M, but contraction could be completely restored by reconstituting the system with a "wild-type" slow isoform of troponin-C. If, however, troponin-C was replaced by mutated troponin-C, the preparation was unable to contract (Fig. 10.10). Obviously, the first residue of the calcium binding loop II (aspartic acid at residue 65) is indeed essential for specific calcium binding and the regulation of contraction. In addition, however, residues 3, 5, 7, 9, and 12 of the EF-loop region are also critical in that they must contain carboxylic and hydroxylic residues engaged in calcium binding.

In fast skeletal muscle, the calcium binding site I also contains carboxylic residues at the critical sites, and it is therefore calcium-bindings as well. As al-

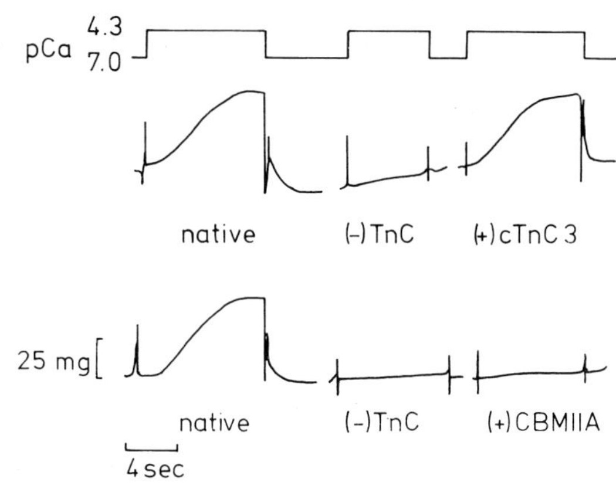

Fig. 10.10. Calcium activation of skinned fibres reconstituted with mutated recombinant TnC. After a control contraction *TnC* was extracted (as described by Güth and Potter 1987) and then substituted with a recombinant troponin-C resembling native troponin-C (cTnC 3), or with a troponin-C isoform, in which the calcium binding site 2 is mutated so that it does not bind Ca^{2+} (denoted as *CBMIIA*). In the latter case, there is no contractile response to calcium. Note that cTnC3 is a bacterially expressed protein, which is virtually identical to troponin-C with some minor changes (e.g. methionine-1 replaced by alanine). In the TnC-CBMIIA mutant, the residues 65, 67 and 70 of calcium binding site 2 are exchanged for alanine, asparagine and glycine, respectively. Conditions: rabbit soleus muscle fibre suspended in ATP salt solution. (From Putkey et al. 1989)

ready mentioned, this site is, however, mutated in troponin-C of slow skeletal and cardiac muscle (Van Eerd and Takahashi 1976; Wilkinson 1980) so that it no longer binds calcium specifically. Mutation involves replacement of the first and third residues of the loop motif by glycine and alanine, respectively, thereby causing the loss of the calcium binding capacity of site I. This effect could be reversed by site-directed mutagenesis. Thus, Putkey and colleagues (1989) were able to replace the glycine and alanine residues of the loop by aspartic acid residues. In this way, they succeeded in transforming the inactive calcium binding site I of vertebrate slow muscle troponin-C or cardiac troponin-C into a specific calcium binding site. Then, they investigated how this mutation would affect the contractile activation of skinned cardiac or slow muscle fibres. The result was remarkable. In these fibres, the force/pCa curves are rather shallow and calcium binding is non-cooperative (Pan and Solaro 1987). However, when skinned slow muscle fibres or cardiac fibres were freed from troponin-C and then reconstituted by mutated troponin, the steepness of the calcium-force relationship of the preparation increased and now resembled the activation characteristics of fast muscle. Obviously, the cooperativity of calcium binding had been increased as the slow muscle troponin-C containing one specific binding site was replaced by mutated troponin-C that contained *two* calcium-specific sites. A similar result was obtained when cardiac or slow muscle troponin-C was replaced by the fast muscle isoform of troponin-C (Moss et al. 1986).

Site-directed mutagenesis has also been used to inhibit troponin-C activity and Ca^{2+}-induced conformational changes by introducing a "stabilizing" intradomain disulphide bridge between helix B and D (Grabarek et al. 1990) that would impede intramolecular movements. Thus, the ATPase activity of myofibrils could not be Ca^{2+}-activated when native TnC was replaced by the mutant TnC. Activity could be restored, however, simply by disrupting the disulphide bonds by reducing agents. In contrast, mutations within the central helix (D/E linker, cf. Fig. 4.1) had little if any effect on the calcium binding properties (Xu and Hitchcock 1988; cf. also Reinach and Karlsson 1988). Rather, they altered the interaction of troponin-C with troponin-I in such a way that it became impossible to inhibit the Ca-activated ATPase of troponin-reconstituted actomyosin with the Ca-chelator EGTA.

Interestingly, the calcium-induced conformational changes of troponin-C just discussed do not require the troponin-C molecule as a whole, but they can be mimicked by small peptide fragments of the protein. Thus, a synthetic 34-residue peptide comprising the calcium binding site III of troponin-C has been shown to bind Ca^{2+} with high affinity (KD 3 μM; cf. Shaw et al. 1990), thus changing its conformation from a "random coil" to a helix-loop-helix motif (Shaw et al. 1990). Moreover, as also revealed by 1 H-NMR spectroscopy, these short synthetic peptides form – by cooperative interaction – dimers that have a tertiary structure similar to that of the C-terminal domain of troponin-C (Shaw et al. 1991 b). Clearly, synthetic peptide approaches have been most fruitful in answering questions regarding the mechanism of calcium regulation. Further examples of this approach will be highlighted in the following sections on thin filament proteins.

10.2.1.2 Troponin-I: Molecular Structure and Function Relationship as Probed by Synthetic Peptides

As mentioned already, troponin-I (TnI) strongly interacts with actin in relaxed muscle thereby repressing the actin-activated myosin-ATPase. Derepression occurs when calcium binds to troponin-C, thus increasing its interaction with troponin-I. Thus, one might visualize troponin-I as being part of a kind of molecular flip-flop mechanism, switching between binding sites on troponin-C and actin, thereby turning contraction "on" and "off". However, this concept may be an oversimplification as there may, in fact, be a dynamic equilibrium between protein conformations corresponding to "on" and "off" states. This equilibrium may then be shifted in the direction of "on" and "off" states, depending on the degree of calcium activation. The sites of interactions are the N-terminal residues of actin (Levine et al. 1988), residues 87–100 of troponin-C and 104–115 of troponin-I (Fig. 10.11). These binding sites could be identified in studies using synthetic peptides comprising the critical amino acid residues of actin or troponin-I. Thus, Talbot and Hodges (1979, 1981) showed that a synthetic TnI peptide containing residues 104–115 of TnI is the minimum sequence required to inhibit the acto-S1-ATPase. Its structure is quite conserved, and almost identical in cardiac and skeletal muscle troponin-I. However, as pointed out by Trayer et al. (1991), one should not assume a priori that the isolated peptides would adopt the same three-dimensional structure in solution as in the parent protein molecule, but rather that they adopt the "right" configuration by "induced fit" when binding to a partner protein, such as actin or troponin-C in the case of the TnI peptides.

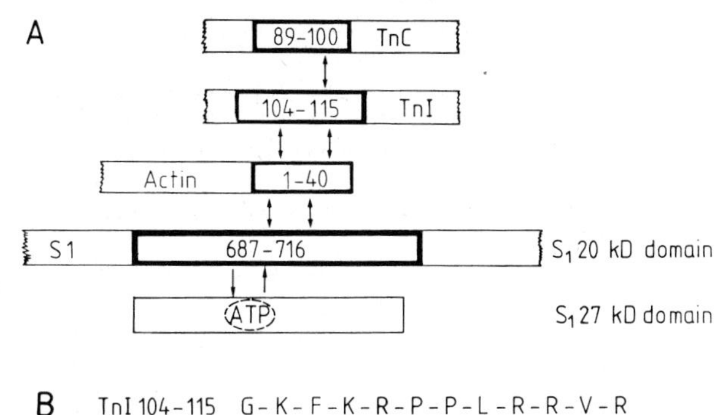

Fig. 10.11. A Interaction of regulatory and contractile proteins in muscle activation; *numbers* refer to amino acid residues of the peptide segments involved; for explanation, see text. *B* Sequence of inhibitory region of skeletal muscle *TnI*. One letter amino acid code, see legend to Fig. 4.14. In the cardiac TnI, proline 110 is replaced by threonine

Further, a peptide comprising the N-terminal region of actin bound to a troponin-I affinity column may be competitively displaced by the TnI peptide (TnI 104–115). This means that the peptide competes with TnI for the TnI-binding site on actin. Clearly, these are reductionist approaches, showing that a small, but essential part of troponin-I could mimic an important biological effect of the whole protein molecule (Van Eyk et al. 1991 a).

The inhibition induced by a troponin-I peptide (TnI 104–115) could be released by troponin-C, and, indeed, the TnI peptide and the inhibitory region of TnI is capable of binding to the C-domain of troponin-C near helix C, which is the N-terminal helix of the calcium binding loop (Cachia et al. 1983; Leszyk et al. 1987, 1988, 1990; Sykes 1990 b). By binding to the protein the peptide acquires a specific structure (Campbell and Sykes 1990 a) and causes an increase in the helical content of troponin-C as well as in the calcium affinity of the calcium binding sites (Rüegg et al. 1989; Van Eyk et al. 1991 b). Furthermore, experiments with 11 analogues of troponin-I, in which a single residue had been replaced by glycine, showed that al residues seem to be essential for full inhibitory activity of the peptide (Van Eyk et al. 1988). Some residues, such as lysine 105, however, were particularly important and if these residues were replaced by glycine, biological activity was lost altogether.

As mentioned earlier, binding of troponin-I to troponin-C also increases the calcium affinity of troponin-C (Zot et al. 1983) and this effect may also be inhibited by the troponin-I peptide. This is because the peptide competes with troponin-I for the TnI-recognition site of troponin-C (Rüegg et al. 1989). Again, all residues of the TnI peptide were required to achieve maximal peptide competition, suggesting that all of them are involved in the process of molecular recognition. These effects of the TnI peptide and various analogues could be assayed in competition experiments using skinned fibres. Thus, the TnI peptide inhibited contractile responses of skinned psoas fibres that were submaximally activated with micromolar concentrations of Ca^{2+} (Fig. 10.12) and it also reduced the apparent calcium sensitivity (Rüegg et al. 1989). This means that the pCa required for half-maximal activation was reduced. Incidentally, a similar effect was also produced by the wasp venom peptide, mastoparan (cf. Cachia et al. 1986) and by a calmodulin binding peptide fragment of a calcium transport ATPase (Sheng et al. 1991). As all these peptides inhibited TnI-TnC interaction, their effect on cal-

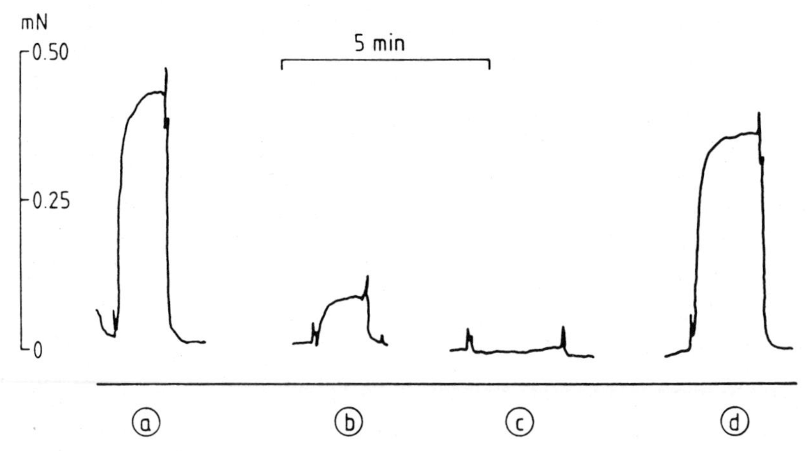

Fig. 10.12. Effect of a peptide fragment from the troponin-I inhibitory region on skinned fibres of rabbit psoas. The peptide has the sequence TnI (104–115, see Fig. 10.11) and inhibits contraction reversibly. Fibre bundles of rabbit psoas fibres were skinned with Triton X-100, attached to an AME-801 force transducer, and relaxation was induced in a well-stirred bathing solution containing (in mM): imidazole, 30; ATP, 10; creatine phosphate, 10; NaN_3, 5; EGTA, 5; $MgCl_2$, 12.5; creatine kinase, 380 U ml^{-1}; dithiothreitol, 5; P_i, 5. Ionic strength was adjusted to 100 mM with KCl, pH 6.7, 20 °C. Isometric contractile responses of a single skinned fibre were elicited by 2.5 μM Ca^{2+} (pCa^{2+} 5.6). Relaxation was induced by immersion into relaxing solution. At *b* and *c* the responses were inhibited by 50 and 100 μM of the peptide (note that fibres were also exposed to the peptide in relaxing solution for 15 min before eliciting the contraction with Ca^{2+}); *a* and *d* are control contractions before addition of the peptide and after its wash out. (From Rüegg et al. 1989)

cium sensitivity clearly points to the importance of TnI-TnC interaction in regulating calcium responsiveness of skinned fibres (Rüegg 1990). These experiments also demonstrate that peptides may be used much like antibodies and both as specific inhibitors and as molecular probes to explore the cognitive interfaces between interacting proteins and their functional significance.

Surely, *peptide competition* is a general approach, which may be used to probe and inhibit many kinds of protein-protein interactions including for instance those of cell adhesion proteins, G-proteins (Hamm et al. 1988) or the interaction of immunoproteins. Indeed, this approach was initiated originally by immunologists, notably Atassi (1975), who showed that the interaction between antigen and antibody may be inhibited by competing peptides that structurally resemble the antigenic site. However, muscle tissue is particularly useful to explore the power of this approach, since the effects of altering protein-protein interactions seem to be so obvious and so readily detectable even by mechanical measurements in skinned fibres.

10.2.1.3 Troponin I: Phosphorylation and Isoforms

In addition to peptide competition experiments, there is further evidence showing that troponin-I may have a role in modulating calcium responsiveness. For instance, in the case of cardiac troponin-I, phosphorylation of the serine 20 residue by cyclic AMP-dependent protein kinase is known to decrease calcium responsiveness as discussed in a previous section (p. 194).

Calcium sensitivity may also be modulated by the expression of various isoforms of troponin-I. Thus, skinned fibres prepared from neonatal dog hearts and expressing neonatal TnI are more calcium-sensitive than those of adult hearts, in particular at slightly acid pH, whereas at neutral pH, the difference is much less (Solaro et al. 1986). A slight acidosis, therefore, seems to reduce the calcium responsiveness of the contractile system of adult hearts, but in neonatal hearts, this pH effect is negligible. Thus, the pH-dependence of calcium sensitivity in cardiac myofibres must obviously be ascribed to a special feature of adult cardiac troponin-I. The nature of troponin-C seems to be less critical, as it is identical in neonatal and adult hearts. For this very reason, it is also unlikely that H-ions reduce calcium sensitivity simply by competing with calcium ions at their TnC-binding sites. Rather, they exert an indirect effect by altering, perhaps, TnI-TnC interaction. The latter may also play a role in cooperative Ca^{2+} binding. Thus, calcium binding of fast skeletal muscle troponin-C is cooperative only when it interacts with other thin filament proteins (Grabarek et al. 1983). Suppose that calcium binds to one of the two specific sites on TnC. Then, the interaction between troponin-C and troponin-I would be increased, which, in turn, would enhance the calcium affinity of the other, non-occupied Ca-binding site. Indeed, such an enhancing effect of troponin-I or troponin-I peptides on the calcium affinity of isolated troponin-C in solution could be demonstrated experimentally (Zot et al. 1983; Van Eyk et al. 1991 b).

To date, the functional significance of the diversity of troponin-I isoforms in cardiac and skeletal muscle is still poorly understood. However, as it is possible to extract TnI selectively from muscle tissue (Strauss et al. 1992), this technique offers the prospect of replacing, in skinned fibres, the native proteins with mutant isoforms and even peptides (Van Eyk et al. 1992) in order to study structure-function relationships.

10.2.1.4 Troponin-T: Isoforms May Also Determine Myofibrillar Calcium Sensitivity

Troponin-T (TnT) has been shown to exist in different classes of isoforms characteristic of fast and slow skeletal muscle as well as cardiac muscle (Dhoot et al. 1979; Anderson and Oakeley 1989; Hartner et al. 1989). Interestingly, the steepness of the Ca-force relation and the different calcium sensitivities of fast, slow and intermediate skinned skeletal muscle fibres also correlated with and were possibly due to different isoforms of troponin-T (Schachat et al. 1987). The contributions of troponin-T isoforms to the regulation of calcium sensitivity of actomyosin ATPase can, in principle, be studied in vitro using an assay system containing the contractile proteins actin and myosin reconstituted with regulatory proteins, including various isoforms of troponin-T.

As already mentioned in a previous section (p. 83), "natural" actomyosin exhibits calcium-sensitive magnesium-dependent ATPase activity, as it is contaminated by troponin and tropomyosin. After purification of actomyosin, calcium sensitivity of the crude ATPase is lost, but can be restored, quite simply, by readdition of tropomyosin and/or the different components of the troponin complex in order to study the differential effects of troponin isoforms. According to

Tobacman and Lee (1987), bovine *cardiac troponin-T* could be separated into two isoforms of slightly different molecular weight. Interestingly, reconstituted actomyosin containing the larger isoform required a lower pCa for half-maximal activation and was thus less calcium-sensitive than the preparation containing the alternate isoform. Thus, it would appear that specific molecular properties of various TnT isoforms are at least partly involved in conferring a different Ca^{2+} sensitivity to the myofilaments. They may contribute, for instance, to the different calcium sensitivities of myofilaments in neonatal and adult hearts (McAuliffe et al. 1990). Naturally, the question arises as to how – during embryogenesis and post-natal periods – specific muscles "know" which isoforms to "choose" and how the expression of different isoforms is switched during the development, thereby altering calcium responsiveness.

The classes of isoforms specific for cardiac and fast and slow skeletal muscle result from the expression of different genes. In skeletal and cardiac muscle, however, many more different subtypes of troponin-T also arise from "*alternative splicing*" during the processing of the primary transcript of the troponin-T gene (Medford et al. 1984; Breitbart et al. 1985; Cooper and Ordahl 1985). Remember that most eucaryotic genes often contain several different exons that are separated by intervening introns. The former are coding sequences of DNA, which determine the primary sequence of the protein expressed, whereas introns are non-coding. This is because exons and intervening introns are transcribed into the primary messenger-RNA, but during further processing and splicing of this pre-messenger-RNA, introns are removed. The exons are then joined together to form the final messenger-RNA, which is then translated into the primary sequence of the peptide chain.

Now consider the case of troponin-T of fast skeletal muscle, which is the product of a gene containing 18 exons, separated by intervening introns. Eleven of these exons are always retained after splicing, but four exons may or may not be expressed or they may be joined together in various combinations after removal of intervening introns, thus giving rise to 32 possible exon arrangements. An additional variability in the final messenger-RNA occurs as it contains either exon 16 or exon 17, but not both. Thus, one might theoretically expect 64 different isoforms produced by alternative splicing, but only few of these have been identified as yet on the protein level (e.g. Sabry and Dhoot 1991). These have been separated by gel electrophoresis, transferred to nitrocellulose sheets and identified by specific antibodies on these "Western blots". However, the number of protein isoforms expressed may possibly be much larger and closer to the theoretically expected one.

Interestingly, most of the diversity of TnT occurs in a variable region near the N-terminal end, thus affecting TnT-tropomyosin interaction, rather than TnT-TnC interaction (Smillie et al. 1988). Remember that troponin-T is a very elongated molecule containing two domains: an N-terminal domain binding to tropomyosin and the C-terminal domain which binds to troponin-C. Thus, the enormous diversity of TnT may well contribute to the functional diversity of muscle, but its understanding on a molecular level remains a challenge for future research.

10.2.1.5 Actin: The N-Terminal Region Has a Regulatory Role

The three-dimensional structure of globular actin has been resolved by X-ray crystallography at 2.3 Å resolution (Kabsch et al. 1990). The molecule consists of a single polypeptide chain which forms a flat structure with 35 Å width, containing two domains, composed of two subdomains (Fig. 10.13). The globular actin monomers are polymerized to F-actin within the thin filaments and tropomyosin polypeptide chain strands follow a long pitch helix, as they run in the groove between subdomains 3 and 4 with actin contacts around residues lys-215 and pro-307. The N-terminal region of actin interacts with both troponin-I (Levine et al. 1988) and myosin (Moir et al. 1987). Myosin heads (crossbridges) attach to actin in a tangential manner, binding mainly at actin residues 1–4 and 24–28, as

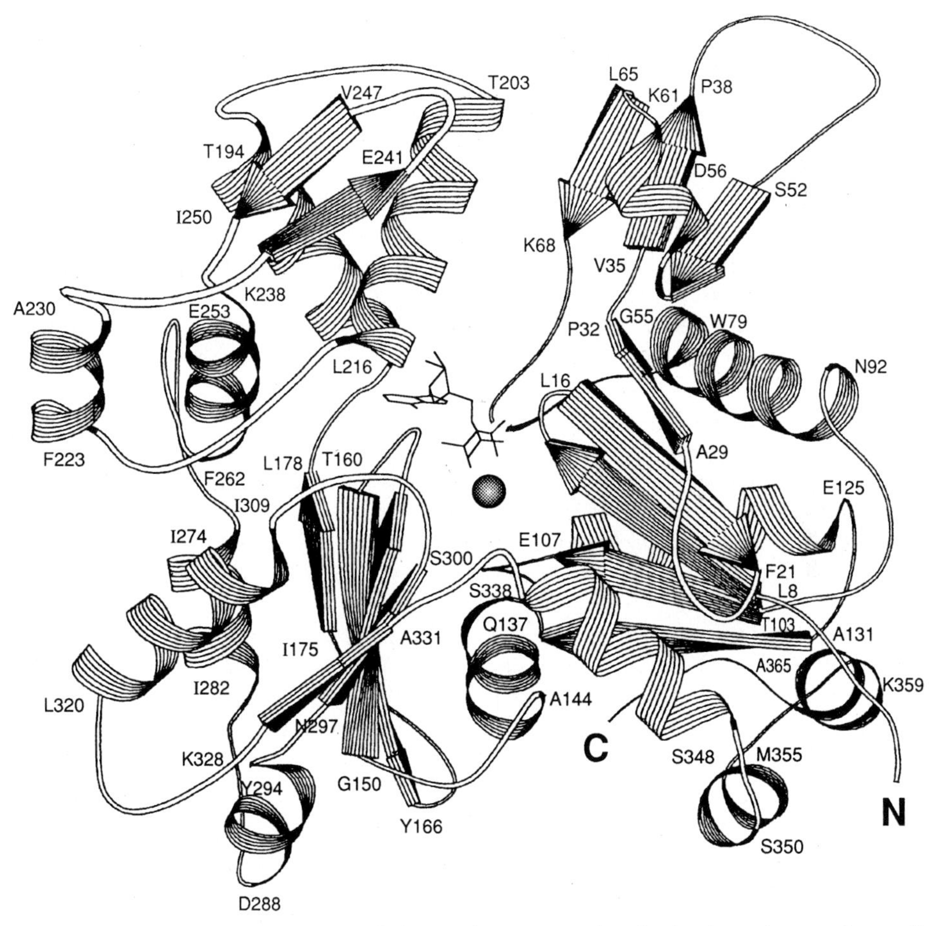

Fig. 10.13. Schematic representation of the atomic structure of actin showing a large and a small domain, each assembled from two subdomains. Note that the bound ATP and calcium ion are found in the cleft between the domains at the center of the molecule. The troponin-I binding region (residues 1–12) and the myosin binding regions (residues 1–7; 24–29 and 92–96) are all located in subdomain I of the small domain. (From Kabsch and Holmes 1991; Courtesy of W. Kabsch)

well as at residues 93 and 95 (Holmes and Kabsch 1991), thereby possibly altering the conformation of F-Actin (cf. Borovikow 1991).

As is well known, muscle contraction involves the cyclic interaction of myosin heads with actin. One should not, however, visualize filamentous actin solely as an inert "backbone" on which myosin heads attach, rotate and pull. Rather, it is a dynamic structure and, importantly, it is also a chemical switch with the capability to turn on the myosin ATPase when interacting with the 20 kDa domain (Keane et al. 1990) and/or possibly also with the 50 kDa domain of myosin subfragment 1 (cf. Chaussepied and Morales 1988). Thus, actin activation of myosin may be partly mimicked by an N-terminal fragment of actin (e.g. actin 1–44) (cf. Kögler et al. 1991) or by a synthetic peptide comprising the stretch of actin that involves residues 1–28 (Van Eyk et al. 1991; Van Eyk and Hodges 1991) and which binds to the 20 kDa domain of myosin subfragment S1 around the SH-1 thiol region (Keane et al. 1990). The importance of that region of myosin has also been clearly demonstrated by showing that a heptapeptide corresponding to a sequence around SH-1 would competitively interfere with actin-myosin interaction, thereby inhibiting acto-S1-ATPase, whereas the activity of subfragment 1 alone was not inhibited (Suzuki et al. 1990; cf. Keane et al. 1990). Moreover, these findings suggested to Morita and her colleagues that the structure around SH-1 may be involved in the activation process induced by actin. In the absence of actin, it would interact with and inhibit the active site of the ATPase located in the 23 kDa domain of S1. However, when the S1-thiol region of myosin interacts with actin rather than with the 23 kDa domain, inhibition will be derepressed (Morita et al. 1991). Accordingly, actin-myosin interaction would be prevented, if actin binds to the inhibitory region of troponin-I (residues TnI 104–115) instead of S1.

The model proposed (Fig. 10.11), therefore, has the N-terminal actin residues 1–28, switching as it were between the inhibitory region of troponin-I and the 20 kDa domain of myosin by a kind of flip-flop mechanism. Such a movement, of course, would require a certain flexibility of that region of actin, which would be in dynamic equilibrium with myosin and, indeed, it has been shown that actin 1–8 is not α-helical, but rather extended. An extended shape is also taken up by the synthetic peptide containing that region (actin 1–28), but, if this peptide is interacting with troponin-I, it acquires a more helical structure in residues 1–4, as demonstrated by 2D-nuclear magnetic resonance spectroscopy (Van Eyk et al. 1991 a). As the TnI-peptide bears six positive charges and the actin peptide four negative ones, electrostatic forces may well play a major role in this type of interaction.

To summarize, then, N-terminal actin is a kind of dynamic "molecular switch" which, by mechanically interacting with myosin, may activate the acto-S1-ATPase, and the degree of activation would depend on a dynamic equilibrium between on and off states. As discussed in Chapter 1, actin activation involves an acceleration of the rate of decomposition of the enzyme product complex, in particular the release of inorganic phosphate bound to S1. This event, then, is probably associated with the process of force generation and with the transition of weakly bound bridges into strongly bound states (Brenner 1990). In muscle activation, this process is triggered by calcium, as will be discussed next.

10.2.1.6 Tropomyosin Structure and Function

In the model proposed above (Fig. 10.11), the possible role of tropomyosin (Tm) has so far not been considered. It is clear, however, that this protein greatly enhances the inhibitory effect of troponin-I (Eaton et al. 1975), as it interacts with up to seven monomers in filamentous actin. Tropomyosin contains two identical α-helical peptide chains of 33 kDa molecular weight oriented in parallel and organized into 14 negatively charged repeats. Each molecule spans seven actin monomers along the thin filament, hence, involving two repeats in the binding of one actin molecule. The critical role of these repeats has been demonstrated by Hitchcock-De Gregory and colleagues who investigated actin binding of tropomyosin that had been altered by site-directed mutagenesis (Hitchcock-De Gregory 1989; Hitchcock-De Gregory and Varnell 1990).

For many years, it has been generally held that calcium activation of muscle contraction was simply due to crossbridge recruitment (Teichholz and Podolsky 1970) caused by a shift of tropomyosin from a position physically or sterically interfering with crossbridge attachment into a non-blocking position (see p. 95). The steric block hypothesis, however, was abandoned – at least in its original version – when it became clear that crossbridges were attached to actin, albeit weakly, even in the relaxed state. The functional significance of calcium-dependent movements of tropomyosin, then, and its interaction with actin remains enigmatic. Perhaps at low Ca-concentration the high affinity of troponin for actin holds the latter in a state which prevents strong crossbridge binding, while still permitting weak interaction (Phillips et al. 1986). Upon binding of calcium, troponin would release tropomyosin from the inhibitory position along the outer surface of the actin helix. This would allow crossbridges to attach in the strongly bound state as already discussed above (p. 96, Fig. 4.6). In turn, attaching crossbridges might also affect tropomyosin conformation and even induce an extremely "turned-on" configuration or potentiated state, in particular at low ATP concentration (Phillips et al. 1986).

10.2.2 Contractile Activation: Crossbridges Called into Action

In the relaxed state of muscle, detached crossbridges are in rapid equilibrium with weakly attached ones and calcium activation seems to promote the transition of these states into strongly bound, force-generating crossbridge states rather than "recruiting" active crossbridges (Brenner 1988). During maintained contraction, these crossbridges are continuously cycling between strongly bound states and various forms of weakly bound states. The apparent rate constants at which crossbridges enter and leave force-generating crossbridge states will be denoted as f_{app} and g_{app} (cf. Sect. 1.3.3.5).

10.2.2.1 Calcium Increases the Apparent Rate Constant of Strong Crossbridge Binding

According to Brenner (1988), it is mainly the increase in the apparent rate constant (f_{app}) of crossbridge attachment in the strongly bound state, which is in-

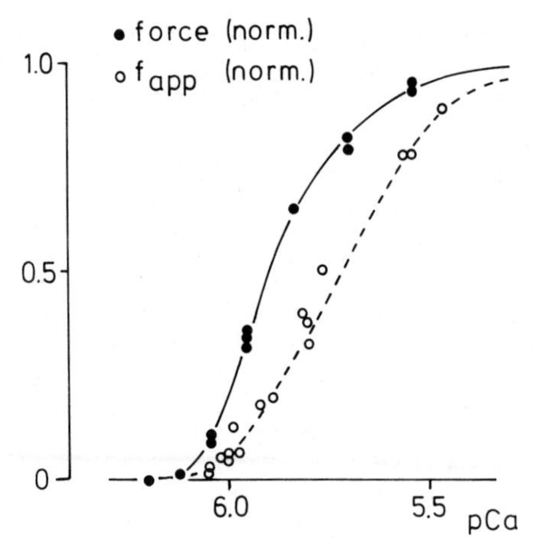

Fig. 10.14. Contractile force (●) and apparent rate constant f (○, crossbridge attachment) as a function of pCa (neg. log of free Ca concentration) in skinned fibres from rabbit psoas muscle in ATP salt solution. Note that rate constant g of crossbridge detachment is assumed to be constant. For further explanation, see text (Brenner 1988, with permission)

duced by calcium activation and leads to an enhanced force development and contractility in muscle. Since the rate constant governing crossbridge detachment (g_{app}) is apparently independent of calcium and since force is related to f/f+g, there must be a curvilinear relationship between f and force in these fibres. Further, the curve relating force and calcium is steeper and shifted towards lower calcium concentrations or higher pCa values than the relationship between f_{app} and calcium (Fig. 10.14). In this respect, it is interesting to note that the force-calcium relationship is also steeper and is shifted to the left with respect to the calcium binding curve, at least in cardiac skinned fibres (Pan and Solaro 1987).

In view of the importance of these concepts, developed by Brenner, some methodological remarks may be appropriate. In skinned or permeabilized rabbit psoas fibres, the apparent rate constant g could be obtained simply from the ratio of ATP splitting rate and force production. This is because force is proportional to f/f+g and ATPase activity is proportional to fg/f+g. In this way, it was found that g is constant at different calcium ion concentrations (Brenner 1988; but cf. Kerrick et a. 1991). Rate constant f could then be readily derived by determining the rate of tension recovery after a kind of large "quick release" followed by restretch of skinned fibres. Note that in these experiments the "quick release" has to be carefully controlled so that tension does not drop to zero, but to a constantly low level, allowing the rapid shortening of the nearly unloaded muscle fibres and crossbridge detachment. As soon as shortening of the fibre is stopped by restretching it to the initial length (isometric conditions), crossbridges reattach and tension recovers with a rate that is related to the sum of f_{app} and g_{app}.

10.2.2.2 Crossbridge Kinetics May Also Be Altered by Calmodulin-Dependent Myosin Light Chain Phosphorylation

Phosphorylation of the regulatory light chain of myosin may alter the conformation of myosin and/or the configuration of the crossbridges (cf. Padron et al. 1991). This effect may, just as calcium activation, result in an increase in the apparent attachment rate f (Sweeney and Stull 1990). Since force is related to f/f+g, as we have seen, it will be increased, both by enhancing either free calcium or the level of phosphorylation at a given free calcium ion concentration. Calcium responsiveness or apparent calcium sensitivity for that matter will therefore be increased by myosin phosphorylation, resulting in a leftward shift of the calcium-force relationship (Morano et al. 1985; Persechini et al. 1985). Activation by myosin phosphorylation is important in the regulation of many invertebrate striated muscle fibres and in particular in vertebrate smooth muscle (see Chap. 8).

In skeletal muscle, myosin light chain phosphorylation not only increases calcium responsiveness, but it is itself controlled by calcium. This is because it is catalyzed by the calcium-calmodulin-dependent *myosin light chain kinase* which is also the case in smooth muscle. Thus, a given level of phosphorylation results from a balance of light chain kinase activity and protein phosphatase activity. While little is known about the regulation of phosphatase, control of myosin light chain kinase by calcium and calmodulin is now understood on a molecular level. Interestingly, a sequence of residues similar to that of the myosin light chain residues interacting as substrate with myosin light chain kinase is also contained within the peptide chain of myosin light chain kinase. This sequence has therefore been called the *pseudo-substrate sequence* (Kemp et al. 1987; cf. Hardie 1988). It is quite flexible and obviously capable of interacting with the catalytic site of the enzyme in an inhibitory manner as it competes with the natural substrate. During Ca^{2+} activation this kind of intramolecular peptide competition is released when the pseudo-substrate sequence interacts with the calcium-calmodulin complex, rather than with the catalytic site. Indeed, there is a partial overlap between the calmodulin binding sites and the pseudo-substrate sequence, so that this part of the molecule may be regarded as a dynamic flip-flop mechanism, switching or possibly oscillating between a calmodulin and substrate binding state (Kemp et al. 1987). This model has been confirmed. Thus, if the pseudo-substrate sequence is removed by controlled proteolysis (Ikebe et al. 1987), or in a recombinant enzyme by site-directed mutagenesis (Ito et al. 1991), myosin light chain kinase is no longer inhibited in the absence of calcium and calmodulin. This constitutively active enzyme, however, can be inhibited by the addition of a synthetic analogue of the inhibitory pseudo-substrate sequence (peptide Ala 789-Gly 804; cf. Ikebe et al. 1987). The peptide obviously competes with the natural substrate for the catalytic site of the enzyme.

In *vertebrate smooth muscle fibres* myosin phosphorylation levels may also influence crossbridge kinetics; however, they may not only be increased by a Ca- and calmodulin-dependent enhancement of myosin light chain kinase activity, but also Ca-independently by inhibiting the activity of myosin phosphatase. There is tentative evidence that this regulation may be brought about by G-proteins (Kitazawa et al. 1991 a). Thus, it could be shown, for instance, that force and myosin light chain phosphorylation can be increased by the addition of the stable G-protein binding nucleotide GTPγS, even at constant levels of free Ca^{2+}. Further research should be directed towards the identification of the nature and mechanism of the G-proteins involved (cf. Sect. 8.4.4).

10.2.2.3 Kinetic Properties of Crossbridges Affect the Relationship Between Free Calcium and Steady State Force

Ultimately, the calcium signal processed by the thin filament proteins affects, as we have seen, the rate constants of critical steps in the crossbridge cycle. But the reverse may also be true, as these rate constants in turn influence calcium responsiveness and even the calcium affinity of the calcium binding sites in troponin-C. Thus, attaching crossbridges that enhance calcium affinity of TnC (Güth and Potter 1987; but cf. Morano and Rüegg 1991) may account for the cooperativity of calcium binding in skinned fibres from fast skeletal muscle. This effect may be mediated perhaps via an alteration of tropomyosin conformation causing a potentiated state as discussed already in a preceding section (cf. Phillips et al. 1986).

In skinned fibres from cardiac muscle the situation, however, is different as there is no cooperative calcium binding (Pan and Solaro 1987); here, the steep calcium-force relation cannot even be ascribed to an enhancement in calcium affinity by increasingly attaching crossbridges, as the affinity has been shown to be constant at different free calcium ion concentrations and force levels.

A plausible explanation for this discrepancy of calcium binding and force generation in skeletal and cardiac muscle is offered by Brenner's hypothesis. As force is related to the quotient of $f/(f+g)$ and f is presumably directly related to calcium binding, the observed displacement of the dependence of force and calcium binding on Ca^{2+} and the different steepness of these curves would be readily accounted for. Brenner (1988) also commented on the importance of the ratio of f_{app} and g_{app} (at pCa 4) in shaping the calcium-force relation curves: any increase in the ratio would increase both the apparent calcium sensitivity and the steepness of the calcium-force relationship. Thus, a steep curve or a high Hill coefficient of cooperativity (e.g. $n=2-4$) would not necessarily be an expression of a large number of cooperatively interacting calcium binding sites (cf. also Shiner and Solaro 1984). Rather it would, according to the model, reflect a high ratio of the rate constants f and g. Of course, lowering constant g would then also enhance the calcium responsiveness, the apparent calcium sensitivity and cooperativity and even the maximal force development simply by slowing crossbridge detachment, without necessarily altering the calcium occupancy of troponin-C.

Incidentally, a force-enhancing mechanism involving altered crossbridge kinetics may also play a role in the mode of action of novel, calcium-sensitizing, positive inotropic drugs to be discussed later (Sect. 10.2.3). The thiadiazinone derivative EMD 53998, for instance, may lower the rate constant g, as indicated by the lower ratio of ATPase/force in skinned fibres of cardiac muscle, and it also induced a leftward shift of the calcium-force relationship of skinned fibres (Beier et al. 1991). In intact cardiac fibres that are only half-maximally activated by myoplasmic calcium in the absence of inotropic interventions, the drug increased force without enhancing the intracellular calcium transient (Allen and Lee 1989); in this way, it improved the contractile performance of the intact heart. In fact, the drug may counteract the effect of inorganic phosphate, which accumulates under hypoxic conditions and increases the tension cost, while decreasing the calcium responsiveness without, however, decreasing calcium affinity of troponin-C. Alternatively, both phosphate and EMD 53998 may affect the size of the crossbridge powerstroke; the latter may well be variable (cf. Lombardi et al. 1992).

10.2.2.4 Exploring the Temporal Relationship Between Calcium Binding and Force Production

In living fibres, the free calcium rises much more quickly than force, but the nature of this delay in force development is still not fully understood. Experiments with intact muscles indicate that the thin filaments bind calcium rapidly and undergo a rapid conformational change, whereas crossbridges attach more slowly (Kress et al. 1986; see Sect. 4.2.3). Similarly, in myocytes of the heart, the rise in myoplasmic calcium levels is fast and calcium may bind and unbind equally fast to troponin-C, as suggested by electron-probe analysis, while contraction is slow (Wendt-Gallitelli and Isenberg 1991). All these findings imply that it may be slow kinetics of crossbridges rather than intracellular calcium handling which determines the relatively long time required to reach peak contraction.

Recently, the delay between Ca^{2+} binding and contraction was also studied in more detail using skinned fibres in which the calcium ion concentration may be increased suddenly and in a stepwise fashion by using "caged" calcium compounds such as Nitr5 (e.g. Ashley et al. 1987b) that release calcium upon photolysis by means of a light flash. The rapid release of calcium following flash photolysis causes a fast isometric contraction, the rate of which is a function of the free calcium ion concentration and presumably of TnC-Ca occupancy (cf. Ashley et al. 1991). How does the calcium occupancy of TnC influence the rate of rise of force development of muscle fibres or pressure change (dp/dt-max) in the heart and even the normalized rate of pressure development [dP/dt (max)/P_{max}]? The maximal rate of pressure change in the ventricle of the heart is, as mentioned earlier, a measure of *contractility*, and understanding its calcium dependence is therefore clearly of great physiological and pathophysiological importance. Ashley and colleagues (1991) suggested a model based on a reaction scheme involving six different states of functional contractile units, each unit comprising one troponin-tropomyosin complex and its associated actin monomeres as well as attached myosin heads. These units are assumed to be calcium-free or occupied with one or two calcium ions, which are either force-generating or non-force-generating. An increase in calcium occupancy is assumed to predominantly affect the transition into force-generating states and computer modelling (using Monte-Carlo simulation) predicted accurately the time course of force development after flash photolysis of caged compounds. However, there are alternative models of contractile activation, as will be discussed below. Clearly, these new concepts will also be of importance regarding a molecular understanding of cardiac contractility.

10.2.3 Molecular Aspects of Contractility in the Myocardium

Contractility of the myocardium is usually defined operationally as that contractile property of heart muscle that determines the maximal rate of pressure development (dP/dt) in the left ventricle or, for that matter, the rate (and extent) of tension generation in the myocardium. Note that, according to the concepts developed here, contractility is assumed to be related to the rate of net attachment of force-generating crossbridges (cf. Rüegg 1990) rather than to the (maximal) shortening velocity (V_{max}) of contractile elements that stretch serious elastic elements, thereby developing tension (cf. Opie 1991 and Hill's model, Sect. 1.3.1). Contractility is a clinically relevant parameter as it is often diminished in diseased hearts, e.g. in congestive heart failure.

10.2.3.1 Calcium Crossbridge Kinetics and Rate of Force Generation (dP/dt)

The hypothesis (cf. Ashley et al. 1991) just discussed in the previous section accounts for the calcium dependence of the rate (and extent) of force generation in skeletal muscle and possibly also in the myocardium. An alternative and possibly simpler proposition is based on the hypothesis of Brenner (1988). In this model, the rate of the rise in tension depends on the sum of f_{app} and g_{app}, the rate constants for forming and leaving the force-generating crossbridge states. As any increase in free calcium increases f_{app}, it is clear that the rate of rise in tension and the maximal rate of pressure change in the left ventricle of the heart (dP/dt_{max}) or *contractility* and even the normalized rate of pressure change must also be calcium-dependent. Of course, this rate could also be increased by switching the isoform of myosin from slow to fast type. This is because a fast-type myosin would be characterized by a large detachment rate constant g, whereas the slow form would have a low g-value. A three-fold difference in g-values for "fast" cadiac α-myosin heavy chain (in atria) and "slow" β-myosin heavy chain (in ventricle) has been determined by measuring the tension cost in skinned fibres (Morano et al. 1991). Subsequently, Ng et al. (1991) showed that in mice heart expressing α-myosin heavy chain, dP/dt or "contractility" is much higher than in mice hearts expressing β-mysoin heavy chain following treatment with thiouracil. Interesting as it is, this finding of course may not necessarily indicate a causal relationship between isozyme type and contractility, as the calcium pump of the sarcoplasmic reticulum is also slowed at low levels of thyroid hormone (Rodgers et al. 1986), and there may be other alterations as well. Molecular genetic experiments using the hearts of transgenic mice expressing selectively altered myosin heavy chains would probably be more conclusive.

10.2.3.2 Duration of Calcium Transients and Contractile Force

In the myocardium, contractile force and contractility are related. Thus, it is clear that the increase in the rate of force generation should increase the twitch force of a heart muscle. The reason is that force-generating crossbridge attachment takes much longer than calcium binding and the development of the active state in the thin filament. Thus, during a brief calcium transient activation, such as it occurs in a twitch, of say, a papillary muscle in vitro, there is simply not sufficient time to develop the maximal force, unless, of course, the rate of force development (i.e. contractility) were to be increased by altering the kinetic properties of the crossbridges. However, a prolongation of the calcium transient would probably be equally effective in increasing twitch force. This is because such an alteration in calcium handling would surely prolong the active state of the thin filaments, thus allowing more time for crossbridges to attach. In tetanic contraction, therefore, contractile force would be expected to be much greater than in a twitch contraction of the heart, and, indeed, this has been confirmed experimentally by Marban et al. (1986). Under the circumstances, it is interesting to note that in the heart, the activity of the sarcoplasmic reticulum calcium pump and, hence, the duration of the intracellular calcium transient may be variable. For instance, it may depend on the phosphorylation state of the pump-associated protein phospholamban (see p. 54). In chronic heart disease, the intracellular calcium transient is prolonged (Gwathmey et al. 1987), as the calcium pumping activity of the sarcoplasmic reticulum is impaired (Limas et al. 1987; Unverferth et al. 1988) and relaxation is delayed. Perhaps these biochemical alterations in calcium handling reflect an adaptation and "remodelling" of the failing heart allowing it to better cope with contractile dysfunction.

10.2.3.3 Calcium Responsiveness in Heart Disease

As mentioned earlier (Sect. 7.4), the calcium sensitivity of the myofilaments is reduced in hypoxia, ischemia and in the stunned myocardium. A widely used animal model of chronic heart failure is the myocardium of aged spontaneously hypertensive rats. Here, contractility is depressed and the response to catecholamine stimulation is diminshed and even abolished. Yet, *catecholamines* increase the amplitude of the intracellular calcium transient, suggesting that the failing contractile response must be ascribed to a diminished calcium responsiveness of the myofilaments (Bing et al. 1991) rather than, for instance, to a diminished synthesis of cyclic AMP following adrenergic intervention. A similar desensitization to myoplasmic free calcium is apparently also found in the failing human heart, which is paced at high frequency (Morgan et al. 1990). Unlike hearts of healthy subjects, in the hearts of cardiomyopathic patients, contractility decreases when the frequency of stimulation is enhanced. As the size of the calcium transient increases rather than decreases under these conditions, the contractility loss must be ascribed to calcium desensitization. This may possibly be caused by the accumulation of metabolites, such as inorganic phosphate as in the case of hypoxia (cf. Allen et al. 1983 b; Hajjar and Gwathmey 1990).

In the long term, the detrimental effects causing Ca-desensitization may be compensated, at least partly, by adaptive alterations. In this respect, it is interesting to note that myofibrillar calcium sensitivity of permeabilized hypertrophic atria is increased when compared with normal tissue (Wankerl et al. 1990). In the heart ventricle, responsiveness to intracellular Ca^{2+} may also be enhanced by the peptide *endothelin* (Kelly et al. 1990). This local hormone is released from the endothelium of blood vessels as well as from the endocardium, particularly in diseased hearts, and it is known to be one of the most potent vasoconstrictors and positive inotropic compounds. It will activate protein kinase C by a G-protein-mediated mechanism (Krämer et al. 1991), thus causing an activation of the H-ion-sodium exchanger of the cell membrane. Such an effect would then lead to intracellular alkalosis with accompanying calcium sensitization, as in the case of α-adrenergic stimulation (cf. Endoh and Blinks 1988; Puceat et al. 1990).

In view of these findings, positive inotropy by increasing calcium sensitivity rather than by increasing intracellular free calcium might constitute a logical pharmacological approach to counteract the effects of pathological calcium desensitization. As pointed out by Morgan et al. (1990), the design of new, positive inotropic drugs should therefore focus on the development of agents that specifically increase calcium responsiveness of the contractile apparatus (called *"calcium sensitizers"*) rather than on compounds enhancing the level of Ca^{2+} or of cyclic AMP, for instance, by inhibiting the enzyme phosphodiesterase (PDE). Clearly, the latter approach would not be reasonable in view of the finding that cyclic AMP synthesis is inhibited in the failing heart (Feldman et al. 1987) and that high cyclic AMP levels may be detrimental, since they may cause calcium overload and, as a consequence, arrhythmias (cf. Opie 1991). The possibility of manipulating the calcium responsiveness of the diseased heart raises, of course, many questions, particularly with regard to the mode of action of novel calcium sensitizers.

10.2.3.4 Cardiotonic Drugs That Increase Myofibrillar Calcium Responsiveness

Enhancement of calcium sensitivity is an important physiological, regulatory principle and may also be a useful therapeutic concept (cf. also Sect. 7.4). *The mechanisms of calcium sensitization* of myofilaments are as yet only partly understood. Remember that the relationship between free calcium levels, calcium occupancy of troponin and force production is rather complex, as it may be affected not only by the calcium affinity of troponin-C, but also by mechanisms "downstream" of troponin-C. The latter may involve, for instance, alterations in the interactions between different isoforms of troponin subunits, different crossbridge kinetics or both. Therefore, the targets of calcium-sensitizing cardiotonic drugs may be many.

First of all, the *calcium affinity* of troponin-C may be increased. Indeed, Solaro and Rüegg (1982) were able to show that at a given intermediate calcium ion concentration, the calcium sensitizer, sulmazole, would increase the amount of calcium bound to myofilaments in vitro, at any given calcium ion concentration. Accordingly, an increase in myofibrillar activity could then be achieved simply by increasing the calcium occupancy of troponin-C, regardless of whether this goal is reached by increasing the free calcium ion concentration or the calcium affinity of troponin-C at a given free calcium level. However, the target of drug action must not necessarily be troponin-C itself, as indeed suggested by the experiments of Smith and England (1990). In fact, all of the subunits of troponin as well as actin and myosin could be target sites. This is because specific interactions between these proteins have been shown to be involved in calcium activation of the contractile machinery. Thus, calcium binding to troponin-C strengthens the interaction between troponin-C and troponin-I, whereas troponin-I-actin interaction is loosened. However, the reverse is also true. If troponin-I interacts with troponin-C, the calcium affinity of the latter increases, but it is reduced again, if actin is allowed to interact with troponin-I. Thus, it is tempting to speculate that drugs binding to troponin subunits and affecting its interactions would also modulate calcium affinity of calcium binding sites, and the calcium sensitivity of the myofilaments.

More interesting, however, is that so-called calcium sensitizer drugs may even increase calcium responsiveness without enhancing calcium affinity of troponin-C (Smith and England 1990), possibly by binding to the contractile proteins. For instance, they might affect the *crossbridge turnover kinetics*, in particular, the ratio of the apparent crossbridge attachment and detachment rate constants f and g, which influence, as we have seen, the relationship of contractile force and intracellular free calcium. Drugs that have this mode of action may even have competitive advantages over compounds which only increase calcium affinity of binding sites and therefore the calcium occupancy of troponin-C. This is because an increase in calcium occupancy would have to be paid for by the release of extra calcium from the sarcoplasmic reticulum, which would be making Ca^{2+} available for binding to troponin-C. In fact, if extra calcium were not released and handled by the sarcoplasmic reticulum, the free calcium level would actually fall, since the calcium ions in myoplasm would be avidly bound to troponin-C when its calcium affinity increases. Cardiotonic drugs affecting crossbridge kinetics rather than TnC-calcium affinity would thus actually be energy sparing, first, because crossbridge turnover might be slowed and, second, because a positive inotropic effect would be achieved without increasing the amount of calcium that has to be released and recuperated by the sarcoplasmic reticulum.

Alteration in calcium responsiveness by changing crossbridge kinetics is a novel concept (cf. Brenner 1988). This mechanism may well involve regulatory subdomains within the myosin molecules themselves. Evidence for such structures has been obtained recently (Keane et al. 1990; Rüegg et al. 1991) by showing that certain peptide fragments from the 20 kDa domains of myosin that are known to interact with actin will also increase the calcium responsiveness of skinned skeletal and cardiac muscle fibres, at least at submaximal levels of calcium activation.

As mentioned already, calcium-sensitizing drugs, that are similar to the thiadiazinone derivative EMD 53998 (Merck, Darmstadt FRG), may have an effect on crossbridge action rather than on the Ca-binding affinity of troponin-C or – by way of phosphodiesterase (PDE) inhibition – on the level of cyclic AMP. Thus, the compound EMD 53998 increases contractile force at submaximal levels of calcium in both permeabilized muscle fibres (Beier et al. 1991) and in intact cardiac muscle fibres; in the latter case without any increase in free calcium levels (Lee and Allen 1991). In the intact heart, such drugs should therefore also increase the expected "isometric contractions maximum", and therefore the extent of shortening in an afterloaded contraction. An increase in fibre segment shortening of hearts of anesthetized, instrumated open-chest dogs could indeed be demonstrated experimentally after systemic application of the calcium-sensitizing compound EMD 57033 (the + enantiomer of the racemic compound 53998) under conditions in which other inotropic mechanisms such as PDE-inhibition could be ruled out (Schelling et al. 1991). These experiments probably provided the first evidence that a pharmacologically induced increase in the sensitivity of the contractile elements to calcium is functional as a positive inotropic mechanism in a living animal. In conclusion, therefore, calcium sensitization of myofilaments may well represent a very promising therapeutic approach to improve the performance of the cardiac pump, in particular in diseased states, in which calcium responsiveness of the myofilaments may be actually decreased, e.g. the hypoxic or stunned myocardium.

10.3 Concluding Remarks and Future Prospects

The preceding chapters gave an account of the pathways leading from excitation to contraction in different kinds of muscle, fast and slow, smooth muscle and striated fibres of skeletal muscle and of the heart. Modern research in this field began in the 1950s. At that time, A. Sandow (1952) coined the term excitation-contraction coupling, but the comparative biologist G. Hoyle (1957) pointed out that "The stages of the processes of excitation and contraction which follow the electrical membrane potential change and lead to the shortening of the contractile elements are ones about which virtually nothing is known." Since then, progress has been enormous, and the flood of information on muscle cell calcium still increases at a rate of several hundred papers per year. Indeed, we are now approaching a molecular level of understanding of the mechanisms that underlie calcium handling in muscle cells, calcium activation of the contractile proteins and how these basic processes have been adapted in various kinds of muscle, such as the myocardium or smooth and skeletal muscle.

Perhaps we will agree with Prosser (1973), who stated: "No other tissue illustrates so well as muscle the theme of a common mechanism with variations adapted to specific functions." Yes, there is a common motif in excitation-contraction coupling. As we have seen, all muscles have calcium channels that act as senders of calcium signals which are received and decoded by specific calcium binding proteins and translated for the user, the contractile elements, composed of actin and myosin. Yet there are differences. In skeletal muscle, for instance, calcium is handled entirely within the cell, whereas in the myocardium extracellular calcium also seems to be important as a determinant of contractility. These differences in excitation-contraction coupling are now understood on a molecular level, as they may be ascribed to a comparatively small structural alteration of calcium channels in the T-tubular systems of the two types of muscle. The also affect calcium release from the sarcoplasmic reticulum.

Thus, in skeletal muscle the calcium signal is usually supramaximal, giving all-or-none responses, while the situation is more complex in cardiac and smooth muscle. Here, calcium signals usually evoke submaximal responses, the generally accepted paradigm being that contractility may depend on the signal amplitude. Regarding the heart, for instance, the accepted dogma was (Opie 1991) that "the changing concentration of calcium ions in the myocardial cell plays the predominant role in the physiological regulation of cardiac function." It has now become clear, however, that the calcium sensitivity or Ca^{2+} responsiveness of the myofilaments may also vary over a wide range in both cardiac and smooth muscle and may have a paramount effect on the contractile response. Improved intracellular calcium probes and calcium-imaging techniques as well as the development of even more sophisticated "skinned" or permeabilized muscle preparations may all have contributed to this new insight.

The skinned fibre method is powerful, as it allows the environment surrounding the myofilaments or the calcium-storing sarcoplasmic reticulum to be influenced at will. It is possible, furthermore, to study in these preparations depolarization-contraction coupling as well as receptor-response coupling, and it is even feasable to replace native regulatory or structural proteins by mutated proteins in order to study structure-function relationships. Mutation of proteins by site-directed mutagenesis and in muscle fibres from transgenic animals is and will be used in these new experimental approaches, while antibodies and specific peptides may serve as inhibitors targeting particular protein structures involved (for instance) in actin-myosin interaction. Indeed, peptides may be employed in muscle research much like antibodies as molecular probes for exploring structure-function relationships (Rüegg 1990 a). In principle, peptides mimicking cognitive protein interfaces should exhibit absolute specificity, which it should be possible to enhance or diminish through "engineering". This may offer the prospect of very selective intervention in cellular mechanisms and may be a basis for the development of a new generation of drugs.

As we have seen, the new concepts regarding calcium regulation and sensitivity modulation may be of pathophysiological and pharmacological relevance, particularly in the cardiovascular field, and may thus have exciting impacts. Already, pharmacologists begin to recognize contractile and regulatory proteins as targets for the actions of drugs, such as cardiotonic calcium sensitizers or antihypertensive calmodulin antagonists, which may act in a manner complementary to compounds that modulate myoplasmic Ca^{2+} levels, such as calcium channel blockers. Some of these drugs may act intracellularly, but they do so differently in cardiac and smooth muscle, because the intracellular regulatory mechanisms of these two kinds of muscles are not the same. The recognition of such diversity in various kinds of muscle, therefore, seems to be of practical importance, quite apart from telling us what features of calcium regulation may be common or fundamental to all muscles and which of them are probably specializations that have evolved because of their suitability for specific functions. Research leading to the discovery of so many different regulatory mechanisms in muscle has been fascinating and somtimes bewildering. Yet there is now also a growing awareness of unity in this diversity as we learn more about the common principles that govern calcium signalling in the regulation of muscle contraction and cell motility.

References

Abbot RH, Cage PE (1984) A possible mechanism of length activation in insect fibrillar flight muscle. J Muscle Res Cell Mot 5:387–397

Achazi RK (1979 a) 5-HT induced accumulation of 3′,5′-AMP and the phosphorylation of paramyosin in the ABRM of *Mytilus edulis*. Malacologia 18:465–468

Achazi RK (1979 b) Phosphorylation of molluscan paramyosin. Pflügers Arch Eur J Physiol 379:197–201

Achazi RK, Dölling B, Haakshorst R (1974) 5-HT-induzierte Erschlaffung und cyclisches AMP bei einem glatten Molluskenmuskel. Pflügers Arch Eur J Physiol 349:19–27

Adams BA, Tanabe T, Mikami A, Numa S, Beam KG (1990) Intramembrane charge movement restored in dysgenic skeletal muscle by injection of dihydro-pyridine receptor cDNAs. Nature (Lond) 346:569–572

Adams S, DasGupta G, Chalovich JM, Reisler E (1990) Immunochemical evidence for the binding of caldesmon to the NH_2-terminal segment of actin. J Biol Chem 265:19652–19657

Adelstein RS, Hathaway DR (1979) Role of calcium and cyclic adenosine 3′,5′ monophosphate in regulating smooth muscle contraction. Am J Cardiol 44:783–787

Adelstein RS, Klee CB (1981) Purification and characterization of smooth muscle myosin light chain kinase. J Biol Chem 256:7501–7509

Adrian RH, Peachey LD (1973) Reconstruction of the action potential of frog sartorius muscle. J Physiol (Lond) 235:103–131

Adrian RH, Chandler WK, Rakowski RF (1976) Charge movement and mechanical repriming in skeletal muscle. J Physiol (Lond) 254:361–388

Ahlquist RP (1948) A study of adrenotropic receptors. Am J Physiol 153:586–600

Ahlquist RP (1962) The adrenotropic receptor-detector. Arch Int Pharmacodyn Ther 139:38–41

Aidley DJ (1965) The effect of calcium ions on potassium contracture in a locust leg muscle. J Physiol (Lond) 177:94–102

Aidley DJ, White DCS (1969) Mechanical properties of glycerinated fibres from the tymbal muscles of a Brazilian cicada. J Physiol (Lond) 205:179–192

Aksoy MO, Williams D, Sharkey EM, Hartshorne DJ (1976) A relationship between Ca^{2+} sensitivity and phosphorylation of gizzard actomyosin. Biochem Biophys Res Commun 69:35–41

Aksoy MO, Mras S, Kamm KE, Murphy RA (1983) Ca^{2+}, cAMP, and changes in myosin phosphorylation during contraction of smooth muscle. Am J Physiol 245:C255–C270

Akster HA, Granzier HLM, Keurs HEDJ ter (1985) A comparison of quantitative ultrastructural and contractile characteristics of muscle fibre types of the perch, *Perca fluviatilis* L. J Comp Physiol B 155:685–691

Allen DG, Blinks JR (1978) Calcium transients in aequorin-injected frog cardiac muscle. Nature (Lond) 273:509–513

Allen DG, Blinks JR (1979) The interpretation of light signals from aequorin-injected skeletal and cardiac muscle cells: a new method of calibration. In: Ashley CC, Campbell AK (eds) Detection an measurement of free Ca^{2+} in cells. Elsevier/North-Holland Biomedical Press, Amsterdam, pp 159–174

Allen DG, Kentish JC (1985) The cellular basis of the length-tension relation in cardiac muscle. J Mol Cell Cardiol 17:821 840

Allen DG, Kurihara S (1980) Calcium transients in mammalian ventricular musle. Eur Heart J 1 (Suppl A):5–15

Allen DG, Kurihara S (1982) The effect of muscle length on intracellular calcium transients in mammalian cardiac muscle. J Physiol (Lond) 327:79–94

Allen DG, Lee JA (1989) EMD 53998 increases tension with little effect on the amplitude of calcium transients in isolated ferret ventricular muscle. J Physiol (Lond) 416:43P

Allen DG, Orchard CH (1983 a) The effects of changes of pH on intracellular calcium transients in mammalian cardiac muscle. J Physiol (Lond) 335:555–567

Allen DG, Orchard CH (1983 b) Intracellular calcium concentration during hypoxia and metabolic inhibition in mammalian ventricular muscle. J Physiol (Lond) 339:107–122

Allen DG, Orchard CH (1987) Myocardial contractile function during ischemia and hypoxia. Circ Res 60:153–168

Allen DG, Kurihara S, Orchard CH (1982) The effects of hypoxia on intracellular calcium transients in mammalian cardiac muscle. J Physiol (Lond) 328:22P

Allen DG, Eisner DA, Orchard CH (1984) Factors influencing free intracellular calcium concentration in quiescent ferret ventricular muscle. J Physiol (Lond) 350:615–630

Allen DG, Morris PG, Orchard CH, Pirolo JS (1985) A nuclear magnetic resonance study of metabolism in the ferret heart during hypoxia and inhibition of glycolysis. J Physiol (Lond) 361:185–204

Allen DG, Lee JA, Smith GL (1989) The consequences of simulated ischemia on intracellular Ca^{2+} and tension in isolated ferret ventricular muscle. J Physiol (Lond) 410:297–323

Allshire A, Piper HM, Cuthbertson KSR, Cobbold PH (1987) Cytosolic free Ca^{2+} in single rat heart cells during anoxia and reoxygenation. Biochem J 244:381–385

Almers W, Palade PT (1981) Slow calcium and potassium currents across frog muscle membrane: measurements with a vaseline-gap technique. J Physiol (Lond) 312:159–176

Almers W, Fink RHA, Palade PT (1981 b) Ca^{++}-depletion in frog muscle tubules: the decline of Ca^{++}-current under maintained depolarization. J Physiol (Lond) 312:177–207

Altringham JD, Johnston IA (1982) The pCa-tension and force-velocity characteristics of skinned fibres isolated from fish fast and slow muscles. J Physiol (Lond) 333:421–449

Ambrogi-Lorenzini C, Colomo F, Lombardi V (1983) Development of force-velocity relation, stiffness and isometric tension in frog single muscle fibres. J Muscle Res Cell Mot 4:177–189

Anderson M (1982) Striated myoepithelial cells. In: Twarog BM, Levine RJC, Dewey MM (eds) Basic biology of muscles: a comparative approach. Raven Press, New York, pp 309–322

Anderson PA, Oakeley AE (1989) Immunological identification of five troponin T isoforms reveals an elaborate maturational troponin T profile in rabbit myocardium. Circ Res 65:1087–1093

Antoni H, Jakob R, Kaufmann R (1969) Mechanische Reaktionen des Frosch- und Säugermyokards bei Veränderung der Aktionspotentialdauer durch konstante Gleichstromimpulse. Pflügers Arch Eur J Physiol 306:33–57

Armstrong CM, Bezanilla F, Horowicz P (1972) Twitches in the presence of ethylene glycol gis (β-aminoethylether)-N,N'-tetraacetic acid. Biochim Biophys Acta 267:605–608

Arner A, Hellstrand P (1983) Activation of contraction and ATPase activity in intact and chemically skinned smooth muscle of rat portal vein. Dependence on calcium and muscle length. Circ Res 53:695–702

Ascadi G, Dickson G, Love DR, Jani A, Walsh FS, Gurusinghe A, Wolff JA, Davies KE (1991) Human dystrophin expression in mdx mice after intramuscular injection of DNA constructs. Nature (Lond) 352:815–818

Ashhurst DE (1967) The fibrillar flight muscles of giant waterbugs: an electron-microscope study. J Cell Sci 2:435–444

Ashiba G, Asada T, Watanabe S (1980) Calcium regulation in clam foot muscle. Calcium sensitivity of clam foot myosin. J Biochem 88:837–846

Ashley CC (1967) The role of calcium in the contraction of single cannulated muscle fibres. Am Zool 7:647–659

Ashley CC (1969) Aequorin-monitored calcium transients in single *Maia* muscle fibres. J Physiol (Lond) 203:32P–33P

Ashley CC, Campbell AK (1978) Free-calcium and tension responses in single barnacle muscle fibres following the application of L-glutamate. Biochim Biophys Acta 512:429–435

Ashley CC, Campbell AK (1979) Detection and measurement of free Ca^{2+} in cells. Elsevier/North-Holland Biomedical Press, Amsterdam

Ashley CC, Griffiths PJ (1983) The effect of injection of parvalbumins into single muscle fibres from the barnacle *Balanus nibulus*. J Physiol (Lond) 345:105P

Ashley CC, Moisescu DG (1977) Effect of changing the composition of the bathing solutions upon the isometric tension-pCa relationship in bundles of crustacean myofibrils. J Physiol (Lond) 270:627–652

Ashley CC, Ridgway EB (1970 a) Simultaneous recordings of membrane potential, calcium transient and tension in single muscle fibres. Nature (Lond) 219:1168–1169
Ashley CC, Ridgway EB (1970 b) On the relationship between membrane potential, calcium transient and tension in single barnacle muscle fibres. J Physiol (Lond) 209:105–130
Ashley CC, Caldwell PC, Lowe AG (1972) The efflux of calcium from single crab and barnacle muscle fibres. J Physiol (Lond) 223:735–755
Ashley CC, Ellory JC, Hainaut K (1974 a) Calcium movements in single crustacean muscle fibres. J Physiol (Lond) 242:255–272
Ashley CC, Moisescu DG, Rose RM (1974 b) Aequorin-light and tension responses from bundles of myofibrils following a sudden change in free calcium. J Physiol (Lond) 241:104P–106P
Ashley CC, Rink TJ, Tsien RY (1978) Changes in free Ca during muscle contraction, measured with an intracellular Ca-selective electrode. J Physiol (Lond) 280:27P
Ashley CC, Potter JD, Strang P, Godber J, Walton A, Griffiths PJ (1985) Kinetic investigations in single muscle fibres using luminescent and fluorescent Ca^{2+} probes. Cell Calcium 6:159–181
Ashley CC, Griffiths PJ, Rakowski RF (1986) Gating current in giant barnacle (*Balanus nubilus*) muscle fibres. J Physiol (Lond) 371:268P
Ashley CC, Barsotti R, Ferenczi MA, Lea TJ, Mulligan IP, Tsien RY (1987) Caged-calcium photolysis activates demembranated muscle fibres from the rabbit. J Physiol (Lond) 390:144P
Ashley CC, Mulligan IP, Trevor JL (1991) Ca^{2+} and activation mechnisms in skeletal muscle. Q Rev Biophys 24:1–73
Atassi MZ (1975) Antigenic structure of myoglobin: the complete immunochemical anatomy of a protein and conclusions relating to antigenic structure of proteins. Immunochemistry 12:423–438
Atsumi S, Sugi H (1976) Localization of calcium-accumulating structures in the anterior byssal retractor muscle of *Mytilus edulis* and their role in the regulation of active and catch contractions. J Physiol (Lond) 257:549–560
Atwood HL, Hoyle G, Smyth T Jr (1965) Mechanical and electrical responses of single innervated crab-muscle fibres. J Physiol (Lond) 180:449–482
Babu YS, Sack JS, Greenhough TJ, Bugg CE, Means AR, Cook WJ (1985) Three-dimensional structure of calmodulin. Nature (Lond) 315:37–40
Bagshaw CR (1980) Divalent metal ion binding and subunit interactions in myosins: a critical review. J Muscle Res Cell Mot 1:255–277
Bagshaw CR (1982) Muscle contraction. Chapman and Hall, London
Bagshaw CR, Kendrick-Jones J (1979) Characterization of homologous divalent metal ion binding sites of vertebrate and molluscan myosins using electron paramagnetic resonance spectroscopy. J Mol Biol 130:317–336
Bagshaw CR, Eccleston JF, Eckstein F, Goody RS, Gutfreund H, Trentham DR (1974) The magnesium ion-dependent adenosine triphosphatase of myosin. Biochem J 141:351–364
Baguet F (1973) The catch-state in glycerol extracted fibers from a lamellibranch smooth muscle (ABRM). Pflügers Arch Eur J Physiol 340:19–34
Baguet F, Gillis JM (1968) Energy cost of tonic contraction in a lamellibranch catch muscle. J Physiol (Lond) 198:127–143
Bailey K (1942) Myosin and adenosinetriphosphatase. Biochem J 36:121–139
Bailey K (1948) Tropomyosin: a new asymmetric protein component of the muscle fibril. Biochem J 43:271–273
Bailey K (1957) Invertebrate tropomyosin. Biochim Biophys Acta 24:612–619
Bárány M (1967) ATPase activity of myosin correlated with speed of muscle shortening. J Gen Physiol 50 (Suppl):197–218
Bárány M, Bárány K (1981) Protein phosphorylation in cardiac and vascular smooth muscle. Am J Physiol 241 (Heart Circ Physiol 10):H117–H128
Bárány K, Bárány M, Gillis JM, Kushmerick MJ (1979) Phosphorylation-dephosphorylation of the 18000-dalton light chain of myosin during the contraction-relaxation cycle of frog muscle. J Biol Chem 254:3617–3623
Barnard RJ, Edgerton VR, Furukawa T, Peter JB (1971) Histochemical, biochemical, and contractile properties of red, white, and intermediate fibers. Am J Physiol 220:410–414
Baron G, Demaille J, Dutruge M (1975) The distribution of parvalbumins in muscle and other tissues. FEBS Lett 56:156–160
Barsotti RJ, Ikebe M, Hartshorne DJ (1987) Effects of Ca^{2+}, Mg^{2+} and myosin phosphorylation on skinned smooth muscle fibres. Am J Physiol 252:C543–C554

Barton PJR, Robert B, Cohen A, Garner I, Sassoon D, Weydert A, Buckingham ME (1988) Structure and sequence of the myosin alkali light chain gene expressed in adult cardiac atria and fetal striated muscle. J Biol Chem 263:12669–12676
Bauer SF, Schwarz K, Rüegg JC (1989) Gluthathione alters calcium responsiveness of cardiac skinned fibers. Basic Res Cardiol 84:591–596
Baylor SM, Chandler WK, Marshall MW (1982) Optical measurements of intracellular pH and magnesium in frog skeletal muscle fibres. J Physiol (London) 331:105–137
Baylor SM, Chandler WK, Marshall MW (1984) Calcium release and sarcoplasmic reticulum membrane potential in frog skeletal muscle fibres. J Physiol (Lond) 348:209–238
Baylor SM, Hollingworth S (1988) Fura-2 calcium transients in frog skeletal muscle fibres. J Physiol (Lond) 403:151–192
Beam KG, Knudson CM, Powell JA (1986) A lethal mutation in mice eliminates the slow calcium current in skeletal muscle cells. Nature (Lond) 320:168–170
Bean BP (1985) Two kinds of calcium channels in canine atrial cells. Differences in kinetics, selectivity, and pharmacology. J Gen Physiol 86:1–30
Beeler GW Jr, Reuter H (1970 a) Membrane calcium current in ventricular myocardial fibres. J Physiol (Lond) 207:191–209
Beeler GW Jr, Reuter H (1970 b) The relation between potential, membrane currents and activation of contraction in ventricular myocardial fibres. J Physiol (Lond) 207:211–229
Beier N, Harting J, Jonas R, Klockow M, Lues I, Haeusler G (1991) The novel cardiotonic agent EMD 53998 is a potent "calcium sensitizer". J Cardiovasc Pharm (in press)
Beinbrech G, Kuhn HJ, Herzig JW, Rüegg JC (1976) Evidence for two attached myosin cross-bridge states of different potential energy. Cytobiologie 12:385–396
Bendall JR (1952) Effect of the Marsh-factor on the shortening of muscle fibre models in the presence of adenosinetriphosphate. Nature (Lond) 170:1058–1060
Benham CD, Tsien RW (1987) A novel receptor-operated Ca^{2+}-permeable channel activated by ATP in smooth muscle. Nature (Lond) 328:275–278
Benham CD, Tsien RW (1988) Noradrenaline modulation of calcium channels in single smooth muscle cells from rabbit ear artery. J Physiol (Lond) 404:767–784
Bennett AJ, Patel N, Wells C, Bagshaw CR (1984) 8-Anilino-1-naphthalenesulphonate, a fluorescent probe for the regulatory light chain binding site of scallop myosin. J Muscle Res Cell Mot 5:165–182
Bennet PM, Elliot A (1989) The "catch" mechanism in molluscan muscle: an electron microscopy study of freeze-substituted anterior byssus retractor muscle of *Mytilus edulis*. J Muscle Res Cell Mot 10:297–311
Benninger C, Einwächter HM, Haas HG, Kern R (1976) Calcium-sodium antagonism on the frog's heart: a voltage clamp-study. J Physiol (Lond) 259:617–645
Berchtold MW, Rowlerson AM, Heizmann CW (1982) Correlation of parvalbumin concentration with relaxation speed in mammalian muscles. J Muscle Res Cell Mot 3:459–460
Berridge MJ (1985) The molecular basis of communication within the cell. Sci Am 253:124–134
Berridge MJ (1987) Inositol trisphosphate and diacylglycerol: two interacting second messengers. Annu Rev Biochem 56:159–193
Berridge MJ, Irvine RF (1984) Inositol trisphosphate, a novel second messenger in cellular signal transduction. Nature (Lond) 312:315–321
Bers DM (1985) Ca influx and sarcoplasmic reticulum Ca release in cardiac muscle activation during postrest recovery. Am J Physiol 248:H366–H381
Bevan JA, Hwa JJ, Owen MP, Winquist RJ (1985) Calcium und myogenic or stretch-dependent vascular tone. In: Rubin RP, Weiss GB, Putney JW Jr (eds) Calcium in biological systems. Plenum Press, New York, pp 391–398
Bialojan C, Rüegg JC, DiSalvo J (1985) Influence of a polycation-modulated phosphatase on actin-myosin interactions in smooth muscle preparations. In: Merlevede W, DiSalvo J (eds) Advances in protein phosphatases II. Leuven University Press, Leuven, pp 105–121
Biajolan C, Rüegg JC, Takai A (1988) Effects of okadaic acid on isometric tension and myosin phosphorylation of chemically skinned guinea-pig taenia coli. J Physiol (Lond) 398:81–95
Bianchi CP, Shanes AM (1959) Calcium influx in skeletal muscle at rest, during activity, and during potassium contracture. J Gen Physiol 42:803–815

Bing OHL, Brooks WW, Conrad CH, Sen S, Perreault CL, Morgan JP (1991) Intracellular calcium transients in myocardium from spontaneously hypertensive rats during the transition to heart failure. Circ Res 68:1390–1400

Blanchard EM, Bo-Sheng P, Solaro RJ (1984) The effect of acidic pH on the ATPase activity and troponin Ca^{2+} binding of rabbit skeletal myofilaments. J Biol Chem 259:3181–3186

Blaustein MP (1974) The interrelationship between sodium and calcium fluxes across cell membranes. Rev Physiol Biochem Pharmacol 70:33–82

Blaustein MP (1977) Sodium ions, calcium ions, blood pressure regulation, and hypertension: a reassessment and a hypothesis. Am J Physiol 232:C165–C173

Blinks JR, Endoh M (1984) Sulmazole (AR-L 115 BS) alters the relation between Ca^{2+} and tension in living canine ventricular muscle. J Physiol (Lond) 353:63P

Blinks JR, Endoh M (1986) Modification of myofibrillar responsiveness to Ca^{2+} as an inotropic mechanism. Circulation 73 (Suppl III):85–98

Blinks JR, Rüdel R, Taylor SR (1978) Calcium transients in isolated amphibian skeletal muscle fibres: detection with aequorin. J Physiol (Lond) 277:291–323

Blinks JR, Wier WG, Hess P, Prendergast FG (1982) Measurement of Ca^{2+} concentrations in living cells. Prog Biophys Mol Biol 40:1–114

Block BA, Imagawa T, Campbell KP, Franzini-Armstrong C (1988) Structural evidence for direct interaction between the molecular components of the transverse tubule/SR junction in skeletal muscle. J Cell Biol 107:2587–2600

Bloomquist E, Curtis BA (1975 a) ^{45}Ca efflux from anterior byssus retractor muscle in phasic and catch contraction. Am J Physiol 229:1237–1243

Bloomquist E, Curtis BA (1975 b) Net calcium fluxes in anterior byssus muscle with phasic and catch contraction. Am J Physiol 229:1244–1248

Blumenthal D, Stull J (1980) Activation of skeletal muscle myosin light chain kinase by Ca^{2+} and calmodulin. Biochemistry 19:5608–5614

Boels PJ, Troschka M, Rüegg JC, Pfitzer G (1991) Higher Ca^{2+}-sensitivity of triton-skinned guinea pig mesenteric microarteries as compared with large arteries. Circ Res 69:989–996

Boettiger EG (1957) The machinery of insect flight. In: Scheer BT (ed) Recent advances in invertebrate physiology. Univ Oregon Press, Eugene, pp 117–142

Bolli R, Jeroudi MO, Patel BS, Aruoma OI, Halliwell B, Lai EK, McCay PB (1989) Marked reduction of free radical generation and contractile dysfunction by antioxidant therapy begun at the time of reperfusion. Circ Res 65:607–622

Bond M, Somlyo AV, Shuman H, Somlyo AP (1983) In situ measurement of cytoplasmic Ca in vascular smooth muscle by electron probe (EPMA) and electron energy loss analysis (EELS). Proc Int Union Physiol Sci XV:261

Bond M, Kitazawa T, Somlyo AP, Somlyo AV (1984) Release and recycling of calcium by the sarcoplasmic reticulum in guinea-pig portal vein smooth muscle. J Physiol (Lond) 355:677–695

Bone Q (1964) Patterns of muscular innervation in the lower chordates. Int Rev Neurobiol 6:99–147

Borle AB (1981) Control, modulation and regulation of cell calcium. Rev Physiol Biochem Pharmacol 90:14–153

Borovikov YS, Kuleva NV, Khoroshev MI (1991) Polarization microfluorimetry study of interaction between myosin head and F-actin in muscle fibers. Gen Physiol Biophys 10:441–459

Bossen EH, Sommer JR, Waugh RA (1978) Comparative stereology of the mouse and finch left ventricle. Tissue Cell 10:773–784

Bowers KC, Allshire AP, Cobbold PH (1992) Bioluminescent measurement in single cardiomyocytes of sudden cytosolic ATP depletion coincident with rigor. J Mol Cell Cardiol 24:213–218

Bozler E (1953) The mechanism of muscular relaxation. Experientia 9:1–6

Bozler E (1972) Feedback in the contractile mechanism of the frog heart. J Gen Physiol 60:239–247

Bradley AB, Morgan KG (1987) Alterations in cytoplasmic calcium sensitivity during porcine coronary artery contractions as detected by aequorin. J Physiol (Lond) 385:437–448

Brandl CJ, Green NM, Korczak B, MacLennan DH (1986) Two Ca^{2+} ATPase genes: homologies and mechanistic implications of deduced amino acid sequences. Cell 44:597–607

Brandt NR, Caswell AH, Wen SR, Talvenheimo JA (1990) Molecular interactions of the junctional foot protein and dihydropyridine receptor in skeletal muscle triads. J Membr Biol 113:237–252

Brandt PW, Cox RN, Kawai M, Robinson T (1982) Regulation of tension in skinned muscle fibers. Effect of cross-bridge kinetics on apparent Ca^{2+} sensitivity. J Gen Physiol 79:997–1016

Breemen C van (1976) Transmembrane calcium transport in vascular smooth muscle. In: Bevan JA (ed) Vascular neuroeffector mechanisms. Karger, Basel, pp 67–79

Breemen C van, Farinas BR, Casteels R, Gerba P, Wuytack F, Deth R (1973) Factors controlling cytoplasmic Ca^{2+} concentration. Philos Trans R Soc Lond B 265:57–71

Breemen C van, Aaronson P, Loutzenhiser R, Meisheri K (1980) Ca^{2+} movements in smooth muscle. Chest 78 (Suppl):157–165

Breitbart RE, Nguyen HT, Medford RM, Destree AT, Mahadavi V, Nadel-Ginard B (1985) Intricate combinatorial patterns of exon splicing generate multiple regulated troponin T isoforms from a single gene. Cell 41:67–82 Z456

Bremel RD, Weber A (1972) Cooperation within actin filament in vertebrate skeletal muscle. Nature (New Biol) 238:97–101

Bremel RD, Murray TM, Weber A (1972) Manifestations of cooperative behaviour in the regulated actin filament during actin activated ATP-hydrolysis in the presence of calcium. Cold Spring Harbor Symp Quant Biol 37:267–275

Brenner B (1988) Effect of Ca^{2+} on crossbridge turnover kinetics in skinned single rabbit psoas fibers: implications for regulation of muscle contraction. Proc Natl Acad Sci USA 85:3265–3269

Brenner B (1990) Muscle mechanics and biochemical kinetics. In: Squire JM (ed) Molecular mechanisms in muscular contraction. Macmillan, London

Brenner B, Schoenberg M, Chalovich JM, Greene LE, Eisenberg E (1982) Evidence for cross-bridge attachment in relaxed muscle at low ionic strength. Proc Natl Acad Sci USA 79:7288–7291

Brenner B, Leepo CY, Chalovich JM (1991) Parallel inhibition of active force and relaxed fiber stiffness in skeletal muscle by caldesmon: implications for the pathway to force generation. Proc Natl Acad Sci USA 88:5739–5743

Bretscher A (1986) Thin filament regulatory proteins of smooth- and non-muscle cells. Nature (Lond) 321:726–727

Brozovich FV, Lawrence DY, Gordon AM (1988) Muscle force and stiffness during activation and relaxation. J Gen Physiol 91:399–420

Brum G, Flockerzi V, Hofmann F, Osterrieder W, Trautwein W (1983) Injection of catalytic subunit of cAMP-dependent protein kinase into isolated cardiac myocytes. Pflügers Arch Eur J Physiol 398:147–154

Buddenbrock WV (1911) Untersuchungen über die Schwimmbewegungen und die Statocysten der Gattung *Pecten*. Ber Heidelberg Akad Wiss 28:1–24

Bülbring E (1955) Correlation between membrane potential, spike discharge, and tension in smooth muscle. J Physiol (Lond) 128:200–221

Bülbring E, Hertog A den (1980) The action of isoprenaline on the smooth muscle of the guinea-pig taenia coli. J Physiol (Lond) 304:277–296

Bullard B (1984) A large troponin in asynchronous insect flight muscle. J Muscle Res Cell Mot 5:196

Bullard B, Dabrowska R, Winkelman L (1973) The contractile and regulatory proteins of insect flight muscle. Biochem J 135:277–286

Bullard B, Leonard K, Larkins A, Butcher G, Karlik C, Fyrberg E (1988) Troponin of asynchronous flight muscle. J Mol Biol 204:621–637

Buller AJ, Eccles JC, Eccles RM (1960) Interactions between motoneurones and muscles in respect of the characteristic speeds of their responses. J Physiol (Lond) 150:417–439

Buss JE, Stull JT (1977) Calcium binding to cardiac troponin and the effect of cyclic AMP dependent protein kinase. FEBS Lett 73:101–104

Butler TM, Siegman MJ (1983) Chemical energy usage and myosin light chain phosphorylation in mammalian smooth muscle. Fed Proc 42:57–61

Butler TM, Siegman MJ, Mooers SU, Barsotti RJ (1983) Myosin light chain phosphorylation does not modulate crossbridge cycling rate in mouse skeletal muscle. Science 220:1167–1169

Butler TM, Siegman MJ, Mooers SU, Narayan SR (1990) Myosin-product complex in the resting state and during relaxation of smooth muscle. Am J Physiol 258:C1092–C1099

Cachia PJ, Van Eyk J, Ingraham RH, McCubbin WD, Kay CM, Hodges RS (1986) Calmodulin and troponin C: a comparative study of the interaction of mastoparan and troponin I inhibitory peptide [104–115]. Biochemistry 25:3553–3562

Caldwell PC, Walster G (1963) Studies on the micro-injection of various substances into crab muscle fibres. J Physiol (Lond) 169:353–372

Campbell AP, Sykes BD (1990 a) Interaction of troponin I and troponin C: use of the two-dimensional transferred nuclear Oberhauser effect to determine the structure of the inhibitory troponin I peptide when bound to turkey skeletal troponin C. J Mol Biol (submitted)
Campbell AP, Sykes BD (1990 a) Interaction of troponin I and troponin C: ^{19}F NMR studies of the binding of the inhibitory troponin I peptide to turkey skeletal troponin C. J Biol Chem (submitted)
Campbell KP, Shamoo AE (1980) Chloride-induced release of actively loaded calcium from light and heavy sarcoplasmic reticulum vesicles. J Membr Biol 54:73–80
Campbell KP, Franzini-Armstrong C, Shamoo AE (1980) Further characterization of light and heavy sarcoplasmic reticulum vesicles. Identification of the sarcoplasmic reticulum feet associated with heavy sarcoplasmic reticulum vesicles. Biochim Biophys Acta 602:97–116
Campbell KP, Knudson CM, Imagawa T, Leung AT, Sutko JL, Kahl SD, Raab CR, Madson L (1987) Identification and characterization of the high affinity [^{3}H] ryanodine receptor of the junctional sarcoplasmic reticulum Ca^{2+} release channel. J Biol Chem 262:6460–6463
Campbell KP, Leung AT, Sharp AH (1988) The biochemistry and molecular biology of the dihydropyridine-sensitive calcium channel. Trends Neurosci 11:425–430
Canfield SP (1971) The mechanical properties and heat production of chicken latissimus dorsi muscles during tetanic contractions. J Physiol (Lond) 219:281–302
Cannell MB, Allen DG (1984) Model of calcium movements during activation in the sarcomere of frog skeletal muscle. Biophys J 45:913–925
Caputo C, Bolanos P (1978) Effect of external sodium and calcium on calcium efflux in frog striated muscle. J Membr Biol 41:1–14
Caputo C, Gottschalk G, Lüttgau HC (1981) The control of contraction activation by the membrane potential. Experientia 37:580–581
Carafoli E (1982) The transport of calcium across the inner membrane of mitochondria. In: Carafoli E (ed) Membrane transport of calcium. Academic Press, London, pp 109–139
Carafoli E (1991) Calcium pump of the plasma membrane. Physiol Rev 71:129–152
Carafoli E, Tiozzo R, Rossi CS, Lugli G (1972) Mitochondrial Ca^{2+} uptake and heart relaxation. In: Bolis L, Keynes RD, Wilbrandt W (eds) Role of membranes in secretory processes. Elsevier/North-Holland, Amsterdam, pp 175–181
Carlson FD, Wilkie DR (1974) Muscle physiology. Prentice-Hall, Englewood Cliffs
Caroni P, Reinlib L, Carafoli E (1980) Charge movements during the Na^{+}-Ca^{+} exchange in heart sarcolemmal vesicles. Proc Natl Acad Sci USA 77:6354–6358
Casteels R, Breemen C van (1975) Active and passive Ca^{2+} fluxes across cell membranes of the guinea-pig taenia coli. Pflügers Arch Eur J Physiol 359:197–207
Casteels R, Raeymaekers L (1979) The action of acetylcholine and catecholamines on an intracellular calcium store in the smooth muscle cells of the guinea-pig taenia coli. J Physiol (Lond) 294:51–68
Castellani L, Cohen C (1987) Rod phosphorylation favors folding in a catch muscle myosin. Proc Natl Acad Sci USA 84:4058–4062
Castellani L, Franzini-Armstrong C, Loesser K (1989) Shape, size and disposition of feet in junctions between transverse tubules and sarcoplasmic reticulum of Bivalvia, Insecta, Crustacea and Arachnida. J Physiol (Lond) 418:118P
Caswell AH, Brandt NR (1989) Does muscle activation occur by direct mechanical coupling of transverse tubules to sarcoplasmic reticulum? TIBS 14: 161–165
Cavero I, Shepperson N, Lefèvre-Borg F, Langer SZ (1983) Differential inhibition of vascular smooth muscle responses to α_1- and α_2-adrenoreceptor agonists by diltiazem and verapamil. Circ Res 52 (Suppl I):169–176
Cecchi G, Colomo F, Lombardi V (1978) Force velocity relation in normal and nitrate treated frog single muscle fibres during rise of tension in an isometric tetanus. J Physiol (Lond) 285:257–273
Cecchi G, Lombardi V, Menchetti G (1984) Development of force-velocity relation and rise of isometric tetanic tension measure the time course of different processes. Pflügers Arch Eur J Physiol 401:396–401
Cecchi G, Griffiths PJ, Taylor S (1986) Stiffness and force in activated frog skeletal muscle fibers. Biophys J 49:437–451
Chacko S (1981) Effects of phosphorylation, calcium ion, and tropomyosin on actin-activated adenosine 5′-triphosphatase activity of mammalian smooth muscle myosin. Biochemistry 20:702–707

Chacko S, Eisenberg E (1990) Cooperativity of actin-activated ATPase of gizzard heavy meromyosin in the presence of gizzard tropomyosin. J Biol Chem 265:2105–2110
Chacko S, Rosenfeld A (1982) Regulation of actin-activated ATP hydrolysis by arterial myosin. Proc Natl Acad Sci USA 79:292–296
Chadwick CC, Inui M, Fleischer S (1988) Identification and purification of a transverse tubule coupling protein which binds to the ryanodine receptor of terminal cisternae at the triad junction in skeletal muscle. J Biol Chem 263:10872–10877
Chadwick CC, Saito A, Fleischer S (1990) Isolation and characterization of the inositol trisphosphate receptor from smooth muscle. Proc Natl Acad Sci USA 87:2132–2136
Chalovich JM, Chock PB, Eisenberg E (1981) Mechanism of action of troponin-tropomyosin. J Biol Chem 256:575–578
Chalovich JM, Chantler PD, Szent-Györgyi AG, Eisenberg E (1983) Regulation of molluscan actomyosin ATPase by Ca^{2+}. Biophys J 41:153 a
Chantler PD (1983) Biochemical and structural aspects of molluscan muscle. In: Wilbur KM (ed) The mollusca 4 (Physiology, part 1). Academic Press, London, pp 78–154
Chantler PD, Szent-Györgyi AG (1980) Regulatory light-chains and scallop myosin: full dissociation, reversibility and co-operative effects. J Mol Biol 138:473–492
Chapman RA (1983) Control of cardiac contractility at the cellular level. Am J Physiol 245 (Heart Circ Physiol 14):H535–H552
Chatterjee M, Murphy RA (1983) Calcium-dependent stress maintenance without myosin phosphorylation in skinned smooth muscle. Science 221:464–466
Chaussepied P, Morales MF (1988) Modifying preselected sites on proteins: the stretch of residues 633–642 of the myosin heavy chain is part of the actin-binding site. Proc Natl Acad Sci USA 85:7471–7475
Cheung WY (1970) Cyclic 3′,5′ nucleotide phosphodiesterase. Demonstration of an activator. Biochem Biophys Res Commun 38:533–538
Cheung WY (1980) Calmodulin plays a pivotal role in cellular regulation. Science 207:19–27
Clarke DM, Loo TW, Inesi G, MacLennan DH (1989) Location of high affinity Ca^{2+}-binding sites within the predicted transmembrane domain of the sarcoplasmic reticulum Ca^{2+}-ATPase. Nature (Lond) 339:476–478
Close RI (1964) Dynamic properties of fast and slow skeletal muscles of the rat during development. J Physiol (Lond) 173:74–95
Close RI (1965) The relation between intrinsic speed of shortening and duration of the active state of muscle. J Physiol (Lond) 180:542–559
Close RI (1972) Dynamic properties of mammalian skeletal muscles. Physiol Rev 52:129–197
Cobbold PH, Bourne PK, Cuthbertson KSR (1984) Evidence from aequorin for injury of metabolically inhibited myocytes independently of free Ca^{2+}. Nature (Lond) 312:444–446
Cohen C (1982) Matching molecules in the catch mechanism. Proc Natl Acad Sci USA 79:3176–3178
Cole HA, Patchell VB, Perry SV (1983) Phosphorylation of chicken gizzard myosin and the Ca^{2+}-sensitivity of the actin-activated Mg^{2+}-ATPase. FEBS Lett 158:17–20
Collins JH (1991) Myosin light chains and troponin C: structural and evolutionary relationships revealed by amino acid sequence comparisons. J Muscle Res Cell Mot 12:3–25
Collins JH, Johnson JD, Szent-Györgyi AG (1983) Purification and characterization of a scallop sarcoplasmic calcium-binding protein. Biochemistry 22:341–345
Colucci WS, Wright RF, Braunwald E (1986 a) New positive inotropic agents in the treatment of congestive heart failure. Mechanisms of action and recent clinical developments. First of two parts. N Engl J Med 314:290–349
Colucci WS, Wright RF, Braunwald E (1986 b) New positive inotropic agents in the treatment of congestive heart failure. Mechanisms of action and recent clinical developments. Second of two parts. N Engl J Med 314:349–358
Connor JA, Kreuglen D, Prosser CL, Weigel R (1977) Interaction between longitudinal and circular muscle in intestine of cat. J Physiol (Lond) 273:665–689
Conti MA, Adelstein RS (1981) The relationship between calmodulin binding and phosphorylation of smooth muscle kinase by the catalytic subunit of 3′:5′-cAMP-dependent protein kinase. J Biol Chem 256:3178–3181
Cooke R, Franks K, Stull JT (1982) Myosin phosphorylation regulates the ATPase activity of permeable skeletal muscle fibers. FEBS Lett 144:33–37

Cooley LB, Johnson WH, Krause S (1979) Phosphorylation of paramyosin and its possible role in the catch mechanism. J Biol Chem 254:2195–2198
Cooper TA, Ordahl CP (1985) A single cardiac troponin T gene generates embryonic and adult isoforms via developmentally regulated alternative splicing. J Biol Chem 260:11140–11148
Coray A, Fry CH, Hess P, McGuigan JAS, Weingart R (1980) Resting calcium in sheep cardiac tissue and in frog skeletal muscle measured with ion-selective micro-electrodes. J Physiol (Lond) 305:60P–61P
Cornelius F (1982) Tonic contraction and the control of relaxation in a chemically skinned molluscan smooth muscle. J Gen Physiol 79:821–834
Costantin LL (1970) The role of sodium current in the radial spread of contraction in frog muscle fibers. J Gen Physiol 55:703–715
Costantin LL, Podolsky RJ, Tice LW (1967) Calcium activation of frog slow muscle fibres. J Physiol (Lond) 188:261–271
Cota G, Stefani E (1981) Effects of external calcium reduction on the kinetics of potassium contractures in frog twitch muscle fibres. J Physiol (Lond) 317:303–316
Coutance A (1878) De l'énergie de la structure musculaire chez les mollusques acéphales. Paris
Cox JA, Wnuk W, Stein EA (1977) Regulation of calcium binding by magnesium. In: Wasserman RH, Corradino RA, Carafoli E, Kretsinger RH, MacLennan DH, Siegel FL (eds) Calcium-binding proteins and calcium function. Elsevier/North-Holland, Amsterdam, pp 266–269
Craig R, Szent-Györgyi AG, Beese L, Flicker P, Vibert P, Cohen C (1980) Electron microscopy of thin filaments decorated with a Ca^{2+}-regulated myosin. J Mol Biol 140:35–55
Craig R, Smith R, Kendrick-Jones J (1983) Light-chain phosphorylation controls the conformation of vertebrate non-muscle and smooth muscle myosin molecules. Nature (Lond) 302:436–439
Crow MT, Kushmerick MJ (1982a) Myosin light chain phosphorylation is associated with a decrease in the energy cost for contraction in fast twitch mouse muscle. J Biol Chem 257:2121–2124
Crow MT, Kushmerick MJ (1982b) Phosphorylation of myosin light chains in mouse fast-twitch muscle associated with reduced actomyosin turnover rate. Science 217:835–837
Crow MT, Kushmerick MJ (1982c) Chemical energetics of slow- and fast-twitch muscles of the mouse. J Gen Physiol 79:147–166
Crow MT, Kushmerick MJ (1983) Correlated reduction of velocity of shortening and the rate of energy utilization in mouse fast-twitch muscle during a continuous tetanus. J Gen Physiol 82:703–720
Cummins P (1983) Contractile proteins in muscle disease. J Muscle Res Cell Mot 4:5–24
Dabrowska R, Aromatorio D, Sherry JMF, Hartshorne DJ (1977) Composition of the myosin light chain kinase from chicken gizzard. Biochem Biophys Res Commun 78:1263–1272
Dantzig JA, Hibberd MG, Trentham DR, Goldman YE (1991) Cross-bridge kinetics in the presence of MgADP investigated by photolysis of caged ATP in rabbit psoas muscle fibres. J Physiol (Lond) 432:639–680
Davies RE (1964) Adenosine triphosphate breakdown during single muscle contractions. Proc R Soc (Lond) B 160:480–485
Dawson MJ, Gadian DG, Wilkie D (1977) Contraction and recovery of living muscles studied by ^{31}P nuclear magnetic resonance. J Physiol (Lond) 267:703–735
DeCouet HC, Mazander KD, Gröschel-Stewart U (1980) A study of invertebrate actins by isoelectric focusing and immunodiffusion. Experientia 36:404–405
Dedman JR, Potter JD, Jackson RL, Johnson JD, Means AR (1977) Physicochemical properties of rat testis Ca^{2+} dependent regulator protein of cyclic nucleotide phosphodiesterase: relationship of Ca^{2+} binding, conformational changes and phosphodiesterase activity. J Biol Chem 252:8415–8422
DeLanerolle P, Nishikawa M, Yost DA, Adelstein RS (1984) Increased phosphorylation of myosin light chain kinase after an increase in cyclic AMP in intact smooth muscle. Science 223:1415–1417
Demaille J, Dutruge E, Capony J-P, Pechère J-F (1974) Muscular parvalbumins: a family of homologous calcium-binding proteins. Their relation to the calcium-binding troponin component. In: Drabikowski W, Strzelecka-Golaszewska M, Carafoli E (eds) Calcium binding proteins. Elsevier/North-Holland, Amsterdam, pp 643–677
Deth R, Breemen C van (1974) Relative contributions of Ca^{2+} influx and cellular Ca^{2+} release during drug induced activation of the rabbit aorta. Pflügers Arch Eur J Physiol 348:13–22
Devine CE, Somlyo AV, Somlyo AP (1972) Sarcoplasmic reticulum and excitation-contraction coupling in mammalian smooth muscles. J Cell Biol 52:690–718

Dewey MM, Colflesh D, Brink P, Fan S-F, Gaylinn B, Gural N (1982) Structural, functional and chemical changes in the contractile apparatus of *Limulus* striated muscle as a function of sarcomere shortening and tension development. In: Twarog B, Levine RJC, Dewey MM (eds) Basic biology of muscles: a comparative approach, vol 37. Raven Press, New York, pp 53–72

D'Haese J (1980) Regulatory light chains of myosin from the obliquely-striated body wall muscle of *Lumbricus terrestris*. FEBS Lett 121:243–245

D'Haese J, Carlhoff D (1987) Localization and histochemical characterization of myosin isoforms in earthworm body wall muscle. J Comp Physiol B 157:171–179

Dhoot GK, Perry SV (1980) The components of the troponin complex and development in skeletal muscle. Exp Cell Res 127:75–87

Dhoot GK, Frearson N, Perry SV (1979) Polymorphic forms of troponin T and troponin C and their localization in striated muscle cell types. Exp Cell Res 122:339–350

Diebler H, Eigen M, Hammes GG (1960) Relaxations-spektrometrische Untersuchungen schneller Reaktionen von ATP in wäßriger Lösung. Z Naturforsch 15b:554–560

Dillon PF, Aksoy MO, Driska SP, Murphy RA (1981) Myosin phosphorylation and the cross-bridge cycle in arterial smooth muscle. Science 211:495–497

DiSalvo J, Gifford D, Bialojan C, Rüegg JC (1983) An aortic spontaneously active phosphatase dephosphorylates myosin and inhibits actin-myosin interaction. Biochem Biophys Res Commun 111:906–911

Donahue MJ, Michnoff CA, Masaracchia RA (1985) Calcium-dependent muscle contraction in obliquely striated *Ascaris suum* muscle. Comp Biochem Physiol 82B:395–403

Dorsett DA, Roberts JB (1980) A transverse tubular system and neuromuscular junctions in a molluscan unstriated muscle. Cell Tissue Res 206:251–260

Droogmans G, Casteels R (1979) Sodium and calcium interactions in vascular smooth muscle cells of the rabbit ear artery. J Gen Physiol 74:57–70

Droogmans G, Raeymaekers L, Casteels R (1977) Electro- and pharmacomechanical coupling in the smooth muscle cells of the rabbit ear artery. J Gen Physiol 70:129–148

Droogmans G, Himpens B, Casteels R (1985) Ca-exchange, Ca-channels and Ca-antagonists. Experientia 41:895–900

Drummond DR, Peckham M, Sparrow JC, White DCS (1990) Alteration in crossbridge kinetics caused by mutations in actin. Nature (Lond) 348:440–442

Dubyak GR, Scarpa A (1982) Sarcoplasmic Ca^{2+} transients during the contractile cycle of single barnacle muscle fibres: measurements with arsenazo III-injected fibres. J Muscle Res Cell Mot 3:87–112

Dulhunty AF (1985) Excitation-contraction coupling and contractile properties in denervated rat EDL and soleus muscles. J Muscle Res Cell Mot 6:207–225

Dulhunty AF, Gage PW (1983) Asymmetrical charge movement in slow- and fast-twitch mammalian muscle fibres in normal and paraplegic rats. J Physiol (Lond) 341:213–231

Dulhunty AF, Gage PW (1985) Excitation-contraction coupling and charge movement in denervated rat extensor digitorum longus and soleus muscles. J Physiol (Lond) 358:75–89

Duncan CJ (1978) Role of intracellular calcium in promoting muscle damage: a strategy for controlling the dystrophic condition. Experientia 34:1531–1672

Dunn JF, Radda GK (1991) Total ion content of skeletal and cardiac muscle in the mdx mouse dystrophy: Ca^{2+} is elevated at all ages. J Neurol Sci 103:226–231

Eastwood AB, Franzini-Armstrong C, Peracchia C (1982) Structure of membranes in crayfish muscle: comparison of phasic and tonic fibres. J Muscle Res Cell Mot 3:273–294

Eaton BL, Kominz DR, Eisenberg E (1975) Correlation between the inhibition of the acto-heavy meromyosin ATPase and the binding of tropomyosin to F-actin: effects of Mg^{2+}, KCl, troponin I and troponin C. Biochemistry 14:2718–2725

Ebashi S (1963) Third component participating in the superprecipitation of "natural actomyosin". Nature (Lond) 200:1010–1012

Ebashi S (1980) Regulation of muscle contraction. Proc R Soc (Lond) B 207:259–286

Ebashi S (1983) Regulation of contractility. In: Stracher A (ed) Muscle and nonmuscle motility. Academic Press, London, pp 217–232

Ebashi S (1991) Excitation-contraction coupling and the mechanism of muscle contraction. Annu Rev Physiol 53:1–16

Ebashi S, Endo M (1968) Calcium ion and muscle contraction. Prog Biophys Mol Biol 18:125–183

Ebashi S, Kodama A (1966) Native tropomyosin-like actin of troponin on trypsin-treated myosin B. J Biochem 60:733–734
Ebashi S, Lipmann F (1962) Adenosine triphosphate-linked concentration of calcium ions in a particulate fraction of rabbit muscle. J Cell Biol 14:389–400
Ebashi S, Kodama A, Ebashi F (1968) Troponin. I. Preparation and physiological function. J Biochem (Tokyo) 64:465–477
Ebashi S, Endo M, Ohtsuki I (1969) Control of muscle contraction. Q Rev Biophys 2:351–384
Edman KAP (1975) Mechanical deactivation induced by active shortening in isolated muscle fibres of the frog. J Physiol (Lond) 246:255–275
Eftimie R, Brenner HR, Buonanno A (1991) Myogenin and myoD join a family of skeletal muscle genes regulated by electrical activity. Proc Natl Acad Sci USA 88:1349–1353
Egelman EH (1985) The structure of F-actin. J Muscle Res Cell Mot 6:129–151
Eggermont JA, Vrolix M, Raeymaekers L, Wuytack F, Casteels R (1988) Ca^{++}-transport ATPases of vascular smooth muscle. Circ Res 62:266–278
Eigen M, Hammes GG (1963) Elementary steps in enzyme reactions (as studied by relaxation spectrometry). Adv Enzymol 25:1–38
Eisenberg BR, Eisenberg RS (1968) Selective disruption of the sarcotubular system in frog sartorius muscle. A quantitative study with exogenous peroxidase as a marker. J Cell Biol 39:451–467
Eisenberg BR, Eisenberg RS (1982) The T-SR junction in contracting single skeletal muscle fibres. J Gen Physiol 79:1–19
Eisenberg BR, Kuda AM (1975) Stereological analysis of mammalian skeletal muscle. II. White vastus muscle of the adult male guinea pig. J Ultrastruct Res 51:176–187
Eisenberg BR, Kuda AM (1976) Discrimination between fiber populations in mammalian skeletal muscle by using ultrastructural parameters. J Ultrastruct Res 54:76–88
Eisenberg BR, Salmons S (1981) The reorganization of subcellular structure in muscle undergoing fast-to-slow type transformation. Cell Tissue Res 220:449–471
Eisenberg BR, Kuda AM, Peter JB (1974) Stereological analysis of mammalian skeletal muscle. I. Soleus muscle of the adult guinea pig. J Cell Biol 60:732–754
Eisenberg E, Hill TL (1985) Muscle contraction and free energy transduction in biological systems. Science 227:999–1006
Eisenberg RS, McCarthy RT, Milton RL (1983) Paralysis of frog skeletal muscle fibres by the calcium antagonist D-600. J Physiol (Lond) 341:495–505
Ellington WR (1983) The extent of intracellular acidification during anoxia in the catch muscles of two bivalve molluscs. J Exp Zool 227:313–317
Elliot A, Bennett PM (1982) Structure of the thick filaments in molluscan adductor muscle. In: Twarog BM, Levine RJC, Dewey MM (eds) Basic biology of muscles: a comparative approach. Raven Press, New York, pp 11–28
El-Saleh SC, Warber KD, Potter JD (1986) The role of tropomyosin-troponin in the regulation of skeletal muscle contraction. J Muscle Res Cell Mot 7:387–404
Endo M (1977) Calcium release from the sarcoplasmic reticulum. Physiol Rev 57:71–108
Endo M, Kitazawa T, Yagi S (1980) Different features of responses of the sarcoplasmic reticulum in cardiac and smooth muscles. In: Ebashi S, Maruyama K, Endo M (eds) Muscle contraction. Its regulatory mechanisms. Japan Sci Soc Press, Tokyo, pp 447–463
Endo M, Yagi S, Ishizuka T, Horiuti K, Koga Y, Amaha K (1983) Changes in the Ca-induced Ca-release mechanism in the sarcoplasmic reticulum of the muscle from a patient with malignant hyperthermia. Biomed Res 4:83–92
Endo T, Naka M, Hidaka H (1982) Ca^{2+}-phospholipid dependent phosphorylation of smooth muscle myosin. Biochem Biophys Res Commun 105:942–948
Endoh M, Blinks JR (1988) Actions of sympathomimetic amines on the Ca^{2+}-transients and contractions of rabbit myocardium: reciprocal changes in myofibrillar responsiveness to Ca^{2+}-mediated through α- and β-adrenoceptors. Circ Res 62:247–265
Endoh M, Yanagisawa T, Taira N, Blinks JR (1986) Effects of new inotropic agents on cyclic nucleotide metabolism and calcium transients in canine ventricular muscle. Circulation 73 (Suppl III):117–133
Epstein HF, Aronow BJ, Harris HE (1976a) Myosin-paramyosin cofilaments: enzymatic interactions with F-actin. Proc Natl Acad Sci USA 73:3015–3019
Epstein HF, Isachsen MM, Suddleson EA (1976b) Kinetics of movement of normal and mutant nematodes. J Comp Physiol 110:317–322

Erdman R, Lüttgau HC (1989) The effect of the phenylalkylamine D888 (Devapamil) on force and Ca^{2+}-current in isolated frog skeletal muscle fibres. J Physiol (Lond) 413:521–541
Ettienne EM (1970) Control of contractility in Spirostomum by dissociated calcium ions. J Gen Physiol 56:168–179
Eusebi F, Miledi R, Takahashi T (1980) Calcium transients in mammalian muscles. Nature (Lond) 284:560–561
Eusebi F, Miledi R, Takahashi T (1983) Aequorin-calcium transients in frog twitch muscle fibres. J Physiol (Lond) 340:91–106
Fabiato A (1981) Myoplasmic free calcium concentration reached during the twitch of an intact isolated cardiac cell and during calcium-induced release of calcium from the sarcoplasmic reticulum of a skinned cardiac cell from the adult rat or rabbit ventricle. J Gen Physiol 78:457–497
Fabiato A (1982a) Fluorescence and differential light absorption recordings with calcium probes and potential-sensitive dyes in skinned cardiac cells. Can J Physiol Pharmacol 60:556–567
Fabiato A (1982b) Calcium release in skinned cardiac cells: variations with species, tissues, and development. Fed Proc 41:2238–2244
Fabiato A (1983) Calcium-induced release of calcium from the cardiac sarcoplasmic reticulum. Am J Physiol 245 (Cell Physiol 14):C1–C14
Fabiato A (1985a) Rapid ionic modifications during the aequorin-detected calcium transient in a skinned canine cardiac Purkinje cell. J Gen Physiol 85:189–246
Fabiato A (1985b) Time and calcium dependence of activation and inactivation of calcium-induced release of calcium from the sarcoplasmic reticulum of a skinned canine cardiac Purkinje cell. J Gen Physiol 85:247–289
Fabiato A (1985c) Stimulated calcium current can both cause calcium loading in and trigger calcium release from the sarcoplasmic reticulum of a skinned canine cardiac Purkinje cell. J Gen Physiol 85:291–320
Fabiato A, Fabiato F (1979) Calculator programs for computing the composition of the solutions containing multiple metals and ligands used for experiments in skinned muscle cells. J Physiol (Paris) 75:463–505
Fatt P, Ginsborg BL (1958) The ionic requirements for the production of action potentials in crustacean muscle fibres. J Physiol (Lond) 142:516–543
Fatt P, Katz B (1953) The electrical properties of crustacean muscle fibres. J Physiol (Lond) 120:171–204
Fawcett DW, McNutt NS (1969) The ultrastructure of the cat myocardium. I. Ventricular papillary muscle. J Cell Biol 42:1–45
Fawcett DW, Revel JP (1961) The sarcoplasmic reticulum of a fast-acting fish muscle. J Biophys Biochem Cytol 10:89–109
Fay FS, Shlevin HH, Granger VC, Taylor SR (1979) Aequorin luminescence during activation of single smooth muscle cells. Nature (Lond) 280:506–508
Fay FS, Fogarty K, Fujiwara K, Tuft R (1982) Contractile mechanism of single isolated smooth muscle cells: In: Twarog BM, Levine RJC, Dewey MM (eds) Basic biology of muscles: a comparative approach. Raven Press, New York, pp 143–157
Feldman MD, Copelas L, Gwathmey JK, Phillips P, Warren SE, Schoen FJ, Grossman W, Morgan JP (1987) Deficient production of cyclic AMP: pharmacologic evidence of an important cause of contractile dysfunction in patients with end-stage heart failure. Circulation 1987:331–339
Feldmeyer D, Lüttgau HC (1988) The effect of perchlorate on Ca currents and mechanical force in skeletal muscle fibres. Pflügers Arch Eur J Physiol 411:R190
Feldmeyer D, Melzer W, Pohl B (1990a) Effects of gallopamil on calcium release and intramembrane charge movements in frog skeletal muscle fibres. J Physiol (Lond) 421:343–362
Feldmeyer D, Melzer W, Pohl B, Zöllner P (1990) Fast gating kinetics of the slow Ca^{2+} current in cut skeletal muscle fibres of the frog. J Physiol (Lond) 425:347–367
Fenn WO (1924) The relation between the work performed and the energy liberated in muscular contraction. J Physiol (Lond) 58:373–395
Ferenczi MA, Simmons RM, Sleep JA (1982) General considerations of cross bridge models in relation to the dependence on MgATP concentration of mechanical parameters of skinned fibers from frog muscle. In: Twarog BM, Levine RJC, Dewey MM (eds) Basic biology of muscles: a comparative approach. Raven Press, New York, pp 91–107

Ferenczi MA, Goldman YE, Simmons RM (1984) The dependence of force and shortening velocity on substrate concentration in skinned muscle fibres from *Rana temporaria*. J Physiol 350:519–543
Ferguson DG, Franzini-Armstrong C (1988) The Ca^{2+}-ATPase-content of slow and fast twitch fibers of guinea pig. Muscle Nerve 11:561–570
Ferguson DG, Schwartz HW, Franzini-Armstrong C (1984) Subunit structure of junctional feet in triads of skeletal muscle: a freeze-drying, rotary-shadowing study. J Cell Biol 99:1735–1742
Fiehn W, Peter JB (1971) Properties of the fragmented sarcoplasmic reticulum from fast-twitch and slow-twitch muscles. J Clin Invest 50:570–573
Fill M, Coronado R, Mickelson JR, Vilven J, Ma J, Jacobson BA, Louis CF (1990) Abnormal ryanodine receptor channels in malignant hyperthermia. Biophys J 57:471–475
Filo RS, Rüegg JC, Bohr DF (1963) Actomyosin like protein of arterial wall. Am J Physiol 205:1247–1252
Filo RS, Bohr DF, Rüegg JC (1965) Glycerinated skeletal and smooth muscle. Calcium and magnesium dependence. Science 147:1581–1583
Fink RHA (1988) Calcium movements and counter-ion conductance in the sarcoplasmic reticulum of cardiac and skeletal muscle. Proc Aust Physiol Pharmacol Soc 19:100–106
Fink RHA, Hillman GM (1989) Increased calcium release from the sarcoplasmic reticulum of normal and dystrophic isolated murine skeletal muscle in the presence of 4-aminopyridine. J Physiol (Lond) 418:156P
Fink RHA, Stephenson DG (1987) Ca^{2+}-movements in muscle modulated by the state of K^+-channels in the sarcoplasmic reticulum membranes. Pflügers Arch Eur J Physiol 409:374–380
Fink RHA, Stephenson DG, Williams DA (1990) Physiological properties of skinned fibres from normal and dystrophic (Duchenne) human muscle activated by Ca^{2+} and Sr^{2+}. J Physiol (Lond) 420:337–353
Finol HJ, Lewis DM, Owens R (1981) The effects of denervation on contractile properties in rat skeletal muscle. J Physiol (Lond) 319:81–92
Fischer W, Pfitzer G (1989) Rapid myosin phosphorylation transients in phasic contractions in chicken gizzard smooth muscle. FEBS Lett 258(1):59–62
Fisher EH, Becker JU, Blum HE, Byers B, Heizmann C, Kerrick GW, Lehky P, Malencik DA, Pocinwong S (1976) Concerted regulation of glycogen metabolism and muscle contraction. In: Heilmeyer LMG Jr, Rüegg JC, Wieland TH (eds) Molecular basis of motility. Springer, Berlin Heidelberg New York, pp 137–158
Fleckenstein A (1983) Calcium antagonism in heart and smooth muscle. Wiley, New York
Fleischer S, Ogunbunmi EM, Dixon MC, Fleer EA (1985) Localization of Ca^{2+} release channels with ryanodine in junctional terminal cisternae of SR of fast skeletal muscle. Proc Natl Acad Sci USA 82:7256–7259
Flitney FW (1971) The volume of the T-system and its association with the sarcoplasmic reticulum in slow muscle fibres of the frog. J Physiol (Lond) 217:243–257
Flitney FW, Johnston IA (1979) Mechanical properties of isolated fish red and white muscle fibres. J Physiol (Lond) 295:49P–50P
Flockerzi V, Oeken HJ, Hofmann F, Pelzer D, Cavalié A, Trautwein W (1986) Purified dihydropyridine-binding site from skeletal muscle T-tubules is a functional calcium channel. Nature (Lond) 323:66–68
Floyd K, Smith ICH (1971) The mechanical and thermal properties of frog slow muscle fibres. J Physiol (Lond) 213:617–631
Ford LE, Podolsky RJ (1972a) Calcium uptake and force development by skinned muscle fibres in EGTA buffered solutions. J Physiol (Lond) 223:1–19
Ford LE, Podolsky RJ (1972b) Intracellular calcium movements in skinned muscle fibres. J Physiol (Lond) 223:21–33
Ford LE, Huxley AF, Simmons RM (1977) Tension responses to sudden length change in stimulated frog muscle fibres near slack length. J Physiol (Lond) 269:441–515
Ford LE, Huxley AF, Simmons RM (1986) Tension transients during the rise of tetanic tension in frog muscle fibres. J Physiol (Lond) 372:595–609
Forssmann WG, Girardier L (1970) A study of the T-system in rat heart. J Cell Biol 42:1–19
Fosset M, Jaimovich E, Delpont E, Lazdunski M (1983) [^{3}H]nitrendipine receptors in skeletal muscle. J Biol Chem 258:6086–6092
Frado LL, Craig R (1992) Electron microscopy of the actin-myosin head complex in the presence of ATP. J Mol Biol 223:391–397

Franco A Jr, Lansman JB (1990) Calcium entry through stretch-inactivated ion channels in mdx myotubes. Nature (Lond) 344:670–673

Frank GB (1982) Roles of extracellular and "trigger" calcium ions in excitation-contraction coupling in skeletal muscle. Can J Physiol Pharmacol 60:427–439

Franzini-Armstrong C (1970) Studies of the triad. I. Structure of the junction in frog twitch fibers. J Cell Biol 47:488–499

Franzini-Armstrong C (1975) Membrane particles and transmission at the triad. Fed Proc 34:1382–1389

Franzini-Armstrong C, Nunzi G (1983) Junctional feet and particles in the triads of a fast twitch muscle fibre. J Muscle Res Cell Mot 4:233–252

Franzini-Armstrong C, Peachey LD (1981) Striated muscle – contractile and control mechanisms. J Cell Biol 91:166s–186s

Fuchs F (1977) The binding of calcium to glycerinated muscle fibers in rigor. The effect of filament overlap. Biochim Biophys Acta 491:523–531

Fuchs F (1985) The binding of calcium to detergent-extracted rabbit psoas muscle fibres during relaxation and force generation. J Muscle Res Cell Mot 6:477–486

Fuchs F, Fox C (1982) Parallel measurements of bound calcium and force in glycerinated rabbit psoas muscle fibers. Biochim Biophys Acta 679:110–115

Fujii J, Otsu K, Zorzato F, de Leon S, Khanna VK, Weiler JE, O'Brien PJ, MacLennan DH (1991) Identification of a mutation in the porcine ryanodine receptor that is associated with malignant hyperthermia. Science 253:448–451

Fujiwara T, Itoh T, Kubota Y, Kuriyama H (1988) Actions of a phorbol ester on factors regulating contraction in rabbit mesenteric artery. Circ Res 63:893–902

Fujiwara T, Itoh T, Kubota Y, Kuriyama H (1989) Effects of guanosine nucleotides on skinned smooth muscle tissue of the rabbit mesenteric artery. J Physiol (Lond) 408:535–547

Furchgott RF (1983) Role of endothelium in responses of vascular smooth muscle. Circ Res 53:557–573

Gabella G (1984) Structural apparatus for force transmission in smooth muscle. Physiol Rev 64:455–477

Galvas DE, DiSalvo J (1983) Concentration and time-dependent relationship between isosorbide dinitrate-induced relaxation and formation of cyclic GMP in coronary arterial smooth muscle. J Pharmacol Exp Ther 224:373–378

Garcia MC, Gonzalez-Serratos H, Morgan JP, Perreault CL, Rozycka M (1991) Differential activation of myofibrils during fatigue in phasic skeletal muscle cells. J Muscle Res Cell Mot 12:412–424

Geeves MA, Goody RS, Gutfreund H (1984) Kinetics of acto-S1 interaction as a guide to a model for the crossbridge cycle. J Muscle Res Cell Mot 5:351–361

Gerday C, Gillis JM (1976) The possible role of parvalbumins in the control of contraction. J Physiol (Lond) 258:96P–97P

Gerday C, Collin S, Gerardin-Otthiers N (1981) The soluble calcium-binding protein from muscle of the sandworm, *Nereis virens*. J Muscle Res Cell Mot 2:225–238

Gerthoffer WT, Trevethick MA, Murphy RA (1984) Myosin phosphorylation and cyclic adenosine 3′,5′-monophosphate in relaxation of arterial smooth muscle by vasodilators. Circ Res 54:83–89

Gerzer R, Hofmann F, Schultz G (1981) Purification of a soluble, sodium-nitroprusside-stimulated guanylate cyclase from bovine lung. Eur J Biochem 116:479–486

Gesser H, Mangor-Jensen A (1984) Contractility and ^{45}Ca fluxes in heart muscle of flounder at a lowered extracellular NaCl concentration. J Exp Biol 109:201–207

Gillard EF, Otsu K, Fujii J, Khanna VK, de Leon S, Derdemezi J, Britt BA, Duff CL, Worton RG, MacLennan DH (1991) A substitution of cysteine for arginine 614 in the ryanodine receptor is potentially causative of human malignant hyperthermia. Genomics 11:751–755

Gillis JM (1980) The biological significance of muscle parvalbumins. In: Siegel FL, Carafoli E, Kretsinger RH, MacLennan DH, Wasserman RH (eds) Calcium binding proteins: structure and function. Elsevier/North-Holland, Amsterdam, pp 309–311

Gillis JM (1985) Relaxation of vertebrate skeletal muscle. A synthesis of the biochemical and physiological approaches. Biochim Biophys Acta 811:97–145

Gillis JM, Thomason D, Lefèvre J, Kretsinger RH (1982) Parvalbumins and muscle relaxation: a computer simulation study. J Muscle Res Cell Mot 3:377–398

Gilly WMF, Hui CS (1980a) Mechanical activation in slow and twitch skeletal muscle fibres of the frog. J Physiol (Lond) 301:137–156
Gilly WMF, Hui CS (1980b) Voltage-dependent charge movement in frog slow muscle fibres. J Physiol (Lond) 301:175–190
Gilly WMF, Scheuer T (1984) Contractile activation in scorpion striated muscle fibers: dependence on voltage and external calcium. J Gen Physiol 84:321–345
Ginsborg BL (1960) Some properties of avian skeletal muscle fibres with multiple neuromuscular junctions. J Physiol (Lond) 154:581–598
Godt RE, Lindley BD (1982) Influence of temperature upon contractile activation and isometric force production in mechanically skinned muscle fibres of the frog. J Gen Physiol 80:279–297
Goldman YE, Hibberd MG, McCray JA, Trentham DR (1982) Relaxation of muscle fibres by photolysis of caged ATP. Nature (Lond) 300:701–705
Gomolla M, Gottschalk G, Lüttgau HC (1983) Perchlorate-induced alterations in electrical and mechanical parameters of frog skeletal muscle fibres. J Physiol (Lond) 343:197–214
González-Serratos H (1971) Inward spread of activation in vertebrate muscle fibres. J Physiol (Lond) 212:777–799
Goodman M (1980) Molecular evolution of the calmodulin family. In: Siegel FL, Carafoli E, Kretsinger RH, MacLennan DH, Wasserman RH (eds) Calcium binding proteins: structure and function. Elsevier/North-Holland, Amsterdam, pp 347–354
Goody RS, Holmes KC (1983) Cross-bridges and the mechanism of muscle contraction. Biochim Biophys Acta 726:13–39
Gordon AM, Huxley AF, Julian FJ (1966) The variation in isometric tension with sarcomere length in vertebrate muscle fibres. J Physiol (Lond) 184:170–192
Gordon AM, Ridgway EB, Martyn DA (1984) Calcium sensitivity is modified by contraction. In: Pollack GH, Sugi H (eds) Contractile mechanisms in muscle (Adv Exp Med Biol 170). Plenum Press, New York, pp 553–563
Gordon AR (1978) Contraction of detergent-treated smooth muscle. Proc Natl Acad Sci USA 75:3527–3530
Gottschalk G, Lüttgau HC (1986) The effect of D600 and Ca^{2+} deprivation on force kinetics in short toe muscle fibres of the frog. J Physiol (Lond) 371:170P
Grabarek Z, Grabarek J, Leavis PC, Gergely J (1983) Cooperative binding to the Ca^{2+}-specific sites of troponin C in regulated actin and actomyosin. J Biol Chem 258:14098–14102
Grabarek Z, Tan RY, Wang J, Tao T, Gergely J (1990) Inhibition of mutant troponin C activity by an intra-domain disulphide bond. Nature (Lond) 345:132–135
Graf F, Schatzmann HJ (1984) Some effects of removal of external calcium on pig striated muscle. J Physiol (Lond) 349:1–13
Grand RJA, Perry SV (1979) Calmodulin-binding proteins from brain and other tissues. Biochem J 183:285–295
Greaser ML, Gergely J (1971) Reconstitution of troponin activity from three protein components. J Biol Chem 246:4226–4233
Greaser ML, Gergely J (1973) Purification and properties of the components from troponin. J Biol Chem 248:2125–2133
Greaser ML, Moss RL, Reiser PJ (1988) Variations in contractile properties of rabbit single muscle fibres in relation to troponin T isoforms and myosin light chains. J Physiol (Lond) 406:85–98
Greene LE, Eisenberg E (1980) Cooperative binding of myosin subfragment-1 to the actin-troponin-tropomyosin complex. Proc Natl Acad Sci USA 77:2616–2620
Grocki K (1982) The fine structure of the deep muscle lamellae and their sarcoplasmic reticulum in *Branchiostoma lanceolatum*. Eur J Cell Biol 28:202–212
Gröschel-Stewart U (1980) Immunochemistry of cytoplasmic contractile proteins. Int Rev Cytol 65:193–254
Grundfest H (1966) Comparative electrobiology of excitable membranes. Adv Comp Physiol Biochem 2:1–116
Grupp I, Wook-Bin I, Chin OL, Shin-Woong L, Pecker MS, Schwartz A (1985) Relation of sodium pump inhibition to positive inotropy at low concentrations of ouabain in rat heart muscle. J Physiol (Lond) 360:149–160
Güth K, Junge J (1982) Low Ca^{2+} impedes cross-bridge detachment in chemically skinned *Taenia coli*. Nature (Lond) 300:775–776

Güth K, Potter JD (1987) Effect of rigor and cycling crossbridges on the structure of troponin-C and on the calcium affinity of the calcium specific regulatory sites in skinned rabbit psoas fibers. J Biol Chem 262:13627–13635

Güth K, Kuhn HJ, Tsuchiya T, Rüegg JC (1981) Length dependent state of activation-length change dependent kinetics of cross bridges in skinned insect flight muscle. Biophys Struct Mech 7:139–169

Güth K, Gagelmann M, Rüegg JC (1984) Skinned smooth muscle: time course of force and ATPase activity during contraction cycle. Experientia 40:174–176

Gwathmey JK, Copelas L, MacKinnon R, Schoen FJ, Feldman MD, Grossman W, Morgan JP (1987) Abnormal intracellular calcium handling in myocardium from patients with end-stage heart failure. Circ Res 61:70–76

Haeberle JR, Hathaway DR, DePaoli-Roach AA (1985) Dephosphorylation of myosin by the catalytic subunit of a type-2 phosphatase produces relaxation of chemically skinned uterine smooth muscle. J Biol Chem 260:9965–9968

Hagiwara S, Hayashi H, Takahashi K (1969) Calcium and potassium currents of the membrane of a barnacle muscle fibre in relation to the calcium spike. J Physiol (Lond) 205:115–129

Hagiwara S, Henkart MP, Kidokoro Y (1971) Excitation-contraction coupling in amphioxus muscle cells. J Physiol (Lond) 219:233–251

Hai CM, Murphy RA (1988) Cross-bridge phosphorylation and regulation of latch state in smooth muscle. Am J Physiol 254:C99–C106

Hainaut K, Desmedt JE (1974) Effects of dantrolene sodium on calcium movements in single muscle fibres. Nature (Lond) 252:728–730

Hajjar RJ, Gwathmey JK (1990) Direct evidence of changes in myofilament responsiveness to Ca^{2+} during hypoxia and reoxygenation in myocardium. Am J Physiol 259 (Heart Circ Physiol 28):H784–H795

Hamill OP, Marty A, Neher E, Sakmann B, Sigworth FJ (1981) Improved patch-clamp techniques for high-resolution current recording from cells and cell-free membrane patches. Pflügers Arch Eur J Physiol 391:85–100

Hamlyn JM, Blaustein MP, Bova S, DuCharme DW, Harris DW, Mandel F, Mathews WR, Ludens JH (1991) Identification and characterization of a ouabain-like compound from human plasma. Proc Natl Acad Sci USA 88:6259–6263

Hamm HE, Deretic D, Arendt A, Hargrave PA, Koenig B, Hofmann KP (1988) Site of G-protein binding to rhodopsin mapped with synthetic peptides from the α subunit. Science 241:832–835

Hamoir G, Gerardin-Otthiers N, Focant B (1980) Protein differentiation of the superfast swimbladder muscle of the toadfish *Opsanus tau*. J Mol Biol 143:155–160

Hansford RG (1987) Relation between cytosolic free Ca^{2+} concentration and the control of pyruvate dehydrogenase in isolated cardiac myocytes. Biochem J 241:145–151

Harder D, Belardinelli L, Sperelakis N, Rubio R, Berne RM (1979) Differential effects of adenosine and nitroglycerin on the action potentials of large and small coronary arteries. Circ Res 44:176–182

Hardie G (1988) Pseudosubstrates turn off protein kinases. Nature (Lond) 335:592–593

Hardwicke PMD, Szent-Györgyi AG (1985) Proximity of regulatory light chains in scallop myosin. J Mol Biol 183:203–211

Hardwicke PMD, Wallimann T, Szent-Györgyi AG (1983) Light-chain movement and regulation in scallop myosin. Nature (Lond) 301:478–482

Harrison SM, Lamont C, Miller DJ (1985) Carnosine and other natural imidazoles enhance muscle Ca sensitivity and are mimicked by caffeine and AR-L 115 BS. J Physiol (Lond) 371:197P

Hartner KT, Kirschbaum BJ, Pette D (1989) The multiplicity of troponin T isoforms. Distribution in normal rabbit muscles and effects of chronic stimulation. Eur J Biochem 179:31–38

Hartshorne DJ, Gorecka A (1980) Biochemistry of the contractile proteins of smooth muscle. In: Bohr DF, Somlyo AP, Sparks HV (eds) The cardiovascular system. Am Physiol Soc, Bethesda, Maryland, pp 93–120 (Handbook of physiology, sect 2, vol II)

Hartshorne DJ, Mrwa U (1982) Regulation of smooth muscle actomyosin. Blood Vessels 19:1–18

Hartshorne DJ, Theiner M, Mueller H (1969) Studies on troponin. Biochim Biophys Acta 175:320–330

Harvey DJ, Godber JF, Timmerman MP, Castell LM, Ashley CC (1985) Measurement of free Ca^{2+} changes and total Ca^{2+} release in a single striated muscle fibre using the fluorescent indicator quin 2. Biochem Biophys Res Commun 128:1180–1189

Haselgrove JC (1973) X-ray evidence for a conformational change in the actin containing filaments of vertebrate striated muscle. Cold Spring Harbor Symp Quant Biol 37:341–352
Haselgrove JC, Rodger CD (1980) The interpretation of X-ray diffraction patterns from vertebrate striated muscle. J Muscle Res Cell Mot 1:371–390
Hashimoto T, Hirata M, Itoh T, Kanmura Y, Kuriyama H (1986) Inositol 1,4,5-trisphosphate activates pharmacomechanical coupling in smooth muscle of the rabbit mesenteric artery. J Physiol (Lond) 370:605–618
Hasselbach W (1964) Relaxing factor and the relaxation of muscle. Prog Biophys Mol Biol 14:167–222
Hasselbach W, Makinose M (1961) Die Calciumpumpe der „Erschlaffungsgrana" des Muskels und ihre Abhängigkeit von der ATP-Spaltung. Biochem Z 333:518–528
Hasselbach W, Makinose M (1962) ATP and active transport. Biochem Biophys Res Commun 7:132–136
Hasselbach W, Makinose M (1963) Über den Mechanismus des Calciumtransports durch die Membranen des Sarkoplasmatischen Reticulums. Biochem Z 339:99–111
Hasselbach W, Oetliker H (1983) Energetics and electrogenicity of the sarcoplasmic reticulum calcium pump. Annu Rev Physiol 45:325–339
Haynes DH (1983) Mechanism of Ca^{2+} transport by Ca^{2+}-Mg^{2+}-ATPase pump: analysis of major states and pathways. Am J Physiol 244:G3–G12
Heilbrunn LV, Wiercinski FJ (1947) The action of various cations on muscle protoplasm. J Cell Comp Physiol 29:15–32
Heilmann C, Pette D (1979) Molecular transformations in sarcoplasmic reticulum of fast-twitch muscle by electro-stimulation. Eur J Biochem 93:437–446
Heilmann C, Brdiczka D, Nickel E, Pette D (1977) ATP-ase activities, Ca^{2+} transport and phosphoprotein formation in sarcoplasmic reticulum subfractions of fast and slow rabbit muscles. Eur J Biochem 81:211–222
Heinl P, Kuhn HJ, Rüegg JC (1974) Tension responses to quick length changes of glycerinated skeletal muscle fibres from the frog and tortoise. J Physiol (Lond) 237:243–258
Heiny JA, Jong DS (1990) A nonlinear electrostatic potential change in the T-system of skeletal muscle detected under passive recording conditions using potentiometric dyes. J Gen Physiol 95:147–157
Heizmann CW (1984) Parvalbumin, an intracellular calcium-binding protein; distribution, properties and possible roles in mammalian cells. Experientia 40:910–921
Hellam DC, Podolsky RJ (1969) Force measurements in skinned muscle fibres. J Physiol (Lond) 200:807–819
Hernandez-Nicaise M-L, Nicaise G, Malaval L (1984) Giant smooth muscle fibers of the ctenophore *Mnemiopsis leydii*: ultrastructural study of in situ and isolated cells. Biol Bull 167:210–228
Herrmann-Frank A, Meissner G (1989) Isolation of a Ca^{2+}-releasing factor from caffeine-treated skeletal muscle fibres and its effect on Ca^{2+} release from sarcoplasmic reticulum. J Muscle Res Cell Mot 10:427–436
Herrmann-Frank A, Darling E, Meissner G (1991) Functional characterization of the Ca^{2+}-gated Ca^{2+} release channel of vascular smooth muscle sarcoplasmic reticulum. Pflügers Arch Eur J Physiol 418:353–360
Herzberg O, James MNG (1985) Structure of the calcium regulatory muscle protein troponin-C at 2.8 Å resolution. Nature (Lond) 313:653–659
Herzig JW (1978) A cross-bridge model for inotropism as revealed by stiffness measurements in cardiac muscle. Basic Res Cardiol 73:273–286
Herzig JW (1984) Contractile proteins: possible targets for drug action. Trends Pharmacol Sci 5:296–300
Herzig JW, Rüegg JC (1977) Myocardial cross-bridge activity and its regulation by Ca^{2+}, phophate and stretch. In: Riecker G, Weber A, Goodwin J (eds) Myocardial failure. Springer, Berlin Heidelberg New York, pp 41–51
Herzig JW, Rüegg JC (1980) Investigations on glycerinated cardiac muscle fibres in relation to the problem of regulation of cardiac contractility – effects of Ca^{2+} and c-AMP. Basic Res Cardiol 75:26–33
Herzig JW, Feile K, Rüegg JC (1981 a) Activating effects of AR-L 115 BS on the Ca^{2+} sensitive force, stiffness and unloaded shortening velocity (V_{max}) in isolated contractile structures from mammalian heart muscle. Arzneim Forsch/Drug Res 31:188–191

Herzig JW, Köhler G, Pfitzer G, Rüegg JC, Wölffle G (1981 b) Cyclic AMP inhibits contractility of detergent treated glycerol extracted cardiac muscle. Pflügers Arch Eur J Physiol 391:208–212
Herzig JW, Yamamoto T, Rüegg JC (1981 c) Dependence of force and immediate stiffness and sarcomere length and Ca^{2+} activation in frog skinned muscle fibres. Pflügers Arch Eur J Physiol 389:97–103
Hess A (1965) The sarcoplasmic reticulum, the T-system, and the motor terminals of slow and twitch muscle fibers in the garter snake. J Cell Biol 26:467–476
Hess A (1970) Vertebrate slow muscle fibres. Physiol Rev 50:40–62
Hess A, Pilar G (1963) Slow fibres in the extraocular muscles of the cat. J Physiol (Lond) 169:780–798
Hess P, Wier WG (1984) Excitation-contraction coupling in cardiac purkinje fibers. Effects of caffeine on the intracellular Ca^{2+} transient, membrane currents, and contraction. J Gen Physiol 83:417–433
Hess P, Metzger P, Weingart R (1982) Free magnesium in sheep, ferret and frog striated muscle at rest measured with ion-selective micro-electrodes. J Physiol (Lond) 333:173–188
Hess WR (1948) Die funktionelle Organisation des vegetativen Nervensystems. Benno Schwabe, Basel
Heumann HG (1969) Calciumakkumulierende Strukturen in einem glatten Wirbellosenmuskel. Protoplasma 67:111–115
Heumann HG, Zebe E (1967) Über Feinbau und Funktionsweise der Fasern aus dem Hautmuskelschlauch des Regenwurms *Lumbricus terrestris*. Z Zellforsch 78:131–150
Hibberd MG, Jewell Br (1982) Calcium- and length-dependent force production in rat ventricular muscle. J Physiol (Lond) 329:527–540
Hidaka H, Yamak T, Totsuka T, Asano M (1979) Selective inhibition of Ca^{2+}-binding modulator of phosphodiesterase produces vascular relaxation and inhibits actin myosin interaction. Mol Pharmacol 15:49–59
Hidalgo C, Jaimovich E (1989) Inositol trisphosphate and E-C coupling in skeletal muscle. J Bionerg Biomembr 21:267–281
Higgins WJ, Greenberg MJ (1974) Intracellular actions of 5-hydroxytryptamine on the bivalve myocardium. II. Cyclic nucleotide-dependent protein kinases and microsomal calcium uptake. J Exp Zool 190:305–316
Higuchi H, Goldman YE (1991) Sliding distance between actin and myosin filaments per ATP molecule hydrolysed in skinned muscle fibres. Nature (Lond) 352:352–354
Hill AV (1938) The heat of shortening and the dynamic constants of muscle. Proc R Soc (Lond) B 126:136–195
Hill AV (1948) On the time required for diffusion and its relation to processes in muscle. Proc R Soc (Lond) B 135:446–453
Hill AV (1950) The development of the active state of muscle fibres during the latent period. Proc R Soc (Lond) B 137:320–329
Himpens B, Somlyo AP (1988) Free-calcium and force transients during depolarization and pharmacomechanical coupling in guinea-pig smooth muscle. J Physiol (Lond) 395:507–530
Hincke MT, McCubbin WD, Kay CM (1979) The interaction between beef cardiac troponin T and troponin I as demonstrated by ultraviolet absorption difference spectroscopy, circular dichroism, and gel filtration. Can J Biochem 57:768–775
Hinrichsen C, Dulhunty AF (1982) The contractile properties, histochemistry, ultrastructure and electrophysiology of the cricothyroid and posterior cricoarytenoid muscles in the rat. J Muscle Res Cell Mot 3:169–190
Hirata K, Kikuchi A, Sasaki T, Kuroda S, Kaibuchi K, Matsuura Y, Seki H, Saida H, Takai Y (1992) Involvement of *rho* p21 in the GTP-enhanced calcium ion sensitivity of smooth muscle contraction. J Biol Chem 267:8719–8722
Hirata M, Kuriyama H (1980) Does activation of cyclic AMP dependent phosphorylation induced by beta-adrenergic agent control the tone of vascular muscle? J Physiol (Lond) 307:143–161
Hirst GDS (1977) Neuromuscular transmission in arterioles of guinea-pig submucosa. J Physiol (Lond) 273:263–275
Hirst GDS, Neild TO (1980 a) Evidence for two populations of excitatory receptors for noradrenaline on arteriolar smooth muscle. Nature (Lond) 283:767–768
Hirst GDS, Neild TO (1980 b) Some properties of spontaneous excitatory junction potentials recorded from arterioles of guinea-pigs. J Physiol (Lond) 303:43–60

Hitchcock SE (1975) Regulation of muscle contraction: Binding of troponin and its components to actin and tropomyosin. Eur J Biochem 52:255–263
Hitchcock-De Gregori SE (1989) Structure-function analysis of thin filament proteins expressed in *Escherichia coli*. Cell Motil Cytoskel 14:12–20
Hitchcock-De Gregori SE, Varnell TA (1990) Tropomyosin has discrete actin-binding sites with sevenfold and fourteenfold periodicities. J Mol Biol 214:885–896
Hochachka PW, Somero GN (1973) Strategies of biochemical adaptation. Saunders, Philadelphia
Hodgkin AL (1964) The conduction of the nervous impulse. The Sherrington Lectures VII. University Press, Liverpool
Hodgkin AL, Horowicz P (1960) Potassium contractures in single muscle fibres. J Physiol (Lond) 153:386–403
Hodgkin AL, Nakajima S (1972) Analysis of the membrane capacity in frog muscle. J Physiol (Lond) 221:121–136
Hoffman EP, Kunkel LM (1989) Dystrophin abnormalities in Duchenne/Becker muscular dystrophy. Neuron 2:1019–1029
Hofmann PA, Metzger JM, Greaser ML, Moss RL (1990) Effects of partial extraction of light chain 2 on the Ca^{2+}-sensitivities of isometric tension, stiffness, and velocity of shortening in skinned skeletal muscle fibers. J Gen Physiol 95:477–498
Hoh JFY, Yeoh GPS (1979) Rabbit skeletal myosin isoenzymes from fetal, fast-twitch and slow-twitch muscles. Nature (Lond) 280:321–323
Hoh JFY, McGrath PA, White RI (1976) Electrophoretic analysis of multiple forms of myosin in fast-twitch and slow-twitch muscles of the chick. Biochem J 157:87–95
Holmes KC, Kabsch W (1991) Muscle proteins: actin. Curr Opinion Struct Biol 1:270–280
Holmes KC, Goody RS, Mannherz HG, Barrington Leigh J, Rosenbaum G (1976) An investigation of the cross-bridge cycle using ATP analogues and low-angle X-ray diffraction from glycerinated fibres of insect flight muscle. In: Heilmeyer LMG Jr, Rüegg JC, Wieland Th (eds) Molecular basis of motility. Springer, Berlin Heidelberg New York, pp 26–41
Holroyde MJ, Potter JD, Solaro RJ (1979) The calcium binding properties of phosphorylated and unphosphorylated cardiac and skeletal myosins. J Biol Chem 254:6478–6482
Holroyde MJ, Robertson SP, Johnson JD, Solaro RJ, Potter JD (1980) The calcium and magnesium binding sites on cardiac troponin and their role in the regulation of myofibrillar adenosine triphosphatase. J Biol Chem 255:11688–11693
Holzmann S (1982) Endothelium-induced relaxation is induced with larger rises in cyclic GMP in coronary arterial strips. J Cycl Nucl Res 5:211–224
Hondeghem LM, Katzung BG (1986) Control of vascular smooth muscle contractility and the action of calcium channel blockers. In: Rupp H (ed) Regulation of heart function. Basic concepts and clinical applications. Thieme, Stuttgart, pp 38–52
Hoyle G (1957) Comparative physiology of the nervous control of muscular contraction. Univ Press, Cambridge
Hoyle G (1968) Correlated physiological and ultrastructural studies on specialized muscles. Ia. Neuromuscular physiology of the levator of the eyestalk of *Podophthalmus vigil* (Weber). J Exp Zool 167:471–486
Hoyle G (1978) Distributions of nerve and muscle fibre types in locust jumping muscle. J Exp Biol 73:205–233
Hoyle G (1983) Muscles and their neural control. Wiley, New York
Hoyle G, McNeill PA, Walcott B (1966) Nature of invaginating tubules in Felderstruktur muscle fibers of the garter snake. J Cell Biol 30:197–201
Huber F, Kleindienst HU, Moore Te, Schildberger K, Weber T (1990) Acoustic communication in periodical cicadas: neuronal responses to songs of sympatric species. In: Gribakin FG, Wiese K, Popov AV (eds) Advances in life sciences. Birkhäuser, Basel, pp 217–228
Hui CS, Milton RL (1987) Suppression of charge movement in frog skeletal muscle by D600. J Muscle Res Cell Mot 3:195–208
Huxley AF (1957) Muscle structure and theories of contraction. Prog Biophys Chem 7:255–318
Huxley AF (1974) Muscular contraction. J Physiol (Lond) 243:1–43
Huxley AF (1980) Reflections on muscle. The Sherrington Lectures XIV, University Press, Liverpool
Huxley AF (1983) The Fenn effect and theories of muscle contraction. Proc Int Union Physiol Sci 15:270

Huxley AF, Niedergerke R (1954) Interference microscopy of living muscle fibres. Nature (Lond) 173:971–973

Huxley AF, Simmons RM (1973) Mechanical transients and the origin of muscular force. Cold Spring Harbor Symp Quant Biol 37:669–680

Huxley AF, Straub RW (1958) Local activation and interfibrillar structures in striated muscle. J Physiol (Lond) 143:40–41P

Huxley AF, Taylor RE (1955) Function of Krause's membrane. Nature (Lond) 176:1068

Huxley AF, Taylor RE (1958) Local activation of striated muscle fibres. J Physiol (Lond) 144:426–441

Huxley HE (1957) The double array of filaments in cross-striated muscle. J Biophys Biochem Cytol 3:631–648

Huxley HE (1963) Electron microscope studies on the structure of natural and synthetic protein filaments from striated muscle. J Mol Biol 7:281–308

Huxley HE (1964) Evidence for continuity between the central elements of the triads and extracellular space in frog sartorius muscle. Nature (Lond) 202:1067–1071

Huxley HE (1969) The mechanism of muscular contraction. Science 164:1356–1366

Huxley HE (1973) Structural changes in the actin and myosin containing filaments during contraction. Cold Spring Harbor Symp Quant Biol 37:361–376

Huxley HE (1979) Time resolved X-ray diffraction studies on muscle. In: Sugi H, Pollack GH (eds) Cross-bridge mechanism in muscle contraction. University Park Press, Baltimore, pp 391–405

Huxley HE, Hanson J (1954) Changes in the cross-striations of muscle during contraction and stretch and their structural interpretation. Nature (Lond) 173:973–976

Huxley HE, Kress M (1985) Crossbridge behaviour during muscle contraction. J Muscle Res Cell Mot 6:153–161

Huxley HH (1990) Sliding filaments and molecular motile systems. J Biol Chem 265:8347–8350

Hymel L, Inui M, Fleischer S, Schindler H (1988) Purified ryanodine receptor of skeletal muscle sarcoplasmic reticulum forms Ca^{2+}-activated oligomeric Ca^{2+}-channels in planar bilayers. Proc Natl Acad Sci USA 85:441–445

Ibraghimov-Beskrovnaya O, Ervasti JM, Leveille CJ, Slaughter CA, Sernett SW, Campbell KP (1992) Primary structure of dystrophin-associated glycoproteins linking dystrophin to the extra-cellular matrix. Nature (Lond) 355:696–702

Ikebe M, Hinkins S, Hartshorne DJ (1983) Correlation of enzymatic properties and conformation of smooth muscle myosin. Biochemistry 22:4580–4587

Ikebe M, Slepinska M, Kemp BE, Means AR, Hartshorne DJ (1987) Proteolysis of smooth muscle myosin light chain kinase. J Biol Chem 260:13828–13834

Imagawa T, Smith JS, Coronado R, Campbell KP (1987) Purified ryanodine receptor from skeletal muscle sarcoplasmic reticulum is the Ca^{2+}-permeable pore of the calcium release channel. J Biol Chem 262:16636–16643

Inesi G (1985) Mechanism of calcium transport. Annu Rev Physiol 47:573–601

Inesi G, Sumbilla C, Kirtley ME (1990) Relationships of molecular structure and function in Ca^{2+}-transport ATPase. Physiol Rev 70:749–760

Inui M, Saito A, Fleischer S (1987) Purification of the ryanodine receptor and identity with feet structures of junctional terminal cisternae of sarcoplasmic reticulum from fast skeletal muscle. J Biol Chem 262:1740–1747

Irving M, Lombardi V, Piazzesi G, Ferenczi MA (1992) Myosin head movements are synchronous with the elementary force-generating process in muscle. Nature (Lond) 357:156–158

Isenberg G (1977) Cardiac Purkinje fibers. Resting, action, and pacemaker potential under the influence of $[Ca^{2+}]_i$, as modified by intracellular injection techniques. Pflügers Arch Eur J Physiol 371:51–59

Isenberg G (1982) Ca entry and contraction as studied in isolated bovine ventricular myocytes. Z Naturforsch 37:502–512

Isenberg G (1984) Contractility of isolated bovine ventricular myocytes is enhanced by intracellular injection of cardioactive glycosides. Evidence for an intracellular mode of action. In: Erdmann I (ed) Cardiac gylcoside receptors and positive inotropy. Steinkopff, Darmstadt, pp 56–71

Isenberg G, Beresewicz A, Mascher D, Valenzuela F (1985) The two components in the shortening of the unloaded ventricular myocytes: their voltage dependence. Basic Res Cardiol 80:117–122

Ishii N, Simpson AWM, Ashley CC (1989a) Free calcium at rest during "catch" in single smooth muscle cells. Science 243:1367–1368

Ishii N, Simpson AWM, Ashley CC (1989b) Effects of 5-hydroxytryptamine (serotonin) and forskolin on intra-cellular free calcium in isolated fura-2 loaded smooth-muscle cells from the anterior byssus retractor (catch) muscle of *Mytilus edulis*. Pflügers Arch Eur J Physiol 414:162–170

Ishii N, Mitsumori F, Takahashi K (1991) Changes in sarcoplasmic metabolite concentrations and pH associated with the catch contraction and relaxation of the anterior byssus retractor muscle of *Mytilus edulis* measured by phosphorus-31 nuclear magnetic resonance. J Muscle Res Cell Mot 12:242–246

Ishijima A, Doi T, Sakurada K, Yanagida T (1991) Sub-piconewton force fluctuations of actomyosin in vitro. Nature (Lond) 352:301–306

Ito M, Guerriero V, Chen X, Hartshorne DJ (1991) Definition of the inhibitory domain of smooth muscle myosin light chain kinase by site-directed mutagenesis. Biochemistry 30:3498–3503

Ito Y, Kitamura K, Kuriyama H (1979) Effects of acetylcholine and catecholamines on the smooth muscle cell of the porcine coronary artery. J Physiol (Lond) 294:595–611

Itoh T, Kajiwara M, Kitamura K, Kuriyama H (1982a) Roles of stored calcium on the mechanical response evoked in smooth muscle cells of the porcine coronary artery. J Physiol (Lond) 322:107–125

Itoh T, Izumi H, Kuriyama H (1982b) Mechanisms of relaxation induced by activation of β-adrenoceptors in smooth muscle cells of the guinea-pig mesenteric artery. J Physiol (Lond) 326:475–493

Itoh T, Kuriyama H, Ueno H (1983) Mechanisms of the nitroglycerine-induced vasodilation in vascular smooth muscles of the rabbit and pig. J Physiol (Lond) 343:233–252

Itoh T, Ikebe M, Kargacin GJ, Hartshorne DJ, Kemp BE, Fay FS (1989) Effects of modulators of myosin light-chain kinase activity in single smooth muscle cells. Nature (Lond) 338:164–167

Itoh Y, Kimura S, Suzuki T, Ohashi K, Maruyama K (1986) Native connectin from porcine cardiac muscle. J Biochem 100:439–447

Jackson DC, Heisler N (1982) Plasma ion balance of submerged anoxic turtles at 3° C: the role of calcium lactate formation. Resp Physiol 49:159–174

Jahromi SS, Atwood HL (1969) Correlation of structure, speed of contraction, and total tension in fast and slow abdominal muscle fibers of the lobster (*Homarus americanus*). J Exp Zool 171:25–37

Jahromi SS, Atwood HL (1971) Structural and contractile properties of lobster leg-muscle fibers. J Exp Zool 176:475–486

Jewell BR (1959) The nature of the phasic and the tonic responses of the anterior byssal retractor muscle of *Mytilus*. J Physiol (Lond) 149:154–177

Jewell BR, Rüegg JC (1966) Oscillatory contraction of insect fibrillar muscle after glycerol extraction. Proc R Soc (Lond) B 165:428–459

Jewell BR, Wilkie DR (1960) The mechanical properties of relaxing muscle. J Physiol (Lond) 152:30–47

Jöbsis FF, O'Connor MJ (1966) Calcium release and reabsorption in the sartorius muscle of the toad. Biochem Biophys Res Commun 25:246–252

Johansson B, Somlyo AP (1980) Electrophysiology and excitation-contraction coupling. In: Bohr DF, Somlyo AP, Sparks HV (eds) The cardiovascular system. Am Physiol Soc, Bethesda, Maryland, pp 301–323 (Handbook of physiology, sect 2, vol II)

Johnson JD, Collins JH, Potter JD (1978) Dansylaziridine-labeled troponin C. A fluorescent probe of Ca^{2+} binding to the Ca^{2+}-specific regulatory sites. J Biol Chem 253:6451–6458

Johnson JD, Charlton SC, Potter JD (1979) A fluorescence stopped-flow analysis of Ca^{2+} exchange with troponin-C. J Biol Chem 254:3497–3502

Johnson JD, Robinson DE, Robertson SP, Schwartz A, Potter JD (1981) Ca^{2+} exchange with troponin and the regulation of muscle contraction. In: Grinnell AD, Brazien MAB (eds) The regulation of muscle contraction: excitation-contraction coupling. Academic Press, New York, pp 241–259

Johnston IA (1982) Biochemistry of myosins and contractile properties of fish skeletal muscle. Mol Physiol 2:15–29

Johnston IA, Brill R (1984) Thermal dependence of contractile properties of single skinned muscle fibres from Antarctic and various warm water marine fishes including skipjack tuna (*Katsuwonus pelamis*) and Kawakawa (*Euthynnus affinis*). J Comp Physiol 155:63–70

Josephson RK (1975) Extensive and intensive factors determining the performance of striated muscle. J Exp Zool 194:135–154

Josephson RK, Young D (1985) A synchronous insect muscle with an operating frequency greater than 500 Hz. J Exp Biol 118:185–208
Julian FJ, Morgan DL (1979) Comparison of tension transients in frog twitch and slow fibres. J Physiol (Lond) 301:72P–73P
Julian FJ, Moss RL (1981) Effects of calcium and ionic strength on shortening velocity and tension development in frog skinned muscle fibres. J Physiol (Lond) 311:179–199
Kabsch W (1991) Structure of actin. In: Rüegg JC (ed) Peptides as probes in muscle research. Springer, Berlin Heidelberg New York, pp 7–14
Kabsch W, Mannherz HG, Suck D, Pai EF, Holmes KC (1990) Atomic structure of actin: DNAse 1 complex. Nature (Lond) 347:37–44
Kameyama M, Hescheler J, Hofmann F, Trautwein W (1986) Modulation of Ca current during the phosphorylation cycle in the guinea pig heart. Pflügers Arch Eur J Physiol 407:123–128
Kaminski EA, Chacko S (1984) Effects of Ca^{2+} and Mg^{2+} on the actin-activated ATP hydrolysis by phosphorylated heavy meromyosin from arterial smooth muscle. J Biol Chem 259:9104–9108
Kamm KE, Stull JT (1985a) The function of myosin and myosin light chain kinase phosphorylation in smooth muscle. Annu Rev Pharmacol Toxicol 25:593–620
Kamm KE, Stull JT (1985b) Myosin phosphorylation, force, and maximal shortening velocity in neurally stimulated tracheal smooth muscle. Am J Physiol (Cell Physiol 18) 249:C238–C247
Kamm KE, Stull JT (1986) Activation of smooth muscle contraction: relation between myosin phosphorylation and stiffness. Science 232:80–82
Katz AM (1979) Role of the contractile proteins and sarcoplasmic reticulum in the response of the heart to catecholamines: a historical review. Adv Cyclic Nucleotide Res II:303–343
Kawai M, Schachat FH (1984) Differences in the transient response of fast and slow skeletal muscle fibers. Correlations between complex modulus and myosin light chains. Biophys J 45:1145–1151
Kawai M, Cox RN, Brandt PW (1981) Effect of Ca ion concentration on cross-bridge kinetics in rabbit psoas fibers: evidence for the presence of two Ca-activated states of thin filament. Biophys J 35:375–384
Keane AM, Trayer IP, Levine BA, Zeugner C, Rüegg JC (1990) Peptide mimetics of an actin binding site on myosin span two functional domains on actin. Nature (Lond) 344:265–268
Keith RA, Burkman AM, Sokoloski TD, Fertel RH (1982) Vascular tolerance to nitroglycerine and cyclic GMP generation in rat aortic smooth muscle. J Pharmacol Exp Ther 221:525–531
Kelly RA, Eid H, Krämer BK, O'Neill M, Liang BT, Reers M, Smith TW (1990) Endothelin enhances the contractile responsiveness of adult rat ventricular myocytes to calcium by a pertussis toxin-sensitive pathway. J Clin Invest 86:1164–1171
Kemp BE, Pearson RB, Guerriero V, Bagchi IC, Means AR (1987) The calmodulin binding domain of chicken smooth muscle myosin light chain kinase contains a pseudosubstrate sequence. J Biol Chem 262:2542–2548
Kendrick-Jones J, Scholey JM (1981) Myosin-linked regulatory systems. J Muscle Res Cell Mot 2:347–372
Kendrick-Jones J, Cohen C, Szent-Györgyi AG, Longley W (1969) Paramyosin: molecular length and assembly. Science 163:1196–1198
Kendrick-Jones J, Lehman W, Szent-Györgyi AG (1970) Regulation in molluscan mucles. J Mol Biol 54:313–326
Kendrick-Jones J, Cande WZ, Tooth PJ, Smith RC, Scholey JM (1983) Studies on the effect of phosphorylation of the 20000 M_r light chain of vertebrate smooth muscle myosin. J Mol Biol 165:139–162
Kentish JC (1986) The effects of inorganic phosphate and creatine phosphate on force production in skinned muscles from rat ventricle. J Physiol (Lond) 370:585–604
Kerrick WGL, Hoar PE (1981) Inhibition of smooth muscle tension by cyclic AMP-dependent protein kinase. Nature (Lond) 292:253–255
Kerrick WGL, Hoar PE, Cassidy PS (1980) Calcium activated tension: the role of myosin light chain phosphorylation. Fed Proc 39:1558–1563
Kerrick WGL, Potter JD, Hoar PE (1991) The apparent rate constant for the dissociation of force generating myosin crossbridges from actin decreases during Ca^{2+}-activation of skinned muscle fibres. J Muscle Res Cell Mot 12:53–60
Kilarski W (1967) The fine structure of striated muscles in teleosts. Z Zellforsch 79:562–580

Kim DH, Ikemoto N (1983) Kinetic resolution of the components involved in Ca release from sarcoplasmic reticulum. Biophys J 41:232a

Kim HW, Steenaart NAE, Ferguson DG, Kranias EG (1990) Functional reconstitution of the cardiac sarcoplasmic reticulum Ca^{2+}-ATPase with phospholamban in phospholipid vesicles. J Biol Chem 265:1702–1709

Kirchberger MA, Tada M (1976) Effects of adenosine 3′:5′-monophosphate-dependent protein kinase on sarcoplasmic reticulum isolated from cardiac and slow and fast contracting skeletal muscles. J Biol Chem 251:725–729

Kirchberger MA, Tada M, Katz AM (1974) Adenosine 3′:5′-monophosphate-dependent protein kinase-catalyzed phosphorylation reaction and its relationship to calcium (transport) in cardiac sarcoplasmic reticulum. J Biol Chem 249:6166–6173

Kitazawa T (1984) Effect of extracellular calcium on contractile activation in guinea-pig ventricular muscle. J Physiol (Lond) 355:635–659

Kitazawa T, Kobayashi S, Horiuti K, Somlyo AV, Somlyo AP (1989) Receptor-coupled, permeabilized smooth muscle. J Biol Chem 264:5339–5342

Kitazawa T, Gaylinn BD, Denney GH, Somlyo AP (1991 a) G-protein-mediated Ca^{2+}-sensitization of smooth muscle contraction through myosin light chain phosphorylation. J Biol Chem 266:1708–1715

Kitazawa T, Masuo M, Somlyo AP (1991 b) G protein-mediated inhibition of myosin light-chain phosphatase in vascular smooth muscle. Proc Natl Acad Sci USA 88:9307–9310

Klee CB, Crouch TH, Richman PG (1980) Calmodulin. Annu Rev Biochem 49:489–515

Klöckner U, Isenberg G (1985) Calcium currents of cesium loaded isolated smooth muscle cells (urinary bladder of the guinea pig). Pflügers Arch Eur J Physiol 405:340–348

Knapp MF, Mill PJ (1971) The contractile mechanism in obliquely striated body wall muscle of the earthworm, *Lumbricus terrestris*. J Cell Sci 8:413–425

Koch-Weser J, Blinks JR (1962) Analysis of the relation of the positive inotropic action of cardiac glycosides to the frequency of contraction of heart muscle. J Pharmacol Exp Ther 136:305–317

Koch-Weser J, Blinks JR (1963) The influence of the interval between beats on myocardial contractility. Pharmacol Rev 15:601–652

Kögler H, Moir AJG, Trayer IP, Rüegg JC (1991) Peptide competition of actin activation of myosin-subfragment 1 ATPase by an amino terminal actin fragment. FEBS Lett 294:31–34

Kometani T, Kasai M (1978) Ionic permeability of sarcoplasmic reticulum vesicles measured by light scattering method. J Membr Biol 41:295–308

Kometani K, Sugi H (1978) Calcium transients in a molluscan smooth muscle. Experientia 34:1469–1470

Kondo N, Shibata S (1984) Calcium source for excitation-contraction coupling in myocardium of non-hibernating and hibernating chipmunks. Science 225:641–643

Konishi M, Kurihara S, Sakai T (1984) The effects of caffeine on tension development and intracellular calcium transients in rat ventricular muscle. J Physiol (Lond) 355:605–618

Kovács L, Szücs G (1983) Effect of caffeine on intramembrane charge movement and calcium transients in cut skeletal muscle fibres of the frog. J Physiol (Lond) 341:559–578

Kovács L, Rîos E, Schneider MF (1979) Calcium transients and intramembrane charge movement in skeletal muscle. Nature (Lond) 279:391–396

Kovács L, Schümperli RA, Szücs G (1983) Comparison of birefringence signals and calcium transients in voltage-clamped cut skeletal muscle fibres of the frog. J Physiol (Lond) 341:579–593

Krämer BK, Smith TW, Kelly RA (1991) Endothelin and increased contractility in adult rat ventricular myocytes – role of intracellular alkalosis induced by activation of the protein kinase C-dependent Na^+-H^+ exchanger. Circ Res 68:269–279

Kramer GL, Hardman JG (1980) Cyclic nucleotides and blood vessel contraction. In: Bohr DF, Somlyo AP, Sparks HV (eds) The cardiovascular system. Am Physiol Soc, Bethesda, Maryland, pp 179–199 (Handbook of physiology, sect 2, vol II)

Krause SM (1988) Myocardial "stunning" does not alter the Ca^{2+}-sensitivity of the myofibrillar ATPase. Circulation 78:Suppl II–77

Krebs H (1975) The August Krogh principle "For many problems there is an animal on which it can be most conveniently studied." J Exp Zool 194:221–226

Kress M, Huxley HE, Faruqi AR, Hendrix J (1986) Structural changes during activation of frog muscle studied by time-resolved X-ray diffraction. J Mol Biol 188:325–342

Kretsinger RH (1980) Structure and evolution of calcium-modulated proteins. CRC Crit Rev Biochem 8:119–174
Krieter H, Schwarz K, Bauer SF, Brückner UB, Rüegg JC (1991) Improvement of postischemic fiber-shortening in dog hearts by infusion of glutathione. Pflügers Arch Eur J Physiol 419:R108
Krisanda JM, Paul R (1984) Energetics of isometric contraction in porcine carotid artery. Am J Physiol 246 (Cell Physiol 15):C510–C519
Krogh A (1939) Osmotic regulation in aquatic animals. Cambridge University Press. Repr 1965, Dover Publications, New York, p 208
Kuffler SW (1946) The relation of electric potential changes to contracture in skeletal muscle. J Neurophysiol 9:367–377
Kuffler SW, Vaughan Williams EM (1953 a) Small-nerve junctional potentials. The distribution of small motor nerves to frog skeletal muscle, and the membrane characteristics of the fibres they innervate. J Physiol (Lond) 121:289–317
Kuffler SW, Vaughan Williams EM (1953 b) Properties of the "slow" skeletal muscle fibres of the frog. J Physiol (Lond) 121:318–340
Kukovetz WR, Holzmann S, Wurm A, Pöch G (1979) Evidence for cyclic GMP-mediated relaxant effects of nitro-compounds in coronary smooth muscle. Naunyn-Schmiedeberg's Arch Pharmacol 310:129–138
Kumagai H, Ebashi S, Takeda F (1955) Essential relaxing factor in muscle other than myokinase and creatine phosphokinase. Nature (Lond) 176:166
Kushmerick MJ (1983) Energetics of muscle contraction. In: Peachey LD, Adrian RH, Geiger SR (eds) Skeletal muscle. Am Physiol Soc, Bethesda, Maryland, pp 189–236 (Handbook of physiology, sect 10)
Kushmerick MJ, Krasner B (1982) Force and ATPase rate in skinned skeletal muscle fibers. Fed Proc 41:2232–2237
Kushmerick MJ, Meyer RA (1985) Chemical changes in rat leg muscle by phosphorous nuclear magnetic resonance. Am J Physiol 248:C542–549
Kusuoka H, Porterfield JK, Weisman HF, Weisfeldt ML, Marban E (1987) Pathophysiology and pathogenesis of stunned myocardium. J Clin Invest 79:950–961
Kusuoka H, Koretsune Y, Chacko VP, Weisfeldt ML, Marban E (1990) Excitation-contraction coupling in postischemic myocardium. Circ Res 66:1268–1276
Lacerda AE, Kim HS, Ruth P, Perez-Reyes E, Flockerzi V, Hofmann F, Birnbaumer L, Brown AM (1991) Normalization of current kinetics by interaction between the α_1 and β subunits of the skeletal muscle dihydropyridine-sensitive Ca^{2+}-channel. Nature (Lond) 352:527–530
Lai FA, Meissner G (1989) The muscle ryanodine receptor and its intrinsic Ca^{2+}-channel activity. J Bioenerg Biomembr 21:227–246
Lai FA, Erickson HP, Rousseau E, Liu QY, Meissner G (1988) Purification and reconstitution of the calcium release channel from skeletal muscle. Nature (Lond) 331:315–319
Lakey A, Ferguson C, Labeit S, Reedy M, Larkins A, Butcher G, Leonard K, Bullard B (1990) Identification and localization of high molecular weight proteins in insect flight and leg muscle. EMBO 9:3459–3467
Lamb GD (1991) Ca^{2+} channels or voltage sensors? Nature (Lond) 352:113
Lamb GD, Stephenson DG (1989) Calcium release in skinned muscle fibres of the toad by transverse tubule depolarization or by direct stimulation. J Physiol (Lond) 423:495–517
Lamb GD, Stephenson DG (1990) Effect of Mg^{2+} on the control of Ca^{2+}-release in skeletal muscle fibres of the toad. J Physiol (Lond) 434:507–528
Langer GD, Frank JS, Philipson KD (1982) Ultrastructure and calcium exchange of the sarcolemma, sarcoplasmic reticulum and mitochondria of the myocardium. Pharmacol Ther 16:331–376
Lännergren J (1978) The force-velocity relation of isolated twitch and slow muscle fibres of *Xenopus laevis*. J Physiol (Lond) 283:501–521
Lännergren J (1979) An intermediate type of muscle fibre in *Xenopus laevis*. Nature (Lond) 279:254–256
Lansman JB, Franco A Jr (1991) What does dystrophin do in normal muscle? J Muscle Res Cell Mot 12:409–411
Laszt L, Hamoir G (1961) Etude par électrophorèse et ultracentrifugation de la composition protéinique de la couche musculaire des carotides de bovidé. Biochim Biophys Acta 50:430–449

Laxson DD, Homans DC, Dai XZ, Sublett E, Bache RJ (1989) Oxygen consumption and coronary reactivity in postischemic myocardium. Circ Res 64:9–20

Lea TJ, Ashley CC (1989) Ca^{2+}-induced Ca^{2+} release from the sarcoplasmic reticulum of isolated myofibrillar bundles of barnacle muscle fibres. Pflüg Arch Eur J Physiol 413:401–406

Lee HC, Smith N, Mohabir R, Clusin WT (1987) Cytosolic calcium transients from the beating mammalian heart. Proc Natl Acad Sci USA 84:7793–7797

Lee JA, Allen DG (1991 a) EMD 53998 sensitizes the contractile proteins to calcium in intact ferret ventricular muscle. Circ Res 69:927–936

Lee JA, Allen DG (1991 b) Mechanisms of acute ischemic contractile failure of the heart – role of intracellular calcium. J Clin Invest 88:361–367

Lee JA, Westerblad H, Allen DG (1991) Changes in tetanic and resting $[Ca^{2+}]_i$ during fatigue and recovery of single muscle fibres from *Xenopus laevis*. J Physiol (Lond) 433:307–326

Lehman W (1982) The location and periodicity of a troponin-T-like protein in the myofibril of the horseshoe crab *Limulus polyphemus*. J Mol Biol 154:385–391

Lehman W (1983) The ionic requirements for regulation by molluscan thin filaments. Biochim Biophys Acta 745:1–5

Lehman W (1991) Calponin and the composition of smooth muscle thin filaments. J Muscle Res Cell Mot 12:221–224

Lehman W, Szent-Györgyi AG (1972) Activation of the adenosine triphosphatase of *Limulus polyphemus* actomyosin by tropomyosin. J Gen Physiol 59:375–387

Lehman W, Szent-Györgyi AG (1975) Regulation of muscular contraction. Distribution of actin control and myosin control in the animal kingdom. J Gen Physiol 66:1–30

Lehman W, Kendrick-Jones J, Szent-Györgyi AG (1973) Myosin-linked regulatory systems: comparative studies. Cold Spring Harbor Symp Quant Biol 37:319–330

Lehman W, Bullard B, Hammond K (1974) Calcium-dependent myosin from insect flight muscles. J Gen Physiol 63:553–563

Lehman W, Head JF, Grand PW (1980) The stoichiometry and location of troponin I- and troponin C-like proteins in the myofibril of the bay scallop, *Aequipecten irradians*. Biochem J 187:447–456

Léoty C, Léauté M (1982) Membrane potential and contractures in segments cut from rat fast and slow twitch muscles. Pflügers Arch Eur J Physiol 395:42–48

Leszyk J, Collins JH, Leavis PC, Tao T (1987) Cross-linking of rabbit skeletal muscle troponin with the photoactive reagent 4-maleimidobenzophenone: identification of residues in troponin-I that are close to cysteine-98 of troponin-C. Biochemistry 26:7042–7047

Leszyk J, Collins JH, Leavis PC, Tao T (1988) Cross-linking of rabbit skeletal muscle troponin subunits: labeling of cysteine-98 of troponin-C with 4-maleimidobenzophenone and analysis of products formed in the binary complex with troponin-T and the ternary complex with troponin I and T. Biochemistry 27:6983–6987

Leszyk J, Grabarek Z, Gergely J, Collins JH (1990) Characterization of zero-length cross-links between rabbit skeletal muscle troponin C and troponin I: evidence for direct interaction between the inhibitory region of troponin I and the NH_2-terminal, regulatory domain of troponin C. Biochemistry 29:299–304

Leung AR, Imagawa T, Block B, Franzini-Armstrong C, Campbell KP (1988) Biochemical and ultrastructural characterization of the dihydropyridine receptor from rabbit skeletal muscle. J Biol Chem 263:994–1001

Levine BA, Moir AJG, Perry SV (1988) The interaction of troponin-I with the N-terminal region of actin. Eur J Biochem 172:389–397

Levine RJC, Elfvin M, Dewey MM, Walcott B (1976) Paramyosin in invertebrate muscles. II. Content in relation to structure and function. J Cell Biol 71:273–279

Levine RJC, Davidheiser S, Kelly AM, Kensler RW, Leferovich J, Davies RE (1989) Fibre types in *Limulus telson* muscles: morphology and histochemistry. J Muscle Res Cell Mot 10:53–66

Levitsky DO, Benevolensky DS, Levchenko TS, Smirnov VN, Chazov EI (1981) Calcium-binding rate and capacity of cardiac sarcoplasmic reticulum. J Mol Cell Cardiol 13:785–796

Lewartowski B, Pytkowski B, Janczewski A (1984) Calcium fraction correlating with contractile force of ventricular muscle of guinea-pig heart. Pflügers Arch Eur J Physiol 401:198–203

Limas CJ, Olivari MT, Goldenberg IF, Levine TB, Benditt DG, Simon A (1987) Calcium uptake by cardiac sarcoplasmic reticulum in human dilated cardiomyopathy. Cardiovasc Res 21:601–605

Liu J, Prosser CL, Job DD (1969) Ionic dependence of slow waves and spikes in intestinal muscle. Am J Physiol 217:1542–1547

Lombardi V, Menchetti G (1984) The maximum velocity of shortening during the early phases of the contraction in frog single muscle fibres. J Muscle Res Cell Mot 5:503–513
Lombardi V, Piazzesi G, Linari M (1992) Rapid regeneration of the actin-myosin power stroke in contracting muscle. Nature (Lond) 355:638–641
Lomo T, Westgaard RH, Dahl HA (1974) Contractile properties of muscle: control by pattern of muscle activity in the rat. Proc R Soc (Lond) B 187:99–103
Lopez JR, Alamo L, Caputo C, Wikinski J, Ledezma D (1985) Intracellular ionized calcium concentration in muscles from humans with malignant hyperthermia. Muscle Nerve 8:355–358
Lorenzini S (1678) Osservazio in intorno alle Torpedim. Onofri, Firenze
Lorkovíc H (1983) Potassium contractures in mouse limb muscles. J Physiol (Lond) 343:569–576
Lowey S (1971) Myosin: molecule and filament. In: Timasheff S, Fasman G (eds) Biological macromolecules. Subunits in biological systems, vol 5. Dekker, New York, pp 201–259
Lowey S (1980) An immunological approach to the isolation of myosin isoenzymes. In: Pette D (ed) Plasticity of muscle. Walter de Gruyter, Berlin, pp 69–81
Lowy J, Poulsen FR (1982) Time-resolved X-ray diffraction studies of the structural behaviour of myosin heads in a living contracting unstriated muscle. Nature (Lond) 299:308–312
Lowy J, Millman BM, Hanson J (1964) Structure and function in smooth tonic muscles of lamellibranch molluscs. Proc R Soc (Lond) B 160:525–536
Loxdale HD, Tregear RT (1985) Dissociation between mechanical performance and the cost of isometric tension maintenance in *Lethocerus* flight muscle. J Muscle Res Cell Mot 6:163–175
Luff AR (1981) Dynamic properties of the inferior rectus, extensor digitorum longus, diaphragm and soleus muscles of the mouse. J Physiol (Lond) 313:161–171
Luff AR, Atwood HR (1971) Changes in sarcoplasmic reticulum and transverse tubular system of fast and slow skeletal muscles of the mouse during postnatal development. J Cell Biol 51:369–383
Lüttgau HC, Niedergerke R (1958) The antagonism between Ca and Na ions on the frog's heart. J Physiol (Lond) 143:486–505
Lüttgau HC, Spiecker W (1979) The effects of calcium deprivation upon mechanical and electrophysiological parameters in skeletal muscle fibres of the frog. J Physiol (Lond) 296:411–429
Lüttgau HC, Stephenson GD (1986) Ion movements in skeletal muscle in relation to the activation of contraction. In: Andreoli TE, Hoffmann JF, Fanestil DD, Schultz SG (eds) Physiology of membrane disorders. Plenum, New York, pp 449–468
Lüttgau HC, Gottschalk G, Berwe D (1986) The role of Ca^{2+} in inactivation and paralysis of excitation-contraction coupling in skeletal muscle. Fortschr Zool 33:195–203
Ma J, Fill M, Knudson MC, Campbell KP, Coronado R (1988) Ryanodine receptor of skeletal muscle is a gap junction-type channel. Science 242:99–102
Machin KEW, Pringle JWS (1959) The physiology of insect fibrillar muscle. II. Mechanical properties of a beetle flight muscle. Proc R Soc (Lond) B 151:204–225
Machin KEW, Pringle JWS (1960) The physiology of insect fibrillar muscle. III. The effect of sinusoidal changes of length on a beetle flight muscle. Proc R Soc (Lond) B 152:311–330
MacLennan DH, Wong PTS (1971) Isolation of a calcium-sequestring protein from sarcoplasmic reticulum. Proc Natl Acad Sci USA 68:1231–1235
Maéda Y (1979) X-ray difraction patterns from molecular arrangements with 38-nm periodicities around muscle thin filaments. Nature (Lond) 277:670–672
Maéda Y, Matsubara I, Yagi N (1979) Structural changes in thin filaments of crab striated muscle. J Mol Biol 127:191–201
Maier L, Rathmayer W, Pette D (1984) pH lability of myosin ATPase activity permits discrimination of different muscle fibre types in crustaceans. Histochemistry 81:75–77
Maita T, Chen J-I, Matsuda G (1981) Amino-acid sequence of the 20000-molecular-weight light chain of chicken gizzard-muscle myosin. Eur J Biochem 117:417–424
Makinose M (1969) The phosphorylation of the membrane protein of the sarcoplasmic vesicles during active calcium transport. Eur J Biochem 10:74–82
Mandel F, Kranias EG, Schwartz A (1983) The effect of cAMP-dependent protein kinase phosphorylation on the external Ca^{2+}-binding sites of cardiac sarcoplasmic reticulum. J Bioenerg Biomembr 15:179–194
Marban E, Rink T, Tsien RW, Tsien RY (1980) Free calcium in heart muscle at rest and during contraction measured with Ca^{2+}-sensitive microelectrodes. Nature (Lond) 286:845–850

Marban E, Kusuoka H, Yue DT, Weisfeldt ML, Wier WG (1986) Maximal Ca^{2+}-activated force elicited by tetanization of ferret papillary muscle and whole heart: mechanism and characteristics of steady contractile activation in intact myocardium. Circ Res 59:262–269

Margulis BA, Bobrova IF, Mashanski VF, Pinaev GP (1979) Major myofibrillar protein content and the structure of mollusc adductor contractile apparatus. Comp Biochem Physiol A 64:291–298

Marsh BB (1951) The effects of adenosine triphosphatase on the fibre volume of a muscle homogenate. Nature (Lond) 167:1065–1066

Marston SB (1982) The regulation of smooth muscle contractile proteins. Prog Biophys Mol Biol 41:1–41

Marston SB (1983) Myosin and actomyosin ATPase: kinetics. In: Stephens NL (ed) Biochemistry of smooth muscle, vol I. CRC Press, Boca Raton, pp 167–191

Marston SB, Lehman W (1985) Caldesmon is a Ca^{2+}-regulatory component of native smooth-muscle thin filaments. Biochem J 231:517–522

Marston SB, Smith CWJ (1984) Purification and properties of Ca^{2+}-regulated thin filaments and F-actin from sheep aorta smooth muscle. J Muscle Res Cell Mot 5:559–575

Marston SB, Tregear RT (1974) Calcium binding and the activation of fibrillar insect flight muscle. Biochim Biophys Acta 347:311–318

Marston SB, Rodger CD, Tregear RT (1976) Changes in muscle crossbridges when β,γ-imido-ATP binds to myosin. J Mol Biol 104:263–276

Martonosi AN (1984) Mechanisms of Ca^{2+} release from sarcoplasmic reticulum of skeletal muscle. Physiol Rev 64:1240–1320

Mathias RT, Levis RA, Eisenberg RS (1980) Electrical models of excitation-contraction coupling and charge movement in skeletal muscle. J Gen Physiol 76:1–31

Matsubara I, Millman BM (1974) X-ray diffraction patterns from mammalian heart muscle. J Mol Biol 82:527–536

Matsubara I, Yagi N (1978) A time-resolved X-ray diffraction study of muscle during twitch. J Physiol (Lond) 278:297–307

Matsuda T, Podolsky RJ (1984) X-ray evidence for two structural states of the actomyosin crossbridge in muscle fibres. Proc Natl Acad Sci USA 81:2364–2368

Matsumura M, Ochi K (1983) Calcium transients in normal and potentiated twitch of frog skeletal muscle examined by arsenazo-III. Proc Int Union Physiol Sci XV:152

Maylie J, Morad M (1984) A transient outward current related to calcium release and development of tension in elephant seal atrial fibres. J Physiol (Lond) 357:267–292

McArdle HJ, Johnston IA (1981) Ca^{2+}-uptake by tissue sections and biochemical characteristics of sarcoplasmic reticulum isolated from fish fast and slow muscles. Eur J Cell Biol 25:103–107

McArdle HJ, Johnston IA (1982) Temperature adaptation and the kinetics of the Ca^{2+}-independent and Ca^{2+}-dependent ATPases of fish sarcoplasmic reticulum. J Therm Biol 7:63–67

McAuliffe JJ, Lizhu G, Solaro RJ (1990) Changes in myofibrillar activation and troponin C Ca^{2+}-binding associated with troponin-T isoform switching in developing rabbit heart. Circ Res 66:1204–1216

McDonald TF, Cavalié A, Trautwein W, Pelzer D (1986) Voltage-dependent properties of macroscopic and elementary calcium channel currents in guinea pig ventricular myocytes. Pflügers Arch Eur J Physiol 406:437–448

McLennan DH, Duff C, Zorzato F, Fujii J, Phillips M, Korneluk R, Frodis W, Britt BA, Worton RG (1990) Ryanodine receptor gene is a candidate for predisposition to malignant hyperthermia. Nature (Lond) 343:559–561

Means AR, Tash JS, Chafouleas JG (1982) Physiological implications of the presence, distribution and regulation of calmodulin in eukaryotic cells. Physiol Rev 62:1–39

Mechmann S, Pott L (1986) Identification of Na-Ca exchange current in single cardiac myocytes. Nature (Lond) 319:597–599

Medford RM, Nguyen HT, Destree AT, Summers E, Nadel-Ginard B (1984) A novel mechanism of alternative RNA splicing for the developmentally regulated generation of troponin T isoforms from a single gene. Cell 38:409–421

Meisheri KD, Breemen C van (1982) Effects of Beta-adrenergic stimulation on calcium movements in rabbit aortic smooth muscle: relationship with cyclic AMP. J Physiol (Lond) 331:429–441

Meisheri KD, Rüegg JC (1983) Dependence of cyclic-AMP-induced relaxation on Ca^{2+} and calmodulin in skinned smooth muscle of guinea pig taenia coli. Pflügers Arch Eur J Physiol 399:315–320

Meisheri KD, Hwang O, Breemen C van (1981) Evidence for two separate Ca^{2+} pathways in smooth muscle plasmalemma. J Membr Biol 59:19–25

Meissner G (1975) Isolation and characterization of two types of sarcoplasmic reticulum vesicles. Biochim Biophys Acta 389:51–68

Melzer W (1982 a) Electrical membrane properties of the muscle lamellae in *Branchiostoma* myotomes. Eur J Cell Biol 28:213–218

Melzer W (1982 b) Twitch activation in Ca^{2+} free solutions in the myotomes of the lancelet (*Branchiostoma lanceolatum*). Eur J Cell Biol 28:219–225

Melzer W, Rios E, Schneider MF (1984) Time course of calcium release and removal in skeletal muscle fibers. Biophys J 45:637–641

Melzer W, Schneider MF, Simon BJ, Szucs G (1986) Intramembrane charge movement and calcium release in frog skeletal muscle. J Physiol (Lond) 373:481–511

Melzer W, Rios E, Schneider MF (1987) A general procedure for determining the rate of calcium release from the sarcoplasmic reticulum in skeletal muscle fibres. Biophys J 51:849–863

Mendelson M (1969) Electrical and mechanical characteristics of a very fast lobster muscle. J Cell Biol 42:548–563

Menke A, Jockusch H (1991) Decreased osmotic stability of dystrophin-less muscle cells from the mdx mouse. Nature (Lond) 349:69–71

Merkel L, Meisheri KD, Pfitzer G (1984) The variable relation between myosin light-chain phosphorylation and actin-activated ATPase activity in chicken gizzard smooth muscle: modulation by tropomyosin. Eur J Biochem 138:429–434

Michell RH, Kirk CJ, Jones LM, Downes CP, Creba JA (1981) The stimulation of inositol lipid metabolism that accompanies calcium mobilization in stimulated cells: defined characteristics and unanswered questions. Phil Trans R Soc (Lond) B 296:123–137

Mignery GA, Südhof TC, Takei K, De Camilli P (1989) Putative receptor for Ins(1,4,5)P_3 similar to ryanodine receptor. Nature (Lond) 342:192–195

Mikami A, Imoto K, Tanabe T, Niidome T, Mori Y, Takeshima H, Narumiya S, Numa S (1989) Primary structure and functional expression of the cardiac dihydropyridine-sensitive calcium channel. Nature (Lond) 340:230–233

Mikawa T (1979) "Freezing" of Ca-regulated conformation of reconstituted thin filament of skeletal muscle by glutaraldehyde. Nature (Lond) 278:473–474

Miledi R, Parker I, Schalow G (1977) Calcium transients in the frog slow muscle fibres. Nature (Lond) 268:750–752

Miledi R, Parker I, Schalow G (1981) Calcium transients in normal and denervated slow muscle fibres of the frog. J Physiol (Lond) 318:191–206

Miledi R, Parker I, Zhu PH (1982) Calcium transients evoked by action potentials in frog twitch muscle fibres. J Physiol (Lond) 333:655–679

Miledi R, Parker I, Zhu PH (1983 a) Calcium transients in frog skeletal muscle fibres following conditioning stimuli. J Physiol (Lond) 339:223–242

Miledi R, Parker I, Zhu PH (1983 b) Calcium transients studied under voltage-clamp control in frog twitch muscle fibres. J Physiol (Lond) 340:649–680

Mill PJ, Knapp MF (1970) The fine structure of obliquely striated body wall muscle in the earthworm *Lumbricus terrestris* Linn. J Cell Sci 7:233–261

Miller JR, Silver PJ, Stull JT (1983) The role of myosin light chain kinase phosphorylation in beta-adrenergic relaxation of tracheal smooth muscle. Mol Pharmacol 24:235–242

Mines GR (1913) On functional analysis by the action of electrolytes. J Physiol (Lond) 46:188–235

Mobley BA, Eisenberg BR (1975) Sizes of components in frog skeletal muscle measured by methods of stereology. J Gen Physiol 66:31–46

Moir AJG, Levine BA, Goodearl AJ, Trayer JP (1987) The interaction of actin with myosin subfragment 1 and with pPDM-cross-linked S1:a ^{1}HNMR investigation. J Muscle Res Cell Mot 8:68–69

Moisescu DG (1976) Kinetics of reaction in Ca^{2+}-activated skinned muscle fibres. Nature (Lond) 262:610–613

Molloy JE, Kyrtatas V, Sparrow JC, White DCS (1987) Kinetics of flight muscles from insects with different wingbeat frequencies. Nature (Lond) 328:449–451

Moore G, Johnston IA, Goldspink G (1983) The pCa-tension characteristics of single skinned fibres isolated from the anterior and posterior latissimus dorsi muscles of the chicken. J Exp Biol 105:411–416

Moore RL, Houston ME, Iwamoto GA, Stull JT (1985) Phosphorylation of rabbit skeletal muscle myosin in situ. J Cell Physiol 125:301–305

Morad M, Goldman Y (1973) Excitation-contraction coupling in heart muscle: membrane control of development of tension. Prog Biophys Mol Biol 27:257–313

Morad M, Orkand RK (1971) Excitation-contraction coupling in frog ventricle: evidence from voltage clamp studies. J Physiol (Lond) 219:167–189

Morano I (1992) Molecular biology of smooth muscle. J Hypertens 10:411–416

Morano I, Rüegg JC (1991) What does TnC_{DANZ}fluorescence reveal about the thin filament state? Pflügers Arch Eur J Physiol 418:333–337

Morano I, Hofmann F, Zimmer M, Rüegg JC (1985) The influence of P-light chain phosphorylation by myosin light chain kinase on the calcium sensitivity of chemically skinned heart fibres. FEBS Lett 189:221–224

Morano I, Rösch J, Arner A, Rüegg JC (1990) Phosphorylation and thiophosphorylation by myosin light chain kinase: different effects on mechanical properties of chemically skinned ventricular fibers from the pig. J Mol Cell Cardiol 22:805–813

Morano I, Bletz C, Wojciechowski R, Rüegg JC (1991) Modulation of crossbridge kinetics by myosin isoenzymes in skinned human heart fibers. Circ Res 68:614–618

Morgan DL, Proske U (1984) Vertebrate slow muscle: its structure, pattern of innervation, and mechanical properties. Physiol Rev 64:103–169

Morgen JP, Blinks JR (1982) Intracellular Ca^{2+} transients in the cat papillary muscle. Can J Physiol Pharmacol 60:520–528

Morgan JP, Morgan KG (1982) Vascular smooth muscle: the first recorded Ca^{2+} transients. Pflügers Arch Eur J Physiol 395:75–77

Morgan JP, Morgan KG (1984a) Stimulus-specific patterns of intracellular calcium levels in smooth muscle of ferret portal vein. J Physiol (Lond) 351:155–167

Morgan JP, Morgan KG (1984b) Alteration of cytoplasmic ionized calcium levels in smooth muscle by vasodilators in the ferret. J Physiol (Lond) 357:539–551

Morgan JP, Gwathmey JK, DeFeo TT, Morgan KG (1986) The effects of amrinone and related drugs on intracellular calcium in isolated mammalian cardiac and vascular smooth muscle. Circulation 73 (Suppl III):65–77

Morgan JP, Erny RE, Allen PD (1990) Abnormal intracellular calcium handling, a major cause of systolic and diastolic dysfunction in ventricular myocardium from patients with heart failure. Circulation 81 (Suppl III):III–21 – III–33

Morita F, Kondo S (1982) Regulatory light chain contents and molecular species of myosin in catch muscle of scallop. J Biochem 92:977–983

Morita F, Katoh T, Suzuki R, Isonishi K, Hori K, Eto M (1991) An actin binding site on myosin. In: Rüegg JC (ed) Peptides as probes in muscle research. Springer, Berlin Heidelberg New York, pp 39–48

Mornet D, Bertrand R, Pantel P, Audemard E, Kassab R (1981) Structure of the actin-myosin interface. Nature (Lond) 292:301–306

Moss RL (1982) The effect of calcium on the maximum velocity of shortening in skinned skeletal muscle fibres of the rabbit. J Muscle Cell Mot 3:295–311

Moss RL (1986) Effects on shortening velocity of rabbit skeletal muscle due to variations in the level of thin-filament activation. J Physiol (Lond) 377:487–505

Moss RL, Giulian GG, Greaser ML (1982) Mechanical effects accompanying the removal of myosin LC_2 from skinned skeletal muscle fibres. J Biol Chem 257:8588–8591

Moss RL, Lauer MR, Giulian GG, Greaser ML (1986) Altered Ca^{2+} dependence of tension development in skinned skeletal muscle fibers following modification of troponin by partial substitution with cardiac troponin C. J Biol Chem 261:6096–6099

Movsesian MA, Nishikawa M, Adelstein RS (1984) Phosphorylation of phospholamban by calcium-activated, phospholipid-dependent protein kinase. Stimulation of cardiac sarcoplasmic reticulum calcium uptake. J Biol Chem 259:8029-8032

Mrwa U, Rüegg JC (1977) The role of the regulatory light chain in pig carotid smooth muscle ATPase. In: Casteels R, Godfraind T, Rüegg JC (eds) Excitation-contraction coupling in smooth muscle. Elsevier/North-Holland, Amsterdam, pp 353–357

Mrwa U, Troschka M, Rüegg JC (1979) Cyclic AMP-dependent inhibition of smooth muscle actomyosin. FEBS Lett 107:371–374

Mueller E, Breemen C van (1979) Role of intracellular Ca^{2+} sequestration in β-adrenergic relaxation of a smooth muscle. Nature (Lond) 281:682–683
Mulieri LA, Alpert NR (1982) Activation heat and latency relaxation in relation to calcium movement in skeletal and cardiac muscle. Can J Physiol Pharmacol 60:529–541
Müller-Beckmann B, Sponer G, Rüegg JC, Freund P, Jäger H, Strein K (1986) Zur Pharmakologie der neuen positiv inotropen Substanz BM 14.478. Z Kardiol 75 (Suppl IV):31
Mullins LJ (1976) A mechanism for Na/Ca transport. J Gen Physiol 70:681–695
Mullins LJ (1979) The generation of electric currents in cardiac fibers by Na/Ca exchange. Am J Physiol 236:C103–C110
Muneoka Y, Twarog BM (1983) Neuromuscular transmission and excitation-contraction coupling in molluscan muscle. In: Wilbur KM (ed) The Mollusca 4 (Physiology, Part 1). Academic Press, London, pp 35–75
Muneoka Y, Cottrell GA, Twarog BM (1977) Neurotransmitter action on the membrane of *Mytilus* smooth muscle. III. Serotonin. Gen Pharmacol 8:93–96
Murase M, Tanaka H, Nishiyama K, Shimizu H (1986) A three-state model for oscillation in muscle: sinusoidal analysis. J Muscle Res Cell Mot 7:2–10
Murphy RA, Herlihy JT, Megerman J (1974) Force-generating capacity and contractile protein content of arterial smooth muscle. J Gen Physiol 64:691–705
Murphy RA, Aksoy MO, Dillon PF, Gerthoffer WT, Kamm KE (1983) The role of myosin light chain phosphorylation in regulation of the cross-bridge cycle. Fed Proc 42:51–56
Murray KJ, England PJ (1980) Contraction in intact pig aortic strips is not always associated with phosphorylation of myosin light chains. Biochem J 192:967–970
Murray JM, Weber A, Bremel RD (1975) Could cooperativity in the actin-filament play a role in muscle contraction. In: Carafoli E, Clementi W, Drabikowski W, Margreth A (eds) Calcium transport in contraction and secretion. Elsevier/North-Holland, Amsterdam, pp 489–496
Nachtigall W, Wilson DM (1967) Neuro-muscular control of dipteran flight. J Exp Biol 47:77–97
Nairn AC, Perry SV (1979) Calmodulin and myosin light chain kinase of rabbit fast skeletal muscle. J Biochem (Tokyo) 179:89–97
Nakajima S, Gilai A (1980) Radial propagation of muscle action potential along the tubular system examined by potential-sensitive dyes. J Gen Physiol 76:751–762
Nakajima Y, Endo M (1973) Release of calcium induced by "depolarization" of the sarcoplasmic reticulum membrane. Nature (Lond) 246:216–218
Natori R (1954) The property and contraction process of isolated myofibrils. Jikeikai Med J 1:119–126
Nauss K, Davies RE (1966) Changes in inorganic phosphate and arginine during the development, maintenance and loss of tension in the anterior byssus retractor muscle of *Mytilus edulis*. Biochem Z 345:173–187
Nayler WG, Poole-Wilson PA, Williams A (1979) Hypoxia and calcium. J Mol Cell Cardiol 11:683–706
Needham DM (1971) Machina carnis. The biochemistry of muscular contraction in its historical development. University Press, Cambridge
Neering IR, Morgan KG (1980) Use of aequorin to study excitation-contraction coupling in mammalian smooth muscle. Nature (Lond) 288:585–587
Negele JC, Dotson DG, Liu W, Sweeney HL, Putkey JA (1992) Mutation of the high affinity calcium binding sites in cardiac troponin C. J Biol Chem 267:825–831
Nelson MT, Worley JF (1989) Dihydropyridine inhibition of single calcium channels and contraction in rabbit mesenteric artery depends on voltage. J Physiol (Lond) 412:65–91
Nelson MT, Standen NB, Brayden JE, Worley JF (1988) Noradrenaline contracts arteries by activating voltage-dependent calcium channels. Nature (Lond) 336:382–385
New W, Trautwein W (1972) The ionic nature of slow inward current and its relation to contraction. Pflügers Arch Eur J Physiol 334:24–38
Newsholme EA, Crabtree B (1976) Substrate cycles in metabolic regulation and heat generation. Biochem Soc Symp 41:61–109
Ng WA, Grupp IL, Subramaniam A, Robbins J (1991) Cardiac myosin heavy chain mRNA expression and myocardial function in the mouse heart. Circ Res 69:1742–1750
Ngai PK, Walsh MP (1984) Inhibition of smooth muscle actin-activated myosin Mg^{2+}-ATPase activity by caldesmon. J Biol Chem 259:13656–13659

Ngai PK, Walsh MP (1985) Properties of caldesmon isolated from chicken gizzard. Biochem J 230:695–707
Nicoll DA, Longoni S, Philipson KD (1990) Molecular cloning and functional expression of the cardiac sarcolemmal Na^+-Ca^{2+}-exchanger. Science 250:562–565
Nielsen KE, Gesser H (1984) Energy metabolism and intracellular pH in trout heart muscle under anoxia and different $[Ca^{2+}]_0$. J Comp Physiol B 154:523–527
Niggli E, Lederer WJ (1991) Molecular operations of the sodium-calcium exchanger revealed by conformation currents. Nature (Lond) 349:621–624
Nilius B, Hess P, Lansman JB, Tsien RW (1985) A novel type of cardiac calcium channel in ventricular cells. Nature (Lond) 316:443–446
Nilsson S (1983) Autonomic nerve function in the vertebrates. Springer, Berlin Heidelberg New York (Zoophysiology, vol 13)
Nimmo HG, Cohen P (1977) Hormonal control of protein phosphorylation. Adv Cycl Nucl Res 8:145–266
Nishikawa M, DeLanerolle P, Lincoln THM, Adelstein RS (1984a) Phosphorylation of mammalian myosin light chain kinases by the catalytic subunit of cyclic AMP-dependent protein kinase and by cyclic GMP-dependent protein kinase. J Biol Chem 259:8429–8436
Nishikawa M, Sellers JR, Adelstein RS, Hidaka H (1984b) Protein kinase C modulates in vitro phosphorylation of the smooth muscle heavy meromyosin by myosin light chain kinase. J Biol Chem 259:8808–8814
Nishikori K, Weisbrodt NW, Sherwood OD, Sanborn BM (1982) Relaxin alters rat uterine myosin light chain phosphorylation and related enzymatic activity. Endocrinology 111:1743–1745
Nishimura J, Kolber M, Breemen C van (1988) Norepinephrine and GTP-c-S increase myofilament Ca^{2+}-sensitivity in α-toxin permeabilized arterial smooth muscle. Biochem Biophys Res Commun 157:677–683
Nishimura J, Khalil RA, Drenth JP, Breemen C van (1990) Evidence for increased myofilament Ca^{2+}-sensitivity in norepinephrine-activated vascular smooth muscle. Am J Physiol 259:H2–H8
Noble D (1984) The surprising heart: a review of recent progress in cardiac electrophysiology. J Physiol (Lond) 353:1–50
Nomura H (1963) The effect of stretching on the intracellular action potential from the cardiac muscle fibre of the marine mollusc, *Dolabella auricula*. Sci Rep Tokyo Daig 11:153–165
Nunzi MG, Franzini-Armstrong C (1981) The structure of smooth and striated portions of the adductor muscle of the valves in a scallop. J Ultrastruct Res 76:134–148
Oetliker H (1982) An appraisal of the evidence for a sarcoplasmic reticulum membrane potential and its relation to calcium release in skeletal muscle. J Muscle Res Cell Mot 3:247–272
Oetliker H, Baylor SM, Chandler WK (1975) Simultaneous changes in fluorescence and optical retardation in single muscle fibres during activity. Nature (Lond) 257:693–696
Ogawa Y, Kurebayashi N (1982) ATP-ADP exchange reaction by fragmented sarcoplasmic reticulum from bull-frog skeletal muscle. J Muscle Res Cell Mot 3:39–56
Ogawa Y, Kurebayashi N, Irimajiri A, Hanai T (1981) Transient kinetics for Ca uptake by fragmented sarcoplasmic reticulum from bull frog skeletal muscle with reference to the rate of relaxation of living muscle. In: Varga E, Kövér A, Kovács T, Kovács L (eds) Molecular and cellular aspects of muscle function. Adv Physiol Sci, vol 5. Pergamon Press, New York, pp 417–435
Ohtsuki I (1975) Distribution of troponin components in the thin filament studied by immunoelectron microscopy. J Biochem (Tokyo) 77:633–639
Onishi H, Wakabayashi T (1982) Electron microscopic studies of myosin molecules from chicken gizzard muscle. I. The formation of the intramolecular loop in the myosin tail. J Biochem 92:871–879
Opie LH (1989) Reperfusion injury and its pharmacologic modification. Circulation 80:1049–1062
Opie LH (1991) The heart – physiology and metabolism, 2nd edn. Raven Press, New York
Orkand RK (1962) The relation between membrane potential and contraction in single crayfish muscle fibres. J Physiol (Lond) 161:143–159
Osterrieder W, Brum G, Hescheler J, Trautwein W, Flockerzi V, Hofmann F (1982) Injection of subunits of cyclic AMP-dependent protein kinase into cardiac myocytes modulates Ca^{2+} current. Nature (Lond) 298:576–578
Otsu K, Khanna VK, Archibald AL, MacLennan DH (1991) Co-segregation of porcine malignant hyperthermia and a probable causal mutation in the skeletal muscle ryanodine receptor gene in backcross families. Genomics 11:744–750

Ovalle WK Jr (1982) Ultrastructural duality of extrafusal fibres in a slow (tonic) skeletal muscle. Cell Tissue Res 222:261–267

Padrón R, Panté N, Sosa H, Kendrick-Jones J (1991) X-ray diffraction study of the structural changes accompanying phosphorylation of tarantula muscle. J Muscle Res Cell Mot 12:235–241

Page E, Surdyk-Droske M (1979) Distribution, surface density, and membrane area of diadic junctional contacts between plasma membrane and terminal cisterns in mammalian ventricle. Circ Res 45:260–267

Page SG (1965) A comparison of the fine structures of frog slow and twitch muscle fibres. J Cell Biol 26:477–497

Page SG (1969) Structure and some contractile properties of fast and slow muscles of the chicken. J Physiol (Lond) 205:131–145

Palade P, Vergara J (1982) Arsenazo III and antipyrylazo III calcium transients in single skeletal muscle fibers. J Gen Physiol 79:679–707

Palmer RMJ, Ferrige AG, Moncada S (1987) Nitric oxide release accounts for the biological activity of endothelium-derived relaxing factor. Nature (Lond) 327:524–526

Pan BS, Solaro RJ (1987) Calcium-binding properties of troponin C in detergent-skinned heart muscle fibers. J Biol Chem 262:7839–7849

Paul RJ (1990) Smooth muscle energetics and theories of cross-bridge regulation. Am J Physiol 258:C369–C375

Paul RJ, Rüegg JC (1976) Biochemistry of vascular smooth muscle: energy metabolism and the proteins of the contractile apparatus. In: Altura BM, Kaley G (eds) Microcirculation, vol II. Univ Park, Baltimore, pp 41–82

Paul RJ, Doermann G, Zeugner C, Rüegg JC (1983) The dependence of unloaded shortening velocity on Ca^{2+}, calmodulin, and duration of contraction in "chemically skinned" smooth muscle. Circ Res 53:342–351

Peachey LD (1965) The sarcoplasmic reticulum and transverse tubules of the frog's sartorius. J Cell Biol 25 (part II):209–231

Peachey LD (1981) Three-dimensional structure of the T-system of skeletal muscle cells. In: Varga E, Kövér A, Kovács T, Kovács L (eds) Molecular and cellular aspects of muscle function. Pergamon Press, New York, pp 299–311 (Adv physiol sci, vol 5)

Peachey LD, Eisenberg BR (1978) Helicoids in the T-system and striations of frog skeletal muscle fibres seen by high-voltage electron microscopy. Biophys J 22:145–154

Peachey LD, Huxley AF (1964) Transverse tubules in crab muscle. J Cell Biol 23:70A–71A

Peckham M, Woledge RC (1986) Labile heat and changes in rate of relaxation of frog muscles. J Physiol (Lond) 374:123–135

Pemrick SM (1980) The phosphorylated LC_2 light chain of skeletal myosin is a modifier of the actomyosin ATPase. J Biol Chem 255:8836–8841

Penner R, Neher E, Takeshima H, Nishimura S, Numa S (1989) Functional expression of the calcium release channel from skeletal muscle ryanodine receptor cDNA. FEBS Lett 259:217–221

Perez-Reyes E, Kim HS, Lacerda AE, Horne W, Wei X, Rampe D, Campbell KP, Brown AM, Birnbaumer L (1989) Induction of calcium currents by the expression of the $alpha_1$-subunit of the dihydropyridine receptor from skeletal muscle. Nature (Lond) 340:233–236

Perry SV, Cole HA, Head JF, Wilson FJ (1973) Localization and mode of action of the inhibitory protein component of the troponin complex. Cold Spring Harbor Symp Quant Biol 37:251–262

Perry SV, Cole HA, Grand RJA, Levine BA (1982) Comparative aspects of the regulation of contraction in vertebrate muscle. In: Twarog BM, Levine RJC, Dewey MM (eds) Basic biology of muscles. A comparative approach. Raven Press, New York, pp 243–254

Perry SV, Levine BA, Moir AJG, Patchell VB (1991) Use of synthetic peptides in the study of the function of dystrophin. In: Rüegg JC (ed) Peptides as probes in muscle research. Springer, Berlin Heidelberg New York, pp 161–170

Persechini A, Hartshorne DJ (1981) Phosphorylation of smooth muscle myosin: evidence for cooperativity between the myosin heads. Science 213:1383–1385

Persechini A, Stull JT (1984) Phosphorylation kinetics of skeletal muscle myosin and the effect of phosphorylation on actomyosin adenosinetriphosphatase activity. Biochemistry 23:4144–4150

Persechini A, Mrwa U, Hartshorne DJ (1981) Effect of phosphorylation on the actin activated ATPase activity of myosin. Biochem Biophys Res Commun 98:800–805

Persechini A, Stull JT, Cooke R (1985) The effect of myosin phosphorylation on the contractile properties of skinned rabbit skeletal muscle fibres. J Biol Chem 260:7951–7954

Pette D (1984) Activity-induced fast to slow transitions in mammalian muscle. Med Sci Sports Exerc 16:517–528

Pfitzer G, Rüegg JC (1982) Molluscan catch muscle: Regulation and mechanics in living and skinned anterior byssus retractor muscle of *Mytilus edulis*. J Comp Physiol 147:137–142

Pfitzer G, Rüegg JC, Flockerzi V, Hofmann F (1982) cGMP-dependent protein kinase decreases calcium sensitivity of skinned cardiac fibers. FEBS Lett 149:171–175

Pfitzer G, Hofmann F, DiSalvo J, Rüegg JC (1984) cGMP and cAMP inhibit tension development in skinned coronary arteries. Pflügers Arch Eur J Physiol 401:277–280

Pfitzer G, Rüegg JC, Zimmer M, Hofmann F (1985) Relaxation of skinned coronary arteries depends on the relative concentrations of Ca^{2+}, calmodulin and active cAMP-dependent protein kinase. Pflügers Arch Eur J Physiol 405:70–76

Pfitzer G, Merkel L, Rüegg JC, Hofmann F (1986) Cyclic GMP-dependent kinase relaxes skinned fibers from guinea pig *Taenia coli* but not from chicken gizzard. Pflügers Arch Eur J Physiol 407:87–91

Pfitzer G, Zeugner C, Chalovich JM (1992) Effect of caldesmon and fragments of caldesmon on isometric force in skinned fibers from chicken gizzard. Biophys J 61:A7

Philipps GN Jr, Fillers JP, Cohen C (1986) Tropomyosin crystal structure and muscle regulation. J Mol Biol 192:111–131

Philipson KD, Bers DM, Nishimoto AY, Langer GA (1980) Binding of Ca^{2+} and Na^{+} sarcolemmal membranes: relation to control of myocardial contractility. Am J Physiol 238:H373–H378

Pifl C, Plank B, Wyskovsky W, Bertel O, Hellmann G, Suko J (1984) Calmodulin $(Ca^{2+})_4$ is the active calmodulin-calcium species activating the calcium-, calmodulin-dependent protein kinase of cardiac sarcoplasmic reticulum in the regulation of the calcium pump. Biochim Biophys Acta 773:197–206

Pitts BJR (1979) Stoichiometry of sodium-calcium exchange in cardiac sarcolemmal vesicles. J Biol Chem 354:6232–6235

Pizarro G, Cleemann L, Morad M (1985) Optical measurement of voltage-dependent Ca^{2+} influx in frog heart. Proc Natl Acad Sci USA 82:1864–1868

Pliszka B, Strzelecka-Golaszewska H (1981) Comparison of myosin isoenzymes from slow-tonic and fast-twitch fibers of frog muscle. Eur J Cell Biol 25:144–149

Podolsky RJ (1968) Membrane systems in muscle cells. In: Miller PL (ed) Aspects of cell motility, vol XXII. Symposia of the Society for Experimental Biology. University Press, Cambridge, pp 87–99

Podolsky RJ, Teichholz LE (1970) The relation between calcium and contraction kinetics in skinned muscle fibres. J Physiol (Lond) 211:19–35

Pollack GH (1983) The cross-bridge theory. Physiol Rev 63:1049–1113

Popescu LM, Nutu O, Panoiu C (1985 a) Oxytocin contracts the human uterus at term by inhibiting the myometrical Ca^{2+}-extrusion pump. Biosci Rep 5:21–28

Popescu LM, Panoiu C, Hinescu M, Nutu O (1985 b) The mechanism of cGMP-induced relaxation in vascular smooth muscle. Eur J Pharmacol 107:393–394

Porter KR, Palade GE (1957) Studies on the endoplasmic reticulum. III. Its form and distribution in striated muscle cells. J Biophys Biochem Cytol 3:269–300

Portzehl H (1957) Die Bindung des Erschlaffungsfaktors von Marsh an die Muskelgrana. Biochim Biophys Acta 26:373–377

Portzehl H, Caldwell PC, Rüegg JC (1964) The dependence of contraction and relaxation of muscle fibres from the crab *Maia squinado* on the internal concentration of free calcium ions. Biochim Biophys Acta 79:581–591

Portzehl H, Zaoralek P, Grieder A (1965) Der Calcium-Spiegel in lebenden und isolierten Muskelfibrillen von *Maia squinado* und seine Regulierung durch die sarkoplasmatischen Vesikel. Pflügers Arch Eur J Physiol 286:44–56

Potter JD, Gergely J (1975) The calcium and magnesium binding sites of troponin and their role in the regulation of myofibrillar adenosinetriphosphatase. J Biol Chem 250:4628–4633

Potter JD, Johnson JD (1982) Troponin. In: Cheung W (ed) Calcium and cell function, vol II. Academic Press, New York, pp 145–173

Potter JD, Johnson JD, Mandel F (1978) Fluorescence stopped flow measurements of Ca^{2+} and Mg^{2+} binding to parvalbumins. Fed Proc 37:1608

Potter JD, Holroyde MJ, Robertson SP, Solaro RJ, Kranias EG, Johnson JD (1982) The regulation of cardiac-muscle contraction by troponin. In: Dowben RM, Shay JW (eds) Cell and muscle motility, vol 2. Plenum, New York, pp 245–255
Potter JD, Sheng Z, Miller T, Strauss W (1991) Both Ca^{2+}-specific sites of skeletal muscles TnC are required for full activity. Biophys J 59:582 a
Powell T (1985) The isolation and characterization of calcium-tolerant myocytes. In: Spieckermann PG, Piper HM (eds) Isolated adult cardiac myocytes. Structure, function, and metabolism. Basic Res Cardiol 80 (Suppl). Steinkopff, Darmstadt, Springer, Berlin Heidelberg New York, pp 15–18
Pringle JWS (1954) The mechanism of the myogenic rhythm of certain insect striated muscles. J Physiol (Lond) 124:269–291
Pringle JWS (1957) Insect flight. Cambridge Univ Press, London
Pringle JWS (1978) Stretch activation of muscle: function and mechanism. The Croonian Lecture, 1977. Proc R Soc (Lond) B 201:107–130
Pringle JWS (1981) The evolution of fibrillar muscle in insects. J Exp Biol 94:1–14
Pritchard K, Ashley CC (1986) Na^{+}/Ca^{2+} exchange in isolated smooth muscle cells demonstrated by the fluorescent calcium indicator fura-2. FEBS Lett 195:23–27
Prosser CL (1960) Comparative physiology of activation of muscles with particular attention to smooth muscles. In: Bourne GH (ed) The structure and function of muscle. Structure, part I, vol 2. Academic Press, New York, pp 387–434
Prosser CL (1967) Ionic analyses and effects of ions on contractions of sponge tissues. Z Vgl Physiol 54:109–120
Prosser CL (1973) Comparative animal physiology, 3rd edn. Saunders, Philadelphia
Puceat M, Clement O, Lechene P, Pelosin JM, Ventura-Clapier R, Vassort G (1990) Neurohormonal control of calcium sensitivity of myofilaments in rat single heart cells. Circ Res 67:517–524
Pulliam DL, Sawyna V, Levine RJC (1983) Calcium sensitivity of vertebrate skeletal muscle myosin. Biochemistry 22:2324–2331
Putkey JA, Sweeney HL, Campbell ST (1989) Site-directed mutation of the trigger calcium-binding sites in cardiac troponin C. J Biol Chem 264:12370–12378
Racker E (1972) Reconstitution of a calcium pump with phospholipids and a purified Ca^{2+} adenosine triphosphatase from sarcoplasmic reticulum. J Biol Chem 247:8198–8200
Raeymaekers L, Hasselbach W (1981) Ca^{2+} uptake, Ca^{2+} ATPase activity, phosphoprotein formation and phosphate turnover in a microsomal fraction of smooth muscle. Eur J Biochem 116:373–378
Rakowski RF, Best PM, James-Kracke MR (1985) Voltage dependence of membrane charge movement and calcium release in frog skeletal muscle fibres. J Muscle Res Cell Mot 6:403–433
Rall JA (1981) Mechanics and energetics of contraction in striated muscle of the sea scallop, *Placopecten magellanicus*. J Physiol (Lond) 321:287–295
Rall JA (1982) Energetics of Ca^{2+} cycling during skeletal muscle contraction. Fed Proc 41:155–160
Rall JA, Schottelius BA (1973) Energetics of contraction in phasic and tonic skeletal muscles of the chicken. J Gen Physiol 62:303–323
Rapoport RM, Murad F (1983) Agonist-induced endothelium-dependent relaxation in rat thoracic aorta may be mediated through cGMP. Circ Res 52:352–357
Rapoport RM, Draznin MB, Murad F (1983) Endothelium-dependent relaxation in rat aorta may be mediated through cyclic GMP-dependent protein phosphorylation. Nature (Lond) 306:174–176
Rasmussen H, Barrett PQ (1984) Calcium messenger system: an integrated view. Physiol Rev 64:938–984
Rathmayer W, Erxleben C (1983) Identified muscle fibers in a crab. I. Characteristics of excitatory and inhibitory neuromuscular transmission. J Comp Physiol 152:411–420
Ray KP, England PJ (1976) Phosphorylation of the inhibitory subunit of troponin and its effects on the calcium dependence of cardiac myofibrillar ATPase. FEBS Lett 70:11–16
Reedy MK, Holmes KC, Tregear RT (1965) Induced changes in orientation of the cross-bridges of glycerinated insect flight muscle. Nature (Lond) 207:1276–1280
Reeves JP, Sutko JL (1979) Sodium-calcium ion exchange in cardiac membrane vesicles. Proc Natl Acad Sci USA 76:590–594
Reeves JP, Sutko JL (1980) Sodium-calcium exchange activity generates a current in cardiac membrane vesicles. Science 208:1461–1464
Regenstein JM, Szent-Györgyi AG (1975) Regulatory proteins of lobster striated muscle. Biochemistry 14:917–925

Reinach FC, Karlsson R (1988) Cloning, expression and site-directed mutagenesis of chicken skeletal muscle troponin C. J Biol Chem 263:2371–2376

Reinach FC, Nagai K, Kendrick-Jones J (1986) Site-directed mutagenesis of the regulatory light-chain Ca^{2+}/Mg^{2+} binding site and its role in hybrid myosins. Nature (Lond) 322:80–83

Reinlib L, Caroni P, Carafoli E (1981) Studies on heart sarcolemma: vesicles of opposite orientation and the effect of ATP on the Na^{+}-Ca^{2+} exchanger. FEBS Lett 126:74–76

Reiser PJ, Moss RL, Giulian GG, Greaser ML (1985) Shortening velocity in single fibers from adult rabbit soleus muscles is correlated with myosin heavy chain composition. J Biol Chem 260:9077–9080

Reiter M, Vierling W, Seibel K (1984) Excitation-contraction coupling in rested-state contractions of guinea-pig ventricular myocardium. Naunyn-Schmiedeberg's Arch Pharmacol 325:159–169

Rembold CM, Murphy RA (1986) Myoplasmic calcium, myosin phosphorylation, and regulation of the crossbridge cycle in swine arterial smooth muscle. Circ Res 58:803–815

Rembold CM, Murphy RA (1988 a) Myoplasmic [Ca^{2+}] determines myosin phosphorylation in agonist-stimulated swine arterial smooth muscle. Circ Res 63:593–603

Rembold CM, Murphy RA (1988 b) Ca^{2+}-dependent myosin phosphorylation in phorboldiester stimulated smooth muscle contraction. J Physiol (Lond) 429:77–94

Reuter H (1983) Calcium channel modulation by neurotransmitters, enzymes and drugs. Nature (Lond) 301:569–574

Reuter H (1984) Ion channels in cardiac cell membranes. Annu Rev Physiol 46:473–484

Reuter H, Scholz H (1977) The regulation of calcium conductance of cardiac muscle by adrenaline. J Physiol (Lond) 264:49–62

Reuter H, Seitz N (1968) The dependence of calcium efflux from cardiac muscle on temperature and external ion composition. J Physiol (Lond) 195:451–470

Reuter H, Blaustein MP, Häusler G (1973) Na-Ca exchange and tension development in arterial smooth muscle. Phil Trans R Soc (Lond) B 265:87–94

Revel JP (1962) The sarcoplasmic reticulum of the bat cricothyroid muscle. J Cell Biol 12:571–588

Richardot M, Wautier J (1971) Une structure intermédiaire entre muscle lisse et muscle strié. La fibre musculaire du bulbe buccal de *Ferrissia wautieri* (Moll Basomm Ancylidae). Z Zellforsch 115:100–109

Ridgway EB, Gordon AM (1984) Muscle calcium transient. Effect of post-stimulus length changes in single fibers. J Gen Physiol 83:75–103

Ridgway EB, Gordon AM, Martyn DA (1983) Hysteresis in the force-calcium relationship in muscle. Science 219:1075–1077

Ringer S (1883) A further contribution regarding the influence of the different constituents of the blood on the contraction of the heart. J Physiol (Lond) 4:29–42

Ríos E, Brum G (1987) Involvement of dihydropyridine receptors in excitation-contraction coupling in skeletal muscle. Nature (Lond) 325:717–720

Ríos E, Pizarro G (1991) Voltage sensor of excitation-contraction coupling in skeletal muscle. Physiol Rev 71:849–908

Ríos E, Ma J, Gonzáles A (1991) The mechanical hypothesis of excitation-contraction (EC) coupling in skeletal muscle. J Muscle Res Cell Mot 12:127–135

Ritchie JM (1954) The duration of the plateau of full activity in frog muscle. J Physiol (Lond) 124:605–612

Robertson JD (1956) Some features of the ultrastructure of reptilian skeletal muscle. J Biophys Biochem Cytol 2:369–380

Robertson SP, Johnson JD, Potter JD (1981) The time-course of Ca^{2+} exchange with calmodulin, troponin, parvalbumin, and myosin in response to transient increases in Ca^{2+}. Biophys J 34:559–569

Robison GA, Butcher RW, Sutherland EW (1971) Cyclic AMP. Academic Press, New York

Rodgers RL, Black S, Katz S, McNeill JH (1986) Thyroidectomy of SHR: effects on ventricular relaxation and on SR calcium uptake activity. Am J Physiol 250:H861–865

Röhrkasten A, Meyer HE, Nastainczyk W, Sieber M, Hofmann F (1988) cAMP-dependent protein kinase rapidly phosphorylates serine-687 of the skeletal muscle receptor for Ca channel blockers. J Biol Chem 263:15325–15329

Rome LC, Funke RP, Alexander RM, Lutz G, Aldridge H, Scott F, Freadman M (1988) Why animals have different muscle fibre types. Nature (Lond) 335:824–827

Rosenbluth J (1965) Ultrastructural organization of the obliquely striated muscle fibres in *Asacaris lumbricoides*. J Cell Biol 25:495–515
Rosenbluth J (1968) Obliquely striated muscle. IV. Sarcoplasmic reticulum, contractile apparatus, and endomysium of the body muscle of a polychaete, *Glycera*, in relation to its speed. J Cell Biol 36:245–259
Rosenbluth J (1969) Sarcoplasmic reticulum of an unusually fast-acting crustacean muscle. J Cell Biol 42:534–547
Rosenbluth J (1972) Obliquely striated muscle. In: Bourne GH (ed) The structure and function of muscle, vol 1, part 1. Academic Press, New York, pp 389–419
Rougier O, Vassort G, Garnier D, Gargouil YM, Coraboeuf E (1969) Existence and role of a slow inward current during the frog arterial action potential. Pflügers Arch Eur J Physiol 308:91–110
Rowlerson A, Scapolo PA, Mascarello F, Carpenè E, Veggetti A (1985) Comparative study of myosins present in the lateral muscle of some fish: species variations in myosin isoforms and their distribution in red, pink and white muscle. J Muscle Res Cell Mot 6:601–640
Rüegg JC (1961) On the tropomyosin-paramyosin system in relation to the viscous tone of lamellibranch catch muscle. Proc R Soc (Lond) B 154:224–249
Rüegg JC (1963) Die Unabhängigkeit des plastischen Tonus von der Kontraktilität bei glatten Muskeln von Säugern. Pflügers Arch Ges Physiol 278:R18–19
Rüegg JG (1964) Tropomyosin-paramyosin system and "prolonged contraction" in a molluscan smooth muscle. Proc R Soc (Lond) B 160:536–542
Rüegg JC (1965) Physiologie und Biochemie des Sperrtonus. Helvetica Physiologica et Pharmacologica Acta (Suppl XVI):1–76
Rüegg JC (1968a) Contractile mechanisms of smooth muscle. In: Miller PL (ed) Aspects of cell motility. Cambridge Univ Press, Cambridge, pp 45–66
Rüegg JC (1968b) Oscillatory mechanism in fibrillar insect flight muscle. Experientia 24:529–536
Rüegg JC (1971) Smooth muscle tone. Physiol Rev 51:201–248
Rüegg JC (1986) Effects of new inotropic agents on Ca^{2+} sensitivity of contractile proteins. Circulation 73 (Suppl III):78–84
Rüegg JC (1990a) Muscle protein interaction; competition by peptide mimetics. J Muscle Res Cell Mot 11:189–190
Rüegg JC (1990b) Towards a molecular understanding of contractility. Cardioscience 1:163–167
Rüegg JC, Paul RJ (1982) Vascular smooth muscle: calmodulin and cyclic AMP-dependent protein kinase alter calcium sensitivity in porcine carotid skinned fibres. Circ Res 50:394–399
Rüegg JC, Pfitzer G (1985) Modulation of calcium sensitivity in guinea pig *Taenia coli*: skinned fiber studies. Experientia 41:997–1001
Rüegg JC, Strassner E (1963) Sperrtonus und Nucleosidtriphosphate. Z Naturforsch 18b:133–138
Rüegg JC, Tregear RT (1966) Mechanical factors affecting the ATPase activity of glycerol-extracted insect fibrillar flight muscle. Proc R Soc (Lond) B 165:497–512
Rüegg JC, Weber HH (1963) Kontraktionszyklus und Sperrtonus. In: Cori CF, Foglia VG, Leloir LF, Ochoa S (eds) Perspectives in biology. Elsevier/North-Holland, Amsterdam, pp 301–320
Rüegg JC, Steiger GJ, Schädler M (1970) Mechanical activation of the contractile system in skeletal muscle. Pflügers Arch Eur J Physiol 319:139–145
Rüegg JC, Sparrow MP, Mrwa U (1981) Cyclic-AMP mediated relaxation of chemically skinned fibers of smooth muscle. Pflügers Arch Eur J Physiol 390:198–201
Rüegg JC, Pfitzer G, Zimmer M, Hofmann F (1984a) The calmodulin fraction responsible for contraction in an intestinal smooth muscle. FEBS Lett 170:383–386
Rüegg JC, Pfitzer G, Eubler D, Zeugner C (1984b) Effect on contractility of skinned fibres from mammalian heart and smooth muscle by a new benzimidazole derivative, [4,5-dihydro-6-2-(4-methoxyphenyl)-1H-benzimidazol-5-yl]-5-methyl-3(2H)-pyridazinone. Arzneim-Forsch/Drug Res 34, 12:1736–1738
Rüegg JC, Kuhn HJ, Güth K, Pfitzer G, Hofmann F (1984c) Tension transients in skinned muscle fibres of insect flight muscle and mammalian cardiac muscle: effect of substrate concentration and treatment with myosin light chain kinase. In: Pollack G, Sugi H (eds) Contractile mechanisms in muscle. Advances in experimental medicine and biology, vol 170. Plenum Press, New York, pp 605–615

Rüegg JC, Zeugner C, Van Eyk J, Kay CM, Hodges RS (1989) Inhibition of TnI-TnC interaction and contraction of skinned muscle fibres by the synthetic peptide TnI [104–115]. Pflügers Arch Eur J Physiol 414:430–436

Rüegg JC, Zeugner C, Van Eyk JE, Hodges RS, Trayer IP (1991) Myosin and troponin peptides affect calcium sensitivity of skinned muscle fibres. In: Rüegg JC (ed) Peptides as probes in muscle research. Springer, Berlin Heidelberg New York, pp 95–110

Rumberger E, Reichel H (1972) The force-frequency relationship: a comparative study between warm- and cold-blooded animals. Pflügers Arch Eur J Physiol 332:206–217

Rupp H (1982) Polymorphic myosin as the common determinant of myofibrillar ATPase in different haemodynamic and thyroid states. Basic Res Cardiol 77:34–46

Sabry MA, Dhoot GK (1991) Identification of and pattern of transitions of cardiac, adult slow and slow skeletal muscle-like embryonic isoforms of troponin T in developing rat and human skeletal muscles. J Muscle Res Cell Mot 12:262–270

Saida K, Breemen C van (1983) A possible Ca^{2+} induced Ca^{2+} release mechanism mediated by norepinephrine in vascular smooth muscle. Pflügers Arch Eur J Physiol 397:166–167

Saida K, Breemen C van (1984) Characteristics of the norepinephrine-sensitive Ca^{2+} store in vascular smooth muscle. Blood Vessels 21:43–52

Saito A, Inui M, Radermacher M, Frank J, Fleischer S (1988) Ultrastructure of the calcium release channel of sarcoplasmic reticulum. J Cell Biol 107:211–219

Saito A, Chadwick CC, Fleischer S (1990) Ultrastructure of the inositol triphosphate receptor (IP_3REC) from smooth muscle. Biophys J 57:285a

Sakmann B, Neher E (1983) Single-channel recording. Plenum Press, New York

Salviati G, Sorenson MM, Eastwood AB (1982) Calcium accumulation by the sarcoplasmic reticulum in two populations of chemically skinned human muscle fibers. J Gen Physiol 79:603–632

Salviati G, Betto R, Danieli Betto DD, Zeviani M (1984) Myofibrillar-protein isoforms and sarcoplasmic-reticulum Ca^{2+}-transport activity of single human muscle fibres. Biochem J 224:215–225

Sánchez JA, Stefani E (1983) Kinetic properties of calcium channels of twitch muscle fibres of the frog. J Physiol (Lond) 337:1–17

Sandow A (1952) Excitation-contraction coupling in muscular response. Yale J Biol Med 25:176–201

Sandow A (1965) Excitation-contraction coupling in skeletal muscle. Pharmacol Rev 17:265–320

Sanger JW (1971) Sarcoplasmic reticulum in the cross-striated adductor muscle of the bay scallop, *Aequipecten irridians*. Z Zellforsch 118:156–161

Sarkar S, Sréter FA, Gergely J (1971) Light chains of myosins from white, red and cardiac muscles. Proc Natl Acad Sci USA 68:946–950

Satyshur KA, Rao ST, Pyzalska D, Drendel W, Greaser M, Sundaralingam M (1988) Refined structure of chicken skeletal muscle troponin C in the two-calcium state at 2-Å resolution. J Biol Chem 263:1628–1647

Sayers ST, Bárány K (1983) Myosin light chain phosphorylation during contraction of turtle heart. FEBS Lett 154:305–310

Scarpa A, Graziotti P (1973) Mechanism for intracellular calcium regulation in heart. J Gen Physiol 62:756–772

Schachat FH, Diamond MS, Brandt PW (1987) Effect of different troponin T-tropomyosin combinations on thin filament activation. J Mol Biol 198:551–554

Schädler M (1967) Proportionale Aktivierung von ATPase-Aktivität und Kontraktionsspannung durch Calciumionen in isolierten contractilen Strukturen verschiedener Muskelarten. Pflügers Arch Eur J Physiol 296:70–90

Schatzmann HJ (1985) Calcium extrusion across the plasma membrane by the calcium-pump and the Ca^{2+}-Na^{+} exchange system. In: Marmé D (ed) Calcium and cell physiology. Springer, Berlin Heidelberg New York, pp 19–52

Schaub MC, Brunner UT, Huber PAJ (1990) Adaptive changes in sarcomere proteins of heart muscle. In: Pette D (ed) The dynamic state of muscle fibers. Walter de Gruyter, Berlin

Scheid CR, Honeyman TW, Fay FS (1979) Mechanism of β-adrenergic relaxation of smooth muscle. Nature (Lond) 277:32–36

Schelling P, Becker KH, Lues I, Minck KO, Schliep HJ, Weygandt H, Wolf HP (1991) In vivo characterization of the cardiotonic action of EMD 57033, a novel drug that increases left ventricular contractility via sensitization of the myofibrils to calcium. J Mol Cell Cardiol 23 (Suppl V):69

Schenk DB, Johnson LK, Schwartz K, Sista H, Scarborough RM, Lewicki JA (1985) Distinct atrial natriuretic factor receptor sites on cultured bovine aortic smooth muscle and endothelial cells. Biochem Biophys Res Commun 127:433–442

Schiaffino S, Margreth A (1969) Coordinated development of the sarcoplasmic reticulum and T-system during postnatal differentiation in rat skeletal muscle. J Cell Biol 41:855–875

Schirmer RH (1965) Die Besonderheiten des contractilen Proteins der Arterien. Biochem Z 343:269–282

Schneider M, Sparrow M, Rüegg JC (1981) Inorganic phosphate promotes relaxation of chemically skinned smooth muscle of guinea-pig *Taenia coli*. Experientia 37:980–982

Schneider MF, Chandler WK (1973) Voltage dependent charge movement in skeletal muscle: a possible step in excitation-contraction-coupling. Nature (Lond) 242:244–246

Scholz H (1984) Inotropic drugs and their mechanisms of action. In: Katz AM (ed) Basic concepts in cardiology. JACC 4:389–397

Schouten VJA, Deen van JK, De Tombe P, Verveen AA (1987) Force-interval relationship in heart muscle of mammals. Biophys J 51:13–26

Schultz KD, Böhme E, Kreye VAW, Schultz G (1979) Relaxation of hormonally stimulated smooth muscular tissues by the 8-bromo derivative of cyclic GMP. Naunyn-Schmiedeberg's Arch Pharmacol 306:1–9

Schulz GE, Schirmer RH (1978) Principles of protein structure. Springer, Berlin Heidelberg New York

Schumacher T (1972) Zum Mechanismus der ökonomischen Halteleistung eines glatten Muskels (Byssus retractor anterior, *Mytilus edulis*). Pflügers Arch Eur J Physiol 331:77–89

Schwartz LM, McCleskey EW, Almers W (1985) Dihydropyridine receptors in muscle are voltage-dependent but most are not functional calcium channels. Nature (Lond) 314:747–750

Schwarzenbach G, Senn H, Anderegg G (1957) Komplexone. XXIX. Ein großer Chelateffekt besonderer Art. Helv Chim Acta 40:1886–1900

Sellers JR (1981) Phosphorylation-dependent regulation of *Limulus* muscle myosin. J Biol Chem 256:9274–9278

Sellers JR (1985) Mechanism of the phosphorylation-dependent regulation of smooth muscle heavy meromyosin. J Biol Chem 260:15815–15819

Sellers JR, Chantler PD, Szent-Györgyi AG (1980) Hybrid formation between scallop myofibrils and foreign regulatory light-chains. J Mol Biol 144:223–245

Sellers JR, Spudich JA, Sheetz MP (1985) Light chain phosphorylation regulates the movement of smooth muscle myosin on actin filaments. J Cell Biol 101:1897–1902

Shaw GS, Hodges RS, Sykes BD (1990) Calcium-induced peptide association to form an intact protein domain: ^{1}H NMR structural evidence. Science 249:280–283

Shaw GS, Golden LF, Hodges RS, Sykes BD (1991 a) Interactions between paired calcium-binding sites in proteins: NMR determination of the stoichiometry of calcium binding to a synthetic troponin-C peptide. J Am Chem Soc 113:5557–5563

Shaw GS, Hodges RS, Sykes BD (1991 b) Probing the relationship between α-helix formation and calcium affinity in troponin-C: ^{1}H NMR studies of calcium binding to synthetic and variant site III helix-loop-helix peptides. Biochemistry 30:8339–8347

Sheetz MP, Spudich JA (1983) Movement of myosin-coated fluorescent beads on actin cables in vitro. Nature (Lond) 303:31–35

Sheng Z, Pan B, Francois JM, Penniston JT, Potter JD (1991) Evidence for the mechanism of inhibition of TnC-regulation of muscle contraction by a calmodulin binding peptide. Biophys J 59:581 a

Shepherd AP, Mao CC, Jacobson ED, Shanbour LL (1973) The role of cyclic AMP in mesenteric vasodilation. Microvascular Res 6:332–341

Shimomura O, Johnson FH, Saiga Y (1962) Extraction, purification and properties of aequorin, a bioluminescent protein from the luminous hydromedusan, *Aequorea*. J Cell Comp Physiol 59:223–239

Shiner JS, Solaro RJ (1982) Activation of thin-filament-regulated muscle by calcium ions: considerations based on nearest neighbor lattice statistics. Proc Natl Acad Sci USA 79:4637–4641

Shiner JS, Solaro RJ (1984) The Hill-coefficient for the Ca^{2+}-activation of striated muscle contraction. Biophys J 46:541–543

Siegman MJ, Butler TM, Mooers SU, Davies RE (1980) Chemical energetics of force development, force maintenance, and relaxation in mammalian smooth muscle. J Gen Physiol 76:609–629

Siegman MJ, Butler TM, Mooers SU, Michalek A (1984) Ca^{2+} can affect V_{max} without changes in myosin light chain phosphorylation in smooth muscle. Pflügers Arch Eur J Physiol 401:385–390
Siegman MJ, Butler TM, Mooers SU (1985) Energetics and regulation of crossbridge states in mammalian smooth muscle. Experientia 41:1020–1025
Silver PJ, DiSalvo J (1979) Adenosine 3′:5′-monophosphate-mediated inhibition of myosin light chain phosphorylation in bovine aortic actomyosin. J Biol Chem 254:9951–9954
Silver PJ, Schmidt-Silver C, DiSalvo J (1982) β-Adrenergic relaxation and cAMP kinase activation in coronary arterial smooth muscle. Am J Physiol 242:H177–H184
Silver PJ, Maximilian Buja L, Stull JT (1986) Frequency-dependent myosin light chain phosphorylation in isolated myocardium. J Mol Cell Cardiol 18:31–37
Simmons RM, Szent-Györgyi AG (1980) Control of tension development in scallop muscle fibres with foreign regulatory light chains. Nature (Lond) 286:626–628
Simmons RM, Szent-Györgyi AG (1985) A mechanical study of regulation in the striated adductor muscle of the scallop. J Physiol (Lond) 358:47–64
Simon W, Ammann D, Oehme M, Morf WE (1977) Calcium-selective electrodes. Ann NY Acad Sci 307:52–70
Skaer HLB (1974) The water balance of the serpulid polychaete *Mercierella enigmatica* (Fauvel). IV. The excitability of the longitudinal muscle cells. J Exp Biol 60:351–370
Sleep JA, Hutton RL (1980) Exchange between inorganic phosphate and adenosine 5′-triphosphate in the medium by actomyosin subfragment 1. Biochemistry 19:1276–1283
Small JV (1974) Contractile units in vertebrate smooth muscle cells. Nature (Lond) 249:324–327
Small JV, Squire JM (1972) Structural basis of contraction in vertebrate smooth muscle. J Mol Biol 67:117–149
Small JV, Fürst DO, De Mey J (1986) Localization of filamin in smooth muscle. J Cell Biol 102:210–220
Smillie LB, Golosinska K, Reinach F (1988) Sequences of complete cDNAs encoding four variants of chicken skeletal muscle troponin T. J Biol Chem 263:18816–18820
Smith DS (1966) The organization and function of the sarcoplasmic reticulum and T-system of muscle cells. Prog Biophys Molec Biol 16:107–142
Smith ICH (1972) Energetics of activation in frog and toad muscle. J Physiol (Lond) 220:583–599
Smith JS, Coronado R, Meissner G (1985) SR contains adenine nucleotide-activated calcium channels. Nature (Lond) 316:446–449
Smith JS, Coronado R, Meissner G (1986) Single channel measurements of the Ca^{2+}-release channel from skeletal muscle SR. J Gen Physiol 88:573–588
Smith JS, Imagawa T, Ma J, Fill M, Campbell KP, Coronado R (1988) Purified ryanodine receptor from rabbit skeletal muscle is the calcium-release channel of sarcoplasmic reticulum. J Gen Physiol 92:1–26
Smith JS, Rousseau E, Meissner G (1989) Calmodulin modulation of single sarcoplasmic reticulum Ca^{2+}-release channels from cardiac and skeletal muscle. Circ Res 64:352–359
Smith SJ, England PJ (1990) The effects of reported Ca^{2+} sensitisers on the rates of Ca^{2+} release from cardiac troponin-C and the troponin-tropomyosin complex. Br J Pharmacol 100:779–785
Sobieszek A (1973) The fine structure of the contractile apparatus of the anterior byssus retractor muscle of *Mytilus edulis*. J Ultrastruct Res 43:313–343
Sobieszek A (1977) Ca^{2+}-linked phosphorylation of a light chain of vertebrate smooth muscle myosin. Eur J Biochem 73:477–483
Sobieszek A, Small JV (1976) Myosin-linked calcium regulation in vertebrate smooth muscle. J Mol Biol 102:75–92
Sobue K, Sellers JR (1991) Caldesmon, a novel regulatory protein in smooth muscle and nonmuscle actomyosin systems. J Biol Chem 266:12115–12118
Sobue KM, Morimoto K, Kanda M, Fukunaga E, Myamoto E, Kakiuchi S (1982) Interaction of 135000 M_r calmodulin binding protein (myosin kinase) and F-actin: another Ca^{2+}- and calmodulin dependent flip flop switch. Biochem Intern 5:503–510
Sohma H, Yazawa M, Morita F (1985) Phosphorylation of regulatory light chain a (RLC-a) in smooth muscle myosin of Scallop, *Patinopecten yessoensis*. J Biochem 98:569–572
Sohma H, Inoue K, Morita F (1988) A cAMP-dependent regulatory protein for RLC-a myosin kinase catalyzing the phosphorylation of scallop smooth muscle myosin light chain. J Biochem (Tokyo) 103:431–435

Solaro RJ, Rüegg JC (1982) Stimulation of Ca^{2+} binding and ATPase activity of dog cardiac myofibrils by AR-L 115 BS, a novel cardiotonic agent. Circ Res 51:290–294

Solaro RJ, Moir AGJ, Perry SV (1976) Phosphorylation of troponin I and the inotropic effect of adrenaline in the perfused rabbit heart. Nature (Lond) 262:615–616

Solaro RJ, Robertson SP, Johnson JD, Holroyde MJ, Potter JD (1981) Troponin-I phosphorylation: a unique regulator of the amounts of calcium required to activate cardiac myofibrils. Cold Spring Harbor Conf Cell Proliferation 8:901–911

Solaro RJ, Kumar P, Blanchard EM, Martin AF (1986) Differential effects of pH on calcium activation of myofilaments of adult and perinatal dog hearts: evidence for developmental differences in thin filament regulation. Circ Res 58:721–729

Somlyo AP (1967) Discussion remarks to Twarog BM. The regulation of catch in molluscan muscle. J Gen Physiol (Suppl) 50:157–169

Somlyo AP (1985a) The messenger across the gap. Nature (Lond) 316:298–299

Somlyo AP (1985b) Excitation-contraction coupling and the ultrastructure of smooth muscle. Circ Res 57:497–507

Somlyo AP, Himpens B (1989) Cell calcium and its regulation in smooth muscle. FASEB J 3:2266–2276

Somlyo AP, Somlyo AV (1968) Vascular smooth muscle. I. Normal structure, pathology, biochemistry, and biophysics. Pharmacol Rev 20:197–272

Somlyo AP, Somlyo AV, Shuman H, Endo M (1982) Calcium and monovalent ions in smooth muscle. Fed Proc 41:2883–2890

Somlyo AV, Somlyo AP (1968) Electromechanical and pharmacomechanical coupling in vascular smooth muscle. J Pharmacol Exp Ther 159:129–145

Somlyo AV, Ashton FT, Lemanski L, Vallieres J, Somlyo AP (1977a) Filament organization and dense bodies in vertebrate smooth muscle. In: Stephens NL (ed) Biochemistry of smooth muscle. Univ Park, Baltimore, pp 445–471

Somlyo AV, Shuman H, Somlyo AP (1977b) Elemental distribution in striated muscle and the effects of hypertonicity. Electron probe analysis of cryo sections. J Cell Biol 74:828–857

Somlyo AV, González-Serratos H, Shuman H, McClellan G, Somlyo AP (1981) Calcium release and ionic changes in the sarcoplasmic reticulum of tetanized muscle. An electron probe study. J Cell Biol 90:577–594

Somlyo AV, Bond M, Somlyo AP, Scarpa A (1985) Inositol trisphosphate-induced calcium release and contraction in vascular smooth muscle. Proc Natl Acad Sci USA 82:5231–5235

Sommer JR, Johnson EA (1969) Cardiac muscle: A comparative ultrastructural study with special reference to frog and chicken hearts. Z Zellforsch 98:437–468

Sommer JR, Johnson EA (1979) Ultrastructure of cardiac muscle. In: Berne RM, Sperelakis N, Geiger SR (eds) The cardiovascular system. Am Physiol Soc, Bethesda, Maryland, pp 113–186 (Handbook of physiology, sect 2, vol I)

Sparrow MP, Bockxmeer FM van (1972) Arterial tropomyosin and a relaxing protein fraction from vascular smooth muscle. Comparison with skeletal tropomyosin and troponin. J Biochem Tokyo 72:1075–1080

Sparrow MP, Maxwell LC, Rüegg JC, Bohr DF (1970) Preparation and properties of a calcium ion-sensitive actomyosin from arteries. Am J Physiol 219:1366–1372

Sparrow MP, Mrwa U, Hofmann F, Rüegg JC (1981) Calmodulin is essential for smooth muscle contraction. FEBS Lett 125:141–145

Sperelakis N, Caulfield JB (eds) (1984) Calcium antagonists. Mechanisms of action on cardiac muscle and vascular smooth muscle. Martinus Nijhoff, Boston

Squire JM (1981) The structural basis of muscle contraction. Plenum Press, New York

Srihari T, Tuchschmid CR, Schaub MC (1982) Isoforms of heavy and light chains of cardiac myosins from rat and rabbit. Basic Res Cardiol 77:599–609

Stafford WF, Szent-Györgyi AG (1978) Physical characterization of myosin light chains. Biochemistry 17:607–614

Stafford WF, Szentkiralyi EV, Szent-Györgyi AG (1979) Regulatory properties of single-headed fragments of scallop myosin. Biochemistry 24:5273

Stedman HH, Sweeney HL, Shrager JB, Maguire HC, Panettieri RA, Petrof B, Narusawa M, Leferowich JM, Sladky JT, Kelly AM (1991) The mdx mouse diaphragm reproduces the degenerative changes of Duchenne muscular dystrophy. Nature (Lond) 352:536–539

Steenbergen C, Murphy E, Levy L, London RE (1987) Elevation in cytosolic free calcium concentration early in myocardial ischemia in perfused rat heart. Circ Res 60:700–707

Steenbergen C, Murphy E, Watts JA, London RE (1990) Correlation between cytosolic free calcium, contracture, ATP, and irreversible ischemic injury in perfused rat heart. Circ Res 66:135–146

Steiger GJ (1977) Stretch activation and tension transients in cardiac, skeletal and insect flight muscle. In: Tregear RT (ed) Insect flight muscle. Elsevier/North-Holland, Amsterdam, pp 221–268

Steiger GJ (1979) Kinetic analysis of isometric tension transients in cardiac muscle. In: Sugi H, Pollack GH (eds) Cross-bridge mechanism in muscle contraction. University Press Tokyo, pp 259–274

Steiger GJ, Rüegg JC (1969) Energetics and "efficiency" in the isolated contractile machinery of an insect fibrillar muscle at various frequencies of oscillation. Pflügers Arch Eur J Physiol 307:1–21

Stein LA, Schwarz RP, Chock PB, Eisenberg E (1979) Mechanism of actomyosin adenosine triphosphatase. Evidence that adenosine 5′-triphosphate hydrolysis can occur without dissociation of the actomyosin complex. Biochemistry 18:3895–3909

Stein P, Palade P (1988) Sarcoballs: direct access to sarcoplasmic reticulum Ca^{2+}-channels in skinned frog muscle fibers. Biophys J 54:357–363

Steinmeyer K, Klocke R, Ortland C, Gronemeier M, Jockusch H, Gründer S, Jentsch TJ (1991) Inactivation of muscle chloride channel by transposon insertion in myotonic mice. Nature (Lond) 354:304–308

Sten-Knudsen O (1960) Is muscle contraction initiated by internal current flow? J Physiol (Lond) 151:363–384

Stephenson DG, Wendt IR (1984) Length dependence of changes in sarcoplasmic calcium concentration and myofibrillar calcium sensitivity in striated muscle fibres. J Muscle Res Cell Mot 5:243–272

Stephenson DG, Williams DA (1980) Activation of skinned arthropod muscle fibres by Ca^{2+} and Sr^{2+}. J Muscle Res Cell Mot 1:73–87

Stephenson DG, Williams DA (1981) Calcium-activated force responses in fast- and slow-twitch skinned muscle fibres of the rat at different temperatures. J Physiol (Lond) 317:281–302

Stephenson DG, Williams DA (1982) Effects of sarcomere length on the force-pCa relation in fast- and slow-twitch skinned muscle fibres from the rat. J Physiol (Lond) 333:637–653

Stephenson DG, Williams DA (1983) Slow amphibian muscle fibres become less sensitive to Ca^{2+} with increasing sarcomere length. Pflügers Arch Eur J Physiol 397:248–250

Stephenson DG, Wendt IR, Forrest QG (1981) Non-uniform ion distribution and electrical potentials in sarcoplasmic regions of skeletal muscle fibres. Nature (Lond) 289:690–692

Stephenson EW (1978) Properties of chloride-stimulated ^{45}Ca flux in skinned muscle fibers. J Gen Physiol 71:411–430

Stephenson EW (1982) The role of free calcium ion in calcium release in skinned muscle fibers. Can J Physiol Pharmacol 60:417–426

Stephenson EW (1985) Excitation of skinned muscle fibers by imposed ion gradients. I. Stimulation of ^{45}Ca efflux at constant [K] [Cl] product. J Gen Physiol 86:813–832

Stokes DL (1991) P-type ion pumps: structure determination may soon catch up with structure predictions. Current Opinion Struct Biol 1:555–561

Stössel W, Zebe E (1968) Zur intracellulären Regulation der Kontraktionsaktivität. Pflügers Arch Eur J Physiol 302:38–56

Strauss JD, Zeugner C, Bletz C, Rüegg JC (1992) Calcium insensitive contraction in skinned procine cardiac muscle fibers following troponin-I (TnI) extraction. Biophys J 61:A18

Streb H, Irvine RF, Berridge MJ, Schulz I (1983) Release of Ca^{2+} from a nonmitochondrial intracellular store in pancreatic acinar cells by inositol-1,4,5-trisphosphate. Nature (Lond) 306:67–69

Strosberg AM, Katzung BG, Lee JC (1972) Glycerol removal treatment of guinea-pig cardiac muscle. J Mol Cell Cardiol 4:39–48

Stuhlfauth I, Reininghaus J, Jockusch H, Heizmann CW (1984) Calcium-binding protein, parvalbumin, is reduced in mutant mammalian muscle with abnormal contractile properties. Proc Natl Acad Sci USA 81:4814–4818

Stull JT, Blumenthal DK, Cooke R (1980) Regulation of contraction by myosin phosphorylation. A comparison between smooth and skeletal muscles. Biochem Pharmacol 29:2537–2543

Suarez-Kurtz G (1982) The role of calcium in excitation-contraction coupling in crustacean muscle fibres. Can J Physiol Pharmacol 60:446–458

Suematsu E, Hirata J, Hasimoto T, Kuriyama H (1984) Inositol 1,4,5-trisphosphate releases Ca^{2+} from intracellular store sites in skinned single cells of porcine coronary artery. Biochem Biophys Res Commun 120:481–485

Suematsu E, Resnick M, Morgan KG (1991) Change of Ca^{2+} requirement for myosin phosphorylation by prostaglandin $F_{2\alpha}$. Am J Physiol 261:C253–258

Sugi H, Suzuki S (1978) Ultrastructural and physiological studies on the longitudinal body wall muscle of *Dolabella auricularia*. I. Mechanical response and ultrastructure. J Cell Biol 79:454–466

Sugi H, Yamaguchi T (1976) Activation of the contractile mechanism in the anterior byssal retractor muscle of *Mytilus edulis*. J Physiol (Lond) 257:531–547

Sugi H, Suzuki S, Daimon T (1982) Intracellular calcium translocation during contraction in vertebrate and invertebrate smooth muscles as studied by pyroantimonate method. Can J Physiol Pharmacol 60:576–587

Sutherland C, Walsh MP (1989) Phosphorylation of caldesmon prevents its interaction with smooth muscle myosin. J Biol Chem 264:578–583

Sütsch G, Brunner UT, Schulthess C von, Hirzel HO, Hess OM, Turina M, Krayenbuehl HP, Schaub MC (1992) Hemodynamic performance and myosin light chain-1 expression of the hypertrophied left ventricle in aortic valve disease before and after valve replacement. Circ Res 70:1035–1043

Suzuki H, Stafford WF III, Slayter HS, Seidel JC (1985) A conformational transition in gizzard heavy meromyosin involving the head-tail junction, resulting in changes in sedimentation coefficient, ATPase activity, and orientation of heads. J Biol Chem 260:14810–14817

Suzuki R, Morita F, Nishi N, Tokura S (1990) Inhibition of actomyosin subfragment 1 ATPase activity by analog peptides of the actin-binding site around the Cys (SH1) of myosin heavy chain. J Biol Chem 265:4939–4943

Swanson CJ (1971) Isometric response of the paramyosin smooth muscle of *Paragordius varius* (Leidy) (Aschelminthes, Nematomorpha). Z Vgl Physiol 74:403–410

Sweeney HL, Stull JT (1990) Alteration of cross-bridge kinetics by myosin light chain phosphorylation in rabbit skeletal muscle: implications for regulation of actin-myosin interaction. Proc Natl Acad Sci USA 87:414–418

Swynghedauw B (1986) Developmental and functional adaptation of contractile proteins in cardiac and skeletal muscles. Physiol Rev 66:710–771

Swynghedauw B, Delcayre C, Moalic J-M, LeCarpentier Y, Ray A, Mercadier JJ, Lompre A-M, Aumont M-C, Schwartz K (1982) Isoenzymic changes in myosin and hypertrophy; adaptation during chronic mechanical overload. Eur Heart J 3 (Suppl A):75–82

Szent-Györgyi A (1953) Chemical physiology of contraction in body and heart muscle. Academic Press, New York

Szent-Györgyi AG (1975) Calcium regulation of muscle contraction. Biophys J 15:707–723

Szent-Györgyi AG, Chantler PD (1986) Control of contraction by myosins. In: Engel AJ, Danker BA (eds) Myology. McGraw-Hill, New York, pp 589–612

Szent-Györgyi AG, Szentkiralyi EM (1973) The light chains of scallop myosin as regulatory subunits. J Mol Biol 74:179–203

Szent-Györgyi AG, Cohen C, Kendrick-Jones J (1971) Paramyosin and the filaments of molluscan "catch" muscles. II. Native filaments: isolation and characterization. J Mol Biol 56:239–258

Szentkiralyi EM (1984) Tryptic digestion of scallop S1: evidence for a complex between the two light-chains and a heavy-chain peptide. J Muscle Res Cell Mot 5:147–164

Tada M, Kirchberger MA, Repke DI, Katz AM (1974a) The stimulation of calcium transport in cardiac sarcoplasmic reticulum by adenosine 3′:5′-monophosphate-dependent protein kinase. J Biol Chem 249:6174–6180

Tada M, Kirchberger MA, Katz AM (1974b) Phosphorylation of a 22K dalton component of the cardiac sarcoplasmic reticulum by adenosine 3′,5′ monophosphate dependent protein kinase. J Biol Chem 250:2641–2647

Tada M, Yamamoto T, Tonomura Y (1978) Molecular mechanism of active calcium transport by sarcoplasmic reticulum. Physiol Rev 58:1–79

Takahashi K, Hiwada K, Kokubu T (1988) Vascular smooth muscle calponin. A novel troponin T-like protein. Hypertension 11:620–626

Takahashi M, Sohma H, Morita F (1988) The steady state intermediate of scallop smooth muscle myosin ATPase and effect of light chain phosphorylation. A molecular mechanism for catch contraction. J Biochem 104:102–107

Takai A, Bialojan C, Troschka M, Rüegg JC (1987) Smooth muscle myosin phosphatase inhibition and force enhancement by black sponge toxin. FEBS Lett 217:81–84

Takeshima H, Nishimura S, Matsumoto T, Ishida H, Kangawa K, Minamino N, Matsuo H, Ueda M, Hanaoka M, Hirose T, Numa S (1989) Primary structure and expression form complementary DNA of skeletal muscle ryanodine receptor. Nature (Lond) 339:439–445

Takisawa H, Makinose M (1981) Occluded bound calcium on the phosphorylated sarcoplasmic transport ATPase. Nature (Lond) 290:271–273

Talbot JA, Hodges RS (1979) Synthesis and biological activity of an icosapeptide analog of the actomyosin ATPase inhibitory region of troponin I. J Biol Chem 254:3720–3723

Talbot JA, Hodges RS (1981) Comparative studies on the inhibitory region of selected species of troponin-I. J Biol Chem 256:12374–12378

Tanabe T, Takeshima H, Mikami A, Flockerzi V, Takahashi H, Kangawa K, Kojima M, Matsuo H, Hirose T, Numa S (1987) Primary structure of the receptor for calcium channel blockers from skeletal muscle. Nature (Lond) 328:313–328

Tanabe T, Beam KG, Powell JA, Numa S (1988) Restoration of excitation-contraction coupling and slow calcium current in dysgenic muscle by dihydropyridine receptor complementary DNA. Nature (Lond) 336:134–139

Tanabe T, Mikami A, Numa S, Beam KG (1990a) Cardiac-type excitation-contraction in dysgenic skeletal muscle injected with cardiac dihydropyridine receptor cDNA. Nature (Lond) 344:451–453

Tanabe T, Beam KG, Adams BA, Niidome T, Numa S (1990b) Regions of the skeletal muscle dihydropyridine receptor critical for excitation-contraction coupling. Nature (Lond) 346:567–569

Tanabe T, Adams BA, Numa S, Beam KG (1991) Repeat I of the dihydropyridine receptor is critical in determining calcium channel activation kinetics. Nature (Lond) 352:800–803

Taylor EW (1979) Mechanism of actomyosin ATPase and the problem of muscle contraction. Crit Rev Biochem 6:103–164

Taylor KA, Amos LA (1981) A new model for the geometry of the binding of myosin cross-bridges to muscle thin filaments. Biophys J 33:84a

Taylor SR, Rüdel R (1970) Striated muscle fibres: inactivation of contraction induced by shortening. Science NY 167:882–884

Taylor SR, Lopez JR, Griffiths PJ, Trube G, Cecchi G (1982) Calcium in excitation-contraction coupling of frog skeletal muscle. Can J Physiol Pharmacol 60:489–502

Thames MD, Teichholz LE, Podolsky RJ (1974) Ionic strength and the contraction kinetics of skinned muscle fibers. J Gen Physiol 63:509–530

Thomas MV (1982) Techniques in calcium research. Academic Press, London

Thompson RJ, Livingston DR, Zwaan A de (1980) Physiological and biochemical aspects of the valve snap and valve closure responses in the giant scallop *Placopecten magellanicus*. I. Physiology. J Comp Physiol 137:97–104

Thorson J, White DCS (1969) Distributed representations for actin-myosin interaction in the oscillatory contraction of muscle. Biophys J 9:360–390

Tobacman LS, Lee R (1987) Isolation and functional comparison of bovine cardiac troponin-T isoforms. J Biol Chem 262:4059–4064

Toida N, Kuriyama H, Tashiro N, Ito Y (1975) Obliquely striated muscle. Physiol Rev 55:700–756

Trautwein W, Pelzer D (1985) Voltage-dependent gating of single calcium channels in the cardiac cell membrane and its modulation by drugs. In: Marmé D (ed) Calcium and cell physiology. Springer, Berlin Heidelberg New York, pp 53–93

Trautwein W, McDonald TF, Tripathi O (1975) Calcium conductance and tension in mammalian ventricular muscle. Pflügers Arch Eur J Physiol 354:55–74

Trayer IP, Trayer HR, Levine BA (1987) Evidence that the N-terminal region of A1-light chain of myosin interacts directly with the C-terminal region of actin. Eur J Biochem 164:259–266

Trayer IP, Keane AM, Murad Z, Rüegg JC, Smith J (1991) The use of peptide mimetics to define the actin-binding sites on the head of the myosin molecule. In: Rüegg JC (ed) Peptides as probes in muscle research. Springer, Berlin Heidelberg New York, pp 57–68

Tregear RT (1983) Physiology of insect flight muscle. In: Peachey LD, Adrian RH, Geiger SR (eds) Skeletal muscle. Am Physiol Soc, Bethesda, Maryland, pp 487–506 (Handbook of physiology, sect 10)

Trueblood CE, Walsh TP, Weber A (1982) Is the steric model of tropomyosin action valid? In: Twarog BM, Levine RJC, Dewey MM (eds) Basic biology of muscles: a comparative approach. Raven Press, New York, pp 223–241

Tsien RW, Tsien RY (1990) Calcium channels, stores, and oscillations. Annu Rev Cell Biol 6:715–760
Tsien RY, Rink TJ (1980) Neutral carrier ion-selective microelectrodes for measurement of intracellular free calcium. Biochim Biophys Acta 599:623–638
Tsien RY, Rink TJ, Peonie M (1985) Measurement of cytosolic free Ca^{2+} in individual small cells using fluorescence microscopy with dual excitation wavelengths. Cell Calcium 6:145–157
Tsokos J, Sans R, Bloom S (1977) Ca^{2+} uptake by hyperpermeable mouse heart cells: effects of inhibitors of mitochondrial function. Life Sci 20:1913–1922
Turner PR, Fong P, Denetclaw WF, Steinhardt RA (1991) Increased calcium influx in dystrophic muscle. J Cell Biol 115:1701–1712
Twarog BM (1954) Responses of a molluscan smooth muscle to acetylcholine and 5-hydroxytryptamine. J Cell Comp Physiol 44:141–163
Twarog BM (1967) Excitation of *Mytilus* smooth muscle. J Physiol (Lond) 192:857–868
Twarog BM (1976) Aspects of smooth muscle function in molluscan catch muscle. Physiol Rev 56:829–838
Uexküll J von (1912) Studien über den Tonus. VI. Die Pilgermuschel. Z Biol 58:305–332
Ulbrich M, Rüegg JC (1971) Stretch induced formation of ATP-^{32}P in glycerinated fibres of insect flight muscle. Experientia 27:45–46
Ulbrich M, Rüegg JC (1976) Is the chemomechanical energy transformation reversible? Pflügers Arch Eur J Physiol 363:219–222
Unverferth DV, Lee SW, Wallick ET (1988) Human myocardial adenosine triphosphatase activities in health and heart failure. Am Heart J 115:139
Uyeda TQP, Kron SJ, Spudich JA (1990) Myosin step size – estimation from slow sliding movement of actin over low densities of heavy meromyosin. J Mol Biol 214:699–710
Uyeda TQP, Warrick HM, Kron SJ, Spudich JA (1991) Quantized velocities at low myosin densities in an in vitro motility assay. Nature (Lond) 352:307–311
Valdeolmillos M, O'Neill SC, Smith GL, Eisner DA (1989) Calcium-induced Ca^{2+} release activates contraction in intact cardiac cells. Pflügers Arch Eur J Physiol 413:676–678
Van Eerd JP, Takahashi K (1976) Determination of the complete amino acid sequence of bovine cardiac troponin C. Biochemistry 15:1171–1180
Van Eyk JE, Hodges RS (1988) The biological importance of each amino acid residue of the troponin I inhibitory sequence 104–115 in the interaction with troponin C and tropomyosin-actin. J Biol Chem 263:1726–1732
Van Eyk JE, Hodges RS (1991) A synthetic peptide of the N-terminus of actin interacts with myosin. Biochemistry 30:11676–11682
Van Eyk JE, Sönnichsen FD, Sykes BD, Hodges RS (1991 a) Interaction of actin 1–28 with myosin and troponin I and the importance of these interactions to muscle regulation. In: Rüegg JC (ed) Peptides as probes in muscle research. Springer, Berlin Heidelberg New York, pp 15–32
Van Eyk JE, Kay CM, Hodges RS (1991 b) A comparative study of the interactions of synthetic peptides of the skeletal and cardiac TnI inhibitory region with skeletal and cardiac Troponin C. Biochemistry 30:9974–9981
Van Eyk JE, Strauss JD, Hodges RS, Rüegg JC (1992) A synthetic peptide replaces troponin I in regulating contractility of skinned fibres from the myocardium. Pflügers Arch Eur J Physiol 420:R103
Varadi G, Lory P, Schultz D, Varadi M, Schwartz A (1991) Acceleration of activation and inactivation by the β-subunit of the skeletal muscle calcium channel. Nature (Lond) 352:159–162
Vergara J, Tsien RY, Delay M (1985) Inositol 1,4,5-trisphosphate: a possible chemical link in excitation-contraction coupling in muscle. Proc Natl Acad Sci USA 82:6352–6356
Vergara J, DiFranco M, Compagnon D, Suarez-Isla BA (1991) Imaging of calcium transients in skeletal muscle fibers. Biophys J 59:12–24
Vibert P, Craig R (1983) Electron microscopy and image analysis of myosin filaments from scallop striated muscle. J Mol Biol 165:303–320
Vibert PJ, Haselgrove JC, Lowy J, Poulsen FR (1972) Structural changes in actin-containing filaments of muscle. J Mol Biol 71:757–767
Vislie T, Fugelli K (1975) Cell volume regulation in flounder (*Platichthys flesus*) heart muscle accompanying an alteration in plasma osmolarity. Comp Biochem Physiol 52A:415–418
Volpe P, Damiani E, Salviati G, Margreth A (1982) Transitions in membrane composition during postnatal development of rabbit fast muscle. J Muscle Res Cell Mot 3:213–230

Wabnitz RW, Wachtendonk D von (1976) Evidence for serotonin (5-hydroxytryptamine) as transmitter in the penis retractor muscle of *Helix pomatia L.* Experientia 32:707–709
Wagenknecht T, Grassucci R, Frank J, Saito A, Inui M, Fleischer S (1989) Three-dimensional architecture of the calcium channel/foot structure of sarcoplasmic reticulum. Nature (Lond) 338:167–170
Wagner PD (1984) Effect of skeletal muscle myosin light chain 2 on the Ca^{2+}-sensitive interaction of myosin and heavy meromyosin with regulated actin. Biochemistry 23:5950–5956
Wagner PD, Giniger E (1981) Calcium-sensitive binding of heavy meromyosin to regulated actin in the presence of ATP. J Biol Chem 256:12647–12650
Walker JW, Somlyo AV, Goldman YE, Somlyo AP, Trentham DR (1987) Kinetics of smooth and skeletal muscle activation by laser pulse photolysis of caged inositol 1,4,5-trisphosphate. Nature (Lond) 327:249–252
Wallimann T, Szent-Györgyi AG (1981 a) An immunological approach to myosin light-chain function in thick filament linked regulation. 1. Characterization, specificity, and cross-reactivity of antiscallop myosin heavy- and light-chain antibodies by competitive, solid-phase radioimmunoassay. Biochemistry 20:1176–1187
Wallimann T, Szent-Györgyi AG (1981 b) An immunological approach to myosin light-chain function in thick filament linked regulation. 2. Effects of anti-scallop myosin light-chain antibodies. Possible regulatory role for the essential light chain. Biochemistry 20:1188–1197
Wallimann T, Hardwicke PMD, Szent-Györgyi AG (1982) Regulatory and essential light-chain interactions in scallop myosin. II. Photochemical cross-linking of regulatory and essential light-chains by heterobifunctional reagents. J Mol Biol 156:153–173
Wallinga-de Jonge W, Gielen FLH, Wirtz P, de Jong P, Broennink J (1985) The different intracellular action potentials of fast and slow muscle fibres. Electroencephalogr Clin Neurophysiol 60:539–547
Walsh MP (1985) Calcium regulation of smooth muscle contraction. In: Marmé D (ed) Calcium and cell physiology. Springer, Berlin Heidelberg New York, pp 170–203
Walsh MP, Bridenbaugh R, Hartshorne DJ, Kerrick WGL (1982a) Phosphorylation-dependent activated tension in skinned gizzard muscle fibers in the absence of Ca^{2+}. J Biol Chem 257:5987–5990
Walsh MP, Dabrowska R, Hinkins S, Hartshorne DJ (1982b) Calcium-independent myosin light chain kinase of smooth muscle. Preparation by limited chymotryptic digestion of the calcium ion dependent enzyme, purification and characterization. Biochemistry 21:1919–1925
Wang K (1984) Cytoskeletal matrix in striated muscle: the role of titin, nebulin and intermediate filaments. In: Pollack GH, Sugi H (eds) Contractile mechanisms in muscle. Plenum, New York, pp 285–305
Wankerl M, Böhm M, Morano I, Rüegg JC, Eichhorn M, Erdmann E (1990) Calcium sensitivity and myosin light chain pattern of atrial and ventricular skinned cardiac fibers from patients with various kinds of cardiac disease. J Mol Cell 22:1425–1438
Wasserstrom JA, Schwartz DJ, Fozzard HA (1983) Relation between intracellular sodium and twitch tension in sheep cardiac Purkinje strands exposed to cardiac glycosides. Circ Res 52:697–705
Weber A (1968) The mechanism of the action of caffeine on sarcoplasmic reticulum. J Gen Physiol 52:760–772
Weber A (1971) Regulatory mechanism of the calcium transport system of fragmented rabbit sarcoplasmic reticulum. I. The effect of accumulated calcium on transport and adenosine triphosphate hydrolysis. J Gen Physiol 57:50–63
Weber A, Murray JM (1973) Molecular control mechanisms in muscle contraction. Physiol Rev 53:612–673
Weber A, Winicur S (1961) The role of calcium in the superprecipitation of actomyosin. J Biol Chem 236:3198–3202
Weber A, Herz R, Reiss I (1966) Study of the kinetics of Ca transport by isolated fragmented sarcoplasmic reticulum. Biochem Z 345:329–369
Weber HH, Portzehl H (1954) The transference of the muscle energy in the contraction cycle. Prog Biophys Mol Biol 4:60–111
Weeds A (1980) Myosin light chains, polymorphism and fibre types in skeletal muscles. In: Pette D (ed) Plasticity of muscle. de Gruyter, Berlin, pp 55–68

Weiss GB (1981) Sites of action of calcium antagonists in vascular smooth muscle. In: Weiss GB (ed) New perspectives on calcium antagonists. Am Physiol Soc, Bethesda, Maryland, pp 83–94
Wells C, Bagshaw CR (1984) The Ca^{2+} sensitivity of the actin-activated ATPase of scallop heavy meromyosin. FEBS Lett 168:260–264
Wells C, Bagshaw CR (1985) Calcium regulation of molluscan myosin ATPase in the absence of actin. Nature (Lond) 313:696–697
Wendt-Gallitelli MF, Isenberg G (1991) Total and free myoplasmic calcium during a contraction cycle: X-ray microanalysis in guinea-pig ventricular myocytes. J Physiol (Lond) 435:349–372
Wendt-Gallitelli MF, Jacob R (1982) Rhythm-dependent role of different calcium stores in cardiac muscle: X-ray microanalysis. J Mol Cell Cardiol 14:487–492
Wendt-Gallitelli MF, Wolburg H (1984) Rapid freezing, cryosectioning, and X-ray microanalysis on cardiac muscle preparations in defined functional states. J Electron Microsc 1:151–174
Westerblad H, Lee JA, Lamb AG, Bolsover SR, Allen DG (1990) Spatial gradients of intracellular calcium in skeletal muscle during fatigue. Pflügers Arch Eur J Physiol 415:734–740
White HD, Taylor EW (1976) Energetics and mechanism of actomyosin adenosine triphosphatase. Biochemistry 15:5818–5826
Wier WG, Hess P (1984) Excitation-contraction coupling in cardiac Purkinje fibers. Effects of cardiotonic steroids on the intracellular Ca^{2+} transient, membrane potential, and contraction. J Gen Physiol 83:395–415
Wier WG, Yue DT (1985) The effects of 1,4-dihydropyridine type Ca^{2+}-channel antagonists and agonists on intracellular $[Ca^{2+}]$ transients accompanying twitch contraction of heart muscle. In: Fleckenstein A, Breemen C van, Groß R, Hoffmeister F (eds) Cardiovascular effects of dihydropyridine-type calcium antagonists and agonists. Springer, Berlin Heidelberg New York, pp 188–197
Wier WG, Yue DT, Marban E (1985) Effects of ryanodine on intracellular Ca^{2+} transients in mammalian cardiac muscle. Fed Proc 44:2989–2993
Wilkinson JM (1980) Troponin C from rabbit slow skeletal and cardiac muscle is the product of a single gene. Eur J Biochem 103:179–188
Wilkinson JM, Grand RJA (1978) Comparison of amino acid sequence of troponin I from different striated muscles. Nature (Lond) 271:31–35
Williams AJ (1992) Ion conduction and discrimination in the sarcoplasmic reticulum ryanodine receptor/calcium-release channel. J Muscle Res Cell Mot 13:7–26
Williams DA, Fogarty KE, Tsien RY, Fay FS (1985) Calcium gradients in single smooth muscle cells revealed by the digital imaging microscope using Fura-2. Nature (Lond) 318:558–561
Williams DA, Head SI, Bakker AJ, Stephenson DG (1990) Resting calcium concentrations in isolated skeletal muscle fibres of dystrophic mice. J Physiol (Lond) 428:243–256
Williams RJP (1970) The biochemistry of sodium, potassium, magnesium and calcium. Q Rev Chem Soc 24:331–365
Wilson DM, Larimer JL (1968) The catch property of ordinary muscle. Proc Natl Acad Sci USA 61:909–916
Winegrad S (1965) Autoradiographic studies of intracellular calcium in frog skeletal muscle. J Gen Physiol 48:455–479
Winegrad S (1970) The intracellular site of calcium activation of contraction in frog skeletal muscle. J Gen Physiol 55:77–88
Winegrad S (1979) Electromechanical coupling in heart muscle. In: Berne RM, Sperelakis N, Geiger SR (eds) The cardiovascular system. Am Physiol Soc. Bethesda, pp 393–428 (Handbook of physiology, sect 2, vol I)
Winkelmann DA, Mekeel H, Rayment I (1985) Packing analysis of crystalline myosin subfragment-1. J Mol Biol 181:487–501
Winkelman L (1976) Comparative studies of paramyosins. Comp Biochem Physiol 55B:391–397
Winkelman L, Bullard B (1980) Phosphorylation of a locust myosin light chain and its effect on calcium regulation. J Muscle Res Cell Mot 1:221–222
Winquist RJ, Faison EP, Waldman SA, Schwartz K, Murad F, Rapoport TM (1984) Atrial natriuretic factor elicits an endothelium-independent relaxation and activates particulate guanylate cyclase in vascular smooth muscle. Proc Natl Acad Sci USA 81:7661–7664
Winton FR (1930) Tonus in mammalian unstriated muscle. J Physiol (Lond) 69:393–410

Winton FR (1937) The changes in viscosity of an unstriated muscle (*Mytilus edulis*) during and after stimulation with alternating, interrupted and uninterrupted direct currents. J Physiol (Lond) 88:492–511

Wnuk W, Schoechlin M, Stein EA (1984) Regulation of actomyosin ATPase by a single calcium-binding site on troponin C from crayfish. J Biol Chem 259:9017–9023

Wohlfart B, Noble MIM (1982) The cardiac excitation-contraction cycle. Pharmacol Ther 16:1–43

Woledge RC, Curtin NA, Homsher E (1985) Energetic aspects of muscle contraction. Academic Press, London

Wood EH, Heppner RL, Weidmann S (1969) Inotropic effects of electric currents. I. Positive and negative effects of constant electric currents or current pulses applied during cardiac action potentials. II. Hypotheses: calcium movements, excitation-contraction coupling and inotropic effects. Circ Res 24:409–445

Wootton RJ, Newman DJS (1979) Whitefly have the highest contraction frequencies yet recorded in non-fibrillar flight muscles. Nature (Lond) 280:402–403

Wray JS (1979a) Structure of a backbone in myosin filaments of muscle. Nature (Lond) 277:37–40

Wray JS (1979b) Filament geometry and the activation of insect flight muscles. Nature (Lond) 280:325–326

Wrogemann K, Nylen EG (1978) Mitochondrial calcium overloading in cardiomyopathic hamsters. J Mol Cell Cardiol 10:185–195

Wuytack F, Raeymaekers L, Casteels R (1985) The Ca^{2+}-transport ATPases in smooth muscle. Experientia 41:900–905

Xu GQ, Hitchcock-De Gregori SE (1988) Synthesis of a troponin C cDNA and expression of wild-type and mutant proteins in *Escherichia coli*. J Biol Chem 263:13962–13969

Xu S, Kress M, Huxley HE (1987) X-ray diffraction studies of the structural state of crossbridges in skinned frog sartorius muscle at low ionic strength. J Muscle Res Cell Mot 8:39–54

Yamada S, Tonomura Y (1972) Reaction mechanism of the Ca^{2+}-dependent ATP-ase of sarcoplasmic reticulum from skeletal muscle. VII. Recognition and release of Ca^{2+} ions. J Biochem 72:417–425

Yamamoto K, Sekine T (1979) Interaction of myosin subfragment-1 with actin. J Biochem 86:1863–1868

Zachar J, Zacharová D (1966) Potassium contractures in single muscle fibres of the crayfish. J Physiol (Lond) 186:596–618

Zange J, Pörtner HO, Hans AWH, Grieshaber MK (1990) The intracellular pH of a molluscan smooth muscle during a contraction-catch relaxation cycle estimated by the distribution of [^{14}C]DMO and by ^{31}P-NMR spectroscopy. J Exp Biol 150:81–93

Zechel K, Weber K (1978) Actins from mammals, bird, fish and slime mold characterized by isoelectric focusing in polyacrylamide gels. Eur J Biochem 89:105–112

Zorzato F, Fujii J, Otsu K, Phillips M, Green NM, Lai FA, Meissner G, MacLennan DH (1990) Molecular cloning of cDNA encoding human and rabbit forms of the Ca^{2+} release channel (ryanodine receptor) of skeletal muscle sarcoplasmic reticulum. J Biol Chem 265:2244–2256

Zot AS, Potter JD (1987) Structural aspects of troponin-tropomyosin regulation of skeletal muscle contraction. Annu Rev Biophys Biophys Chem 16:535–559

Zot HG, Iida S, Potter JD (1983) Thin filament interactions and Ca^{2+} binding to Tn. Chemica Scripta 21:133–136

Zot HG, Güth K, Potter JD (1985) Measurement of fluorescence and tension development in skinned skeletal muscle fibers reconstituted with TnC_{DANZ}. Biophys J 47:473a

Zot HG, Güth K, Potter JD (1986) Fast skeletal muscle skinned fibers and myofibrils reconstituted with N-terminal fluorescent analogues of troponin C. J Biol Chem 261:15883–15890

Zubrzycka-Gaarn E, Korczak B, Osinska H, Sarzala MG (1982) Studies on sarcoplasmic reticulum from slow-twitch muscle. J Muscle Res Cell Mot 3:191–212

Subject Index

Printing: Druckerei Zechner, Speyer
Binding: Buchbinderei Schäffer, Grünstadt